The von Grunebaum Center
for Near Eastern Studies

PROOFS FOR ETERNITY, CREATION AND THE EXISTENCE OF GOD IN MEDIEVAL ISLAMIC AND JEWISH PHILOSOPHY

Proofs for Eternity, Creation and the Existence of God in Medieval Islamic and Jewish Philosophy

HERBERT A. DAVIDSON
University of California, Los Angeles

New York * Oxford
OXFORD UNIVERSITY PRESS
1987

Oxford University Press

Oxford New York Toronto
Delhi Bomby Calcutta Madras Karachi
Petaling Jaya Singapore Hong Kong Tokyo
Nairobi Dar es Salaam Cape Town
Melbourne Auckland
and associated companies in
Beirut Berlin Ibadan Nicosia

Published by Oxford University Press, Inc.,
200 Madison Avenue, New York, New York 10016

Library of Congress Cataloging-in-Publication Data
Davidson, Herbert A. (Herbert Alan)
Proofs for eternity, creation, and the existence
of God in medieval Islamic and Jewish philosophy.
New York: Oxford University Press, 1987.
Includes indexes.
1. God (Islam)—Proof—History of doctrines.
2. God (Judaism)—History of doctrines. 3. Philosophy, Islamic.
4. Philosophy, Jewish. I. Title.
B745.G63D38 1987 212'.1'0902 86-33179
ISBN 0-19-504953-5

Published with the assistance of The Louis and Minna Epstein Fund of
the American Academy for Jewish Research

1 3 5 7 9 10 8 6 4 2

Printed in the United States of America
on acid-free paper

For
Rachel and Jessica

Acknowledgments

Chapters IV and V of the present book are a reworking of an article that appeared in the *Journal of the American Oriental Society*, Volume 89. Chapter VI incorporates material from articles appearing in *Philosophy East and West*, Volume 18, and *Studies in Medieval Jewish History and Literature* (Cambridge, 1979). An expanded version of Appendix A, part 2, appeared in *Studies in Jewish Religious and Intellectual History Presented to Alexander Altmann* (University, Alabama, 1979).

The National Endowment for the Humanities and the UCLA Academic Senate have supported my work; the von Grunebaum Center for Near Eastern Studies, UCLA, took responsibility for preparing a camera-ready copy; a grant from the American Academy for Jewish Research to the von Grunebaum Center helped to defray some of the composition costs. I wish to express my warmest thanks to all those institutions. I also wish to thank Marina Preussner of the von Grunebaum Center for her invaluable aid and irrepressible good cheer.

My wife subjected the book to a painstaking, and often painful, critique. Any clarity that is to be detected is her doing.

The book was complete in 1980.

Contents

I Introduction 1

1. Eternity, creation, and the existence of God 1
2. The present book 6

II Proofs of Eternity from the Nature of the World 9

1. Proofs of eternity 9
2. Proofs of eternity from the nature of the physical world 12
3. Replies to proofs from the nature of the world 30
4. Summary 46

III Proofs of Eternity from the Nature of God 49

1. The proofs 49
2. Replies to proofs from the nature of the cause of the universe 67
3. Summary 85

IV John Philoponus' Proofs of Creation and Their Entry into Medieval Arabic Philosophy 86

1. Philoponus' proofs of creation 86
2. Saadia and Philoponus 95
3. Kindi and Philoponus 106
4. Summary 116

V Kalam Proofs for Creation 117

1. Proofs from the impossibility of an infinite number 117
2. Responses of the medieval Aristotelians to proofs of creation from the impossibility of an infinite number 127
3. The standard Kalam proof for creation: the proof from accidents 134
4. Juwaynī's version of the proof from accidents 143
5. Proofs from composition 146

VI Arguments from the Concept of Particularization 154

1. Inferring the existence of God from creation 154
2. Arguments from the concept of particularization 174
3. Particularization arguments for the existence of God without the premise of creation; particularization arguments for creation 187
4. Ghazali and Maimonides 194
5. Additional arguments for creation in Maimonides and Gersonides 203

VII Arguments from Design 213

1. Cosmological, teleological, and ontological proofs of the existence of God 213
2. Teleological arguments 216
3. Summary 236

VIII The Proof from Motion 237

1. Aristotle's proof from motion 237
2. Maimonides' version of the proof from motion 240
3. Ḥasdai Crescas' critique of the proof from motion 249
4. Another proof from motion 275

IX Avicenna's Proof of the Existence of a Being Necessarily Existent by Virtue of Itself 281

1. First cause of motion and first cause of existence 281
2. The existence of God: a problem for metaphysics 284
3. Necessarily existent being and possibly existent being 289
4. The attributes of the necessarily existent by virtue of itself 293
5. Proof of the existence of the necessarily existent by virtue of itself 298
6. Questions raised by Avicenna's proof 304
7. The version of Avicenna's proof in Shahrastānī and Crescas 307
8. Summary 309

X Averroes' Critique of Avicenna's Proof 311

1. The proof of the existence of God as a subject for physics 312
2. Necessarily existent by virtue of another, possibly existent by virtue of itself 318
3. The nature of the celestial spheres according to Averroes 321
4. Averroes' critique of the body of Avicenna's proof 331
5. Summary 334

XI Proofs of the Existence of God from the Impossibility of an Infinite Regress of Efficient Causes 336

1. The proof from the impossibility of an infinite regress of causes 336
2. Unity and incorporeality 345
3. The proof from the impossibility of an infinite regress of efficient causes and the proof from the concepts possibly existent and necessarily existent 350
4. Resumé 362
5. Crescas on the impossibility of an infinite regress 365
6. Ghazali's critique of Avicenna's proof 366
7. Summary 375

XII Subsequent History of Proofs from the Concept of *Necessary Existence* 378

1. Maimonides and Aquinas 378
2. The influence of Avicenna's proof 385
3. Proofs of the existence of God as a necessarily existent being in modern European philosophy 388
4. Summary 405
5. Concluding remark 406

Appendix A. Two Philosophic Principles 407

1. The principle that an infinite number is impossible 407
2. The principle that a finite body contains only finite power 409

Appendix B. Inventory of Proofs 412

Primary Sources 414

Index of Philosophers 421

Index of Terms 427

PROOFS FOR ETERNITY, CREATION AND THE EXISTENCE OF GOD IN MEDIEVAL ISLAMIC AND JEWISH PHILOSOPHY

I

Introduction

1. Eternity, creation, and the existence of God

One might well expect the existence of God to be the initial issue of natural theology; but such was not the case in the Islamic and Jewish Middle Ages. Medieval Muslim and Jewish philosophers did as a matter of course construct their natural theology and their metaphysics in general on the existence of God. The provability of the deity's existence was, furthermore, disputed; for whereas most philosophers were confident that the deity's existence can be demonstrated rationally, some demurred. When the possibility of demonstrating the existence of God was challenged, the challenge came, however, not from radicals who doubted the proposition, but from conservatives who questioned the competence of human reason to demonstrate it. While the provability of God's existence might, then, be subject to dispute, God's existence never was, and the Middle Ages were free from atheism and agnosticism, at least public atheism and agnosticism, on the philosophic plane. The existence of God could not, as a consequence, be the initial issue for natural theology. The existence of God, as distinct from the provability of God's existence, was not strictly an issue at all.

The initial issue of natural theology for Muslims and Jews, the most fundamental issue where opinions divided, was, it may be ventured, the inquiry concerning whether the world is eternal or had a beginning. Much more is at stake there than chronology or hermeneutics—the age of the universe or the question whether the scriptural account of the genesis of the universe should be taken literally or allegorically. The issue of eternity and creation[1] provided an arena for determining the relationship of God to the universe, for determining, specifically, whether God is a necessary or a voluntary cause. If the world should be eternal, and a deity is recognized, the deity's relationship to the universe would likewise be eternal. Since eternity and necessity are, by virtue of an Aristotelian

[1] I employ the term *creation* to mean the thesis that the world came into existence after not having existed, not the more specific thesis that a creator brought the world into existence. Medieval thinkers who accepted the former thesis were invariably certain that the latter thesis can be inferred from it.

rule, mutually implicative,[2] an eternal relationship is a relationship bound by necessity; and necessity excludes will.[3] The eternity of the world thus would imply that the deity is, as the cause of the universe,[4] bereft of will. A beginning of the world would, by contrast, lead to a deity possessed of will. Should the world be understood to have a beginning, all medieval thinkers agreed, the existence of a creator can be inferred; and the decision on the creator's part to bring a world into existence where no world existed before would constitute a supreme and paradigmatic act of volition. Will in the deity would, therefore, be ruled out by the eternity of the world and entailed by creation.

The issue of eternity and creation often intertwined with the enterprises of proving the existence of God. The majority of Islamic and Jewish proofs for the existence of God take either eternity or creation as a premise, and require a resolution of that issue before their own proper subject can be broached. In the Aristotelian proof of the existence of God as the prime mover and in kindred proofs, the eternity of the world is an indispensable premise and must be established prior to the proof itself. The world is shown to be eternal, and the eternal motion or the eternal existence of the universe is shown to have a cause, which is identified as the deity.[5] In Kalam proofs of the existence of God, the indispensable premise is creation. The Kalam thinkers followed what has been called the Platonic procedure,[6] that is, the procedure of first proving the creation of the world and then inferring therefrom the existence of a creator, again identified as the deity.[7]

The decision to demonstrate the existence of God from the premise of eternity, on the one hand, or from the premise of creation, on the other, was not the result merely of one party's happening to be convinced of the truth of eternity and the other's being convinced of the truth of creation. The decision was connected with the diverse conceptions of the deity which accompany the two premises. The choice of one or the other premise would be reflected in the conclusion of a proof

[2]Aristotle, *De Generatione* II, 11, 338a, 1.

[3]Plotinus speaks of a "necessary free will" (*Enneads*, IV, 8, 5); and Avicenna and Averroes speak of the deity's "eternal will" (Avicenna, *Shifā': Ilāhīyāt*, ed. G. Anawati and S. Zayed [Cairo, 1960], p. 366; Averroes, *K. al-Kashf*, ed. M. Mueller [Munich, 1859], p. 52; German translation, with pagination of the Arabic indicated: *Philosophie und Theologie von Averroes*, trans. M. Mueller [Munich, 1875]). Crescas, *Or ha-Shem*, III, i, 5, also defends the possibility of an eternal will in God. But these are Pickwickian senses of *will*; cf. Maimonides, *Guide to the Perplexed*, II, 21.

[4]Conceivably the deity could, though the universe is eternal, exercise his will in some fashion that does not relate to the universe. But that thought, if it has any meaning at all, was not entertained by the medievals.

[5]Cf. below, p. 239.

[6]Cf. Plato, *Timaeus*, 28; Moses Narboni, *Commentary on Maimonides' Guide* (Vienna, 1853), II, 2; C. Baeumker, *Witelo* (Muenster, 1908), pp. 320 ff.; H. Wolfson, "Notes on Proofs of the Existence of God in Jewish Philosophy," reprinted in his *Studies in the History of Philosophy and Religion*, Vol. I (Cambridge, Mass., 1973), pp. 571–572.

[7]Cf. below, pp. 154 ff.

for the existence of God, since, as already seen, a proof from the premise of eternity would lead to a deity bound by necessity and a proof from the premise of creation would lead to a deity possessed of will. Something more might be involved. Proofs proceeding from the two different premises can differ in their understanding of what constitutes a genuine proof of the existence of God. Every proof of the existence of God must at some stage, whether explicitly or—as occurs far more often—implicitly, presuppose a definition of God, a set of specifications requisite and sufficient for the deity. A proof of the existence of God is a chain of reasoning which concludes with the existence of a being distinguished by certain attributes. Unless the attributes qualify the being possessing them as the deity, no proof, however correctly reasoned, can claim that the being whose existence it arrives at is God. Proofs of the existence of God from the premise of eternity, at least among Islamic and Jewish philosophers, implicitly assume or explicitly state three specifications for the deity. By deity, a being is meant which is, firstly, an uncaused cause; secondly, incorporeal; and thirdly, one.[8] Any chain of reasoning concluding with the existence of a single, uncaused, incorporeal cause would accordingly constitute a successful proof of the existence of God. Volition is not, in proofs from the premise of eternity, included among the specifications. In fact, volition is ruled out, since, as has been seen, the deity would be bound by necessity if the world is eternal. As Ghazali explains the virtue of the proof from the opposite premise, the premise of creation, that proof must be resorted to because of the inadequacy of the definition just given. To be an uncaused cause, incorporeal, and one, Ghazali insists, is a good deal less than to be the deity. For nothing could conceivably be designated as the deity if it is unable to make decisions affecting the course of events in the universe.[9] The specifications presupposed in proofs from the premise of eternity must, on this view, be supplemented with a further attribute, volition.

Ghazali had an additional reason for rejecting any proof of the existence of God not based on the premise of creation. He contends that the notion of eternal causation is intrinsically nonsensical, that what comes into existence after not existing can alone be thought of as having a cause. To advocate eternity would thus be tantamount to denying a cause of the universe, and to countenance eternity would be tantamount to countenancing the causelessness of the universe. Argumentation that does not employ the premise of creation hence would fail to prove the existence of God not merely for those who number volition among the requisite specifications. It would fail to prove the existence of God even on the view of the proponents of the proofs from eternity; for it would fail to establish the

[8]Explicitly in Maimonides, *Guide,* I, 71. Also cf. Alfarabi, *K. Arā' Ahl al-Madīna al-Fāḍila,* ed. F. Dieterici (Leiden, 1895), beginning; German translation: *Der Musterstaat,* trans. F. Dieterici (Leiden, 1900). Avicenna, *Shifā' : Ilāhīyāt,* pp. 37–47.

[9]Ghazali, *Tahāfut al-Falāsifa,* ed. M. Bouyges (Beirut, 1927), III, §§3, 16; English translation in *Averroes' Tahafut al-Tahafut,* trans S. van den Bergh (London, 1954), pp. 89, 96.

existence of anything having the first of the three specifications—the attribute of being the cause of the universe—which the proponents of proofs from eternity themselves deem requisite for the deity.[10]

Besides proofs for the existence of God from one or the other premise, from the premise of the eternity of the world or the premise of creation, proofs were also advanced with neither eternity nor creation as a premise. In some instances, creation is established in the course of proving the existence of God; a single train of reasoning arrives at both the existence of God and the creation of the world.[11] In other instances, the issue of eternity and creation is nowise touched on.[12] Proofs of the latter sort avoid taking a position on the presence or absence of volition in the deity. Nevertheless, they cannot avoid taking a position on the specifications for the deity. By professing to be genuine proofs of the existence of God without demonstrating that God possesses will, they tacitly affirm that will is not a requisite specification.

Still another procedure for proving the existence of God was in evidence. Ibn Ṭufayl and Maimonides, as well as the Scholastic philosopher Thomas Aquinas, do not themselves subscribe to the eternity of the world, yet they put forward proofs of the existence of God resting on the premise of eternity. They demonstrate the existence of God on two parallel, alternative tracks, on both the hypothesis of eternity and the hypothesis of creation; and proofs resting on the premise of eternity serve them hypothetically, as a means for establishing the existence of God on one of the two conceivable alternatives. The world, so their reasoning goes, either is eternal or had a beginning. Should the world be eternal, the Aristotelian proof from motion and other proofs from the premise of eternity establish a first incorporeal cause, who is the deity; whereas if the world should not be eternal, its having come into existence permits the immediate inference of a creator. In either event, the existence of God is established. Ibn Ṭufayl chose the procedure outlined in order to refrain from any stand whatsoever on the issue of eternity and creation. He explains that the issue is unresolvable and that the existence of God will have been demonstrated only if shown to follow from both the hypothesis of the eternity of the world and the hypothesis of creation.[13] The rationale of Maimonides and Aquinas was different. They do take a stand on the issue of eternity and creation, and advocate the latter. They nonetheless wish to

[10]Ghazali, *Tahāfut al-Falāsifa,* III, §§17, 28; X, §1; English translation, pp. 96, 102, 250. The contention that eternal causation is nonsensical is found elsewhere; cf. below, pp. 190, 193, 210.

[11]Below, pp. 149–150, 151, 188, 190, 387.

[12]The notable example is Avicenna's proof, below, Chapter IX. Kalam writers subsequent to Avicenna commonly advance versions of Avicenna's proof side by side with the traditional Kalam arguments.

[13]*Ḥayy ben Yaqdhān,* ed, and trans. L. Gauthier (Beirut, 1936), Arabic text, pp. 81–86; French translation, pp. 62–65; English translation with pagination of the Arabic indicated: *Hayy Ibn Yaqzān,* trans. L. Goodman (New York, 1972).

avoid linking the more fundamental and less problematic doctrine of the existence of God to the less fundamental and more problematic doctrine of creation. They therefore leave the question of creation open while demonstrating the existence of God and, like Ibn Ṭufayl, prove the existence of God on parallel tracks, on both the hypothesis of eternity and the hypothesis of creation.[14] Once having shown that God exists whether or not the world is eternal, they return, however — unlike Ibn Ṭufayl—to the issue of creation and eternity; and Maimonides, for his part, offers "arguments," acknowledged by him to be less than apodictic demonstrations, for creation.[15] In espousing creation, Maimonides and Aquinas clearly espouse, as well, the presence of volition in God. But by postponing the subject of creation until after proving the existence of God, they tacitly affirm that volition is not integral to the concept of the deity, that a proof of the existence of a single incorporeal cause, albeit a necessary cause, would constitute a genuine proof of the existence of God.

Various procedures for proving the existence of God are, in sum, discernible in medieval Islamic and Jewish philosophy, some of which do, whereas others do not, require that the issue of eternity and creation be settled before a proof of the existence of God can be accomplished. The existence of God might be proved through the premise of the world's eternity. In proofs of the sort, the deity is explicitly or implicitly defined as an *uncaused cause, incorporeal,* and *one*. The existence of God might be proved through the contrary premise, creation, the attendant conclusion being a deity possessed of will. Ghazali, who expatiates upon the import of proofs based on the premise of creation, stresses that volition is integral to an adequate concept of the deity, that unless a proof establishes a first cause possessed of will, it is not a genuine proof of the existence of God. The existence of God, creation, and hence the attribute of will, might all be established by means of a single train of reasoning. Here, it is not clear whether volition is viewed as integral to the concept of God. The existence of God might also be proved utilizing neither the premise of eternity nor the premise of creation. The implication now would plainly be that whether or not God does possess will, volition is not part of the irreducible concept of the deity. Finally, a proof from the premise of eternity might be employed by philosophers who do not themselves subscribe to the premise. The proof from eternity would serve a hypothetical function and would be supplemented through a parallel proof from creation. Again, the implication would be that the irreducible concept of the deity does not contain volition.

[14]Maimonides, *Guide,* I, 71; II, 2; Aquinas, *Summa contra Gentiles,* I, 13. Also cf. Aquinas, *Commentary on Physics,* VIII, §970, and *Commentary on Metaphysics,* XII, lectio 5 (end).

[15]Maimonides, *Guide,* II, 19; 22. Aquinas, *Summa Theologiae,* I, q. 46, arts. 1, 2; *Summa contra Gentiles,* II, chap. 38. Also cf. A. Maier, "Problem des aktuell Unendlichen," *Ausgehendes Mittelalter,* Vol. I (Rome, 1964), p. 48.

In a later chapter it will be seen that medieval Islamic and Jewish arguments for the existence of God are, in the main, cosmological; teleological arguments are also found; and no argument is ontological.[16]

2. The present book

The chapters to follow examine medieval Islamic and Jewish arguments for eternity; responses to the arguments for eternity on the part of the proponents of creation; medieval Islamic and Jewish arguments for creation; responses thereto by the proponents of eternity; proofs for the existence of God resting on the premise of creation, the premise of eternity, or neither premise; and refutations of proofs for the existence of God.

Medieval Islamic and Jewish philosophy will be treated here as a single philosophic tradition. Treating the several branches of medieval philosophy—Islamic, Jewish, and Christian—in conjunction with one another is surely justified considering the extent to which they draw sustenance from the same, or similar, sources and are animated by the same spirit.[17] From the standpoint of medieval Jewish philosophy, treating Islamic and Jewish philosophy conjointly has an additional justification, inasmuch as Jewish philosophy is rooted in Islamic philosophy and cannot be properly understood in isolation from it. The reverse is of course not true: Medieval Islamic philosophers barely knew of the existence of the Jewish philosophers. Yet when viewed historically, Jewish philosophy does shed light on medieval Islamic philosophy in a number of areas, notably in two. For the early period, Jewish thinkers who read and wrote Arabic, especially Saadia, complement the available Islamic material and help delineate the beginnings of Arabic philosophy; and for a later period, Hebrew sources are an invaluable aid in the study of Averroes. Many of Averroes' writings are preserved exclusively in Hebrew, and Hebrew commentaries furnish the best, and often the sole, access to the meaning of even those preserved in Arabic.

I have not tried to take account of each and every argument for eternity, creation, and the existence of God, which might be unearthed in medieval Arabic and Hebrew literature, but I have tried to take account of every argument for those doctrines which is of a philosophic character. Kalam reasoning consequently falls within the scope of the present work when it appears to me to be philosophic, but has been disregarded when it does not. The difficulty in drawing up a satisfactory definition of philosophy always lies in the demarcation of the precise boundary between philosophy and other domains. Still, any thoughtful acceptation of the term will distinguish philosophy from pure theology, from speculation whose underlying premises are founded entirely on religious faith, and will also exclude from the domain of philosophy enunciations that fall below some minimal level of plausibility. The Kalam arguments that I have disregarded

[16]Below, pp. 214–216.
[17]Cf. H. Wolfson, *Philo* (Cambridge, Mass., 1948), chap. 14.

are of a purely theological character or so unsubstantial that they cannot, even with magnanimity, be taken seriously.[18]

My discussion purports to be exhaustive as regards Arabic and Hebrew arguments; that is, I have undertaken to examine every medieval Arabic and Hebrew philosophic argument for eternity, creation, and the existence of God. Arguments in the two literatures which are not discussed were either judged to be nonphilosophic, were subsumed under other arguments that are discussed elsewhere in the book,[19] or were simply overlooked. In a number of instances I have pursued the penetration of Islamic and Jewish arguments into medieval Christian philosophy, and in a few instances into modern European philosophy. There, though, I make no pretense at exhaustiveness, and the citations are of a kind that are ready at hand in obvious primary and secondary works.

The discipline to which the present study belongs is what I would venture to call the history of philosophic ideas. I am attempting to trace the history of philosophic constructions rather than reproduce the complete system of any individual philosopher or philosophic school. My concern is chiefly with the history of proofs—of eternity, creation, and the existence of God; but also worthy of attention is the history of the components from which the proofs are fashioned. The starting point both for the history of the proofs and the history of their components is, with rare exceptions, Aristotle. That is hardly surprising; the importance of Aristotle for medieval philosophy is common knowledge. The direction in which the Aristotelian conceptions developed in the Middle Ages was, however, often determined by the late Greek philosophers, and their importance for the Middle Ages is far from common knowledge. In much of the material to be examined, Proclus (5th century) and, in greater measure, John Philoponus (6th century) are responsible for the direction in which Aristotelian conceptions developed. The significance for medieval Islamic and Jewish thought of Proclus, one of the last Greek Neoplatonists, and of Philoponus, a Christian and sometime commentator on Aristotle, is a subordinate theme of the present book.

The proofs for eternity, creation, and the existence of God are admirably suited for an organized presentation. They arrange themselves spontaneously and gracefully into chapters and chapter sections. Unfortunately, the arrangement into chapters and sections according to the proofs tends to obscure the history of the components from which the proofs are fashioned. The components wend their way through one proof after another. A principle or conception may originate in Aristotle; undergo development at the hands of Proclus or Philoponus, or both;

[18]I have, for example, passed over the argument for creation offered by Ash'ari, *K. al-Luma'*, §3 (in *The Theology of al-Ash'arī*, ed. and trans. R. McCarthy [Beirut, 1953]), and also recorded by Shahrastānī, *K. Nihāya al-Iqdām*, ed. A. Guillaume (Oxford and London, 1934), p. 12. H. Wolfson, *Philosophy of the Kalam* (Cambridge, Mass., 1976), pp. 382–383, terms it the "argument from the analogy of things in the world." It is simply an expansion of Quran 23:12–14.

[19]For example, some of the Kalam arguments for creation listed by Wolfson, *Philosophy of the Kalam*, p. 374, will be treated here, Chapters IV, V, and VI, as variations of other arguments.

and repeatedly cross linguistic, confessional, and school boundaries—as well as the chapter headings of the present book—while traveling down fifteen centuries or more. In crossing boundaries, the principle may remain imbedded within the context with which it was originally associated, but may also be transformed and reembodied in a new context. And a principle or conception originally formulated by a given party may be converted by the opposing party into ammunition for attacking the basic tenets of the former.[20] Particularly far-reaching examples will be offered in Appendix A. It traces the peregrinations of two Aristotelian principles—the principle that an infinite number is impossible, and the principle that a finite body can contain only finite power—which make their appearance in several chapters but do not lend themselves to a complete treatment at any single place in the body of the book.[21]

Medieval Islamic and Jewish arguments for eternity and responses made to them by the adherents of creation are discussed in Chapters II and III. Chapters IV and V treat a group of arguments for creation which can be linked with John Philoponus, together with refutations of the arguments. Chapters VI and VII explore two styles of reasoning, argumentation from the concept of *particularization* and teleological argumentation, which were used in the formulating of proofs for both creation and the existence of God. Chapter VIII analyzes the Aristotelian proof of the existence of God as a prime mover, together with Ḥasdai Crescas' refutation. Chapter IX examines Avicenna's proof of the existence of God as a being necessarily existent by virtue of itself, and Chapter X analyzes Averroes' refutation of Avicenna's proof. Chapter XI discusses a family of proofs for the existence of God, all resting on the principle that an infinite regress of causes is impossible; refutations of the proofs are treated in the same chapter. Chapter XII deals with the history, subsequent to Avicenna, of arguments for the existence of God as a necessary being.

All the translations are mine.

[20]This occurs in connection with both principles discussed in Appendix A.

[21]Other principles appearing repeatedly—and employed by different parties, each for its own purpose—are the principle that every possibility must eventually be realized; and the principle that whatever is generated is destructible, and vice versa.

II

Proofs of Eternity from the Nature of the World

1. Proofs of eternity

The writings of medieval Islamic and Jewish philosophers reveal stock proofs of the eternity of the world, proofs that recur repeatedly through the centuries with little change. Proofs of the eternity of the world are, of course, advanced by philosophers espousing eternity. But they are also recorded by philosophers who, on the contrary, espouse creation; and since the advocates of creation were more numerous than the advocates of eternity, more citations of the standard proofs for eternity can be gleaned from the writings of the former than from the writings of the latter. When the advocates of creation record proofs of eternity, their purpose, naturally enough, is to prepare the way for a refutation. In that, John Philoponus may well have served as the model; Philoponus had painstakingly refuted all the arguments for eternity which he had discovered in Aristotle and Proclus, and his refutations of Aristotle and Proclus together with his own arguments for creation were known to, and used by, the Islamic and Jewish philosophers.[1] Not only were the proofs for eternity standardized; the refutations were, as well. A repertoire of stock refutations was thus arrayed against a repertoire of stock arguments. And the refutations, in their turn, often elicited surrejoinders from the proponents of eternity. This chapter and the next examine the philosophic proofs for eternity appearing in the writings of medieval Islamic and Jewish advocates of eternity or creation; rebuttals of the proofs for eternity; and surrejoinders to the rebuttals.

Ancient and medieval proofs of eternity do not all have the same aim. The more comprehensive seek to establish that the world has existed from eternity in the form in which it exists today. But less comprehensive proofs, too, are in evidence. They seek to establish the eternity of the matter of the world, leaving open the possibility that the form of the world is not eternal, that the world was,

[1]Examples are offered in this and the following three chapters.

in other words, created out of a preexistent matter. Unfortunately, both adherents of eternity and adherents of creation often fail to state explicitly what a given proof is intended to accomplish; and it sometimes even happens that proofs ostensibly of the eternity of the world as a whole argue, in fact, for nothing more than the eternity of matter.

As will be seen in later chapters, the doctrine of creation is likewise supported by more comprehensive and less comprehensive proofs. A more comprehensive proof of creation seeks to establish the creation of the world *ex nihilo,* whereas a less comprehensive proof seeks only to establish the creation of the form of the world, leaving open the possibility that the matter of the world is eternal. A less comprehensive proof of eternity, that is, a proof of the eternity of matter, is plainly compatible with a less comprehensive proof of creation, a proof solely of the creation of the world in its present form. Accordingly, a proof of the eternity of matter is compatible as well with volition in the deity; for creation in any mode, even creation from a preexistent matter, would constitute a voluntary act on the part of the creator.

Maimonides drew a dichotomy that is highly helpful, if not indispensable, for any analysis of proofs for eternity. The proofs of eternity deserving consideration, he explained, fall into two categories. There are, on the one hand, arguments formulated by Aristotle which take their departure "from the world." On the other hand, there are arguments "extracted" by subsequent philosophers "from Aristotle's philosophy," these, in contrast to the previous category, taking their departure "from God."[2] The distinction between proofs of eternity from the nature of the world and proofs from the nature of God does not seem to have been articulated before Maimonides, but almost every medieval Islamic and Jewish philosopher who treats the issue of eternity does adduce proofs belonging to both categories. If Maimonides' dichotomy were to be collated with the distinction between proofs of the eternity of the world in its form as well as its matter and proofs of the eternity of matter alone, the result would be as follows: Some proofs from the nature of the world, it turns out, seek to demonstrate the eternity of the universe—or, to be more specific, the eternity of the physical universe[3]—in its entirety, whereas others merely have the aim of demonstrating the eternity of matter;[4] proofs of eternity from the nature of God, however, are invariably of the more comprehensive kind and aim at demonstrating the eternity of the universe in its entirety.

[2]*Guide to the Perplexed,* II, 14. The dichotomy reappears in Albertus Magnus, Bonaventure, Aaron ben Elijah, and Crescas. Aquinas in the *Summa contra Gentiles,* II, 32–34, subdivides proofs from the nature of the world into two categories and thereby reaches a trichotomy. Isaac Abravanel, who knew Aquinas' works, has a similar trichotomy in *Mif'alot* (Venice, 1592), IV–VI.

[3]Proofs from the nature of the physical world can lead, naturally enough, only to the eternity of the physical world or to the eternity of the underlying matter of the world. Proofs from the nature of the deity would lead to the eternity of the entire universe, nonphysical as well as physical.

[4]See below, pp. 29–30.

Maimonides' identification of Aristotle as the author of the proofs for eternity from the nature of the world is not inaccurate. The medieval Islamic and Jewish proofs in that category either are borrowed directly from Aristotle or are adaptations of arguments he put forward. Maimonides' description of proofs from the nature of God as having been "extracted . . . from Aristotle's philosophy," in other words, as having been fashioned out of Aristotelian components, is less exact; for one of the three proofs in this category is animated by a Neoplatonic theme that Aristotle would have found uncongenial. The main source or channel through which proofs for eternity from the nature of God reached the Middle Ages was apparently Proclus.[5]

Heeding Maimonides' dichotomy, I deal in the remainder of the present chapter with proofs of eternity from the nature of the world, and in the next chapter with proofs of eternity from the nature of God. Besides those discussed, arguments for eternity are occasionally referred to in the writings of the proponents of creation which are so flimsy that they can hardly be regarded as more than straw men; they cannot conceivably have been advanced by intelligent, self-respecting proponents of the eternity of the world.[6] Arguments of the sort have been disregarded here except insofar as they can be related to serious philosophic proofs.

In classifying proofs of eternity, it is sometimes difficult to decide whether a proof should be assigned to one general category or the other, to the category that reasons from the nature of the world or to the category that reasons from the nature of God.[7] And when enumerating separate and distinct proofs within each category, arbitrariness is unavoidable. Through the centuries the arguments subdivided and underwent variations; and should all subdivisions and variations be listed as separate proofs, the total number swells. For example, arguments for eternity which Maimonides listed as single proofs[8] were later subdivided by Crescas into two or three proofs, with the result that Crescas' list is several times as long as Maimonides'.[9] Aquinas, by enumerating subdivisions and variations, was able in one work to draw up a list of over twenty-five philosophic proofs for eternity.[10] Whereas some philosophers thus subdivide the arguments, others combine them. Proofs that can stand independently may coalesce to form a single complex chain of reasoning. Or else, what was originally an independent proof for eternity may be advanced by a philosopher not as such, but as a surrejoinder

[5]See below, p. 51.

[6]See, for instance, Ibn Ḥazm, *K. al-Faṣl fī al-Milal* (Cairo, 1964), I, p. 10(5). Spanish translation: *Abenházam de Córdoba y su Historia Crítica de las Ideas Religiosas,* trans. M. Asín Palacios, Vol. II (Madrid, 1928), p. 99.

[7]For example, the argument from motion, which is discussed in this chapter, blends into the argument from the unchangeability of the cause of the universe, which is discussed in the next chapter.

[8]*Guide,* II, 14.

[9]*Or ha-Shem,* III, i, 1. Similarly, Abravanel, *Mif'alot,* IV, 3; V, 1; VI, 1.

[10]*De Potentia,* q.3, art. 17. Aquinas lists, in fact, thirty arguments for eternity, but some are scriptural and not philosophic, and some are duplicates.

to the refutation of another proof of eternity.[11] I found that the requirements of exposition would be best served by distinguishing six separate proofs of eternity from the nature of the world and three separate proofs from the nature of God. Alternative enumerations would undoubtedly be equally justified.

The medieval advocates of eternity who will be cited here are: the authors of a corpus of Arabic writings (precise dates unknown) attributed to Alexander of Aphrodisias, Avicenna (980–1037), Abū al-Barakāt (d. ca. 1160), Averroes (1126–1198), and Moses Narboni (d. 1362). In addition, two philosophers argue for the eternity of matter, although they espouse the creation of the world in its present form; they are Abū Bakr b. Zakarīyā Rāzī (ca. 864–925) and Gersonides (1288–1344). The medieval advocates of creation *ex nihilo* who will be cited are: Saadia (892–942), Bāqillānī (d. 1013), ʻAbd al-Jabbār (d. 1025), Ibn Ḥazm (994–1064), the authors of the Jābir corpus (10th century), Juwaynī (1028–1085), Bazdawī (Pazdawī) (d. 1099), Ghazali (1058–1111), Shahrastānī (1086–1153), Maimonides (1135–1204), Fakhr al-Dīn al-Rāzī (1149–1209/10), Āmidī (1156–1233), Ṭūsī (d. 1274), Albertus Magnus (1206–1280), Bonaventure (1221–1274), Thomas Aquinas (1225–1274), Ījī (d. 1355),[12] Aaron ben Elijah (d. 1369), Ḥasdai Crescas (1340–1410), Joseph Albo (1380–1444), and Isaac Abravanel (1437–1508). A final writer to be cited, Leone Ebreo (Judah Abravanel) (d. 1535), does not state clearly whether he endorses creation from a preexistent matter or creation *ex nihilo*.

2. Proofs of eternity from the nature of the physical world

Six proofs for eternity from the nature of the world can, as was mentioned, be distinguished; and some of the six undergo variations. In every instance Aristotle is either the immediate or ultimate source.

The first four of the six proofs disclose an identical structure. To begin, they all proceed indirectly, establishing their own thesis by focusing on the opponent's position and showing it to be untenable: They argue that the world must be eternal since creation is impossible. Indirect reasoning is far from unique to them, however. The fifth proof from the nature of the world also reasons indirectly. Much of the argumentation in the second category, where eternity is proved from the nature of God, proceeds in the same fashion. And the advocates of creation too had a predilection for indirect reasoning; proofs for creation, as will appear in later chapters, typically focus on the thesis of eternity and argue that it is untenable.

What is unique in the first four proofs for eternity is the manner in which they establish the untenability of creation and thereby indirectly establish the eternity of the world. They all contend that the laws of nature—or, to be more accurate,

[11]For example, below, pp. 75–76.

[12]I have not distinguished between the text of Ījī and the commentary of Jurjānī, which is interwoven with it.

the laws of Aristotelian physics—are such that for creation to have taken place, something would already have had to exist similar to what was supposedly coming into existence for the first time. In order that matter should have come into existence, matter would already have had to exist; and in order that a world should come into existence, a world would already have had to exist. Hence the assumption of an absolute beginning of matter or an absolute beginning of the world is self-contradictory, and the world, or matter, must exist from eternity.

I turn now to the individual proofs.

(a) The argument from the nature of matter

In the course of an exhaustive analysis of the nature of matter, Aristotle argued that the underlying matter of the universe must be eternal, and his procedure reveals the pattern just outlined. He established the eternity of matter by showing the creation of matter to be untenable; and his grounds for the untenability of the creation of matter are that matter could only have come into existence from an already existent matter. Aristotle reasoned: Everything that comes into existence does so from a substratum.[13] If the underlying matter of the universe came into existence, it also would come into existence from a substratum. But the nature of matter is precisely to be the substratum from which other things arise. Consequently, the underlying matter of the universe could have come into existence only from an already existing matter exactly like itself; and to assume that the underlying matter of the universe came into existence would require assuming that an underlying matter already existed. The assumption is thus self-contradictory, and matter must be eternal.[14] The argument that the underlying matter of the universe must be eternal since matter could only come into existence from an already existing, prior matter is adduced by the adherents of eternity in defense of their position, and it is recorded by adherents of creation who have in view a subsequent refutation. The adherents of the eternity of matter adducing the argument are Avicenna,[15] Averroes,[16] and Gersonides;[17] the adherents of creation recording it are Maimonides,[18] Albertus Magnus,[19] Aquinas,[20] Ījī,[21] Aaron ben

[13]*Physics* I, 7.

[14]*Physics* I, 9, 192a, 29–31.

[15]*Shifā': Ilāhīyāt*, ed. G. Anawati and S. Zayed (Cairo, 1960), p. 376.

[16]*Epitome of Physics*, in *Rasā'il Ibn Rushd* (Hyderabad, 1947), p. 10; *Middle Commentary on Physics* (Oxford, Bodleian Library, Hebrew MS. Neubauer 1380 = Hunt. 79), I, iii, 3 and 5; *Long Commentary on Physics*, in *Aristotelis Opera cum Averrois Commentariis*, Vol. IV (Venice, 1562), I, comm. 82; *Long Commentary on De Caelo*, in ibid., Vol. V, I, comm. 22.

[17]*Milḥamot ha-Shem* (Leipzig, 1866), VI, i, 3, pp. 300, 302. In VI, i, 17, p. 364, Gersonides in effect accepts the argument.

[18]*Guide*, II, 14(2).

[19]*Physics*, in *Opera Omnia*, ed. A. Borgnet, Vol. III (Paris, 1890), VIII, tr. 1, chap. 11.

[20]*Compendium of Theology*, chap. 99. Cf. *Summa Theologiae*, I, 46, 1, obj. 3.

[21]*Mawāqif* (Cairo, 1907), VII, p. 228.

Elijah,[22] Crescas,[23] Isaac Abravanel,[24] and Leone Ebreo.[25] In most instances, the force of the argument is sharpened. Matter, in the Aristotelian physical scheme, never exists in actuality devoid of all form, and the generation of any object therefore starts not merely with a previously existing matter but with a previously existing compound of matter and form.[26] Generation is the process wherein the compound of matter and form receives an additional form or exchanges its own form for another. The sharpened medieval version of the argument for the eternity of matter accordingly runs: The underlying matter of the world could only have come into existence from an already existing matter. Matter, however, never exists devoid of form, but solely in a compound, together with form. To assume an absolute beginning of the underlying matter of the universe would, then, not merely require assuming an already existent matter before matter existed. It would require assuming that before matter existed, there already existed a full-blown compound of matter and form.[27]

The heart of the foregoing argument or arguments clearly is the principle that whatever comes into existence does so from a substratum. Once the principle is accepted, discussion is virtually precluded; for if whatever comes into existence does so from an already existing substratum or matter, matter obviously is eternal. As for the critical principle, Aristotle in one passage supported it inductively. We can, he wrote, "always" observe "something underlying, from which the generated object comes, plants and animals, for example, [coming] from seed."[28] Elsewhere, though, Aristotle treated the impossibility "that generation should take place from nothing" as self-evident.[29]

In the Middle Ages, the presupposition upon which the argument for the eternity of matter rests received frequent attention. Generally it was adherents of the opposing position whose attention was roused, their object being to expose the argument's feeble foundation; but an exception to the generalization is Zakarīyā Rāzī. Rāzī, an advocate of the eternity of matter although not of the eternity of the world, is reported to have maintained the eternity of matter on the grounds that "intellect" rejects the coming into existence of something from nothing.[30] In effect, that is to say, Rāzī regarded the presupposition of the argument for the

[22]*'Eṣ Ḥayyim,* ed. F. Delitzsch (Leipzig, 1841), chap. 6.

[23]*Or ha-Shem,* III, i, 1.

[24]*Mif'alot,* V, 1(1).

[25]*Dialoghi d'Amore* (Bari, 1929), p. 237; Hebrew translation: *Wikkuaḥ 'al ha-Ahaba* (Lyck, 1871), p. 55b; English translation: *The Philosophy of Love,* trans. F. Friedeberg-Seeley and Jean H. Barnes (London, 1937), pp. 278–279.

[26]*De Generatione* II, 1, 329a, 24–26; *Physics* IV, 7, 214a, 15.

[27]Such is the implication in Averroes' version. The argumentation is explicit in Maimonides, Albertus,Ījī, Aaron ben Elijah, Crescas, and Abravanel.

[28]*Physics* I, 7, 190b, 3–5.

[29]*Metaphysics* III, 4, 999b, 8; see Ross's note, *ad locum.*

[30]Rāzī, *Opera Philosphica,* ed. P. Kraus (Cairo, 1939), I, p. 221.

eternity of matter as self-evident. Aquinas, Aaron ben Elijah, and Crescas, all advocates of creation *ex nihilo,* similarly record arguments for the eternity of matter which rest on a supposedly self-evident proposition. Aquinas reports an argument for "eternity"—which is in fact an argument for the eternity of matter—resting on the "universal opinion of philosophers that nothing is made from nothing."[31] Aaron and Crescas likewise refer to an argument for the eternity of matter resting on the premise that—in Aaron's words—"something cannot be generated from nothing," or—in Crescas' words—"the generation of a body out of nothing is absurd."[32]

Among Kalam writers and writers affiliated with the Kalam, the belief not merely in the eternity of matter but also in the eternity of the world is commonly traced to an induction or analogy, one that the Kalam, of course, deemed improperly drawn.[33] Saadia, for example, avers that the "strongest argument" of the adherents of the eternity of the world in its entirety (*al-dahr*) is the circumstance that an absolute beginning of the processes of nature has never been witnessed.[34] The Jābir corpus wrestles with, and endeavors to refute, two parallel contentions, one to the effect that the world must be eternal since nobody ever observed the world's coming into existence, and the other to the effect that every man must have been born of woman since man was never observed not to be born of woman.[35] Ibn Ḥazm represents the adherents of eternity (*dahrīya*) as reasoning: "We do not see anything coming into existence except from something or in something, and whoever maintains the contrary is maintaining what neither is now observed nor has ever been observed."[36] 'Abd al-Jabbār records a proof possibly containing an echo of the passage in which Aristotle cited the fact that "plants and animals [come] from seed."[37] According to 'Abd al-Jabbār, the advocates of eternity reasoned: "We do not find a hen except from an egg or an egg except from a hen, and things must therefore always have been so. That implies the eternity of the world."[38] Juwaynī's account is similar to 'Abd al-Jabbār's. He describes the advocates of eternity (*dahrīya*) as arguing: "Since we never observe a hen except from an egg or a man except from the sperm of male and female, the judgment that such takes place must be extended to what is hidden from us [in past time]."[39] Bazdawī reports an argument for the eternity of matter resting

[31] *Summa contra Gentiles,* II, 34(1).

[32] *'Eṣ Ḥayyim,* chap. 6; *Or ha-Shem,* III, i, 1.

[33] See below, pp. 30, 34.

[34] *K. al-Amānāt wa-l-I'tiqādāt,* ed. S. Landauer (Leiden, 1880), I, 3(10), p. 63. English translation, with pagination of the Arabic indicated: *Book of Beliefs and Opinions,* trans. S. Rosenblatt (New Haven, 1948).

[35] Jābir ibn Ḥayyān, *Textes Choisis,* ed. P. Kraus (Cairo, 1935), p. 422.

[36] *K. al-Faṣl fī al-Milal,* I, p. 9; Spanish translation, II, p. 94.

[37] Above, n. 28.

[38] *Sharḥ al-Uṣūl* (Cairo, 1965), p. 117. See below, Chapter V, n. 162.

[39] *K. al-Shāmil* (Alexandria, 1969), p. 224.

on the circumstance that "we never have seen anything created from nothing," and concluding "that the world was not created from nothing."[40] Aaron ben Elijah—who, as already seen, refers to an argument for the eternity of matter based on the supposedly self-evident "premise . . . that something cannot be generated from nothing"[41]—also gives what seems to be an argument by induction: Since what is generated is generated "only from something else . . . and only from something specifically [adapted to the resulting product] . . . it follows *a fortiori* that nothing can be generated out of nothing."[42] Finally, Isaac Abravanel records two closely related arguments for the eternity of matter, one in almost the same language as Aaron, and the other explicitly labeled as "inductive." The latter begins with an examination of "particular things, whether substances or accidents, which come into existence"; it discovers that whatever comes into existence does so from "from something, not from nothing"; and it arrives at the "judgment that everything comes into existence from something, and nothing can come into existence from nothing."[43]

(b) The argument from the concept of possibility

In the Aristotelian physical system, *possibility* and *matter* are closely related concepts, matter being the locus of potentiality, or possibility.[44] The relationship between the concepts led to the development of a proof for the eternity of matter from the concept of possibility, alongside the proof from the nature of matter, examined in the preceding paragraphs. The germ of the proof from the concept of possibility can be discovered easily enough in Aristotle. In one passage, for example, Aristotle established that the process of coming into existence "necessarily" requires "the prior presence of something existent potentially [or: possibly], but not existent in actuality";[45] it follows that nothing whatsoever, including matter, can come into existence from absolute nothingness. In the Middle Ages, the proof takes a distinctive cast, however, and the earliest philosopher to whom I could trace the new and distinctive formulation is Avicenna. Considering that the concept of possibility was a central concern of Avicenna's,[46] he very likely is the author of the argument as it is found in the Middle Ages.

Avicenna lays down the proposition that prior to a thing's coming into actual existence, its existence must have been possible; were its existence necessary, he explains, the thing would already have existed, and were its existence impossible, the thing would never exist.[47] The "possibility (*imkān*) of the existence" of a

[40] Bazdawī (Pazdawī), *K. Uṣūl al-Dīn*, ed. H. Linss (Cairo, 1963) p. 16.
[41] Above, n. 32.
[42] *'Eṣ Ḥayyim*, chap. 6.
[43] *Mif'alot*, IV, 3(1,2).
[44] Cf. *Metaphysics* XIV, 1, 1088b, 1. *Potentiality* and *possibility* both translate the Greek δύναμις.
[45] *De Generatione* I, 3, 317b, 16–17.
[46] See below, Chapter IX.
[47] Cf. Aristotle, *De Generatione* II, 9, 335a, 32 ff.

thing must, moreover, in some sense have its own existence. For if the possibility were completely "nonexistent," it could not legitimately be spoken of as "being prior" to the thing's actual existence, and everyone will surely acknowledge that the possibility is prior. Mere possibility of existence is plainly not a substance. It can only belong to the class of entities that are "present in a subject." Thus whenever anything comes into existence, the possibility of its existence must previously have subsisted in a subject. The "possibility (*imkān*) of existence" may also be termed the "potentiality (*qūwa*) of existence";[48] and the subject in which possibility or potentiality is found is called "matter" (*mādda, hayūlā*). Whenever anything comes into existence, the possibility or potentiality of its existence must, then, previously have subsisted in an already existent subject, or matter. But, Avicenna concludes, if an already existent matter must precede everything coming into existence, clearly nothing, including matter, can come into existence *ex nihilo,* that is, from absolute nothingness. An absolute beginning of the existence of matter is impossible.[49]

The argument for the eternity of matter from the concept of possibility is cited in Ghazali's critique of Avicenna's philosophic system,[50] in Shahrastānī's account of Avicenna's philosophy,[51] in Fakhr al-Dīn al-Rāzī,[52] Averroes,[53] Maimonides,[54] Āmidī,[55] Albertus Magnus,[56] Aquinas,[57] Gersonides,[58] Ījī,[59] Aaron ben Elijah,[60] and Isaac Abravanel.[61]

(c) The argument from the nature of motion

In *Physics* VIII, 1, as a preliminary to his demonstration of the existence of a first mover of the universe, Aristotle undertook to prove the eternity of motion. His proof contains several strands, and they were put to various uses in the Middle Ages.

[48]δύναμις underlies both *imkān* and *qūwa*.

[49]*Najāt* (Cairo, 1938), pp. 219–220; cf. *Shifā': Ilāhīyāt,* pp. 177–178.

[50]*Tahāfut al-Falāsifa,* ed. M. Bouyges (Beirut, 1927), I, §§81–82; English translation in *Averroes' Tahafut al-Tahafut,* trans. S. van den Bergh (London, 1954), p. 57.

[51]*K. al-Milal wa-l-Niḥal,* ed. W. Cureton (London, 1842–1846), p. 371.

[52]*Muḥaṣṣal* (Cairo, 1905), p. 91; *K. al-Arba'īn* (Hyderabad, 1934), p. 49.

[53]*Tahāfut al-Tahāfut,* ed. M. Bouyges (Beirut, 1930), I, pp. 69, 74, 100; English translation, with pagination of the Arabic indicated: *Averroes' Tahafut al-Tahafut,* trans. S. van den Bergh (London, 1954).

[54]*Guide,* II, 14(4).

[55]*Ghāya al-Marām* (Cairo, 1971), p. 267.

[56]*Physics,* VIII, tr. 1, chap. 11.

[57]*Summa Theologiae,* I, 46, 1, obj. 1; *Summa contra Gentiles,* II, 34(3); *De Potentia,* q. 3, art. 17(10).

[58]*Milḥamot ha-Shem,* VI, i, 17, p. 365; he accepts the argument.

[59]*Mawāqif,* VII, p. 230; the argument is not fully developed.

[60]*'Eṣ Ḥayyim,* chap. 6.

[61]*Mif'alot,* IV, 3(3).

Aristotle's proof ran: Motion—taking the term in the broad sense that includes change of all types[62]—obviously can occur only if an object undergoing motion is "present" (ὑπάρχειν).[63] If an absolute beginning of motion should be assumed, the object to undergo the first motion must either (i) have come into existence, or (ii) have been eternal. That is to say, an absolute beginning of motion can be construed in one of two ways: (i) The world came into existence and began to move; or (ii) the world existed in an eternal state of rest before beginning to move. The eternity of motion will be established when both alternatives are ruled out.

Inasmuch as coming into existence is one type of motion,[64] alternative (i) asserts in effect that before the world performed its absolutely first motion, it had already performed another motion, namely the motion of coming into existence. Alternative (i) asserts, therefore, that the absolutely first motion was not after all the first motion, which is a self-contradiction; and the proposition that the world came into existence and thereupon began to move is thus untenable.[65] It is to be noted that Aristotle fails to explain why the absolutely first movement may not have been precisely the coming into existence of the physical universe. On such a theory, motion and the physical universe would have come into existence together; the requirement that motion takes place only if an object undergoing motion is "present" would be met; and an absolute beginning of motion as well as of the world could be defended. The question is taken up by Themistius[66] and Averroes.[67]

After Aristotle had, to his satisfaction, ruled out alternative (i)—an absolute beginning of motion in a world that likewise had a beginning—alternative (ii) remained. Alternative (ii) is the thesis that motion had an absolute beginning but the world is eternal, in other words, the thesis that the world existed in a state of rest for an eternity before starting to move. Aristotle found the thesis to be inadmissible for two reasons.

[62]Aristotle distinguished four sorts of motion or change, each of which takes place in a different category: change in the category of substance, that is to say, coming into existence or destruction; change in the category of quantity, that is to say, growth or diminution; change in the category of quality, that is to say, alteration; and change in the category of place. His statements are by no means consistent, however. See E. Zeller, *Die Philosophie der Griechen,* Vol. II, Part 2, (4th ed.; Leipzig, 1921), pp. 389–390; W. D. Ross, *Aristotle* (London, 1953), pp. 82–83.

[63]Aristotle bases the proposition both on common sense and on his definition of motion.

[64]See n. 62.

[65]*Physics* VIII, 1, 251a, 8–20.

[66]Themistius deals with the question indirectly. He has Aristotle contend not that the object to undergo motion must be "present" in order for motion to take place, but that it must be "previously present." Something would accordingly always have to exist prior to the occurrence of motion. See Themistius, *Paraphrase of Physics,* ed. H. Schenkel, *Commentaria in Aristotelem Graeca,* Vol. V/2 (Berlin, 1900), p. 210.

[67]See below, p. 21.

(α) Just as some factor must be posited as the cause of the motion of whatever undergoes motion, so too, Aristotle writes, must a factor be posited as the cause of the state of rest of whatever is at rest. If the world had been at rest before starting to move, something must have produced its state of rest. But producing the state of rest would itself be a motion. A motion would consequently have preceded the supposed absolutely first motion; and the assumption that the physical universe existed in a state of rest from all eternity before undergoing an absolutely first motion embodies, like the previous alternative, a self-contradiction.[68]

Aristotle's thinking, again, has puzzling aspects. In the first place, once he laid down the rule that everything at rest must be set at rest he could have proceeded more simply. He could have argued that the thesis of a world in eternal rest is in itself self-contradictory because it implies a prior motion whereby the world was set at rest; and the supposedly eternal state of rest would not be eternal after all. In the second place, the rule that everything at rest must have been set at rest by something else is most strange and gratuitous. What grounds can there be for supposing that every state of rest is produced and has a cause? The likelihood—and this addresses only the second of the two puzzling points—is that Aristotle was arguing not in the abstract but *ad hominem*; a few lines earlier he had been discussing Empedocles,[69] and in Empedocles' system the world in its state of rest is indeed set at rest through a prior process. At any rate, the contention that a preexistent state of rest would have had to be brought about by a cause does not, as far as I could discover, recur in the Middle Ages.

(β) Aristotle offered a second reason for ruling out alternative (ii). He submits that the world's beginning to move after having been stationary allows only a single interpretation. The relationship between the world and whatever causes its motion would previously not have permitted motion, whereupon the relationship would have changed, the motion of the world being an outcome of the change. But if that had happened, the supposed first motion would have been preceded by another motion, to wit, the change in relationship between what produces and what undergoes motion. The thesis that the body of the universe existed in a state of rest before undergoing an absolutely first motion is seen once more to imply a prior motion and to embody a self-contradiction.[70]

In sum: If motion should have had a beginning and the world is (i) assumed to be generated, its generation would have constituted a motion prior to the supposed absolutely first motion. If motion should have had a beginning and the world is (ii) assumed to be eternal—if the world is assumed to have existed in an eternal state of rest before starting to move—then (α) the process whereby the world was set in its state of rest would have constituted a motion prior to the supposed absolutely first motion. Moreover, (β) the change in relationship which

[68] *Physics* VIII, 1, 251a, 20–28.

[69] Ibid., 250b, 26–27.

[70] Ibid., 251b, 1–10.

initiated the first motion—that is, the change in the relationship between the world and the cause producing its motion—would also have constituted a motion before the supposed absolutely first motion. Having exposed the inadmissibility of the two possible ways of construing an absolutely first motion, Aristotle concluded that motion cannot have had a beginning but must, together with a world capable of undergoing motion, be eternal.

In the Middle Ages, Aristotle's complex argumentation was sometimes advanced with both alternatives carefully laid out and shown to be untenable. In addition, his reason for ruling out alternative (i) was sometimes advanced as an independent proof either for the eternity of motion or for the eternity of the celestial spheres.

Instances of the complete argument

In one of a collection of Arabic works attributed to Alexander of Aphrodisias, the conceivable ways of construing an absolutely first motion are spelled out as Aristotle had spelled them out: The object to undergo the first motion—that is, the physical universe—would either (i) have to come into existence, or (ii) be eternal. As for alternative (i), the world's "coming into existence would clearly precede [its existing and moving]. . . . Coming into existence occurs, however, only through[71] motion, and consequently motion would have existed before motion existed." That is to say,[72] the body undergoing the first motion would have come into existence through a motion preceding what was, by hypothesis, its first motion. The supposed absolutely first motion would have been preceded by another motion and would not in fact be first.

As for alternative (ii), it is ruled out by two considerations. The first is not the one offered by Aristotle—the strange notion that if the world had been at rest some factor would have had to produce its rest. Instead, a substitution is made and the rhetorical question is posed: (α) "How might anyone explain the world's starting to move now, after having been at rest for an infinite time?"[73] That question reflects a separate, widely utilized proof of the eternity of the world, which is to be taken up in the next chapter.[74] The second consideration advanced by the text attributed to Alexander is a slight expansion[75] of Aristotle's second reason for the inadmissibility of alternative (ii). Here the argument is: (β) If the world were stationary before beginning to move, "either the factor producing motion did not exist previously . . . or else the factor producing motion did not stand in the requisite relation to the object that was to undergo motion." But to assume either that the factor producing motion came into existence or that the

[71] Aristotle reasoned that coming into existence *is* a motion. See Averroes' version of the argument, immediately below.

[72] See previous note.

[73] Strictly speaking, for an Aristotelian there would have been no time without motion.

[74] Below, p. 52.

[75] Aristotle did not mention the possibility that the cause producing motion was not yet in existence.

relationship between it and the object undergoing motion changed would be to assume a motion prior to the supposed first motion. The assumption that the world was eternally at rest and thereupon began to move thus entails, once more, a motion prior to the first motion, which is a self-contradiction. Since the possible ways of conceiving a first motion are untenable, motion, the conclusion is drawn, must be eternal.[76]

Averroes also offers a proof of the eternity of motion which repeats Aristotle's distinction of the two conceivable ways for motion to have begun. Should it be assumed, Averroes writes, that (i) "the object to undergo the first motion came into existence," a prior motion would thereby be implied. In specifying what the implied prior motion would be, Averroes fills in a gap left by Aristotle. Aristotle had failed to explain why the coming into existence of the object to undergo motion—that is, the creation of the world—might not itself be the very first motion ever to have occurred;[77] the deficiency is now made up by Averroes with the aid of a tenet of Aristotelian physics. Four genera of motion or change are recognized in Aristotelian physics, and the most primary of the four is motion in place, a motion in place ultimately lying behind all incidents of motion or change in the other genera.[78] Averroes accordingly completes his elimination of alternative (i) by arguing that the "motion of coming into existence cannot be the first [motion ever to have occurred], since coming into existence is always dependent upon a [prior] motion in place."[79] To assume, therefore, either that the physical universe came into existence and then performed the first motion to have occurred, or alternatively, to assume that the coming into existence of the physical universe was itself the first motion ever to occur, is self-contradictory. In either case, "the motion that was by supposition first would not in fact be first," since a prior motion in place would perforce be implied.

As for the assumption that (ii) "the object to undergo motion was at rest for an eternity and thereupon moved," such an assumption would also imply "a motion prior to the supposed first motion. . . . For if what produces motion and what is to undergo motion exist eternally . . . [and yet motion occurs only at a certain moment], some further factor must come into existence which induced the cause producing motion to produce motion and the object undergoing motion to undergo motion, after not having done so. But what comes into existence[80] is

[76] Alexander of Aphrodisias (?), *Mabādi' al-Kull,* in *Arisṭū 'ind al-'Arab,* ed. A. Badawi (Cairo, 1947), p. 263.

[77] See above, p. 18.

[78] *Physics* VIII, 7, and cf. below, p. 241.

[79] An added point made by Averroes is that when motion is taken in a narrow sense, the process of coming into existence is, although a form of change, not a form of motion. Aristotle is not consistent on the question whether coming into existence should, or should not, be characterized as motion. See references to Zeller and Ross, above, n. 62.

[80] One would expect Averroes to have written: "the *process* of coming into existence is a motion. . . ."

either a motion or the result of a motion."[81] Consequently, the coming into existence of the factor inducing motion would constitute, or at least entail, a prior motion; and the supposedly first motion again turns out to have been preceded by another. The two possible ways of construing an absolute beginning of motion having been ruled out, motion—and hence something capable of undergoing motion—must be eternal.[82]

Versions of the complete Aristotelian proof for the eternity of motion very similar to Averroes' version appear in Albertus Magnus,[83] who could have borrowed the proof from Averroes; in Gersonides,[84] who undoubtedly did; in Crescas,[85] who drew from either Averroes or Gersonides; and in Abravanel,[86] who copies from Gersonides.

Instances where the argument eliminating alternative (i) is offered as an independent proof

Side by side with the composite argument just examined, Averroes offers a proof from motion which has the specific aim of establishing the eternity of the heavens. His proof consists in a slight elaboration of the reasoning whereby Aristotle had eliminated alternative (i) in the complete, composite argument.

In Aristotelian physics, as already mentioned, motion in place is the primary genus of motion and change, every incident of motion and change in the other genera being ultimately traceable to a motion in place. Averroes now reasons: Motion in place is responsible for the other kinds of motion and change. Whenever motion and change occur in the sublunar world, what is ultimately responsible is the circular motion in place of the celestial spheres. Should the celestial spheres themselves have come into existence, there must exist a body "prior to the celestial spheres" which "undergoes motion in place . . . and is thereby responsible for the spheres' coming into existence." Should the body that brought about the existence of the celestial spheres have itself come into existence, its coming into existence must be due to the motion in place of still another body, prior to it. Since an infinite series of these bodies is impossible,[87] the series must "end at [a class of] eternal bodies undergoing a motion in place precisely like

[81] Averroes adds the words "the result of a motion" because he understands that the process of coming into existence is not strictly a motion. See n. 79.

[82] *Middle Commentary on Physics,* VIII, ii, 2, taken together with *Epitome of Physics,* pp. 108–109.

[83] *Commentary on II Sentences,* in *Opera Omnia,* ed. A. Borgnet, Vol. XXVII (Paris, 1894), d. 1, B, art. 10.

[84] *Milḥamot ha-Shem,* VI, i, 3, pp. 299–300.

[85] *Or ha-Shem,* III, i, 1. A few lines have fallen out in the printed editions and I have used one of the manuscripts.

[86] *Mif'alot,* V, 1(2).

[87] Both because of the impossibility of an infinite number of bodies (cf. below, p. 243) and because of the impossibility of an infinite regress of causes (cf. below, p. 241).

[that of] the celestial spheres." Consequently either the celestial spheres are eternal or there exists something prior to them and precisely like them which is eternal.[88]

A proof of the eternity of the celestial spheres framed in the same pattern is recorded by Gersonides,[89] who undoubtedly took it from Averroes; by Crescas,[90] whose source was either Averroes or Gersonides, and who condenses the argument considerably; and by Abravanel,[91] who copies the argument from Crescas.

A somewhat similar argument is recorded by Maimonides. Like the original proof of Aristotle, Maimonides' version undertakes to demonstrate the eternity of motion; but his formulation is plainly designed to simplify Aristotle's, where separate treatment had been given to the alternative that the world was created and the alternative that the world is eternal. Ignoring the distinction between the two alternatives, Maimonides' version reads: "Whatever comes into existence is preceded by a motion, to wit, its passing [from potentiality] to actuality and its coming into existence after not having existed." Therefore, "if motion itself should have come into existence . . . [a prior] motion would already have had to exist, namely the motion through which motion supposedly came into existence." The very coming into existence of motion would, in other words, itself have been a motion. Should, moreover, the motion through which motion came into existence likewise be assumed to have a beginning, its coming into existence too would have been a motion. The assumption of an absolute beginning of motion thus entails the assumption of "an infinite regress" of motions. The assumption is hence contradictory and self-destructive, and motion taken collectively must be eternal.[92] Virtually the same argument is also recorded by Bonaventure.[93]

Moses Narboni subsequently discovered a flaw in the reasoning. Aristotle's *Physics* had established the plausible proposition that "there cannot be motion of motion, or a coming into existence of coming into existence, or, in general, change of change."[94] What the proposition means is that the beginning of a given change is not a change distinct from, and prior to, the given change. Accordingly, Narboni insists, no true philosopher of the Aristotelian school would maintain that motion comes about through a process which is itself a motion; and the argument recorded by Maimonides is neither cogent nor genuinely Aristotelian.[95] In its place, Narboni offers his own adaptation of Aristotle's original composite proof of the eternity of motion.[96]

[88] *Middle Commentary on Physics,* VIII, ii, 2.

[89] *Milḥamot ha-Shem,* VI, i, 3, pp. 298–299.

[90] *Or ha-Shem,* III, i, 1; the printed texts again have to be corrected with the aid of the manuscripts. Crescas omits explicit mention of the celestial spheres, possibly because he was not certain of their existence.

[91] *Mif'alot,* V, 1(2).

[92] *Guide,* II, 14(1).

[93] *Commentary on II Sentençes,* d. 1, p. 1, a. 1, q. 2(2).

[94] *Physics* V, 2, 225b, 15–16.

[95] *Commentary on Guide* (Vienna, 1853), II, 14, p. 30a.

[96] Ibid., 30b–31a.

Another argument of the type we are considering appears in Aaron ben Elijah. Aaron's version opens with the statement that "every change is preceded by a motion," but he furnishes no explanation as to why the statement is true or even what exactly is meant by *change*. If the motion preceding a given change is preceded by another motion, and the latter by still another, an infinite regress ensues, which is absurd. As motions are traced back, therefore, a motion must eventually be reached which never underwent change, which is ever constant, and hence is eternal.[97] The conclusion is stronger than the conclusion of the argument recorded by Maimonides. Maimonides' version established that motion taken collectively and generically must be eternal; Aaron's version establishes that there must exist a single specific motion—the motion of the celestial spheres—which is eternal.

Brief arguments for the eternity of motion also appear in Albertus Magnus,[98] Aquinas,[99] and Leone Ebreo.[100]

(d) Arguments from the nature of time

After Aristotle presented his proof for the eternity of motion, he added that eternity is implied by the nature of time. He brought forward two considerations.

He argued (i) that time must be eternal because there can be no "*before* and *after* without time." That is to say, should time be assumed to have a beginning, what was before time could still legitimately be spoken of. The term *before* has, however, a temporal connotation, signifying *prior in time,* and therefore everyone who assumes a beginning of time inescapably finds himself referring to prior time. An absolute beginning of time is consequently impossible, and time must be eternal. Since time must be eternal, and since time goes hand in hand with motion,[101] there being no "time without motion, . . . motion too must be eternal." And if motion is eternal, something undergoing motion must have always existed. The impossibility of an absolutely first time entails, then, the eternity of time as well as the eternity of motion and some sort of physical world.[102]

Aristotle further argued (ii) that the concept of time and the concept of the moment are interrelated: "Time can neither be, nor be thought of, apart from the moment." But the nature of the moment is to divide past from future; for the moment serves as "a beginning of the future and an end of the past." The assumption of an absolutely first moment would consequently carry with it the implication of a period of time which is terminated by, and prior to, that first moment;

[97] *'Eṣ Ḥayyim,* chap. 6.
[98] *Physics,* VIII, tr. 1, chap. 11.
[99] *Summa Theologiae,* I, 46, 1, obj. 5.
[100] *Dialoghi d'Amore,* p. 238.
[101] *Physics* V, 11, 220a, 24–25: "Time is the number of motion in respect of before and after."
[102] *Physics* VIII, 1, 251b, 10–13.

and the prior time would itself contain moments. The assumption of an absolutely first moment is thus self-contradictory. Inasmuch as a first moment is impossible, time together with motion and a world undergoing motion must be eternal.[103]

These Aristotelian grounds for the eternity of time might strike us as highly dubious, since they rest not so much on the nature of time as on the idiosyncracies of human discourse and on the Aristotelian definitions of time and the moment.[104] Nevertheless, the arguments were taken seriously in the Middle Ages.

(i) A number of philosophers cite the argument that if time should be assumed to have a beginning, what was before time could still legitimately be spoken of. Frequently the reasoning is that, at the least, the "nonexistence (*'adam*) of time," could be described as preceding time. But if anything, even nonexistence, can be spoken of as *before* the assumed first time, time already existed. An absolutely first time is therefore impossible; and time, as well as motion, and a world, are eternal. The argument as thus formulated is to be found in Avicenna,[105] Ibn Ṭufayl,[106] Averroes,[107] Fakhr al-Dīn al-Rāzī,[108] Āmidī,[109] Aquinas,[110] Gersonides,[111] Ījī,[112] Crescas,[113] and Abravanel.[114]

A minor variation consists in the combination of the argument from time with the argument from the concept of possibility, examined earlier.[115] The reasoning now is: To assume that the world was created would be to assume that the world had the possibility of being created before actually being created. But to speak of a possibility before creation is to imply a time before creation; and to imply time before creation is to imply that motion and a world already existed before creation. The assumption of the creation of a world where no world existed before is therefore untenable. This version of the argument, like the previous version,

[103]Ibid., 19–28.

[104]Cf. G. Hourani, "The Dialogue between al-Ghazālī and the Philosophers on the Origin of the World," *Muslim World*, XLVIII (1958), 190.

[105]*Najāt*, p. 117.

[106]*Ḥayy ben Yaqdhān*, ed. and trans. L. Gauthier (Beirut, 1936), Arabic text, p. 81; French translation, p. 62; English translation: *Hayy Ibn Yaqzān*, trans. L. Goodman (New York, 1972), with pagination of the Arabic indicated. Of the writers cited here Ibn Ṭufayl is the only one who does not reason specifically that it is the nonexistence of time which would precede time.

[107]*Long Commentary on Physics*, VIII, comm. 10; *Epitome of Metaphysics*, ed. and Spanish trans. C. Quirós Rodríguez (Madrid, 1919), IV, §4; German translation: *Die Epitome der Metaphysik des Averroes*, trans. S. van den Bergh (Leiden, 1924), p. 106. *K. al-Kashf*, ed. M. Mueller (Munich, 1859), p. 34; German translation: *Philosophie und Theologie von Averroes*, trans. M. Mueller (Munich, 1875), with pagination of the Arabic indicated.

[108]*Muḥaṣṣal*, p. 91; K. *al-Arba'īn*, p. 50.

[109]*Ghāya al-Marām*, p. 266.

[110]*Summa contra Gentiles*, II, 33(6).

[111]*Milḥamot ha-Shem*, VI, i, 3, p. 298.

[112]*Mawāqif*, VII, p. 228.

[113]*Or ha-Shem*, III, i, 1.

[114]*Mif'alot*, V, 1(3).

[115]Above, p. 16.

is found in Avicenna[116] and probably originated with him. It subsequently appears in Ghazali's critique of Avicenna's philosophy,[117] in Shahrastānī, who cites it in the name of Avicenna,[118] in Gersonides,[119] in Aaron ben Elijah,[120] and in Crescas.[121]

Another variation insists that the creator would surely have had to exist *before* any assumed creation of the world. The assumed beginning of time would for that reason be preceded by another time, and time would not after all have an absolute beginning. The philosopher responsible seems again to have been Avicenna.[122] The argument reappears in Ghazali's critique of Avicenna's philosophy[123] and in Aquinas.[124]

Yet a further variation plays on the term *when* rather than the term *before*. The reasoning is that should a beginning of time be assumed, the period *when* there was no time could nonetheless be spoken of; and a time prior to the assumed first time would be acknowledged. The contention appears in Proclus,[125] in the medieval Arabic translation of Proclus' arguments for eternity,[126] in Shahrastānī's paraphrase of Proclus' arguments,[127] and in the Arabic corpus attributed to Alexander of Aphrodisias.[128]

(ii) A number of philosophers also cite Aristotle's second consideration, the argument that since the *moment* by its nature divides the past from the future, every moment is preceded by other moments, and an absolutely first moment is impossible. That line of reasoning is advanced by the Arabic corpus attributed to Alexander[129] and by Averroes.[130] It is recorded by Albertus Magnus,[131] Bonaventure,[132] Aquinas,[133] Gersonides,[134] Crescas,[135] Abravanel,[136] and Leone Ebreo.[137]

[116]See the passage from an unpublished work of Avicenna's quoted by S. Pines, "An Arabic Summary of a Lost Work of John Philoponus," *Israel Oriental Studies,* II (1972), 350.

[117]*Tahāfut al-Falāsifa,* I, §81; English translation, p. 57.

[118]*K. Nihāya al-Iqdām,* ed. A. Guillaume (Oxford and London, 1934), p. 33.

[119]*Milḥamot ha-Shem,* VI, i, 3, p. 299.

[120]*'Eṣ Ḥayyim,* chap. 6.

[121]*Or ha-Shem,* III, i, 1.

[122]*Shifā': Ilāhīyāt,* p. 379.

[123]*Tahāfut al-Falāsifa,* I, §§57, 70; English translation, pp. 37, 48.

[124]*Summa Theologiae,* I, 46, 1, obj. 8; *De Potentia,* q. 3, art. 17(20).

[125]Quoted by John Philoponus, *De Aeternitate Mundi Contra Proclum,* ed. H. Rabe (Leipzig, 1899), p. 103. See Philoponus' exposition, ibid., p. 104.

[126]Published by A. Badawi in *Neoplatonici apud Arabes* (Cairo, 1955), p. 38.

[127]*K. al-Milal wa-l-Niḥal,* p. 339.

[128]*Mabādi' al-Kull,* p. 264.

[129]Ibid.

[130]*Epitome of Physics,* p. 111; *Middle Commentary on Physics,* VIII, ii, 3; *Long Commentary on Physics,* VIII, comm. 11; *Epitome of Metaphysics,* IV, §4.

[131]*Commentary on II Sentences,* d. 1, B, art. 10.

[132]*Commentary on II Sentences,* d. 1, p. 1, a. 1, q. 2.

[133]*Summa Theologiae,* I, 46, 1, obj. 7; *Summa contra Gentiles,* II, 33(5).

[134]*Milḥamot ha-Shem,* VI, i, 3, p. 298.

[135]*Or ha-Shem,* III, i, 1.

[136]*Mif'alot,* V, 1(3).

[137]*Dialoghi d'Amore,* p. 238.

(iii) In addition to Aristotle's two considerations, the following line of thought is to be found: Everything that comes into existence, comes into existence *in* time. If time itself came into existence, it too would have come into existence in time. Time would therefore extend back beyond the assumed first time; and the assumption that time had an absolute beginning turns out, once more, to be self-contradictory. The argument is recorded by 'Abd al-Jabbār,[138] Bonaventure,[139] Gersonides,[140] and Abravanel.[141]

(e) The argument from the vacuum

Aristotle had established that a vacuum, completely empty space, cannot possibly exist, and the impossibility of a vacuum became one of the principles of his physics. In one passage he utilized the principle in constructing a proof of the eternity of matter.

He reasoned: Material objects can come into existence only in place. On the hypothesis of "absolute" generation, that is, the hypothesis that something came into existence from nothing, "the place to be occupied by what comes into existence would previously have been occupied by a vacuum, inasmuch as no body existed." But a vacuum is impossible. Consequently, the generation of something from nothing is impossible, and matter must be eternal.[142] In the Middle Ages the argument from the vacuum is hinted at by Saadia[143] and is given by Averroes,[144] Aquinas,[145] Gersonides,[146] Crescas,[147] and Abravanel.[148]

[138] *K. al-Majmū' fī al-Muḥīṭ bi-l-Taklīf,* ed. J. Houben (Beirut, 1965), pp. 65–66.

[139] *Commentary on II Sentences,* d. 1, p. 1, a. 1, q. 2.

[140] *Milḥamot ha-Shem,* VI, i, 3, p. 298; 20, p. 381. Gersonides seems to attribute the argument to Aristotle. I was not able to find it in Aristotle; but the source could be Averroes' *Middle Commentary on Physics,* VIII, ii, 3, where Averroes writes: "What is generated and destroyed undergoes those processes *in a moment*; and the moment is the beginning of the future and the end of the past." Averroes probably means merely that if time began, its initial terminus would be a moment, that being the sense of Aristotle's *Physics* VIII, 1, 251b, 23, which he is paraphrasing. But Averroes could also be taken as meaning that if time underwent the process of generation, it would have been generated in a moment, hence *in* time.

[141] *Mif'alot,* V, 1(3).

[142] *Physics* IV, 6–9; *De Caelo* III, 2, 301b, 31 ff.

[143] *K. al-Amānāt,* p. 71.

[144] *Long Commentary on De Caelo,* III, comm. 29; *K. al-Kashf,* p. 34.

[145] *Summa Theologiae,* I, 46, 1, obj. 4.

[146] *Milḥamot ha-Shem,* VI, i, 3, p. 301 (bottom); on VI, i, 17, p. 364, Gersonides accepts the argument.

[147] *Or ha-Shem,* III, i, 1.

[148] *Mif'alot,* IV, 3(4).

(f) The argument from the nature of the celestial spheres

In *De Caelo* I, Aristotle set forth an argument for the eternity of the celestial spheres which rests on one of the fundamentals of his physics, his analysis of the process of generation.

Aristotle understood the process of generation to consist in something's losing its previous character and adopting the contrary character. When a substance is generated,[149] a portion of matter loses its previous character, which was the absence, or "privation," of the form being acquired, and it adopts the contrary character, which is the new form.[150] The process of destruction likewise is a passage from one contrary to the other: In the destruction of a substance, the matter loses its form and is left with the absence or privation of the form, which, again, is the contrary of its previous character. The two processes differ in that the outcome is positive in the case of generation and negative in the case of destruction; generation begins with, whereas destruction ends with, the absence of a given form.[151] It follows from Aristotle's analysis that the process of generation as well as the process of destruction can occur only where a substratum is amenable to contraries. For without a substratum amenable to contraries, no substance can be generated through the acquisition of a new form nor destroyed through the loss of a present form.

Aristotle further explained that the nature of a thing expresses itself in the thing's motion, and contrary natures express themselves in contrary motions. Yet he demonstrated, or thought he demonstrated, that one type of motion has no contrary: "No motion is contrary to motion in a circle."[152] Now the celestial spheres do, by their nature, undergo circular motion. Since the motion of the celestial spheres has no contrary, their nature has no contrary; and whatever substratum the spheres have,[153] Aristotle inferred, must be of a type not amenable to contraries. The substratum of the spheres can, consequently, never have passed from a condition wherein it did not have its present nature and form to the contrary condition wherein it does, and the spheres can never have undergone the process of generation. They, together with the sublunar region whose existence is implied by theirs, must have always existed.[154]

[149]Aristotle's analysis also covers cases where a substance lacking a certain characteristic acquires the characteristic, as when an unmusical man becomes a musical man.

[150]*Physics* I, 7. Either one of two things can be thought of as having come into existence: the abiding material substratum together with the contrary character adopted by it; or alternatively, and in a stricter sense, only the new character adopted by the underlying substratum.

[151]Cf. *Physics* V, 1.

[152]*De Caelo* I, 4.

[153]Different positions were taken in the Middle Ages on the question whether the celestial spheres have a material substratum analogous to the substratum of objects in the lower, sublunar world. See H. Wolfson, *Crescas' Critique of Aristotle* (Cambridge, Mass., 1929), pp. 594–598.

[154]*De Caelo* I, 3, 270a, 12–22. Aristotle more or less takes for granted that the existence of the sublunar region is entailed by the existence of the celestial region. See *De Caelo* I, 8; Zeller, *Die Philosophie der Griechen*, II, 2, pp. 432–433.

In the Middle Ages, the foregoing argument, usually much abbreviated, is found in Proclus,[155] Avicenna (?),[156] Averroes,[157] Aquinas,[158] Gersonides,[159] Aaron ben Elijah,[160] Crescas,[161] and Abravanel.[162]

A slightly roundabout variation appears as well. It starts not with the process of generation but with the process of destruction; and it combines Aristotle's analysis of the process of destruction with the principle—also demonstrated by Aristotle in the *De Caelo*[163]—that what is not subject to destruction is not subject to generation. The reasoning is: Since the celestial spheres undergo only circular motion, their motion has no contrary and hence their nature has no contrary. The substratum of the spheres is therefore not amenable to contraries and can never pass from the condition wherein it does have its present nature and form to the contrary condition wherein it no longer will; and the spheres are not subject to destruction. But what is not subject to destruction is not subject to generation. The spheres, together with the sublunar region,[164] must accordingly have existed forever. This variation of the argument is recorded by Shahrastānī,[165] Maimonides,[166] Albertus Magnus,[167] Aquinas,[168] Gersonides,[169] Crescas,[170] and Leone Ebreo.[171]

Another variation is alluded to by Maimonides,[172] and is reported by Aquinas,[173] Aaron ben Elijah,[174] and Leone Ebreo.[175] It runs simply, and perhaps speciously: The celestial spheres cannot have a "beginning" because the shape and motion of the spheres are circular, and the circle has no beginning.

The arguments from (a) the nature of matter, (b) the concept of possibility, and (e) the vacuum, would establish the eternity of matter. They disprove creation

[155]Quoted by Philoponus, *De Aeternitate,* p. 478.

[156]*De Caelo,* chap. 4, in *Opera* (Venice, 1508). This seems to be a paraphrase of Themistius' *De Caelo,* and not a genuine work of Avicenna's. See M. Alonso, "Ḥunayn Traducido al Latin," *Al-Andalus,* XVI (1951), 37–47.

[157]*Epitome of De Caelo,* in *Rasā' il Ibn Rushd* (Hyderabad, 1947), p. 11; *Middle Commentary on De Caelo* (Vatican Library, Hebrew MS. Urb. 40), I, vi; *Long Commentary on De Caelo,* I, comm. 20.

[158]*Summa contra Gentiles,* II, 33(1). Cf. *Summa Theologiae,* I, 46, 1, obj. 3.

[159]*Milḥamot ha-Shem,* VI, i, 3, p. 300.

[160]*'Eṣ Ḥayyim,* chap. 6.

[161]*Or ha-Shem,* III, i, 1.

[162]*Mif'alot,* V, 1(4).

[163]*De Caelo* I, 12. Cf. below, pp. 91, 320.

[164]Cf. above, n. 154.

[165]*K. al-Milal wa-l-Niḥal,* p. 340.

[166]*Guide,* II, 14(3).

[167]*Physics,* VIII, tr. 1, chap. 11.

[168]*Summa contra Gentiles,* II, 33(2). Cf. *Summa Theologiae,* I, 46, 1, obj. 2; *De Potentia,* q. 3, art. 17(2).

[169]*Milḥamot ha-Shem,* VI, i, 3, p. 300.

[170]*Or ha-Shem,* III, i, 1.

[171]*Dialoghi d'Amore,* pp. 237–238.

[172]*Guide,* II, 17. Cf. S. Munk's translation, *Le Guide des Égarés,* Vol II (Paris, 1861), p. 135, n. 2.

[173]*De Potentia,* q. 3, art. 17 (17, 18, and replies).

[174]*'Eṣ Ḥayyim,* chap. 6.

[175]*Dialoghi d'Amore,* p. 238.

ex nihilo, but are compatible with the creation of the present form of the world. The arguments from (c) the nature of motion and (d) the nature of time would establish that a world always existed. Whether or not they would establish specifically that the world we know today has existed forever would depend upon whether or not the eternity of motion implies the eternity, specifically, of the celestial spheres.[176] The argument from (f) the nature of the celestial spheres, would establish the eternity of the physical universe as it exists today.

3. Replies to proofs from the nature of the world.

The arguments we have been examining all reason from the laws of nature to the eternity of the world, and they invite a single overall response. An adherent of creation could maintain that the laws of nature govern the world as it now exists, but need not have governed the process whereby the world would have come into existence. A response more or less along that line was already made by Philoponus when he was dealing with arguments for eternity resting on the rule that "something cannot come from nothing." Philoponus explained that although the rule is in truth inviolable within the natural realm, it does not constrain "God, whose essence and actuality transcend the universe," and therefore it does not prove eternity.[177] In the Middle Ages, the response that the laws of nature operative today would not have governed the process of creation appears in various forms, both as an overall refutation of arguments from the nature of the world taken collectively and also as a refutation of the individual arguments.

Several Kalam writers describe their opponents, the advocates of eternity, as having proceeded from what is "present and perceivable" (*shāhid*), that is, from what can be observed in the world today, to what is "not present and perceivable" (*ghā'ib*), to the conditions that would have obtained when the world came into existence. But the nature of what is present and perceivable, the same writers object, is no infallible guide to the nature of what is not. "The Negro" is sometimes brought as an illustration. It would be not science but the height of foolishness for the African Negro to generalize from his personal experience and to affirm that all mankind is black; it is no less foolish to affirm of a previous state of the world everything that is known about the present state. This response to arguments for eternity is offered by 'Abd al-Jabbār,[178] Juwaynī,[179] and the Jābir corpus.[180]

[176]Aristotle and Averroes were certain that the eternity of motion does imply the eternity specifically of the celestial spheres. See *De Caelo* I, 2, and above, n. 88.

[177]Philoponus, as cited by Simplicius, *Commentary on Physics,* ed. H. Diels, *Commentaria in Aristotelem Graeca,* Vol. X (Berlin, 1895), p. 1141.

[178]*Sharḥ al-Uṣūl,* p. 117.

[179]*K. al-Shāmil,* p. 224. Juwaynī adds a general observation: The adherents of eternity failed to support their analogy by showing that the perceivable and nonperceivable realms are subject to the same rules.

[180]Jābir ibn Ḥayyan, *Textes Choisis,* pp. 422–423.

A similar thought is formulated by Maimonides much more circumstantially and precisely. In a comprehensive refutation of all arguments for eternity from the nature of the world, Maimonides writes: "Whenever something comes into existence after not having existed, even in instances where the matter already existed and merely divests itself of one form to assume another, the nature of the thing after it has already come into existence . . . is different from its nature during the process of coming into existence . . . and different as well from its nature before it began the process, [that is to say, different from the nature of whatever it might have come into existence from]. For example, the nature of the female seed [before pregnancy] . . . is different from its nature during pregnancy . . . and different as well as from the nature of a complete living being after the living being is born." Given the differences between the three stages through which generated objects pass, "no inference can be drawn from the nature of the thing when already existent . . . to its state while progressing towards existence, nor can an inference be drawn from the latter to the state of the thing before it began to move towards existence." Proponents of eternity thus commit a fallacy in citing "the nature of the stable, perfected, actual universe" and concluding therefrom that the universe must have existed forever. The adherents of creation, Maimonides continues, do not believe that the world came into existence under the laws of nature operative today; such, plainly could not have occurred. They believe that "God brought the world, in its totality, into existence after nonexistence," and that the state of the world "when stable and perfected, in no way resembles the state of the world during its coming into existence." Maimonides proceeds to show how each of the arguments from the nature of the world recorded by him is resolved, once the laws now governing the world are understood to be different from those that would have governed the world during the process of its coming into existence.[181] His comprehensive response to arguments from the nature of the world reappears, somewhat condensed, in Albertus Magnus[182] and Aaron ben Elijah.[183]

Crescas, too, recommends it; "Maimonides' comprehensive response," he finds, is "clearly correct, and sufficient to refute the arguments" for eternity from the nature of the world.[184] In discussing it Crescas does not, however, employ Maimonides' own terminology, but borrows a formulation from Gersonides.[185] He has Maimonides distinguish between "general coming into existence," that is, the coming into existence of the world as a whole, and "partial coming into existence," the coming into existence of objects within the world.[186] The error in

[181]*Guide,* II, 17.

[182]*Physics,* VIII, tr. 1, chap. 14; *Commentary on II Sentences,* d. 1, B, art. 10.

[183]*'Eṣ Ḥayyim,* chap. 6.

[184]*Or ha-Shem,* III, i, 4 (beginning).

[185]Cf. below, p. 32 f. In *Milḥamot ha-Shem,* p. 306, Gersonides seems to attribute the formulation to Maimonides, but on p. 366, he quotes Maimonides' genuine formulation.

[186]Maimonides distinguished between the separate stages in the generation of an object, whereas Crescas has him distinguish between the part and the whole.

arguments from the nature of the physical world is, as Crescas explains Maimonides' intent, that they draw an improper "analogy between partial coming into existence and general coming into existence." The laws governing the coming into existence of objects within the world were transferred by the adherents of eternity to a phenomenon of a different kind, the coming into existence of the world as a whole; since the coming into existence of the world as a whole need not have been governed by the laws of partial coming into existence, the arguments in question are all fallacious.[187]

Aquinas similarly offers a comprehensive refutation of arguments for eternity from the nature of the world and he arrives at the same result as the writers already quoted, although I could see no evidence of direct filiation.[188] The distinction Aquinas delineates is that between the realm of "nature," on the one hand, and the realm of "divine" action, on the other. In the realm of nature, there occur "change" (*mutatio*) and "particular" production, that is to say, the production of a particular object from another particular object; whereas in the realm of "divine" action, there occur "creation," "simple emanation," and "universal production," that is to say, the bringing forth of being when nothing at all previously existed. Arguments for eternity from the physical world reason from laws that relate to the phenomenon of particular production, but not necessarily to the phenomenon of universal production or creation. As a consequence, the arguments are intrinsically invalid.[189]

The foregoing refutations of arguments from the nature of the world invalidate every argument of the sort, without exception. Nothing in the present state of the world, the recurrent objection goes, is pertinent to the issue of creation and eternity, since the laws of nature operative today need not have been operative during the process of creation. Gersonides takes a separate tack. He too proposes an overall refutation of arguments for eternity from the nature of the world, but his refutation is qualified and restricted. A certain group of arguments from the nature of the world is, for Gersonides, fallacious, whereas another group is valid.

Gersonides begins with the distinction, which in one passage he appears to attribute to Maimonides,[190] between "partial coming into existence" and "general coming into existence." And he warns against indiscriminate "analogy"—to be specific, against analogies wherein "we affirm of [general] coming into existence, everything that can be affirmed of the coming into existence of each single part of the world." But the distinction between partial and general coming into existence is found by Gersonides to be too loose for a definitive resolution of the issue of creation, inasmuch as some characteristics of partial coming into existence can properly be affirmed of general coming into existence. Only "what

[187] *Or ha-Shem*, III, i, 2.

[188] Aquinas' refutation is very similar to Philoponus', above, n. 177.

[189] *Commentary on Physics*, VIII, §§974, 987; *Summa Theologiae*, I, 45, 2, ad 2.

[190] *Milḥamot ha-Shem*, VI, i, 4, p. 306; cf. above, p. 31, and n. 185.

belongs to the world because it is in its present state (*to'ar*)," he understands, "need not have belonged to the world before it was in that state. What, by contrast, belongs to the world simply because it exists, irrespective of its state, would have had to belong to the world even during the period of its coming into existence."[191] The unqualified distinction between partial and general coming into existence must accordingly be supplemented, or replaced, by a distinction between characteristics that are and those that are not linked to the world's existing in its present state. Characteristics linked to the world's existing in its present state would not appertain to the world's coming into existence as a whole; but characteristics of existent beings, not related to any given state of the world, would. Consequently, any argument for eternity which takes its departure from characteristics of the former sort is invalid, since the characteristics in question would have been absent—as Maimonides and other advocates of creation had insisted—during the world's coming into existence. But arguments from characteristics of the latter sort, from characteristics of existent beings which are not related to any given state of the world, retain their validity; for even at the moment of its generation the world would have been something existent. Gersonides criticizes Maimondes for having failed to take account of the distinction between these two sorts of characteristics and for having therefore rejected, without exception, every inference from the state of the world when already existent to the state of the world during the process of its coming into existence.[192] One may wonder how Gersonides could recognize and identify characteristics of things which are and characteristics of things which are not due to the present state of the world, but he seems to have thought that the identification could be made intuitively.[193] It will appear that his qualified refutation of arguments from the nature of the world has the effect, as he applies it, of sanctioning arguments for the eternity of matter, while at the same time ruling out arguments for the eternity of the form of the world.

Thus far we have seen the comprehensive refutations of arguments for eternity from the nature of the world. In addition to their comprehensive refutations, the advocates of creation offered individual refutations for each of the several arguments.

(a) Responses to the arguments from the nature of matter

The argument from the nature of matter, stated briefly, had been that whatever comes into existence does so from a preexisting substratum, or matter; matter too could only have come into existence from a preexisting substratum, or matter; consequently, to assume an absolute coming into existence of matter is self-contradictory.[194]

[191] *Milḥamot ha-Shem,* VI, i, 4, p. 304; 17, p. 366.
[192] Ibid., 4, p. 306; 17, p. 366.
[193] Cf. ibid., 17, pp. 364–366.
[194] Above, p. 13.

The sole grounds Aristotle provided to support the rule that whatever comes into existence does so from an already existing substratum were inductive; otherwise he treated the rule as self-evident.[195] And as has been seen, writers associated with the Kalam school repeatedly ascribe their adversaries' position to induction or analogy.[196] When Kalam writers take up the task of refuting their opponents, their reply consists largely in exposing the unreliability of inductions and analogies from what is empirically known.

Saadia and Ibn Ḥazm had represented the advocates of "eternity" (*dahr*) as accepting nothing but the reports of the senses and as rejecting the creation of matter and the world because sense perception cannot attest to an instance of creation. In response, Saadia and Ibn Ḥazm point out that no man has ever had sense perception of the eternity of the world or a single one of its parts. Consistency therefore would demand that anyone who relies exclusively on analogy and induction from what he perceives should reject not merely the creation of matter and the world, but also the eternity of matter and the eternity of the world. Saadia goes on to list various items of nonempirical knowledge—such as memories and inferences, including inductions themselves—which are not directly acquired through sense perception, yet are perforce accepted by all mankind including the advocates of eternity. The conclusion he and Ibn Ḥazm reach is that uncompromising empiricism is indefensible and that "analogy" from what is perceived through the senses cannot settle the issue of eternity and creation.[197]

'Abd al-Jabbār, Juwaynī, and the Jābir corpus respond to the argument from the circumstance that no instance of creation *ex nihilo* has ever been observed by pressing their comprehensive refutation of arguments from what is "present and perceivable." The nature of the perceivable, they contend, is no reliable guide to the nature of what is not perceivable, and events need not always have occurred as they now are perceived to occur.[198]

Maimonides and advocates of creation who follow his lead refute Aristotle's argument from the nature of matter by applying their own comprehensive refutation of arguments from the nature of the physical world. Maimonides concedes that the underlying matter of the universe cannot be generated and have come into existence in the manner in which generated-destructible things are generated and come into existence; for, as Aristotle correctly held, all generated-destructible objects are generated from a preexisting substratum. The belief of the advocates of creation, Maimonides explains, is that "God created (*awjada*) matter from nothing" and that the act of creation is entirely different from the process of generation as it takes place within a stable world. Creation consequently does

[195] Above, p. 14.

[196] Above, p. 15.

[197] Saadia, *K. al-Amānāt,* I, 3, pp. 63–65; Ibn Ḥazm, *K. al-Faṣl fī al-Milal,* I, p. 10.

[198] 'Abd al-Jabbār, *Sharḥ al-Uṣūl,* p. 117; Juwaynī, *K. al-Shāmil,* p. 224; Jābir, *Textes Choisis,* p. 422.

not require a preexisting substratum; thus Aristotle's argument from the nature of matter has no bearing on the doctrine of creation *ex nihilo*.[199] Albertus Magnus,[200] Aquinas,[201] Aaron ben Elijah,[202] Crescas,[203] and Leone Ebreo[204] similarly rebut the argument from the nature of matter by distinguishing, through one formula or another, the phenomenon of generation within the world from the completely different phenomenon of the creation of the world in its entirety.

Aaron ben Elijah appends a further consideration in an *ad hominem* mode. He notes that Aristotelian philosophy recognizes instances of things' coming into existence from nothing. When a new object comes into existence within the world, its material side is indeed drawn from already existing matter. But the form of the new object comes neither *from* the already existing matter nor *from* anything else whatsoever; the form comes from nothing. Since the Aristotelian adherents of eternity do not gainsay the constant coming into existence of forms from nothing,[205] how, Aaron marvels, can they balk at the coming into existence of matter from nothing?[206]

In contrast to the foregoing, Gersonides' overall response to arguments for eternity from the nature of the world[207] is formulated in a way that lets the argument from the nature of matter stand. Gersonides excludes any inference regarding the coming into existence of the world as a whole which reasons from characteristics tied to the present state of the world. The impossibility "that a body should come into existence from . . . absolutely nothing" is not, he understands, such a characteristic. It is rather a universal law of physical existence, "unrelated to the state in which the world exists."[208] To assume that the physical universe came into existence from absolutely nothing therefore embodies for Gersonides, as for Aristotle, a self-contradiction, and Gersonides concludes that matter is eternal. His considered position is that the world was created from a preexistent eternal matter.[209]

[199] *Guide,* II, 17.

[200] *Physics,* VIII, tr. 1, chap. 14.

[201] *Summa Theologiae,* I, 46, 1, ad 3; *Summa contra Gentiles,* II, chap. 37(1).

[202] *'Eṣ Ḥayyim,* chap. 6.

[203] *Or ha-Shem,* III, i, 5.

[204] *Dialoghi d'Amore,* p. 239.

[205] In *Metaphysics* VII, 8, 1033b, 5–6, Aristotle stated that forms are not generated; but ibid., 15, 1039b, 26, he conceded that forms, although not generated, sometimes "are" and sometimes "are not." Cf. Ross's note to 1033b, 5–6. The anomalousness of Aristotle's position is underlined by Zeller, *Die Philosophie der Griechen,* II, 2, pp. 347–348, and C. Baeumker, *Das Problem der Materie* (Muenster, 1890), pp. 287–288.

[206] *'Eṣ Ḥayyim,* chap. 6. The same point was made by Philoponus; see *De Aeternitate,* p. 351, and Simplicius, *Commentary on Physics,* p. 1142.

[207] Above, p. 32.

[208] *Milḥamot ha-Shem,* VI, i, 17, pp. 365–366. Abravanel rejoins that the impossibility of something's coming into existence from absolutely nothing is in fact a characteristic tied to the present state of the world; *Mif'alot,* IV, 3.

[209] *Milḥamot ha-Shem,* VI, i, 17, pp. 367–368. To be more precise, Gersonides' position is that

(b) Responses to the argument from the concept of possibility

The argument had been: Prior to something's coming into existence, there is a possibility of its existing. The possibility of existing must subsist somewhere; and it can only subsist in an already existing substratum, hence in an already existing matter. The assumption that matter came into existence from absolutely nothing consequently embodies a self-contradiction, and matter must be eternal.[210] Medieval advocates of the creation of the world developed three responses to the argument.

One response consists in referring the possibility of matter's coming into existence to the agent that produced matter. The premise is accepted according to which the possibility of matter's existing would have to precede the actual existence of matter. But the prior possibility of matter's existing is not located in a substratum, from which matter would have come into existence. It is instead identified with the power of the creator to create. When the possibility of the existence of matter is so construed, the assumption of creation *ex nihilo* no longer contradicts itself by implying the prior existence of a substratum containing the possibility.

This response to the argument from the concept of possibility is mentioned, but not seriously, by Avicenna,[211] the apparent author of the argument that is being rebutted. It is also mentioned in passing by Ghazali[212] and is employed by Aquinas,[213] Aaron ben Elijah,[214] Crescas,[215] and Abravanel.[216] The same response was known as well to Averroes, who attributed it to John Philoponus, and to Maimonides, who attributed it to the later Kalam.[217] Both Averroes and Maimonides reject it because of a certain distinction that had been drawn by Aristotle.

Aristotle had distinguished two δυνάμεις—"powers," or "potentialities," or "possibilities"—in the process of change, namely the power, or possibility, of the agent to effect the change, and the power, or possibility, of the object to undergo the change.[218] Accordingly, Averroes and Maimonides maintain, although one possibility of the existence of matter may properly be identified as the agent's

---

the world was created from a preexistent "body free of all form"; in the passage cited, he attempts to elucidate the concept.

[210] Above, pp. 16–17.

[211] *K. al-Ishārāt wa-l-Tanbīhāt*, ed. J. Forget (Leiden, 1892), p. 151; French translation, with pagination of Arabic indicated: *Livre des Directives et Remarques*, trans. A. Goichon (Beirut and Paris, 1951).

[212] *Tahāfut al-Falāsifa*, I, §85; English translation, p. 59.

[213] *Summa Theologiae*, I, 46, 1, ad 1; *Summa contra Gentiles*, II, chap. 37.

[214] *'Eṣ Ḥayyim*, chap. 6.

[215] *Or ha-Shem*, III, i, 5.

[216] *Mif'alot*, IV, 3.

[217] Averroes, *Long Commentary on Metaphysics*, XII, comm. 18; Averroes writes that he drew his information from Alfarabi's work *On Changeable Beings*. Maimonides, *Guide*, II, 14(4).

[218] *Metaphysics* V, 12, 1019a, 15–22; cf. ibid., IX, 1.

power to create matter, there remains the question of the other possibility of the existence of matter, the possibility from the side of some object that is to undergo the process of becoming matter. Until the latter possibility is explained or explained away, the argument that the prior possibility of the existence of matter would have to be located in an already existing substratum will not have been answered.[219] Perhaps because of the point raised by Averroes and Maimonides, some writers—Aquinas, Aaron ben Elijah, and Abravanel—do not utilize the first response to the argument from the concept of possibility by itself, but always buttress it with additional considerations.[220]

The second response to the argument from the concept of possibility seems to have originated in Ghazali. Ghazali construes the "possibility" of something's coming into existence—along with the impossibility, and the necessity, of something's coming into existence—as nothing but an "intellectual judgment," a judgment on the part of the intellect that the thing may—or that it cannot, or that it must—exist. Since *possibility, impossibility,* and *necessity,* have no objective existence in the external world, they do "not require anything existent" to serve as their substratum. And since the possibility of matter's existing does not require a substratum, the argument from the concept of possibility has no validity.[221]

Following Ghazali, Shahrastānī construes the possibility of matter's coming into existence as a "mental supposition" (*taqdīr*);[222] Fakhr al-Dīn al-Rāzī denies that the possibility of matter's coming into existence is an "existent attribute";[223] Āmidī writes that it is not "a real essence."[224] Aquinas offers an interpretation in the same vein, and even finds support in Aristotle. In the course of analyzing the divers meanings of the term "possible" (δυνατόν), Aristotle had isolated what we should call the logically possible. And, he had stated, the possible in its logical sense carries no implication of a power—or potentiality, or possibility (δύναμις)—in either an agent producing change or an object undergoing change.[225] Aquinas calls attention to Aristotle's statement; and he explains that the coming into existence of matter and of the world was possible "in the way a thing is said to be absolutely possible, that is, not by virtue of any potentiality, but solely from the relation of the terms, which are compatible with one another."[226] The possibility of existence preceding the actual existence of matter was thus not anything with an external existence of its own. A possibility of existence preceded

[219]Averroes, *Tahāfut al-Tahāfut,* I, pp. 100–101; Maimonides, *Guide,* II, 14(4).

[220]See below, p. 38.

[221]*Tahāfut al-Falāsifa,* I, §87; English translation, p. 60.

[222]*K. Nihāya al-Iqdām,* p. 34.

[223]*Muḥaṣṣal,* p. 91; *K. al-Arba'īn,* p. 51.

[224]*Ghāya al-Marām,* p. 272.

[225]*Metaphysics* V, 12, 1019b, 27–29, 34–35. This is the chapter in the *Metaphysics* referred to above in n. 218.

[226]*Summa Theologiae,* I, 46, 1, ad 1.

the actual existence of matter and of the world only inasmuch as the terms in the proposition 'matter exists' or 'the world exists' were logically compatible—only inasmuch as the proposition involved no logical impossibility—before those things did exist. Aquinas combines the previous response to the argument from the concept of possibility with the response we are now examining. When the possibility of the existence of the world is said to have preceded its actual existence, the meaning, Aquinas writes, is either that the creator had the power to create the world, or else that the creation of the world was logically possible.[227]

A third response to the argument from the concept of possibility consists in applying the comprehensive refutation of all arguments from the nature of the world. The requirement that the possibility of existence must be located in a substratum is held to be a characteristic of change and generation within the world, but not necessarily a characteristic of the creation of a world.

Such is Maimonides' response. He states that what comes into existence must be preceded by its possibility "only in our stable universe, where things are generated solely from something existent. When, by contrast, something is created *ex nihilo,* nothing at all existed [previously . . . which might permit the thing coming into existence] to be preceded by possibility."[228] The same interpretation is given by Albertus Magnus[229] and Leone Ebreo.[230] Ṭūsī rejects the previous response to the argument from possibility, the response that construes *possibility* as a "nonreal" (*ghayr wujūdī*) attribute; for, according to Ṭūsī, the possibility of being generated is unquestionably a "disposition" that "requires a subject" in which to inhere. In preference to the previous response, Ṭūsī offers the one now being examined. He distinguishes between the generation of things within the world and creation; and he contends that creation is a completely different phenomenon, that in the case of "created things, no disposition is conceivable prior to their existence."[231] Aaron ben Elijah joins the third response to arguments from the concept of possibility with the first. He writes: The possibility of a thing's coming into existence precedes the actual existence of the thing only in the "stable, settled universe" but would not do so, in the creation of matter *ex nihilo*; moreover, whatever possibility there might be in the case of creation *ex nihilo* is to be referred to the "agent," that is to say, the creator.[232] Abravanel similarly explains that in "first creation" as distinct from "generation," no "possibility in a substratum" precedes the process of coming into existence. The sole true possibility at that stage is the "possibility of action" on the part of God's "infinite power."[233] Aquinas in one passage combines all three solutions. He

[227]Ibid.; *De Potentia,* q. 3, art. 17 (ad 10).
[228]*Guide,* II, 17.
[229]*Physics,* VIII, tr. 1, chap. 14.
[230]*Dialoghi d'Amore,* p. 239.
[231]Glosses to Rāzī's *Muḥaṣṣal,* p. 92.
[232]*'Eṣ Ḥayyim,* chap. 6.
[233]*Mif'alot,* IV, 3. See below, n. 254.

maintains that the phenomenon of creation is radically different from the phenomenon of things' coming into existence by motion; in the former, as distinct from the latter, the prior possibility of existence is to be referred to the agent; alternatively, it is to be construed as nothing other than the logical compatibility of the terms in the proposition "matter exists."[234]

(c) Responses to arguments from the nature of motion

Aristotle's argument from the nature of motion, it was seen, reappears in the Middle Ages in its original complex form as well as in a simplified form. The contention in both forms of the argument was that the supposedly very first motion would, by virtue of the laws of motion, have to be preceded by another motion, and hence the supposedly first motion would not in fact be first.[235] The adherents of creation who respond to the proof in one or the other of its forms all adduce the comprehensive refutation of arguments from the nature of the world. They distinguish between natural processes within the world and creation; and they explain that argumentation from the nature of motion can be valid only where the laws of nature are operative but can have no bearing on God's creating the world *ex nihilo*. This response is given by John Philoponus,[236] and it is repeated in the Middle Ages with minor variations by Maimonides,[237] Albertus Magnus,[238] Bonaventure,[239] Aquinas,[240] Gersonides,[241] Aaron ben Elijah,[242] Crescas,[243] Abravanel,[244] and Leone Ebreo.[245]

(d) Responses to arguments from the nature of time

The proof for eternity from the nature of time reasons in one fashion or another that anyone assuming an absolute beginning of time cannot avoid recognizing a prior time; hence time cannot have a beginning but must be eternal; time, however, involves motion and a moving body; and time being eternal, motion and a world undergoing motion must also be eternal.[246] A response might proceed either by rebutting the premise that time involves motion and a moving body, or else by rebutting the grounds that had been advanced for the eternity of time. The only writer I could find taking the former course is Aaron ben Elijah. In reply to one of the versions of the proof, he acknowledges that time is eternal, but denies that time involves motion. "Time," he asserts, "is something extrinsic

234 *Summa contra Gentiles,* II, 37(3).

235 Above, pp. 18–20.

236 Quoted by Simplicius, *Commentary on Physics,* pp. 1141, 1150.

237 *Guide,* II, 7.

238 *Commentary on II Sentences,* d. 1, B, art. 10.

239 *Commentary on II Sentences,* d. 1, p. 1, a. 1, q. 2, ad. 2.

240 *Commentary on Physics,* VIII, §987. *Summa Theologiae,* I, 46, 1, ad 5; *Summa contra Gentiles,* II, 36(3).

241 *Milḥamot ha-Shem,* VI, i, 4, p. 304; much elaborated and nuanced, ibid., 24, p. 395.

242 *'Eṣ Ḥayyim,* chap. 6.

243 *Or ha-Shem,* III, i, 4 (beginning).

244 *Mif'alot,* V, 3.

245 *Dialoghi d'Amore,* p. 239.

246 Above, pp. 24–25.

to motion" and motion is not entailed by it. Arguments establishing the preexistence of time are therefore irrelevant to the issue of creation.[247]

Most proponents of creation from Plato onwards take the other course. They accept the premise that time involves motion and a moving body, but maintain that time is not eternal, that there was no time before the world was created.[248] These proponents of creation had to refute Aristotle's arguments for the eternity of time.

(i) The first argument for the eternity of time ran: On the assumption of an absolute beginning of time, the period before time and when there was not yet time could still be spoken of; but the terms *before* and *when* imply time; time would thus already have existed before the assumed absolute beginning of time. The tenor of medieval refutations was set, or anticipated, by John Philoponus. Philoponus was addressing the version of the argument according to which the words 'when there was no time' imply that time already existed. He explains that with care the mischievous expression is easily avoided. A careful speaker can avoid mentioning the period "when there was no time" and restrict himself to saying, "simply, that time is not eternal."[249] But the complete solution, Philoponus holds, goes deeper. The complete solution lies in understanding that the argument is "sophistical," since it is concerned with "wording" rather than with what is "meant" thereby. Nothing can be inferred about reality from the "weakness" of human speech, and attention should always be directed to "intent" rather than "words."[250] The expression "when there was no time" and similar expressions consequently shed no light on the issue of creation and eternity.

In the Middle Ages, it is repeatedly stated that language seeming to connote time does not imply actual time. Philoponus' distinction between intent and words does not explicitly appear, but another motif that may have derived from him does. Aristotle had affirmed the finiteness of space side by side with the infiniteness of time, and proponents of creation pounce on the apparent inconsistency. They draw an analogy between temporal and spatial extension and contend that just as space is universally acknowledged to have a terminus beyond which there is no space, so too may time have a beginning before which there was no time. For example, the following critique of the Aristotelian doctrine of eternity is reported in the name of "Yaḥyā," that is to say, Yaḥyā ibn 'Adī or, possibly, John Philoponus:[251] The proposition that time began at a certain moment with no time

[247] *'Eṣ Ḥayyim,* chap. 6. Aaron is answering the argument that at the very least the possibility of the existence of the world would be present *before* the world came into existence, and hence time would already exist before the supposed beginning of time.

[248] See Plato, *Timaeus,* 38; Philo, *De Opificio Mundi,* vii, 26; Augustine, *City of God,* XI, 6; *Confessions,* XI, 30; H. Wolfson, *Philo,* Vol. I (Cambridge, Mass., 1948), p. 311.

[249] *De Aeternitate,* p. 105.

[250] Ibid., pp. 104, 116.

[251] In Arabic, Philoponus is called Yaḥyā al-Naḥwī.

preceding is neither more nor less admissible than the proposition that space terminates at a certain point with nothing beyond. It is a familiar tenet of Aristotelian physics that the world is finite and that "no extension whatsoever," neither "plenum nor vacuum," exists outside the world. By what right then do the Aristotelians reject a parallel theory in regard to time? By what right do they brand as "inconceivable, the thesis that there was no time" before creation, that time extends back to a certain moment with absolutely no time preceding?[252]

Ghazali weaves together two motifs, the contention that the vagaries of language shed no light on the issue of creation, and the analogy between temporal and spatial extension. The former motif is developed in a manner akin to his response to the arguments from the concept of possibility. There he contended that the possibility preceding the creation of the world is merely a judgment of the mind, a logical judgment, representing nothing in the external world;[253] here he contends that the time preceding the creation of the world can represent nothing in the external world because it is purely imaginary.[254]

Ghazali is answering the argument that the statement 'God exists *before* creation' implies a time before the assumed creation of the world. He replies that the issue cannot be settled through the testimony of human language. The statement that God existed before creation can be recast in the form: "God was, without a world and without time, then was, with a world and time"; and the implication of a time before creation vanishes.[255] But in any event, even if a speaker should not trouble, or should be unable, to avoid language with temporal connotations in referring to what was before time, no actual preexistent time is thereby implied. For the creation of the world is preceded by time only in our "imagination"; creation is preceded by nothing more than a "supposition" (*taqdīr*) of time. The situation is exactly analogous to that obtaining in spatial extension.

[252]Abū al-Faraj's (?) comment to *Physics* VIII, 1; in Aristotle, *Ṭabī'a* (medieval Arabic translation of *Physics* together with four medieval Arabic commentaries), ed. A. Badawi (Cairo, 1964), p. 816. It is not clear from the printed text whether Abū al-Faraj is quoting "Yaḥyā" or whether the passage belongs not to Abū al-Faraj's commentary but to the commentary of Yaḥyā ibn 'Adī.

[253]Above, p. 37.

[254]The later Kalam writers who maintained that the *possibility* of matter's coming into existence is not real (above, p. 37), and especially Shahrāstanī (above, n. 222), may well have been following Ghazali's response to the argument from time, rather than his response to the argument from the concept of possibility. They may have meant, in other words, not that the possibility of matter's coming into existence is a logical judgment, but that it is a product of the imagination. An unambiguous instance where the response to the argument from time is applied to the argument from the concept of possibility can be found in Abravanel. After giving the reply to the argument from the concept of possibility quoted above, n. 233, Abravanel adds: "It also can be stated that the subject of this *possibility* is in a certain sense the human intellect. Just as a beginning of time is inconceivable without . . . a time extending beyond . . . and prior thereto, . . . so too it is difficult to conceive of the actual coming into existence of an object without the notion that the possibility of the thing precedes its actual existence. . . . But that is only a mental precedence."

[255]*Tahāfut al-Falāsifa,* I, §58; English translation, p. 38.

Nothing whatsoever lies beyond the boundary where the world and its space terminate, although the human imagination "balks" at the notion and mistakenly insists upon "supposing" an empty space beyond the world. Similarly, no time whatsoever existed before the world and time came into existence, although the human imagination again balks and insists upon "supposing" a prior time.[256]

In rebuttals of arguments for the eternity of time, the phrases *imagination,* or *supposition* of time become catchwords; and the analogy between the termination of space and the beginning of time also recurs. Shahrastānī answers arguments of the type we are examining by explaining that any implication of time prior to the world is "an imaginary supposition, like the supposition of a vacuum beyond the world."[257] Merely imagining that a vacuum exists beyond the world does not establish that a vacuum exists in actuality. By the same token, merely imagining that time existed prior to the world does not establish that time truly existed. Maimonides, who does not include an argument from time in his formal classification of proofs for eternity, forestalls such an argument. The statement "God existed before creating the world" and other phraseology that seemingly implies time prior to creation refer, Maimonides writes, solely to the "supposition of time or to the imagination of time, and not to real time."[258] Fakhr al-Dīn al-Rāzī maintains that "priority is not an existent attribute"[259] or a "positive attribute."[260] Āmidī characterizes the assumption of a temporal extension prior to the world as an "imaginative supposition."[261] Aquinas writes that the "*before* we speak of as preceding time" does not refer to time "in reality, but solely in [our] imagination." The situation, he adds, is analogous to "saying that there is nothing above the heavens." In that sentence, the term *above* does not have a genuine spatial reference but merely an imagined reference, and all will acknowledge that no actual space above the heavens is implied. By the same token, the words "*before* time" should be acknowledged to contain no implications of an actual prior time.[262]

Gersonides' response to the argument from the nature of time is similar to the responses made by his predecessors, but—as Gersonides often does when repeating commonplaces—he appends a small twist of his own. The statement that time did not exist "before" the creation of the world is, he writes, analogous to the statement that "neither a vacuum nor plenum is to be found outside the world." The term *before* in the one instance and the term *outside* in the other are employed

[256] Ibid., §§61–63; English translation, pp. 41–42, and cf. van den Bergh's note.
[257] *K. Nihāya al-Iqdām,* p. 52.
[258] *Guide,* II, 13(1).
[259] *Muḥaṣṣal,* p. 91.
[260] *K. al-Arba'īn,* p. 53.
[261] *Ghāya al-Marām,* p. 272.
[262] *Summa contra Gentiles,* II, 36(6). Cf. *Commentary on Physics,* VIII, §990; *Summa Theologiae,* I, 46, 1, ad 8.

equivocally and not in their ordinary sense, and hence their use cannot serve as grounds for inferring either the infinity of space or the eternity of time. We have here, Gersonides continues, an area where human language, which is in any event "conventional," fails to provide the technical terminology required by philosophy and science; the result is that philosophers must use ordinary words in a "borrowed" sense.[263] A similar reply to the argument is made by Abravanel.[264]

(ii) Aristotle's second argument for the eternity of time was that the *moment* invariably divides past from future; to assume a first moment would therefore amount to assuming a preceding past time that terminates at the first moment; and the assumption of an absolutely first moment is self-contradictory.[265] The argument did not receive as much attention as the previous argument from time, but several responses can be cited.

Philoponus condemns the argument as "begging the question," since in assuming that every moment divides past from future, the eternity of time is presupposed from the outset. To elucidate his meaning, Philoponus draws another analogy between space and time. Aristotle had once characterized the moment in time as analogous to the point on a line,[266] inasmuch as the point divides a line into two segments and the moment divides past time from future time. It clearly would, Philoponus writes, be a begging of the question to presuppose that every point, without exception, divides a line into two segments, to infer therefrom that no point can ever stand at the terminus of a line segment, and to conclude that every line must be infinite. It is no less begging the question to presuppose that every moment, without exception, divides past time from future time, to infer that no moment can stand at the very beginning of time, and to conclude that time must be eternal.[267]

Alfarabi and Averroes knew that the argument from the nature of the moment might be rebutted on the grounds that the moment in time is analogous to the point on a line, and every line segment does begin at a point.[268] Any rebuttal of

[263] *Milḥamot ha-Shem,* VI, i, 21, pp. 384–385.

[264] *Mif'alot,* V, 3.

[265] Above, p. 24.

[266] *Physics* IV, 11, 220a, 9–10; 13, 222a, 13.

[267] Quoted by Simplicius, *Commentary on Physics,* p. 1167.

[268] *Epitome of Metaphysics,* IV, §4, with reference to Alfarabi; *Tahāfut al-Tahāfut,* I, p. 77. It very likely was thanks to Alfarabi and Averroes that Philoponus' rebuttal reached the Middle Ages. Averroes reports that Alfarabi's book *On Changeable Beings,* now lost, laid bare the sophism in any comparison between the possibility of a beginning of time at a moment and the possibility of a beginning of a line at a point. As far as I could see, Averroes does not mention Philoponus in this connection. But refuting Philoponus is known to have been one of Alfarabi's objectives in his book *On Changeable Beings* (see above, n. 217; below, p. 128; M. Steinschneider, *Al-Farabi* [St. Petersburg, 1869], pp. 120–122). It may therefore well have been just Philoponus whom Alfarabi had in mind in the passage referred to by Averroes, that is to say, in the surrejoinder to the rebuttal of the argument from the nature of the moment.

the sort is dismissed by them as a sophism,[269] but several medieval adherents of creation employ it nonetheless and they for their part dismiss their adversaries' imputation of sophistry.

Aquinas makes a reply to the argument from the nature of the moment which is strikingly like Philoponus'. He contends that a moment may stand at the beginning of time, with no past time preceding, in exactly the way a point can stand at the end of a line segment with nothing beyond. It would clearly be circular reasoning to presuppose that every point must have a line segment on each side and then proceed to prove that all lines are infinite. Aristotle's argument from the moment is guilty of the same circularity; the argument presupposes its own conclusion in assuming that every moment must be preceded by time.[270]

Gersonides does not speak of question begging or circular reasoning, but he too uses the analogy between space and time. He compares the argument that a moment is always preceded by time and time is therefore eternal, to a possible argument to the effect that the universe must be infinite since one body comes to an end only where it meets another body. Both arguments are a product of the human "imagination," which refuses to allow a point with nothing beyond or a moment with nothing before; and the imagination is not a reliable guide in the realm of science. Gersonides appends a lengthy disquisition on the nature of the *moment,* wherein he shows that not every moment need have the characteristic of dividing the past from the future. The upshot is that a moment might serve as the absolute beginning of time just as a point can serve as the absolute beginning of a line segment.[271] A highly condensed restatement of Gersonides' reply to the argument from the nature of the moment is offered by Crescas,[272] and Abravanel also offers a restatement of Gersonides' reply.[273]

[269] *Epitome of Metaphysics,* IV, §4; *Tahāfut al-Tahāfut,* I, p. 77. Averroes' exposition of the sophism is extremely subtle. The reason a point can serve as the terminus of a line is, he explains—obviously with Aristotle, *Physics* IV, 13, 222a, 13–14, in mind—that the line is "at rest" and the point at the end of the line enjoys "actual" existence. A moment, by contrast, cannot serve as an absolute terminus of time for the reason that time is not at rest, no moment ever exists in actuality, and therefore no moment can ever exist except as subsequent to the past and antecedent to the future. Hence, Averroes concludes, it is a sophism to infer the possibility of time's having its terminus in a moment from the possibility of a line's having its terminus in a point; and Aristotle's argument for eternity from the nature of the moment accordingly remains intact.

[270] *Commentary on Physics,* VIII, §983; cf. *Summa contra Gentiles,* II, chap. 36. Aquinas dismisses Averroes' imputation of sophistical reasoning as follows: The fact that a line is at rest whereas time is ever flowing can have no bearing on the question whether a line or time might have an initial terminus. Therefore, if a line can begin at a point, time should be able to begin at a moment.

[271] *Milḥamot ha-Shem,* VI, i, 21, pp. 388–390. In an indirect reference to Averroes' strictures (above, n. 269), Gersonides, ibid., p. 389, explains that the potentiality associated with the moment can have nothing to do with the ability, or inability, of a moment to serve as a beginning of time with nothing preceding.

[272] *Or ha-Shem,* III, i, 3.

[273] *Mif'alot,* V, 3.

An alternative answer to the argument for eternity from the nature of the moment consists in applying the comprehensive response to all arguments from the nature of the world. Albertus Magnus,[274] Bonaventure,[275] and Leone Ebreo[276] grant that the moment always divides past from future in the present state of the world; but, they maintain, the moment need not have divided past from future during the process of creation.

(iii) A further argument for the eternity of time had run: Everything that comes into existence does so "in" time; therefore the supposed absolutely first time would have to be preceded by another time, namely, the time *in* which it came into existence. 'Abd al-Jabbār,[277] Bonaventure,[278] Gersonides,[279] and Abravanel,[280] who record the argument, all refute it by rejecting the premise that time would have to come into existence *in* time.

(e) The response to the argument from the vacuum

The argument had been that the doctrine of creation *ex nihilo* implies the prior existence of a vacuum in the location the world was to occupy, whereas a vacuum is impossible.[281] Aquinas responds that on the assumption of creation *ex nihilo*, "there was no place or space prior to the world," and consequently no vacuum, space being created together with the world.[282] Crescas and Abravanel offer the same solution. The argument from the vacuum can, they write, be answered by understanding that "prior to the world, no dimensions existed, and God created them when he created body from nothing."[283] Crescas makes an additional and atypical observation, however. He finds Aristotle's grounds for the impossibility of a vacuum unconvincing[284] and is therefore not in the least discomfited by the implication of a vacuum prior to creation. Prior to the existence of the world, he is quite willing to admit, a vacuum did exist, and in it God created the world.[285]

Gersonides is an adherent of creation who accepts the argument from the vacuum. As was seen, he ruled out only those arguments for eternity which rest on characteristics of the universe tied to its present state, whereas he endorsed arguments which he viewed as resting on universal laws of existence. The argument from the vacuum is understood by him to belong to the latter type; for the

[274] *Commentary on II Sentences,* d. 1, B, art. 10.

[275] *Commentary on II Sentences,* d. 1, p. 1, a. 1, q. 2.

[276] *Dialoghi d'Amore,* p. 239.

[277] *K. al-Majmū',* p. 66.

[278] *Commentary on II Sentences,* d. 1, p. 1, a. 1, q. 2.

[279] *Milḥamot ha-Shem,* VI, i, 20, pp. 382–383.

[280] *Mif'alot,* V, 3.

[281] Above, p. 27.

[282] *Summa Theologiae,* I, 46, 1, ad 4. A similar point is made by Saadia, *K. al-Amānāt,* p. 71.

[283] *Or ha-Shem,* III, i, 5; *Mif'alot,* IV, 3. In Abravanel, the sentence ends: ". . . when he created *something* from nothing."

[284] Or ha-Shem, I, ii, 1.

[285] Ibid., III, i, 5.

proposition that objects come into existence in place states a universal characteristic of physical existence, and the impossibility of a vacuum is also absolute. Gersonides accordingly accepts the conclusion that the place now occupied by the world must always have contained something. His position, as has been mentioned, was that matter is eternal and that the world was created out of a preexistent, eternal formless matter.[286]

(f) The response to the argument from the nature of the celestial bodies

The argument here had been that the process of coming into existence as well as the process of destruction consist in a passage from one contrary to another; but whatever substratum the heavens might have cannot be amenable to contraries; consequently, the heavens cannot have come into existence and must be eternal.[287] The advocates of creation who undertake to refute the argument employ the standard comprehensive response to arguments from the nature of the world. They distinguish between natural processes and creation, and maintain: The law that things come into existence only from their contraries is a law peculiar to the process of generation within the present state of the world, whereas in creation, things may well come into existence in a different manner. This solution is put forward in one version or another by Maimonides,[288] Albertus Magnus,[289] Aquinas,[290] Gersonides,[291] Aaron ben Elijah,[292] Abravanel,[293] and Leone Ebreo.[294]

4. Summary

The argument for eternity from the nature of matter rested on the premise that everything coming into existence does so from a preexistent substratum or matter; matter too, it followed, could have come into existence only from an already existing matter; hence an absolute coming into existence of matter, the coming into existence of matter from absolutely nothing, is impossible. The proponents of creation reply that the critical premise can be justified solely by induction or

[286] *Milḥamot ha-Shem,* VI, i, 17, p. 364. Abravanel, *Mif'alot,* IV, 3, disputes Gersonides' reasoning.

[287] Above, p. 28.

[288] *Guide,* II, 17.

[289] *Commentary on II Sentences,* d. 1, B, art. 10.

[290] *Summa Theologiae,* I, 46, 1, ad 3. Cf. ibid., ad 2; *Summa contra Gentiles,* II, 36(1).

[291] *Milḥamot ha-Shem,* VI, i, 26. Gersonides adds a statement that, I think, misses the point of Aristotle's argument. He contends that not everything is generated from its contrary; forms, for example, are not generated from other forms that are their contraries. Aristotle's contention, however, had not been that a given form is generated from the contrary form. His contention had been that a given form is generated in a situation wherein the substratum lacks the given form; and that is the contrary of the form. See above, p. 28.

[292] *'Eṣ Ḥayyim,* chap. 6.

[293] *Mif'alot,* V, 3.

[294] *Dialoghi d'Amore,* p. 239.

analogy, and inductions and analogies are unreliable. Alternatively, the proponents of creation bring to bear their comprehensive refutation of arguments from the nature of the world. They maintain that although in the present state of the world everything coming into existence does so from a preexistent substratum, no preexistent substratum need be assumed for creation. Creation *ex nihilo*—the creation of the world together with its matter—accordingly remains a viable hypothesis.

The argument from the concept of possibility reasoned that everything coming into existence is preceded by the possibility of its existence; the possibility of existence must inhere in a substratum or matter; matter would therefore have had to exist before matter could come into existence; thus the absolute coming into existence of matter is impossible. The proponents of creation respond in one of three ways. They deny that the possibility of the world's coming into existence is to be located in an already existing matter and refer that possibility instead to the creator. They construe the possibility of the existence of the world as a judgment of the mind and as having no objective existence in the external world. Or else they resort to their comprehensive refutation of arguments from the nature of the world; and they maintain that in the phenomenon of creation, the possibility of existence need not precede actual existence. The upshot is that creation *ex nihilo* is a viable hypothesis.

The argument from motion showed that the assumption of a beginning of motion inescapably involves prior motion; and therefore motion, as well as some sort of world undergoing motion, must be eternal. The proponents of creation respond by again applying their comprehensive refutation of arguments from the nature of the world. They explain that although in the present state of the world, motion cannot occur without a prior motion, in the creation of the world as a whole, an absolute beginning of motion is conceivable. The world and its motion may accordingly have had a beginning.

The argument from time reasoned that a beginning of time cannot be spoken of without implying a prior time, and hence time is eternal. The companion argument from the moment reasoned that moments invariably divide past from future, and therefore a first moment, with no time and no moments preceding, is impossible. Both arguments conclude that since time is eternal, motion and some sort of world undergoing motion must also be eternal. In one instance these arguments are countered with the denial that time does involve motion and a moving body. If time does not involve motion, the eternity of time may be granted without admitting the eternity of the world. The usual response, however, is that neither the nature of time nor the nature of the moment implies, in fact, the eternity of time. The argument from the nature of time—the contention that there is no avoiding the implication of a time prior to the supposed beginning of time—is answered by ascribing the implication to deceptive human language; all that in truth precedes the beginning of time, the explanation goes, is a supposition of time or imaginary time. A supporting consideration is provided by an analogy between time and space. It was universally acknowledged that the physical world is finite and that nothing whatsoever, not even empty space, lies beyond the world; by the same token, the adherents of creation submit, time might be finite with no time whatsoever preceding the beginning of the world. The argument from the nature of the moment is answered by the adherents of creation with the aid of another analogy between space and time. They maintain that a moment might serve as the absolute beginning of time just as a point can serve as the absolute terminus of a line segment. If time need not be eternal and if a first moment is possible, then the beginning of motion, and of a world undergoing motion, remains a viable hypothesis.

The argument from the vacuum rested on another principle of Aristotelian physics, the impossibility of a vacuum or completely empty space. The argument was that the creation of matter is impossible since it would imply the prior existence of a vacuum where the world was to come into existence; and a vacuum is impossible. The proponents of creation respond that the place the world occupies was created together with the world, so that no vacuum would have preceded creation; and in one instance, the premise that a vacuum is impossible is also rejected. It follows once again that creation *ex nihilo* is a viable hypothesis.

The argument from the nature of the celestial spheres ran: Everything coming into existence does so from its contrary; the circular motion of the celestial spheres reveals that the substratum of the spheres is not amenable to contraries; consequently the spheres and the rest of the physical universe—the existence of which is entailed by the existence of the spheres—cannot have come into existence. The proponents of creation respond by applying their comprehensive refutation of arguments from the nature of the world. The rule that everything coming into existence does so from its contrary is, they hold, valid only in the world as it exists today and need not be true of the phenomenon of creation. The hypothesis of the creation of the world thus remains viable.

III

Proofs of Eternity from the Nature of God

1. The proofs

The proofs for eternity discussed in the previous chapter proceeded from the nature of the physical world to the eternity of the physical world, or, in some instances, to the eternity merely of the underlying matter of the world. The proofs to be discussed in the present chapter take their departure not from the world but from its cause; moreover, they cover not merely the physical universe, but the nonphysical universe as well. They argue that the cause of the universe being such as it is, the entire universe, whether physical or nonphysical, must be eternal.

Three basic proofs from the nature of the cause of the universe can be differentiated, and other arguments may be treated as variations of the three. The themes of the basic proofs are clearly distinguishable. The first proof argues that no given moment, as against any other, could have suggested itself to the creator as the proper moment for creating the universe. The second proof argues that the cause of the universe must be unchangeable and could not, therefore, have undertaken the act of creation after having failed to do so. The third proof argues that the cause of the universe possesses certain eternal attributes and that the existence of the universe is an expression of those attributes; since the attributes are eternal, the universe, which they give rise to, must likewise be eternal. While these three lines of reasoning are sufficiently distinct, a difficulty in classification does arise. Sometimes a variation on a basic proof or an argument as an individual philosopher happened to formulate it can plausibly be subsumed under more than one basic proof, and the decision to classify it under one rather than another is to an extent arbitrary.[1]

The three basic proofs differ in a significant respect that can be brought out by considering a possible criticism of any argument for eternity from the nature of the cause of the universe. Does not any argument of the sort, it may be asked, rest on the unwarranted presupposition that the universe in truth has a cause

[1]Cf., for example, below, pp. 59, 64–65.

outside itself? In the case of the first proof, a satisfactory reply is at hand: The standpoint there is, in principle, hypothetical, the intent being to reduce the adversary's position to absurdity. A cause of the existence of the universe is not presupposed. Rather, the reasoning is that on the assumption of creation, the universe would have had to be brought into existence by a cause;[2] the cause supposedly bringing the universe into existence would not have been able to create it at any given moment; hence the assumption of creation is untenable. When stating that the standpoint of the first proof is hypothetical, the qualification has to be added that the proof is so *in principle,* since not every instance of the proof keeps the hypothetical standpoint consciously in view.

In the third proof, the standpoint has plainly shifted away from the hypothetical. Presuppositions now are in evidence to the effect that the universe does have a cause of its existence, that the cause of the existence of the universe possesses certain eternal attributes, and that the attributes express themselves in the existence of the universe. The proof is valid only if the presuppositions are granted.

Most actual instances of the second proof—the proof from the unchangeability of the cause of the universe—similarly appear to rest on a presupposition, namely, that the universe owes its existence to an unchangeable cause. The second proof nevertheless differs from the third in that any argument from the unchangeability of the cause of the universe can easily be read or recast in a hypothetical mode. On the assumption of creation, the reasoning can be construed, some cause must be ultimately responsible for bringing the universe into existence. The cause ultimately responsible for bringing the universe into existence would have to be immune from change; for were it to undergo change, it would be dependent upon whatever produced the change, and it would not, after all, be the ultimately responsible cause.[3] And yet, to bring a universe into existence after having failed to do so, would constitute change. Creation would thus imply an unchangeable cause that undergoes change, and the assumption of creation is self-contradictory.

The standpoint of the first proof, then, is hypothetical; the second proof can easily be read or recast in a hypothetical mode; whereas the standpoint of the third is not hypothetical but dogmatic.

Maimonides, it will be recalled, characterized proofs of eternity from the nature of the cause of the world as having been developed from Aristotle's principles by philosophers following Aristotle.[4] The characterization is apt in respect to the first and second of the three proofs. A trace of the first proof is to be found in Aristotle's physical works, and Aristotle seems to have articulated a version of the proof in an early dialogue.[5] The second proof can be seen as a development of one of the strands from which Aristotle's argument for the eternity of motion,

[2] For the principle that nothing could come into existence without a cause, cf. below, pp. 154 ff.
[3] See further, below.
[4] Above, p. 10.
[5] Below, p. 52.

discussed in the previous chapter, is woven.[6] Furthermore, the conception of the ultimate cause of the universe which is operative in the two proofs accords with Aristotle's conception of the prime mover. On the assumption of creation, it is argued in both proofs, the ultimate cause of the universe's coming into existence would have to be an entity that is unaffected by anything outside itself and that is unchanging; and those are traits of Aristotle's prime mover. Maimonides' characterization is not, by contrast, accurate as regards the third proof from the cause of the universe. The third proof presupposes that there is indeed a being from whom the very existence of the universe flows, something quite foreign to the spirit of genuine Aristotelianism, where a cause only of the motion of the universe is recognized. The third proof would be more accurately characterized as Neoplatonic.

Proclus apparently was the main source or channel through which medieval Arabic philosophers received the three proofs for eternity from the cause of the world. A list of eighteen proofs of eternity had been drawn up by him,[7] and part, if not all, of the list was available in Arabic in the Middle Ages: A medieval Arabic translation of the first nine of the eighteen proofs has been discovered,[8] and in addition Shahrastānī records, in Proclus' name, a paraphrase of seven of the nine, together with an eighth proof taken from the second half of Proclus' list.[9] The three basic arguments for eternity from the nature of the cause of the universe, although not every variation, can be discovered among the proofs of Proclus' which have been preserved in Arabic,[10] and he may for that reason be taken as the medievals' probable source.

(a) The argument that nothing could have led a creator to create the universe at a particular moment

The contention that what exists cannot have come into being goes back at least to Parmenides. In conjunction with other considerations, Parmenides is reported

[6]Below, p. 57.

[7]Preserved in John Philoponus, *De Aeternitate Mundi contra Proclum*, ed. H. Rabe (Leipzig, 1899).

[8]In A. Badawi, ed., *Neoplatonici apud Arabes* (Cairo, 1955), pp. 34–42. Cf. G. Anawati, "Un fragment perdu du *de aeternitate mundi* de Proclus ," *Mélanges de philosophie grecque offerts à Mgr. Diès* (Paris, 1956), pp. 21–25.

[9]*K. al-Milal wa-l-Niḥal*, ed. W. Cureton (London, 1842–1846), pp. 338–340. The eighth proof given by Shahrastānī is Proclus' thirteenth. In *K. Nihāya al-Iqdām*, ed. A. Guillaume (Oxford and London, 1934), pp. 45–46, Shahrastānī records, in Proclus' name, three proofs from the first half of Proclus' list.

[10]I find five of Proclus' proofs—the first, third, fourth, sixteenth, and eighteenth—to be versions of the three proofs discussed in the present chapter. The fifth of Proclus' eighteen proofs is a version of the argument from the nature of time; cf. above, p. 24. Most of the remaining are exegetical. They undertake to show that Plato's philosophy implies the eternity of the world, and that Plato too, despite the apparent sense of the *Timaeus*, believed the world to be eternal in the past as well as the future.

to have posed the rhetorical question: "What need could have made that which exists, exist later rather than sooner?"[11] The purport of the question is that since nothing could have determined a later, rather than an earlier, moment for the existence of the universe, the universe cannot have come into existence at any moment whatsoever, but must have existed from eternity. An intimation of the same thought can be unearthed in Aristotle. In the course of proving that what is indestructible cannot have been generated, Aristotle remarked: "Why . . . after not existing for an infinite time, would the thing be generated . . . at a particular moment?"[12] The argument is known to students of modern philosophy from Kant.[13]

Neither the fragment from Parmenides nor the passage in Aristotle makes reference to whatever it might be that brings the world into existence. But the medieval philosophers as well as Aristotle were certain that nothing could have come into existence spontaneously and without a cause. Should the cause responsible for the supposed creation of the world be taken into account, the argument being examined would run: On the assumption of creation, no given moment in an undifferentiated eternity could, as distinct from any other moment, have recommended itself to the cause bringing the world into existence as the proper time for it to create the world; the cause that would, on the assumption of creation, have had to create the world could not therefore have acted at any moment whatsoever. This train of reasoning, too, was employed by Aristotle, not however in his preserved works but in a lost dialogue. Aristotle, as reported by Cicero, argued for the eternity of the world on the grounds that no "new plan" (*novo consilio*) could have arisen which might have occasioned the world's creation.[14] The mention of a "plan" indicates that the argument has in view not merely the moment at which the world would have come into existence on the assumption of creation, but also a cause, specifically an intelligent agent, that would have had to decide upon and execute the project. Aristotle's meaning must, in other words, have been that creation is impossible because there was no moment at which the creator might have decided to act. Similar formulations appear elsewhere before the medieval period. Cicero, the source of the passage from the forementioned Aristotelian dialogue, represents a member of the Epicurean school as asking rhetorically: "Why should the builders of the world suddenly have sprung into action after innumerable ages of slumber?"[15] Augustine takes up certain unnamed skeptics who had a like predilection for rhetorical questions.

[11] H. Diels, *Fragmente der Vorsokratiker* (Berlin, 1934–1938), Parmenides, fragment 8. English translation in J. Burnet, *Early Greek Philosophy* (London, 1930), p. 175.

[12] *De Caelo* I, 12, 283a, 11–12; cf. *Physics* VIII, 1, 252a, 15–16.

[13] Kant, *Critique of Pure Reason*, A427/B455, proof of the antithesis of the first antinomy. The proof of the thesis of the first antinomy can be traced to John Philoponus; cf. below, p. 88.

[14] Quoted by Cicero, *Academica*, II, xxxviii, 119.

[15] *De Natura Deorum*, I, ix, 21.

"Why," the skeptics asked, "did the eternal God choose at a particular point to make the heavens and earth, which he had not made previously?"[16] "How did the idea of making something come into his mind despite his never having made anything before?"[17]

In the Middle Ages, the argument we are examining—the argument that no given moment could, in preference to any other, have served as the proper moment for the world's coming into existence—invariably contains a reference to the creator. The reasoning invariably is that no given moment could have suggested itself to the creator as the proper moment for him to create the world. Avicenna, for example, defends the eternity of the world by posing, once again, a rhetorical question. He asks: "How within [the stretch of] nonexistence could one time be differentiated for [a creator's] not acting and another time for [his] starting [to act]? How might one time differ from another?"[18] Instances of the argument that no moment could have suggested itself to the creator as the moment for creating the world can be found in Ghazali,[19] Maimonides,[20] Albertus Magnus,[21] Aquinas,[22] Crescas,[23] and Joseph Albo.[24]

A distinction can be drawn—and it is admittedly a fine distinction—between instances of the argument which do not and a larger number of instances which do include an additional element, the element of the creator's motive. Avicenna and the philosophers just mentioned asked: How could any given moment have suggested itself to the creator, in preference to infinite identical moments, as the time for creating the world? In the examples to be examined now the question is: On the assumption that the creator did choose one moment in preference to others, what could possibly have induced him to make the choice? What could his motive have been? The theme can already be detected in the passage where Aristotle spoke of a "new plan,"[25] and in the passage where Augustine's skeptics wondered how the "idea" of creating the world could have come into the mind

[16] *City of God,* XI, 4.

[17] *Confessions,* XI, xxx; cf. XI, x. In *City of God,* XII, 15, Augustine states his own position that God was *not* motivated by "a new plan" (*novo . . .consilio*) in creating the world.

[18] *Shifā': Ilāhīyāt,* ed. G. Anawati and S. Zayed (Cairo, 1960), p. 378.

[19] *Tahāfut al-Falāsifa,* ed. M. Bouyges (Beirut, 1927), I, §28; English translation in Averroes *Tahafut al-Tahafut,* trans. S. van den Bergh (London, 1954), p. 18.

[20] *Guide to the Perplexed,* II, 14(8). Maimonides includes a point also met elsewhere, namely that it is unimaginable that the deity should have remained "idle" for an eternity before creating the world. Cf. Cicero, *De Natura Deorum,* I, ix, 22; Augustine, *City of God,* II, 18; Simplicius, *Commentary on Physics,* ed. H. Diels, *Commentaria in Aristotelem Graeca,* Vol. X (Berlin, 1895), p. 1331. A related notion is that the deity is never "grudging"; cf. below, n. 96.

[21] *Commentary on II Sentences,* in *Opera Omnia,* ed. A. Borgnet, Vol. XXVII (Paris, 1894), d. 1, B, art. 10.

[22] *Summa contra Gentiles,* II, 32(5).

[23] *Or ha-Shem,* III, i, 1.

[24] *'Iqqarim,* I, 23.

[25] Above, n. 14.

of the creator.[26] An especially clear statement of the theme is provided by Maimonides. He portrays the proponents of eternity as reasoning: "An agent acts at one time and not at another because of either preventative factors (*māni'*) or motivating factors (*dā'in*) which occur (*ṭārin*) in him. The former bar an agent from accomplishing what he wills; the latter lead the agent to will what he previously did not will. Since the creator, [who must be absolutely self-sufficient[27]] is subject to neither motivating factors . . . nor preventive factors, . . . it is impossible for him to act at one time but not at another."[28]

The contention that nothing imaginable could have motivated the creator to act at one given moment in preference to others was common. Bāqillānī alludes to it, and forestalls it, when he stresses that in creating the world "after not having done so," the creator was not led to act by a "motivating factor (*dā'in*), . . . or moving factor, . . . or inducing factor (*bā'ith*), . . . or disturbing factor, . . . or new idea (*khāṭir*)."[29] Abū al-Barakāt argues that no new "state" (*ḥāl*), or "inducing factor" (*bā'ith*), or "necessitating factor" (*muqtaḍin*) could conceivably have led the creator to select "a given time as distinct from what preceded and what succeeded, inasmuch as all times were equivalent." The conclusion Abū al-Barakāt draws is that the world must be eternal.[30] With variations in wording, the reasoning recurs in Ibn Ṭufayl,[31] Averroes,[32] Albertus Magnus,[33] Aquinas,[34] Fakhr al-Dīn al-Rāzī,[35] Gersonides,[36] Aaron ben Elijah,[37] Crescas,[38] and Abravanel.[39]

Proclus' version of the argument we are examining exhibits still a further element. His contention is that on the assumption of the creator's acting to create the world, the factors inducing him to act at the critical moment would regress infinitely. For, if the world were created, the creator would, up to the moment of creating the world, have been a "potential creator," and something would have

[26]Above, n. 17.

[27]Cf. below, p. 66.

[28]*Guide*, II, 14(6).

[29]*K. al-Tamhīd*, ed. R. McCarthy (Beirut, 1957), p. 30.

[30]*K. al-Mu'tabar* (Hyderabad, 1939), pp. 33, 43. Abū al-Barakāt includes the point that the deity would have been "idle" for an eternity.

[31]*Ḥayy b. Yaqdhān*, ed. and trans. G. Gauthier (Beirut, 1936), Arabic text, p. 82; French translation, pp. 62–63; English translation with pagination of the Arabic indicated: *Ḥayy Ibn Yaqzān*, trans. L. Goodman (New York, 1972).

[32]*K. al-Kashf*, ed. M. Mueller (Munich, 1859), p. 31; German translation with pagination of the Arabic indicated: *Philosophie und Theologie von Averroes*, trans. M. Mueller (Munich, 1875).

[33]*Physics*, in *Opera Omnia*, ed. A. Borgnet, Vol. III (Paris, 1890), VIII, tr. 1, chap. 11(6).

[34]*De Potentia*, q. 3, art. 17(13).

[35]*Muḥaṣṣal* (Cairo, 1905), p. 91.

[36]*Milḥamot ha-Shem* (Leipzig, 1866), VI, i, 18, p. 371.

[37]*'Es Ḥayyim*, ed. F. Delitzsch (Leipzig, 1841), chap. 7.

[38]*Or ha-Shem*, III, i, 4.

[39]*Mif'alot* (Venice, 1562), VI, 1(1), quoting what appears to be Ghazali, *Tahāfut al-Falāsifa*, I, §5.

had to "activate" him. But the activating factor would, before inducing the creator to create the world, have been a "potential" activating factor, and hence would have stood in need of a prior factor to activate it as well. And the prior factor would in its turn also have had to be activated. The supposition of creation therefore leads to an infinite regress of factors activating the creator. Since an infinite regress of causes is absurd,[40] the world, Proclus concludes, must be eternal.[41]

Argumentation of the same type is alluded to and forestalled, as in a previous instance,[42] by Bāqillānī. He explains that the creator cannot be understood to have created the world "for a reason" (*li-'illa*), since the "reason" for the world's creation would have been absent up until the moment of creation and would only then have come into existence. Its coming into existence at the moment of creation could only have been due to a different reason; and the reasons for the creation of the world would regress infinitely.[43] 'Abd al-Jabbār records an argument for eternity according to which the creator's becoming active after having been inactive would have to be due to a factor (*ma'nā*). The factor's being present and operative precisely at the moment of creation would have to be due to another factor, and it, in turn, to yet another. Therefore, the argument concludes, the assumption of creation implies an infinite regress of causes, and is absurd.[44] Averroes defends the doctrine of eternity, writing: If the creator be assumed to have "acted at a given time and not at another, some cause would have had to assign him the one state as distinct from the other." But then an additional cause would have to be responsible for the cause's being present at the critical moment, and so forth *ad infinitum*. The thesis of creation is consequently absurd.[45] In a separate passage, Averroes contends in a similar vein that to assume a beginning of motion in the universe would imply an infinite regress of movers, which is absurd.[46] That argument is later recorded by Gersonides.[47]

The Kalam writers were fond of the notion that when events might take more than a single course, something must *tip the scales* between the equivalent possibilities and determine the course that is taken.[48] The notion of *tipping the scales* is employed now by several Kalam writers in restating their opponents' argument that creation would imply an infinite regress of motivating factors. Ghazali has

[40]See below, pp. 241, 337.

[41]Quoted by Philoponus, *De Aeternitate*, pp. 42–43.

[42]Above, n. 29.

[43]*K. al-Tamhīd*, pp. 31–32.

[44]*Sharḥ al-Uṣūl* (Cairo, 1965), p. 115.

[45]*K. al-Kashf*, p. 30.

[46]*Epitome of Metaphysics*, ed. and trans. C. Quirós Rodriguez (Madrid, 1919), IV, §3; German translation: *Die Epitome der Metaphysik des Averroes*, trans. S. van den Bergh (Leiden, 1924), pp. 105–106. The argument is closely related to the argument from motion, discussed in the previous chapter.

[47]*Milḥamot ha-Shem*, VI, i, 3, p. 299.

[48]Cf. below, p. 162.

the adherents of eternity reason: The world's being produced at a given moment would require the presence at the given moment of a "factor tipping the scales" (*murajjiḥ*) in favor of the creator's acting. The presence of the factor at just the required moment would demand another factor to tip the scales in favor of it, and so on *ad infinitum*. But an infinite regress of causes is inadmissible; hence the assumption of creation is inadmissible.[49] Similar formulations are recorded by Fakhr al-Dīn al-Rāzī,[50] Āmidī,[51] and Ījī.[52] And Rāzī and Ījī describe the argument that creation would imply an infinite regress of factors tipping the scales as the "pillar" of the adherents of eternity.

Resumé

A progression can be discerned. As far back as Parmenides we find the thought that no given moment in the undifferentiated stretch of eternity could, in preference to any other, have lent itself to the world's coming into existence. The argument thereupon developed that no given moment could have recommended itself to the creator as the proper moment for him to create the world. A number of medieval philosophers offer that version, usually through the medium of a rhetorical question. An even larger number offer a version into which the element of motivation is introduced. The thinking here is that no imaginable motive could have induced the creator to act and create the world at one given moment rather than another. Finally, the element of an infinite regress is added by Proclus and by a line of medieval philosophers who are probably dependent directly or indirectly on him. They contend that the factors, or reasons, or causes, or factors tipping the scales, which would have led the creator to create the world at a given moment would regress infinitely. Since an infinite regress of causes is impossible, the creation of the world is impossible as well.

(b) The argument from the unchangeability of the cause of the universe

Should an unchangeable cause of the existence of the universe be presupposed, an argument for eternity can be framed which runs: An unchangeable cause is known to be responsible for the existence of the universe. But an unchangeable cause would not pass from a state of inaction to a state of action. The cause of the universe cannot, therefore, have acted to bring its effect into existence after having failed to do so, and the universe must be eternal. The argument can also be put in a hypothetical form, although, it must be confessed, actual instances of the hypothetical form are not the rule. Put hypothetically, the argument would go: On the assumption of creation, the universe must have been brought into existence by a cause. The ultimate and true cause of the coming into

[49] *Tahāfut al-Falāsifa*, I, §§3, 5; English translation, p. 1.
[50] *K. al-Arba'īn* (Hyderabad, 1934), p. 42.
[51] *Ghāya al-Marām* (Cairo, 1971), p. 265.
[52] *Mawāqif* (Cairo, 1907), VII, pp. 228–229.

existence of the universe would have to be unchangeable. For if the immediate cause of the universe's coming into existence underwent change, it would be dependent on whatever produced the change; if what produced the change underwent change, it would in its turn be dependent upon something else; and unless an unchangeable cause should be reached, nothing would be truly and ultimately responsible for the universe's coming into existence.[53] The assumption of creation implies, then, an ultimate cause that is unchangeable. Yet the assumption equally implies that the cause ultimately responsible for the universe's coming into existence did change inasmuch as it passed from a state of not acting to a state of acting. Creation thus embodies a self-contradiction and is untenable, and the universe cannot have been created but must be eternal.

The argument from the unchangeability of the cause of the universe, especially when formulated hypothetically, resembles and could well have been suggested by a strand in Aristotle's proof of the eternity of motion, a proof that will be recalled from the preceding chapter. At one stage of that proof Aristotle reduced the assumption of an absolute beginning of motion to absurdity by looking at the relationship between the cause producing, and the body performing, the supposed first motion or change. The relationship between what produces and what performs the supposed first motion would, he reasoned, have had to change before the motion could occur; and the motion or change assumed to be absolutely first would, consequently, not be first after all.[54] The argument we are presently examining, the argument from the unchangeability of the cause of the universe, looks, for its part, not at the relationship between the cause of the universe and the universe, but at the cause alone. The argument reduces the assumption of a beginning of the universe to absurdity by showing that the assumption would imply a change in the cause ultimately responsible for creation, whereas anything subject to change could not after all be an ultimately responsible cause.

The argument from the unchangeability of the cause of the universe can also be seen as an extension of the proof examined under the previous heading, the argument that no given moment could have suggested itself to the creator as the proper moment to act. In the argument from the unchangeability of the cause of the universe, the impossibility of an infinite regress can explain—although other explanations are possible too[55]—why the cause ultimately responsible for the existence of the universe would have to be unchangeable: The unchangeability of the ultimate cause must be posited in order to avoid an infinite regress of changeable causes. Creation is thereupon found to be untenable because it would imply that a cause which must be unchangeable nevertheless changes. In the final version of the proof examined under the previous heading, the contention was, more simply, that a cause of the universe could not act to bring the world into

[53]Cf. below, p. 337.

[54]Above, pp. 19–20.

[55]See, for example, Proclus' argument, immediately below.

existence at a given moment because the factors inducing the cause to act would regress infinitely.[56] The present argument goes beyond the previous one in only a single detail. The unchangeability of the cause of the universe is explicit here but not there.

The argument for eternity from the unchangeability of the cause of the universe is found in Proclus, who defended the eternity of the universe by reasoning: The cause of the universe must be "immovable." For were it subject to motion, it would pass from a state wherein it was "imperfect" to a state wherein it was "perfect," which is inconceivable in regard to the highest being. And it would moreover "stand in need of time," which is inconceivable in the being that is the cause of everything outside itself, including time. But if "an agent is immovable, it is unchangeable; and if unchangeable, it . . . [cannot] pass . . . from [a state of] not acting to [a state of] acting. . . . Therefore, if something . . . is an immovable cause of something else . . . it is so eternally . . . ; and if the cause of the universe is immovable . . . the universe must be eternal."[57] In the Middle Ages, the argument for eternity from the unchangeability of the cause of the universe was employed or recorded by Avicenna,[58] Ibn Ḥazm,[59] Shahrastānī in his paraphrase of Proclus,[60] Shahrastānī in his account of Aristotle's philosophy,[61] and Averroes. Averroes' formulation runs: The assumption of creation entails either that things come into existence spontaneously, which is absurd; or else that the agent bringing the world into existence underwent "change and hence stood in need of an agent apart from itself to bring about the change." On the latter alternative, the second agent would stand in need of a further agent; and we would be left in the end with no "first agent" who is responsible for the world's coming into existence. The creation of the world is accordingly untenable and the world must be eternal.[62]

Sometimes the concepts of potentiality and actuality are called into play. On the hypothesis of creation, the reasoning goes, the creator would have passed from the state of being a potential creator to the state of being an actual creator; such a transition would be a change; but change is impossible for the cause of

[56]Above, pp. 54–55.

[57]Quoted by Philoponus, *De Aeternitate,* pp. 55–56. This is only part of Proclus' argument, and the rest is examined under the next heading. The notion that the unchangeability of the deity implies the eternity of the world also appears in other proofs of Proclus'; see Philoponus, ibid., pp. 42, 604.

[58]*Shifā': Ilāhīyāt,* p. 376.

[59]*K. al-Faṣl fī al-Milal,* (Cairo, 1964), I, p. 20; Spanish translation: ***Abenházam de Córdoba y su Historia Crítica de las Ideas Religiosas,*** trans. M. Asín Palacios, Vol. II (Madrid, 1928), pp. 113 ff.

[60]*K. al-Milal wa-l-Niḥal,* p. 339; *K. Nihāya al-Iqdām,* p. 46.

[61]*K. al-Milal wa-l-Niḥal,* p. 320.

[62]*Tahāfut al-Tahāfut,* I, p. 8.

the universe; and hence creation is impossible.[63] That formulation is recorded by Shahrastānī,[64] Albertus Magnus,[65] Bonaventure,[66] Aquinas,[67] Gersonides,[68] Aaron ben Elijah,[69] Crescas,[70] and Abravanel.[71] Maimonides,[72] and Aquinas[73] record a truncated version, which runs: If the world had been created, the creator would have passed from a state of potentiality to a state of actuality; something would have had to bring about the transition; and nothing could possible bring about a transition from potentiality to actuality in the cause of the universe. The reason why no transition from potentiality to actuality can be brought about in the cause of the universe is not stated. Very likely, the intended reason is that the cause of the universe must be unchangeable.[74] But the reason could equally be—as in the final version of the proof examined under the previous heading[75]—that the activating factors would regress infinitely. And, since the eternal actuality of the ultimate cause had been established by Aristotle and other philosophers,[76] the reason could also be that the cause of the universe is always in a state of pure actuality. In fact, these ostensibly different reasons are interconnected. The impossibility of an infinite regress of activating factors can serve as grounds both for the unchangeability, and the eternal actuality of the ultimate cause;[77] and the unchangeability of the ultimate cause both entails, and is entailed by, its state of eternal actuality.[78] All three principles—the impossibility of an infinite regress, the unchangeability of the cause of the universe, and the eternal state of actuality of the cause of the universe—were commonplace in the Middle Ages, and any or all of them could easily have been taken for granted.

[63]See the argument of Proclus' discussed under the previous heading, above, p. 55. Philoponus, *De Aeternitate*, p. 82, quotes an argument from another work of Proclus' to the effect that if the world were created, the creator would have been in a state of potentiality, and hence imperfect, before becoming actual.

[64]*K. al-Milal wa-l-Niḥal*, p. 339; *K. Nihāya al-Iqdām*, p. 45. Shahrastānī is paraphrasing the argument of Proclus' which I included under the previous heading, above, pp. 54–55.

[65]*Physics*, VIII, tr. 1, chap. 11.

[66]*Commentary on II Sentences*, d. 1, p. 1, a. 1, q. 2.

[67]*De Potentia*, q. 3, art. 17(12).

[68]*Milhamot ha-Shem*, VI, i, 24, p. 393.

[69]*'Eṣ Ḥayyim, chap. 7.*

[70]*Or ha-Shem*, III, i, 1.

[71]*Mif'alot*, VI, 1.

[72]*Guide*, II, 14(5).

[73]*Summa contra Gentiles*, II, 32(2).

[74]Aquinas gives this reason elsewhere; above, n. 67.

[75]Above, p. 55.

[76]Aristotle, *Metaphysics* XII, 6, 1071b, 20; below, p. 347.

[77]To avoid an infinite regress of causes, a first cause must be posited which is neither changeable nor subject to a transition from potentiality to actuality.

[78]A cause that is not subject to change will always be in its state of actuality; and anything that is eternally actual will never change.

The argumentation thus far has been that the ultimate cause of the universe must, in general, be unchangeable and consequently could not pass from inaction to action. Several variations are in evidence as well, each of which directs its attention to a respect wherein the cause of the universe would have to be unchangeable. One variation appears in Augustine,[79] Proclus,[80] Maimonides,[81] Aaron ben Elijah,[82] and Gersonides.[83] Proclus argues, and the others record the argument, that the decision to create the world at a particular moment would constitute a change in God's will. Inasmuch, however, as God's will must be identical with his essence,[84] and his essence is unchangeable, his will must be unchangeable. God consequently could not have decided to create the world after having failed to make the decision, and the world must be eternal.

According to another variation, reported by 'Abd al-Jabbār, creation would imply a change in God's "knowledge," hence "a change in his state"; for the creator would have spent an eternity without knowing that the world exists, whereupon, at the moment of creation, he would acquire a new item of knowledge, the knowledge that the world does exist. But since God is not changeable, his knowledge is not changeable. Therefore the external object of his knowledge, the universe, is unchangeable and must have existed forever.[85]

Still another variation was probably inspired by Aristotle's point that the assumption of a beginning of motion would imply a prior change in the relationship between the cause producing, and the object performing, the supposed absolutely first motion.[86] Crescas records an argument concerned not with the change of relationship which would precede creation but with the change of relationship which would result. The argument is that on the assumption of creation, the cause of the universe, or the deity, would enter a new relationship; for before creation he would not have had a relation to the universe, whereas subsequently he would. But a change in the creator's relationship to the world would entail a change in himself, which is an impossibility. Consequently, creation is an impossibility.[87] The argument is repeated by Abravanel.[88]

[79] *City of God,* XII, 18.

[80] Quoted by Philoponus, *De Aeternitate,* p. 560.

[81] *Guide,* II, 14(6); 18.

[82] *'Eṣ Ḥayyim,* chap. 7.

[83] *Milḥamot ha-Shem,* VI, i, 18, p. 377.

[84] The reason is that the essence of the first cause must be absolutely simple. Cf. Plotinus, *Enneads,* VI, 8, 13. On the question whether an Aristotelian or Plotinian deity can properly be described as having will, see E. Zeller, *Die Philosophie der Griechen,* Vol. II, Part 2 (4th ed.; Leipzig, 1921), pp. 368–370; III, 2 (5th ed.; Leipzig, 1923), pp. 539–540.

[85] *Sharḥ al-Uṣūl,* p. 117. 'Abd al-Jabbār does not explain how the authors of the argument would harmonize the unchangeability of God's knowledge with the constant changes in the objects of his knowledge within an eternal universe. Reasoning similar to that recorded by 'Abd al-Jabbār appears in Khayyāṭ, *K. al-Intiṣār,* ed. and trans. A. Nader (Beirut, 1957), §71.

[86] Above, p. 19.

[87] *Or ha-Shem,* III, i, 1.

[88] *Mif'alot,* VI, 1.

In sum, it was argued that creation would imply a change in the cause of the universe, or a change in his will, or in his knowledge, or in his relationship to the world; but a change of any sort is impossible in the cause of the universe; therefore the creation of the world is impossible.

Arguments from God's eternal attributes[89]

An argument for eternity from the attribute of divine goodness[90] was known to Augustine. He portrays the Stoics as defending their peculiar theory of an eternity of world cycles—as distinct from the eternity of a single world—on the grounds that God's "goodness" could never have been "inoperative";[91] since God's goodness must always have been in operation, the reasoning went, there must always have existed an expression of his goodness, that is to say, a world. A similar argument appears in a more fully developed form in Proclus, who characterizes it as the "most convincing . . . demonstration"[92] of the eternity of the world. The original Greek text of the passage in Proclus is lost, but Philoponus' refutation has been preserved in the original Greek, and a medieval Arabic translation of Proclus has also been preserved. Philoponus' refutation indicates that the argument had originally taken God's goodness—together with his power—as its premise.[93] In the medieval Arabic translation, however, a small change has been made and the proof rests on God's beneficence (*jūd*), together with His power.

Proclus, as refracted through the Arabic translation, lays down the proposition that "when an agent does not act, he fails to act either because he does not wish to or because he is unable to." In the issue at hand, the world's coming into existence, it is impossible that the agent should not have wished to act. For it is unimaginable that the supreme being should be merely "sometimes beneficent (*jawād*), sometimes not." He undoubtedly is "eternally beneficent"; and, being eternally beneficent, he must eternally "wish the universe to resemble him," to exist and be good, even as he exists and is good.[94] The impossibility that the deity should not have wished to bring the world into existence is matched by the

[89]The sense in which the deity can have attributes, and the proper manner of construing terms predicated of him, were of course perennial and much debated questions. Cf. H. Wolfson, *Philo*, Vol. II (Cambridge, Mass., 1948), pp. 149–164; idem, *Studies in the History of Philosophy and Religion*, Vol. I (Cambridge, Mass., 1973), pp. 98–169; idem, *Philosophy of the Kalam* (Cambridge, Mass., 1976), pp. 112–234.

[90]That God's goodness is the source of the existence of things outside of God was a tenet of Platonic and Neoplatonic philosophy. See Plato, *Republic*, 509B; Plotinus, *Enneads*, V, 5, 10; 13; Arabic paraphrase: *Plotinus apud Arabes*, ed. A. Badawi (Cairo, 1955), p. 182; Proclus, *Elements of Theology*, ed. E. Dodds (Oxford, 1963), §12. Aristotle, for his part, characterized the prime mover as supremely good, *Metaphysics* XII, 7 and 9.

[91]*City of God*, XII, 18.

[92]In Badawi, *Neoplatonici apud Arabes*, p. 34.

[93]*De Aeternitate*, p. 13, *et passim*.

[94]Cf. Plato, *Timaeus*, 29E.

impossibility that he should be incapable of accomplishing what he wishes. Inasmuch as the deity must have wished eternally to bring the universe into existence and must, moreover, be eternally able to effect what he wishes, he must have brought the universe into existence eternally.[95] Unlike the Stoic arguments in Augustine's report, Proclus' argument concludes that God's beneficence—which perforce produces something as similar to God as possible—must give rise to a single eternal universe, as distinct from an eternal succession of universes. Virtually all medieval arguments from the eternal divine attributes likewise concluded that God's attributes must give rise to a single eternal universe.

In Arabic, the argument for eternity from the beneficence of the first cause is common. The adherents of eternity, 'Abd Jabbār explains, "base their thesis" on, among other things, the premise "that the creator is eternally beneficent." If God were not eternally beneficent, "he would be sometimes beneficent and at other times not beneficent . . . [but] niggardly,[96] [which is unimaginable]. Once the foregoing is established, the eternal existence of the world follows."[97] The argument from God's "beneficence" is alluded to by Avicenna,[98] and is found in Ibn Ḥazm,[99] in Shahrastānī's paraphrase of Proclus,[100] in Fakhr al-Dīn al-Rāzī,[101] Āmidī,[102] Ījī,[103] and Leone Ebreo.[104] Aquinas records an argument in the same vein which infers the eternity of the world from God's "most perfect" and "infinite . . . goodness."[105]

The proof for eternity from the deity's eternal attributes, like the other proofs for eternity, exhibits a number of variations. In each variation an eternal attribute of God is shown to express itself eternally in the existence of the world, God's knowledge, his wisdom, his perfection in general, and his character as Lord of the universe, all being brought into play as grounds for eternity. 'Abd al-Jabbār, for example, takes up an argument to the effect that God cannot have brought the world into existence through the "inducement of need" (*dā'ī al-ḥāja*), since he is self-sufficient and immune from need.[106] God could only have brought the universe into existence through the "inducement of wisdom" (*dā'ī al-ḥikma*). God's wisdom comprises "his knowledge of the goodness of the world and his knowledge of the benefit others will derive from it. That [knowledge] is firm

[95]In Badawi, *Neoplatonici apud Arabes*, p. 34.

[96]For the doctrine that the deity is free of "grudging" (φθόνος), see Plato, *Phaedrus*, 247A; *Timaeus*, 29E; Aristotle, *Metaphysics* I, 2, 983a, 2; Philoponus, *De Aeternitate*, p. 13.

[97]*K. al-Majmū' fī al-muḥīṭ bi-l Taklīf*, ed. J. Houben (Beirut, 1965), p. 66.

[98]*Shifā': Ilāhīyāt*, p. 380.

[99]*K. al-Faṣl fī al-Milal*, I, p. 20. The passage is quoted below, p. 64.

[100]*K. al-Milal wa-l-Niḥal*, p. 339; *K. Nihāya al-Iqdām*, p. 45.

[101]*K. al-Arba'īn*, p. 50.

[102]*Ghāya al-Marām*, p. 266.

[103]*Mawāqif*, VII, p. 230.

[104]*Dialoghi d'Amore*, p. 238.

[105]*Summa contra Gentiles*, II, 32(7). Cf. *De Potentia*, q. 3, art. 17 (1, 14).

[106]This part of the argument is related to the arguments discussed above, pp. 53–54.

through eternity. And therefore the existence of the world is necessary through eternity."[107] Here the contention is that God's eternal wisdom and knowledge would require him to produce the world eternally. Aquinas has a kindred argument, which intertwines and perhaps confuses two notions, the notion that God's eternal knowledge is the cause of the universe and the notion that true knowledge mirrors objects in the external world. Aquinas' version reads: "God is the cause of things through his knowledge." Knowledge is "relative" to the thing known;[108] for knowledge and the object of knowledge "exist together by nature," true knowledge occurring solely when something actually exists to serve as its object. God's knowledge of what is brought into existence through his knowledge accordingly requires, for it to be true knowledge, that the things in question actually exist. And inasmuch as "God's knowledge is eternal, things apparently are produced by him from eternity."[109]

The notion that true knowledge must mirror objects in the external world also plays a role in a passage where Gersonides, followed by Crescas and Abravanel, speculates about the underlying considerations leading Aristotle to the doctrine of eternity. Gersonides observes that "an intelligible thought corresponding to no object outside the mind would seem necessarily to be false." But "God is the [intelligible] order [*nimus*] of the universe"; that is to say, God's thought, which is identical with his essence, comprises a mental representation of the universe. Aristotle concluded herefrom, in Gersonides' reconstruction, that the actual and external order of the universe must exist whenever God and his thought—which comprises the intelligible order of the universe—exist. God consequently "could not exist without the universe," and since God is eternal, the universe must be eternal.[110]

Another ground adduced for the eternity of the world was the broad attribute of divine perfection. Averroes writes that since God is perfect, he cannot "fail to perform the superior act and perform instead the lesser act; for that would be a defect" in him. It would surely be the "greatest defect" in "the eternal agent, whose existence and act are [in fact] unlimited, . . . were his act to be limited and finite." God's act, the causation he exercises vis-à-vis the universe,[111] must, therefore, be infinite in duration, and the universe, which is the product of his causation, must be eternal.[112]

Eternity could, as has been seen, be inferred from the attribute of divine perfection or from the attribute of divine wisdom. Maimonides, followed by

[107] *Sharḥ al-Uṣūl*, p. 116.

[108] Cf. Aristotle, *Topics* IV, 1, 121a, 1.

[109] *De Potentia*, q. 3, art. 17(19).

[110] Gersonides, *Milḥamot ha-Shem*, VI, i, 3, 302; Crescas, *Or ha-Shem*, III, i, 1 (very much abbreviated); Abravanel, *Mif'alot*, VI, 1 (copying from Crescas).

[111] In Averroes' view, God's causation is primarily in the realm of motion; but, as a cause of motion, God is also the cause of the existence of the universe. See below, pp. 325–326, 341.

[112] *Tahāfut al-Tahāfut*, I, p. 96.

Albertus Magnus, shows how eternity might be inferred from the two attributes conjointly. The adherents of eternity, Maimonides reports, argued that since "God's acts are perfect and contain no defect," and God makes "everything as perfect as can be," our "universe must be the most perfect possible." The existence of the world flows, moreover, from God's wisdom, which undoubtedly maximizes the perfection of the world. God's wisdom is, in its turn, "identical with his essence" and hence "eternal," so that the maximum measure of existence it could bring forth is eternal existence. The world, as the most perfect possible product of God's eternal wisdom, must likewise be eternal.[113]

Proclus' argument for eternity from the deity's goodness or beneficence had cited the attribute of divine power side by side with the attribute of divine goodness. Proclus had reasoned that since God is good and beneficent, he could not fail to wish the existence of the world; and since God is powerful he could not fail to accomplish what he wishes.[114] Ibn Ḥazm knows of a singularly spare argument combining not two but three divine attributes: "[The proponents of eternity] affirm that the cause of the creator's act is his beneficence, wisdom, and power; he is always beneficent, wise, and powerful; and seeing that the cause of the world always exists, the world must always exist."[115] Coincidentally, Aquinas too records an argument combining the attributes of divine knowledge, divine power, and divine goodness. It runs: God is "not ignorant"—not lacking in knowledge—and hence he must know how to produce the world from eternity. He is "not impotent"—not lacking in power—and hence he must be capable of producing the world from eternity. He is not "envious"—not lacking in goodness—and hence he must want to produce the world from eternity. The world must consequently be eternal.[116]

One more variation of the proof from the eternity of God's attributes was suggested to Aquinas by a passage in which Augustine described God as "Lord from eternity."[117] It might be argued, Aquinas observes, that since God is eternally the "Lord," he must eternally have subjects with respect to whom he can be designated as the Lord; and a universe must therefore always exist.[118]

The foregoing are arguments for eternity from God's eternal attributes. A further recurring argument consists in the inference of the eternity of the universe from the deity's being a cause, but the appropriate location of that argument in the classification scheme I am observing is not clear-cut. The argument posits that the deity is, by his nature, eternally a *cause by virtue of itself* or that he is

[113]Maimonides, *Guide,* II, 14(7); Albertus Magnus, *Physics,* VIII, tr. 1, chap. 11.

[114]Above, pp. 61–62.

[115]*K. al-Faṣl fī al-Milal,* I, 20.

[116]*De Potentia,* q. 3, art. 17(22).

[117]Augustine, *City of God,* XII, 16; the quotation is not exact. The editions of *De Potentia* refer to Augustine, *De Trinitate,* V, 16, but I could not find an appropriate passage there.

[118]*De Potentia,* q. 3, art. 17(21).

an *eternally actual cause*. Both those terms are convertible into *unchangeable cause*,[119] and the argument from the deity's being a cause may accordingly be assimilated to the argument from the unchangeability of the cause of the universe. The deity's being a cause by his very nature may, however, also be regarded as a divine attribute, and the argument may accordingly be treated as an added variation of the proof from the eternal attributes of God. It can thus be plausibly assigned to either of two headings. In any event—and wherever the argument is best classified—the earliest instances I could discover are two passages in Proclus where the deity's character as an eternal cause appears not independently, but interwoven with other considerations.

The first passage happens to be the one in which Proclus proved that the cause of the universe must be unchangeable.[120] In the course of working out his proof for eternity from the unchangeability of the cause of the universe, Proclus explains that what is unchangeable cannot be a cause "sometimes." If it is a cause at all, it must be "a cause eternally"; and the cause of the universe, which is indeed known to be unchangeable, must eternally possess the character of being a cause. Having come this far, Proclus proceeds to the inference that is of interest here. He lays down the rule that an eternal cause is perforce "a cause of something eternal." Inasmuch as the cause of the universe is an eternal cause, and an eternal cause is a cause of something eternal, the cause of the universe must, Proclus concludes, brings forth an eternal universe.[121]

The second pertinent passage in Proclus is the one in which he has been seen to contend that creation would imply an infinite regress of factors activating the creator.[122] The complete argument there takes the form of a dilemma. The creator, Proclus writes, was either once potential, whereupon he became actual, or else he was eternally actual. If the former alternative were correct and the creator had once been potential, the factors required to activate him would, as was seen earlier, regress infinitely, which is impossible. The latter alternative, then, remains, and the cause of the universe must be eternally actual. But Aristotle showed that "when . . . the cause is actual, the effect is likewise actual."[123] Given Aristotle's proposition together with the proposition that "the creator is an eternally actual creator," the conclusion "ensues that his effect is likewise eternally actual." An eternally actual cause of the universe must bring an actual universe into existence eternally.[124]

In the Middle Ages, Ibn Ḥazm recorded an argument resembling the second of the two arguments just quoted from Proclus. Ibn Ḥazm's version takes the

[119]If something should exist exclusively by virtue of itself, there would be no factor that could bring it into a new condition; and similarly, if something is purely actual and completely free of potentiality, it could never be brought into a new condition.

[120]Above, p. 58.

[121]Quoted by Philoponus, *De Aeternitate,* p. 56. [122]Cf. above, pp. 54–55.

[123]*Physics* II, 3, 195b, 17–18; 28. Also cf. Plotinus, *Enneads,* IV, 5, 7.

[124]Quoted by Philoponus, *De Aeternitate,* pp. 42–43.

form, again, of a dilemma: On the assumption of creation, the creator brought the universe into existence either "by reason of himself (*li-annahu*) or . . . by reason of a cause [distinct from himself]." The latter alternative is inadmissible since it would imply an infinite regress of causes inducing the creator to act.[125] The former alternative therefore remains, and the creator must have acted to produce the world "by reason of himself." But a "cause is inseparable from its effect"; that is to say, when the cause is present, its effect is present. Inasmuch as the creator is a cause by reason of himself, he is a cause as long as he exists, and inasmuch as he exists eternally, he is a cause eternally. Consequently, his effect, the universe, must also exist eternally.[126] Fakhr al-Dīn al-Rāzī and Ījī record the same argument, formulated with the aid of the same dilemma.[127]

Avicenna is more direct. The most expeditious way of settling the issue of creation and eternity is, he maintains, through understanding that the deity is a *cause by virtue of itself.* "A cause by virtue of itself produces its effect by necessity (*awjaba*)," so that whenever it exists, it acts. "If such a cause exists eternally, it acts to produce its effect eternally." Therefore, the eternal cause of the universe must produce an eternal universe.[128] Averroes drops the qualification "by virtue of itself," and argues: When the agent "is eternal," in other words, when something is eternally an agent, "its action must be . . . eternal, and its effects must be eternal."[129]

Bonaventure and Aquinas know of an argument based on the deity's being a "sufficient" cause, a term akin to *cause by reason of itself* or *cause by virtue of itself.* A sufficient cause is such that it does not require the aid of auxiliary causes or conditions to accomplish its ends, which means that as soon as a "sufficient cause is given, its effect is given." Since God is a sufficient cause and since he exists eternally, his effect, the conclusion goes, must also exist eternally.[130] Besides the argument from God's being a sufficient cause, Aquinas records the contention that God's "action . . . must be eternal" because it "is identical with his substance, which is eternal." The conclusion again is that "the effect" of God's action, the universe, must be eternal.[131]

Fakhr al-Dīn al-Rāzī has yet another version of the argument from the deity's being an eternal cause. He describes the proponents of eternity as maintaining, in astonishingly good Kalam style, that to be an agent (*mu'aththir*) is not a "negative attribute" but an "eternal . . . positive attribute added to the essence." To be an agent, the thinking continues, is a "relative attribute," that is, an attribute implying a correlative.[132] Since being an agent is both a positive and a relative

[125]Cf. above, p. 55.

[126]*K. al-Faṣl fī al-Milal,* I, p. 9.

[127]*K. al-Arba'īn,* pp. 41–42; *Mawāqif,* VII, 228–229.

[128]*Shifā': Ilāhīyāt,* p. 373.

[129]*K. al-Kashf,* p. 30.

[130]Bonaventure, *Commentary on II Sentences,* in *Opera Omnia,* Vol. II (Quaracchi, 1882), d. I, p. 1, a. 1, q. 2; Aquinas, *Summa Theologiae,* I, 46, art. 1, obj. 9; *Summa contra Gentiles,* II, 32(3).

[131]*Summa Theologiae,* I, 46, art. 1, obj. 10; *De Potentia,* q. 3, art. 17(6).

[132]Cf. above, p. 63.

attribute, its presence implies the existence of a positive correlative, that is, an actually existing corresponding effect. God's being an agent eternally thus implies the eternal existence of his effect, the universe.[133]

Gersonides, followed by Crescas and Abravanel, provides still another version. Medieval Aristotelians had been divided on the question whether the first cause of the universe is, or is not, identical with the incorporeal mover of the outermost celestial sphere. Some philosophers, most notably Avicenna, understood that the first cause of the universe—the deity—is a being beyond the incorporeal mover of the outermost sphere;[134] but others understood that the highest being in existence, the deity, is the mover of the outermost sphere, and that beyond the movers of the spheres nothing further exists. The second position was espoused by Averroes,[135] and Gersonides naturally enough takes it to be the genuine position of Aristotle. Gersonides now speculates that one of the underlying considerations which led Aristotle to his belief in eternity must have been his position regarding the mover of the outer sphere. Aristotle's reasoning, as reconstructed by Gersonides, was that since the deity is by his nature the cause of the motion of the outermost celestial sphere, he could not exist without a sphere to move. The eternal existence of the first cause would imply the eternal existence of the sphere moved by him, and hence the eternity of the rest of the world as well, seeing that the existence of the rest of the world is entailed by the existence of the sphere.[136]

2. Replies to proofs from the nature of the cause of the universe

The medieval advocates of creation have no single comprehensive response to arguments for eternity from the nature of the cause of the universe similar to their comprehensive response to arguments from the nature of the world itself.[137] What we do find is that each of the three proofs from the nature of the cause of the universe elicits its own set of responses, and that some ancillary motifs recur in answers to more than one proof. It is, I think, not unfair to add that the responses to arguments from the nature of the cause are less satisfactory than were the responses to arguments from the nature of the world.[138]

[133]*K. al-Arba'īn,* p. 49.

[134]Cf. Avicenna, *Shifā': Ilāhīyāt,* pp. 392–393, 401.

[135]Cf. Averroes, *Long Commentary on Metaphysics,* XII, comm. 44.

[136]Gersonides, *Milḥamot ha-Shem,* VI, i, 3, p. 302; Crescas, *Or ha-Shem,* III, i, 1; Abravanel, *Mif'alot,* VI, 1. Crescas and Abravanel give the argument in a very abbreviated form.

[137]Cf. above, pp. 30–33.

[138]This is recognized by Crescas, *Or ha-Shem,* III, i, 4 and 5, and Abravanel, *Mif'alot,* VI, 3 (end). As a means of escaping the arguments for eternity from the nature of the cause, Crescas inclines towards, and Abravanel embraces, an old rabbinic theory (*Genesis Rabbah,* ix, 2) according to which God continually creates and destroys a succession of worlds. See *Or ha-Shem,* III, i, 5; *Mif'alot,* VII, 3 and 5 (Abravanel distinguishes his own theory of successive worlds from Crescas' theory). It is difficult to see, however, just how the theory of successive worlds answers the first two of the three proofs. Leone Ebreo, *Dialoghi d'Amore,* p. 252, attributes a theory of successive worlds to Plato.

(a) Responses to the argument that nothing could have led a creator to create the world at a particular moment

The proof, as will be recalled, had several variations or stages. The proponents of eternity argued that no given moment in an undifferentiated eternity could, in preference to any other, have suggested itself as the proper moment for the creator to create the universe; that there could be no imaginable motive for the creator to create the universe at a given moment; that the factors motivating the creator to create the universe at a given moment would regress infinitely, whereas an infinite regress is impossible. Medieval responses to the proof in its several variations rest largely on the thesis that the creator brought the world into existence through an exercise of will, the central thesis then being buttressed with secondary considerations. The central thesis as well as the secondary considerations can be discovered in Augustine and John Philoponus, either in contexts where the two were replying to the present proof, or in contexts where they were replying to other proofs for eternity.

Augustine takes up the question how the creator might have created the world at a given moment *in time* as distinct from the infinite other identical moments when the world could have been created. He does not so much answer the question as dismiss it, branding it as a sophism on the grounds that there was "no time" before creation, and consequently the act of creation cannot legitimately be spoken of as having occurred in time at all.[139] As for the question what the new factor might have been which induced the creator to act after not having acted for an eternity, Augustine denies that any factor need be posited. No new circumstances, he submits, induced the creator to create the world at the moment when he did. For God did not create the world "in accordance with a new [plan] . . . but rather in accordance with an eternal plan,"[140] and in accordance with "one and the same eternal and immutable will."[141] From all eternity God's will determined the moment at which the world should come into existence. Through one and the same eternal plan, and one and the same eternal act of will, creatures "previously were not, as long as they were not"; "and they thereupon were,"[142] at what Augustine—despite his insistence that creation did not strictly occur in time at all—allows himself to style a given "time."[143] In addition to thus responding to the arguments on their own merits, Augustine voices an *ad hominem* animadversion. It consists in applying the analogy between space and time, an analogy that advocates of creation were earlier seen to utilize when responding to arguments for eternity from the nature of the world.[144] Augustine is addressing

[139] *City of God*, XI, 6; *Confessions*, XI, 30.
[140] *City of God*, XII, 15.
[141] Ibid., 18.
[142] Ibid.
[143] Ibid., XI, 5.
[144] Above, pp. 41–42.

proponents of eternity who recognize a cause of the existence of the universe, and he contends that they, at least, do not have the right to pose difficulties regarding the moment chosen for creation. Such proponents of eternity must acknowledge that the deity has determined a location in space for the existence of the finite physical world in preference to the infinite alternative places where the world might exist. Consistency would therefore require that they acknowledge the deity's ability to determine a moment in eternity for the beginning of the existence of the world in preference to the infinite other moments when the world might have begun to exist.[145]

Augustine's position, in sum, is that the world cannot legitimately be described as having been created at a moment in time. No new factor motivated the creator to act, for the previous nonexistence, and subsequent existence of the world were determined eternally and immutably through an eternal divine plan and an eternal act of the divine will. Proponents of eternity who recognize a cause of the existence of the universe cannot pose difficulties about the deity's ability to create the world at one given moment in preference to other possible moments, since they must admit something analogous, namely the deity's ability to produce the world—albeit eternally—at a given place in preference to the other possible places where the world might exist.

John Philoponus was confident that the creation of the world was not merely defensible, but demonstrable,[146] and his response to the proof for eternity we are examining builds on his demonstration of creation. It must be recognized, Philoponus stresses, that creation can be demonstrated and that the universe must have come into existence at some moment. It must further be recognized that questions such as "why the universe did not come into existence earlier" can be asked about every moment at which the world might have been created. Since the universe is known to have come into existence at one moment or another, and since questions can be raised about whatever is assumed to have been the moment of creation, the questions, Philoponus holds, are pointless and can be dismissed.[147]

Like Augustine, Philoponus explains that creation took place in conformity with an eternal decision of God's eternal will. Philoponus is answering Proclus, who, for his part, had also affirmed the eternity of God's will, but had inferred therefrom, in another proof for eternity than the one we are now examining, that the world is eternal. Proclus' reasoning had been that since the world exists by virtue of God's will, which is eternal, it must likewise be eternal.[148] In response, Philoponus insists upon the distinction between God's eternally willing that a thing should exist and his willing that the thing should exist eternally. Although

[145] *City of God*, XI, 5.

[146] See below, p. 93.

[147] *De Aeternitate*, pp. 11–12.

[148] Cf. above, p. 60.

the adherents of creation do maintain that God willed the existence of the world from eternity, this is not identical with, nor does it entail, God's having willed the eternal existence of the world.[149] To support the proposition that an eternal will can determine noneternal events, Philoponus develops an *ad hominem* device, different from Augustine's. He is addressing philosophers who acknowledge, as Philoponus understands Proclus to have done, that the divine will is responsible for whatever occurs in the universe. Those philosophers concede, in effect, that through a "single . . . simple . . . act of willing"[150] the deity wills the occurrence of myriad noneternal and changing events at given determinate moments in the course of history. How, Philoponus marvels, can the same philosophers shrink from the proposition that God eternally willed the occurrence of an additional noneternal event at a given moment, to wit, the coming into existence of the world as a whole?[151]

The following motifs, to recapitulate, are brought into play by Augustine and Philoponus either in answer specifically to the proof for eternity we are presently examining or in answer to other proofs. No moment in time can strictly be described as having been chosen for creation, inasmuch as there was no time before creation. The creation of the world at a given moment came about in accordance with an eternal decision on the part of the creator. To will something eternally is not equivalent to willing its eternal existence. Once creation is conclusively demonstrated, the question why one moment rather than another was chosen becomes pointless. At any rate, proponents of eternity who recognize a cause of the existence of the universe perforce acknowledge that the deity is responsible for the existence of the world in a given location, as distinct from every other possible location where the world might exist. They can therefore hardly balk at the thesis that the creator is capable of choosing a given moment for the existence of the universe. Proponents of eternity who, moreover, recognize a being possessed of will which is the cause of everything occurring in the universe perforce acknowledge that an eternal unchanging will can decide upon the occurrence of given events at given moments; they therefore cannot balk at the thesis that an eternal unchanging will could decide upon the creation of the world as well at a given moment. In the Middle Ages, these motifs recur in varying combinations.

The question "Why did God not create the world prior to the time" when he did is taken up by Saadia. And Saadia responds: Prior to creation "there was no time that could be asked about; furthermore, it is of the character of a voluntary agent to act when he wishes."[152]

[149] *De Aeternitate*, p. 566.

[150] Ibid., p. 568; cf. p. 81.

[151] Ibid., pp. 566–567; 580.

[152] Saadia, *K. al-Amānāt wa-l-I'tiqādāt*, ed. S. Landauer (Leiden, 1880), p. 73; English translation with pagination of Arabic indicated: *Book of Beliefs and Opinions*, trans. S. Rosenblatt (New Haven, 1948).

Ghazali addresses the rhetorical question what could differentiate one specific time from earlier and later times as the moment for creation, and responds by referring to the creator's will. It is, he too explains, of the nature of will, and not merely divine will, precisely to differentiate between things that are similar in every respect. Hence by the exercise of sheer will, God could select one from among an infinite number of identical moments and designate it as the moment for creation. To support his assertion, Ghazali has recourse to the analogy between time and space. He observes, much as Augustine had done, that adherents of eternity who recognize a cause of the existence of the universe cannot avoid acknowledging wholly arbitrary determinations in the spatial realm. The location of the north and south poles[153] at a given pair of points on the celestial sphere is, in the example he offers, wholly arbitrary, seeing that any other pair of opposite points would be as suitable. Since Ghazali's adversaries recognize that the first cause, the deity, did arbitrarily determine a pair of points for the poles, how, he remonstrates, can they question the deity's competence arbitrarily to determine a moment for the world to begin to exist?[154]

Ghazali also responds to the argument that the factor "tipping the scales" in favor of creation at a particular moment would have to have the scales tipped in its favor by a previous factor, and the latter by yet another factor, *ad infinitum*. An eternal will, he maintains, is capable of determining that something should remain nonexistent up to a certain stage and only thereupon begin to exist; consequently, no factor need have tipped the scales at the critical moment.[155] To support the assertion, Ghazali observes, as Philoponus had done, that his adversaries also trace myriad temporal events in the universe back to the eternal first cause of the universe. To grant that the eternal first cause is responsible for the occurrence of temporal events, even through intermediate causes, is to grant that the first cause is ultimately responsible for the moment at which each event occurs. How then, Ghazali again remonstrates, can the proponents of eternity question the ability of the eternal first cause to bring about one additional non-eternal event at a determinate moment, namely the coming into existence of the entire universe?[156]

Ghazali's contentions reappear in Abū al-Barakāt, who cites them in the name of "the creationists." The creationists, Abū al-Barakāt informs us, answer their adversaries by maintaining that the divine will[157] is fully competent to differentiate between things similar in every respect, such as identical moments; that the adherent of eternity too has to acknowledge arbitrary determinations made by God's will in the realm of space and should not shrink from arbitrary determinations by God's will in the realm of time; that an eternal will can decide upon

[153]That is to say, the points around which the celestial spheres rotate.

[154]*Tahāfut al-Falāsifa,* I, §§30, 38; English translation, pp. 14, 24.

[155]Ibid., §8; English translation, p. 3.

[156]Ibid., §47; English translation, p. 32.

[157]Ghazali stated this of will in general.

a noneternal effect; that the adherent of eternity traces events occurring at given moments back to the divine will and consequently should not question the competence of the divine will to bring the entire universe into existence at a given moment.[158] Abū al-Barakāt was not in the least swayed by any of these considerations and his belief in the eternity of the world remained unshaken.[159]

Kalam writers coming after Ghazali were, however, confident that the proof for eternity under discussion had been refuted. Fakhr al-Dīn al-Rāzī takes up the question: What tipped the scales in favor of creation at a particular moment? He responds that no factor had to "tip the scales," for the selection of the moment was accomplished exclusively by God's "will." The arbitrary selection of a moment for creation, Rāzī adds, is analogous to the arbitrary selection of a location for the stars on the celestial spheres. Inasmuch as adherents of eternity acknowledge the ability of God's will, in the spatial realm, to assign given locations to the stars in preference to other possible locations, they cannot deny the ability of his will, in the temporal realm, to fix upon a given moment for creation in preference to others.[160] The creation of the world at a given moment is, Rāzī observes in another work, analogous to the emergence of sundry individual events at countless moments through history. The adherents of eternity recognize that sundry temporal events do flow from the unchanging state of actuality of the first cause, albeit through intermediate causes; and they recognize that no factor standing behind the first cause tips the scales to activate him, since to assume such a factor would lead to an infinite regress. The adherents of eternity thereby acknowledge, in effect, that the first cause is ultimately responsible for the moment at which each temporal event occurs. How then can they question the ability of the first cause to determine a moment for one additional event, the coming into existence of the entire universe?[161]

The argument that creation would involve an infinite regress of factors "tipping the scales" is also dealt with by Āmidī and Ījī. Āmidī counters by reference to God's eternal "will," which determined "that the nonexistence of the world would extend up to" a particular moment and that the world would "come into existence at the moment when it did come into existence."[162] Ījī writes: "A voluntary agent" is capable of "tipping the scales . . . in favor of one of two [equipollent] alternatives within his power by pure will, with no need of" an "inducing factor . . . added to himself." And at any rate, the adherent of creation is in no worse a predicament than adherents of eternity who admit that the universe has a cause. The latter acknowledge that individual "things daily coming into existence" are traceable back to the eternal cause of the universe and that no factor intervenes

[158] *K. al-Mu'tabar*, III, pp. 33, 43–44.
[159] Below, p. 75.
[160] *Muḥaṣṣal*, p. 91.
[161] *K. al-Arba'īn*, p. 51.
[162] *Ghāya al-Marām*, p. 268.

to tip the scales and induce the eternal cause to act. They can hardly scruple at attributing the coming into existence of one more thing—the world in its entirety—to an eternal cause, without the intervention of a factor tipping the scales.[163]

Aquinas similarly espouses the theory, which he could have learned from either Augustine or Ghazali, that an eternal will is capable of deciding from eternity when a noneternal product should come into existence. No new factor, he writes, need have activated the creator at the moment of creation. For God produced the world through his "thought and will," which are eternal, and they both eternally determined that the world should come into existence at the moment when it did.[164]

Maimonides affirms that the creator could act at a given moment without any "inducement" (*dāʿin*), prompting him to act, nor any "preventive factor" (*māniʿ*), blocking his action up to the moment he acted; and Maimonides defends that position by setting forth the difference between two kinds of voluntary agent. The voluntary agents that act or fail to act because of inducing or preventative factors are, he explains, agents that seek to attain "a purpose . . . external to the will itself." For example, a man is induced to build a house by factors outside his will, such as the weather or environment; and he may be prevented from building the house by factors outside himself, such as the absence of building materials. By contrast, the agent who acts "exclusively by will" is not affected by external factors. Nothing external to the will motivates such an agent, and nothing outside him prevents him from acting. He acts or refrains from acting with absolute autonomy and exclusively as his will dictates; consequently, without external inducement, and exclusively as his will determined, he would be competent to decide upon the moment for creation.[165] Aaron ben Elijah[166] responds to the argument that nothing could have induced the creator to create the world at a given moment in the same manner as Maimonides.

In the foregoing instances, medieval adherents of creation explain away difficulties regarding the moment chosen for creation. Several medieval adherents of creation take the more radical step of branding their adversaries' questions as illegitimate and disallowing the questions outright.

Part of Augustine's response to difficulties regarding the moment chosen for creation was the consideration that before creation time did not exist.[167] The thought reappears when Ṭūsī—who hardly could have had any link, direct or indirect, to Augustine—addresses the familiar question: How could God have settled upon a moment for creation in preference to the infinite other possible moments when creation might have occurred? To handle the matter properly, Ṭūsī

[163] *Mawāqif*, VII, 229–230.
[164] *Summa contra Gentiles*, II, 35.
[165] *Guide*, II, 18(2).
[166] *Eṣ Ḥayyim*, chap. 7.
[167] Above, p. 68.

asserts, one must understand that time did not precede the existence of the world. The supposed time before creation was "imaginary," not real, and any "differentiation" between moments before creation is equally "imaginary."[168] In no true sense was a choice made between moments, and it is illegitimate even to ask how God could have settled upon a particular moment.[169]

Aquinas, in one passage, combines the consideration that time did not exist before creation with an application of the analogy between time and space. The question why the creator should have chosen a given moment for creation, rather than any other, is compared by him to the question why God should have chosen to locate the world at a given spot in space, rather than any other. Outside the world, Aquinas notes, there is no space, and to speak of places outside the world where the world might have been located is illegitimate. Similarly, there was no time before creation, and to speak of moments before the existence of the world when the world might have come into existence is no less illegitimate. Questions regarding the time before creation, like those regarding space outside the world, may be dismissed.[170]

In Gersonides, the thesis that God's eternal will eternally determined the moment for creation is joined with a consideration found earlier in Philoponus,[171] but presumably reached by Gersonides independently. Gersonides treats the familiar problem: What might have led God to create the world at the moment when he did? And the familiar solution is offered: The choice of a moment may be ascribed to God's eternal unchanging will. But Gersonides understood creation to be demonstrable and not merely defensible, which means that if a world was to exist, it could not be eternal but had to be created.[172] That being so, he continues, the question why creation occurred at a given moment rather than any other loses legitimacy. Inasmuch as the world must have come into existence at some moment, and inasmuch as the question regarding the moment of creation could have been raised regarding whatever moment God should have chosen, the question may be dismissed as pointless.[173]

Thus far the adherents of creation have either answered or disallowed their adversaries' questions regarding the possibility of a moment's being selected for creation. Kalam writers perceived that the thrusts and parries over the possibility of a moment's being selected for creation could also be turned to a constructive end. The defense of the doctrine of creation led almost invariably to the creator's will; an exercise of sheer will had to be assumed in order to explain the selection

[168]Cf. above, pp. 41–42.

[169]Gloss to *Muḥaṣṣal,* p. 92. A similar position is taken by Leibniz in *A collection of Papers which passed between the Late Mr. Leibnitz and Dr. Clarke* (London, 1717), III, §6; V, §55–60.

[170]*Summa contra Gentiles,* II, 35(5).

[171]Above, p. 69.

[172]Cf. below, pp. 209–211.

[173]*Milḥamot ha-Shem,* VI, i, 18, p. 377; cf. ibid., 24, pp. 395–397. Albertus Magnus answers questions about the factors inducing the creator to act (cf. above, n. 33) in a similar fashion; his response is that an eternal world is impossible. See Albertus, *Physics,* VIII, tr. 1, chap. 14.

of a moment for creation. As long as advocates of creation adopted a purely defensive attitude, will in the creator remained nothing more than an assumption offered in support of another assumption, the assumption of creation. The Kalam school was, however, certain that creation can be demonstrated conclusively. And once creation is demonstrated, will in the creator may be seen as its demonstrated corollary. For if the world is known to have been created, one out of an infinite number of identical moments is known to have been selected for creation, and the creator, who selected the moment, is known to be possessed of will. The realization that will in the creator is implied by the doctrine of creation was incorporated into the standard Kalam procedure for proving the existence of God, which consisted in demonstrating the creation of the world and inferring the existence of God from creation.[174] Inasmuch as the selection of a moment for creation would imply will in the creator, Kalam thinkers could, after having proved creation, infer not merely the existence of a creator but the existence of a creator possessed of will. The procedure of proving creation and inferring therefrom not merely a deity, but a deity possessed of will, is employed by Bāqillānī,[175] 'Abd al-Jabbār,[176] Juwaynī,[177] Ghazali,[178] Rāzī,[179] Āmidī,[180] and Ījī.[181]

The theory that a moment for creation could have been selected by God's eternal will elicited surrejoinders from Abū al-Barakāt and Averroes. Abū al-Barakāt makes two points, only the second of which he regards as decisive. Adherents of eternity can, he writes, surrejoin by noting that whenever an agent resolves to do something in the future, "another resolution (*'azīma*) or exercise of will" is required to activate the agent when the awaited moment arrives. The theory of eternal will does not, therefore, exempt the proponents of creation from explaining what might have induced the creator to create the world when the moment did arrive for him to act. The second point made by Abū al-Barakāt—and, in his view, the decisive one—is that effects are never deferred fortuitously. Whether a cause acts "through will or without will," its effect is delayed only because of "a deficiency in the causality," that is to say, either because "knowledge . . . strength, . . . will," happen to be absent, or because "preventative factors" happen to interfere. Inasmuch as the deity lacks nothing needed for action and nothing ever hinders him from acting, his action can nowise be delayed. Whatever he wills must be accomplished forthwith, and if he should will the existence of a universe, he could not help but bring a universe into existence immediately.[182]

[174]Cf. above, p. 2; below, p. 154.
[175]*K. al-Tamhīd*, p. 27.
[176]*Sharḥ al-Uṣūl*, p. 120.
[177]*K. al-Irshād*, pp. 28–29.
[178]*K. al-Iqtiṣād fī al-I'tiqād* (Ankara, 1962), pp. 101–102.
[179]*K. al-Araba'in*, pp. 147–148, with nuances.
[180]*Ghāya al-Marām*, p. 45.
[181]*Mawāqif*, VIII, pp. 82–83
[182]*K. al-Mu'tabar*, III, pp. 34–35; cf. ibid., pp. 47–48.

Both these points reappear in Averroes, who reasons: On the assumption that an eternal will predetermined the moment for the world to come into existence, a new "resolution (*'azm*), . . . which did not exist prior to the moment in question," would still be required in order that the eternal agent and his eternal will should be activated. The theory of *eternal will* consequently does not exempt the adherents of creation from the necessity of explaining what induced the creator to act at the moment when he did.[183] And in another work Averroes writes: No "delay" is conceivable between a "voluntary agent's resolution to act" and his acting,[184] nor is any delay conceivable between the "agent's acting" and the appearance of the effect. Immediately upon God's resolution to produce the world, the world must have existed; and if God's resolution to produce the world is eternal, the world cannot have come into existence later but must have existed from all eternity.[185]

(b) Responses to the argument that creation would imply change in the creator

The second proof for eternity from the nature of the cause of the universe turns on two propositions: the proposition that a cause ultimately responsible for the existence of the universe would have to be unchangeable, and the proposition that should creation be assumed, the cause ultimately responsible for the existence of the universe would have undergone change in passing from inaction to action. If, on the one hand, an ultimate cause of the universe must be unchangeable, whereas creation would, on the other hand, involve change in the ultimate cause, the assumption of creation is untenable.[186] The advocates of creation usually respond by inverting the reasoning. They agree that the ultimate cause, or deity, must be unchangeable. But the line they take is that since the deity must be unchangeable, the act of creating the world would not, in his case, have constituted a change.

That response is offered by Ibn Ḥazm, who imparts to it a Kalam tinge. "Alteration," Ibn Ḥazm writes, "consists in something's coming into existence in the subject of alteration which was not there before, with the result that the subject exchanges its [previous] attribute . . . for another." A situation of the sort is not possible in God inasmuch as "he is never the subject of an attribute." God accordingly acts or does not act "by virtue of his essence" and without change.[187] Bonaventure employs similar language, although historical links with Ibn Ḥazm cannot easily be supposed. When the deity produces something new, Bonaventure

[183] *K. al-Kashf,* p. 31.

[184] Averroes obviously is using the expression *resolution to act* in a sense that excludes a voluntary agent's resolving to do what he is not capable of.

[185] *Tahāfut al-Tahāfut,* I, pp. 7–8; the English translation is not satisfactory. What appears here in Averroes as a surrejoinder is recorded by Aquinas as an independent argument for eternity; *Summa contra Gentiles,* II, 32(4).

[186] Above, pp. 56–57.

[187] *K. al-Faṣl fī al-Milal,* I, p. 20.

maintains, nothing "is added" to him which was not there before; consequently, the act of creation would not constitute change in God.[188]

The same approach is also taken by Maimonides except that, as would be expected, Aristotelian terminology is now used. Maimonides deals with the version of the argument where creation is shown to involve a transition from potentiality to actuality in the agent responsible for creation, while the cause ultimately responsible for creation could not undergo such a transition.[189] And Maimonides replies: A transition from potentiality to actuality occurs, as all Aristotelians recognize,[190] solely in material objects. Therefore when an incorporeal agent does produce something after not having done so, no passage from potentiality to actuality occurs in him.[191] Maimonides' solution is repeated by Aaron ben Elijah.[192] Another formulation is provided by Aquinas, but the burden is still the same. "A new divine effect does not," in Aquinas' words, "signify a new action in God; for God's action is identical with his essence" and hence is, like his essence, eternally unchangeable. God could therefore produce a new effect without undergoing change.[193]

One variation of the proof for eternity from the unchangeability of the cause of the universe concerned itself with will, the contention being that a change specifically in the creator's will would be implied by creation.[194] The advocates of creation respond here as they do to the basic argument. Maimonides and, following him, Aaron ben Elijah lay down the proposition that voluntary agents undergo a change of will only when they are corporeal; for only corporeal voluntary agents act "to attain an external end," hence their wills alone are moved by external inducing and preventative factors.[195] By contrast, the will of an incorporeal agent, which seeks no external end, is immune from change. So far, of course, the advocates of eternity would concur; and they would conclude that an agent who is unaffected by circumstances outside himself and who is immune from change could not possibly decide at a particular moment to bring the world into existence. Maimonides and Aaron, however, conclude not that creation is impossible, but rather that creation would involve no change in the will of an incorporeal agent: Inasmuch as the will of an incorporeal agent is not moved by external factors, when—or if—an incorporeal agent did begin to bring the universe into existence, his action would be accompanied by no change of will.[196]

[188]*Commentary on II Sentences,* d. 1, p. 1, a. 1, q. 2, ad 6.

[189]Above, pp. 58–59.

[190]See Zeller, *Die Philosophie der Griechen,* II, 2, pp. 318–323.

[191]*Guide,* II, 18(1).

[192]*'Eṣ Ḥayyim,* chap. 7

[193]*Summa contra Gentiles,* II, 35(1). Cf. *De Potentia,* q. 3, art. 17 (ad 12).

[194]Above, p. 60.

[195]Cf. above, p. 73.

[196]Maimonides, *Guide,* II, 18(2); Aaron ben Elijah, *'Eṣ Ḥayyim,* chap. 7.

Another version of the argument from the unchangeability of the deity reasoned that creation would involve a change of relationship in the creator.[197] Crescas, who gave that version, responds: "A new relationship does not constitute a change of essence [in the case] of an incorporeal being."[198] An incorporeal, as distinct from a corporeal, agent could as a consequence bring something new into existence and thereby enter a new relationship without undergoing any change in himself.

The line taken by these adherents of creation betrays more than a whiff of question begging. Their adversaries had pressed the plausible thought that if an agent never undergoes change he cannot conceivably do something he did not do before. In responding, the adherents of creation insist, as firmly as their opponents, on the deity's immunity from change. And they nowise reveal how an agent might begin doing something he previously did not do without changing. They circumvent the straightforward and obvious conclusion that the deity never passes from the state of not creating the world to the state of creating it, and they assert instead that when the deity does pass from the one state to the other, he does so without changing.

Philoponus and, in the Middle Ages, Shahrastānī and Gersonides make a different response, which—whether cogent or not—does come to grips with their opponents' argument. All three belong to the circle of philosophers who are confident that the creation of the world can be demonstrated definitively. Taking the truth of creation and the impossibility of eternity as a premise, they explain: The deity eternally wills the creation of the world (Philoponus and Gersonides) and is eternally an "actual creator" (Shahrastānī), but since an eternal world is intrinsically impossible, the world cannot possibly be produced by God from eternity. The world comes into existence as soon as it can, and the soonest it can come into existence is in the finite past. God, thus, eternally and unchangeably wills the existence of the universe and is eternally in the state of creating it; yet all that his eternal willing of the universe and his eternal act of creation gives rise to is a noneternal universe.[199]

The response to one final version of the argument from the unchangeability of the deity may be mentioned. 'Abd al-Jabbār recorded a version according to which creation is inadmissible because it would imply a change in the creator's knowledge; the creator would previously not know, and subsequently would know, that the universe is in existence.[200] The response 'Abd al-Jabbār makes is that knowledge of something not yet existent is nowise different from knowledge

[197] Above, p. 60.

[198] *Or ha-Shem,* III, i, 2. The printed editions are corrupt and have to be corrected with the aid of manuscripts of the text.

[199] Philoponus, *De Aeternitate,* p. 81, taken together with p. 8; Shahrastānī, *K. Nihāya al-Iqdām,* pp. 47–48; Gersonides, *Milḥamot ha-Shem,* VI, i, 18, p. 377.

[200] Above, p. 60.

of the same thing when it exists. The assumption of creation does not, therefore, imply that any change occurred in the creator's knowledge.[201]

(c) Responses to the arguments for eternity from God's eternal attributes

The contention here had been that God's attributes—including his character of being a cause by his very nature—express themselves in the existence of the universe, and therefore the eternity of his attributes entails the eternity of the universe.[202] Rebuttals take three forms. The adherents of creation deny that the attributes concerned are eternal attributes of God; they maintain that although God may possess the attributes eternally, the attributes do not necessarily express themselves by producing a universe eternally; or else they undertake to show that it would be intrinsically impossible for God's attributes to express themselves in an eternal universe.

The first form of response is to be found in ʿAbd al-Jabbār, who refuses to "concede that God is eternally beneficent"; and if God is not eternally beneficent, the eternity of the world plainly cannot be inferred from his beneficence.[203]

Shahrastānī's thinking is more subtle. He denies that beneficence is an "essential attribute added to the essence," and interprets it instead "as an attribute of action," or "a relation." The statement that God is beneficent means no more than that God is an "agent"; and the statement that God "is sometimes beneficent, sometimes not," means no more than that "he is sometimes an agent, sometimes not an agent." Since the term *agent* and the attendant term *beneficent* have their reference not in God but in what he has produced, in the relation of the universe to God, nothing in God's essence could disclose whether he is eternally, or merely sometimes, beneficent. The question whether God is, or is not, eternally beneficent and eternally an agent cannot, then, serve as a starting point for settling the issue of eternity and creation. On the contrary, that question is itself the "locus of dispute" between the proponents of creation and the proponents of eternity, and can be resolved only through settling the issue of eternity and creation on its own merits. The eternity of God's beneficence hinges on the eternity of the universe and not vice versa.[204]

In a similar vein, Āmidī refuses to interpret divine beneficence as an "attribute of perfection" in God, that is to say, as an attribute pertaining to the perfection of God's essence and consequently coeternal therewith. Beneficence is "an attribute of action," reducible to God's "being an agent" with no "end" or "profit" in view. Āmidī gives two reasons for not construing divine beneficence as an attribute of perfection. The proposition that beneficence is an "attribute of perfection"

[201] *Sharḥ al-Uṣūl,* p. 117.
[202] Above, pp. 61–62, 64.
[203] *K. al-Majmūʿ*, p. 66.
[204] *K. Nihāya al-Iqdām,* p. 46. Cf. Ashʿari, *Maqālāt al-Islāmīyīn,* ed. H. Ritter (Istanbul, 1929–1930), p. 182.

in God cannot, he writes, be accepted a priori, since it is not an item "of necessary or immediate [knowledge]";[205] and since there is no way in which it can be known a posteriori either, the proposition is groundless. Furthermore, and more importantly, beneficence could not possibly be an attribute of perfection in God. For if beneficence were such an attribute, God's perfection would be dependent on the existence of his creatures, who are the expression of his beneficence,[206] and the "superior would acquire perfection through the inferior, which is absurd." Seeing that beneficence is not an attribute of perfection and does not pertain to God's essence, the eternity of the world cannot be inferred from it.[207]

In the instances to follow the adherents of creation do not merely refuse to admit the eternity of the attributes from which the eternity of the world had been inferred. They refuse to admit that the attributes or traits are in any sense possessed by God. Ibn Ḥazm recorded an argument for eternity turning on the proposition that God could only have produced the world insofar as he is a cause "by reason of himself." If God produces the world insofar as he is a cause by "reason of himself," the thinking went, his effect must have existed for as long as he has himself existed.[208] The response Ibn Ḥazm makes is that God does not act by "reason of himself" nor, indeed, "by reason of" anything whatsoever. God simply acts "as he wishes" to act. Since the proposition that God acts by reason of himself is false, nothing can be inferred from it.[209]

Gersonides reconstructed and ascribed to Aristotle an argument for eternity from the premise of the deity's being the mover of the outermost sphere. The reconstructed reasoning was that since the deity is by his nature the mover of the outermost sphere, the sphere must exist as long as the deity exists.[210] Gersonides responds by denying that the deity has the trait in question. It can be demonstrated, he maintains, that the ultimate cause of the universe is not in fact the mover of the outermost sphere but an entity distinct from, and beyond the movers of the spheres.[211] Inasmuch as the deity is not by his nature the mover of the sphere, the eternity of the sphere cannot be deduced from the eternity of the deity.[212]

Fakhr al-Dīn al-Rāzī, to take a final example, knew of an argument for eternity from the deity's character of being an "agent." To be an agent, the argument ran,

[205]Cf. Ghazali, *Tahāfut al-Falāsifa,* I, §§13 and 33 (English translation, pp. 6 and 21), where "necessary" knowledge, in the sense of immediate knowledge, is contrasted with "speculative" knowledge. For the term "necessary knowledge," cf. J. van Ess, *Die Erkenntnislehre des 'Aḍudaddīn al-Īcī* (Weisbaden, 1966), pp. 116, 118.

[206]Cf. A. Lovejoy, *The Great Chain of Being* (Cambridge, Mass., 1936), pp. 54, 62.

[207]*Ghāya al-Marām,* p. 270.

[208]Above, p. 66.

[209]*K. al-Faṣl fī al-Milal,* I, p. 10.

[210]Above, p. 67.

[211]This is Avicenna's position; above, n. 134.

[212]*Milḥamot ha-Shem,* VI, i, 4, p. 303.

is an "eternal . . . positive attribute" added to God's essence, and it is moreover a "relative attribute." Since the attribute is eternal, positive, and relative, its presence implies the existence of an eternal positive correlative, an eternal effect.[213] Rāzī's response is that to be an agent is not—and could not possibly be[214]—a positive attribute added to God's essence. No light, therefore, can be shed on the issue of eternity and creation by God's being an agent.[215]

This has been one form of response to arguments for eternity from the eternal divine attributes. A second form of response consists not in denying that the attributes are eternal or that God possesses them, but in denying that they entail the eternity of the world. When answering an argument from the divine attributes of wisdom and knowledge,[216] ʿAbd al-Jabbār agrees that God's wisdom and his knowledge of the goodness of the world are eternal. But, ʿAbd al-Jabbār insists, wisdom "does not necessitate action." Therefore, although God undoubtedly knew from eternity that a world should exist and that he would create one, his knowledge did not necessitate the production of the world from eternity.[217] Ibn Ḥazm, for his part, refuses to acknowledge that God's "beneficence, wisdom, and power" are at all the "cause" of God's creating the world. "There is," he asserts, "no cause of God's bringing things into existence"; God simply created the world, and nothing more can be known or said. Consequently, the attributes of beneficence, wisdom, and power, whatever their status might be, cannot be understood to have necessitated the existence of a world, and certainly not the existence of an eternal world.[218]

The argument from God's eternal beneficence is also answered by Fakhr al-Dīn al-Rāzī. Rāzī does not dispute the premise that God is eternally beneficent. He submits, however, that the attribute of eternal beneficence need not express itself in an eternal world, and to support his position, he employs a device already met in the answer to another argument for eternity.[219] Even on the assumption of the eternity of the world, Rāzī notes, God's beneficence is expressed in the production of myriad objects—"forms and accidents"—that "come into existence" and are not eternal. All parties must thus recognize that God's eternal attribute of beneficence is compatible with noneternal products, and no one should

[213] Above, pp. 66–67.

[214] The reason Rāzī gives is a form of the "third man" argument: If *being an agent* were an attribute added to the essence, it would be an entity of some sort dependent on the essence, and hence it would be a *possibly existent* entity (cf. below, p. 290). As *possibly existent,* it would stand in need of an agent to give it existence. The agent could only be God. But if God's character of being the agent that gave His attribute of *being an agent* existence were, then, itself an attribute added to God's essence, it would, in turn, also stand in need of an agent. And so on.

[215] *K. al-Arbaʿīn,* p. 51.

[216] Above, p. 60.

[217] *Sharḥ al-Uṣūl,* p. 116.

[218] *K. al-Faṣl fī al-Milal,* I, p. 20.

[219] Above, pp. 70, 71–72.

balk at the hypothesis that God's eternal beneficence can give rise to one more noneternal product, namely the universe.[220]

Aquinas replies to the argument from God's "infinite" goodness by observing that God's goodness should be expected to express itself by producing creatures who "represent the divine goodness" in the most accurate possible manner. The relation of the creator to what he creates is surely not a relation of equals but a relation of the superior to the inferior, of the infinite to the finite. And the "superiority of divine goodness over what it creates is best expressed [precisely] in the latter's not being eternal." The eternal attribute of infinite divine goodness has its most appropriate expression in a noneternal, rather than eternal, product; divine goodness, far from entailing eternity, harmonizes better with creation.[221] A similar response to the argument from divine wisdom and beneficence is made by Abravanel.[222] In an alternative and more succinct response to the argument that God's eternal goodness must give rise to an eternal world, Aquinas posits that "God's goodness does not exist . . . for the sake of creatures"; consequently, "divine goodness would not have been idle [and thereby deficient] even if it had never produced a creature."[223]

Maimonides offers a reply to the argument that the eternity of the world is entailed by God's eternal attribute of wisdom. The proposition that God's "wisdom is eternal like his essence" and the proposition that the existence of the world "flows from his eternal unchanging wisdom" are embraced by Maimonides without reservation. But, Maimonides affirms, the human mind "is completely ignorant of" God's essence and wisdom. Therefore, although it is true that God's eternal wisdom did express itself in the existence of the universe, nothing justifies the conclusion that God's eternal wisdom had to express itself in an eternal universe.[224]

An argument in which eternity is inferred from God's knowledge or thought was recorded by Aquinas and by Gersonides. In Aquinas' version, the reasoning was that the world must be eternal because God's knowledge needs an eternal world to serve eternally as its object.[225] In Gersonides' version, the starting premise was that God's thought, which is identical with his essence, comprises "the [intelligible] order of the universe." The reasoning then went: The actual and external order of the universe must exist whenever God and his thought, which comprises the intelligible order of the universe, exist; and since God and

[220] *K. al-Arba'īn*, p. 53. The consideration that God's beneficence can express itself in noneternal objects would only counter an argument to the effect that eternal beneficence must express itself exclusively in something eternal. It would not counter the argument that eternal beneficence must eternally express itself in something.

[221] *Summa contra Gentiles*, II, 35(7).

[222] *Mif'alot*, VII, 4. The notion also appears in Leone Ebreo, *Dialoghi d'Amore*, p. 239.

[223] *Commentary on De Caelo*, I, §66. Cf. *De Potentia*, q. 3, art. 17 (ad 14).

[224] *Guide*, II, 18(3). The notion also appears in Leone Ebreo, *Dialoghi d'Amore*, p. 239.

[225] Above, p. 63.

his thought are eternal, the actual universe must be eternal.[226] Aquinas and Gersonides make virtually identical replies to the argument. God's knowledge and thought, they stress, is in no sense dependent on the world, being, instead, the cause of the world and prior thereto. Consequently, in Aquinas' words: "God can have knowledge even should the object of his knowledge not exist."[227] And in Gersonides' words: The "[intelligible] order may exist even if the existent things, whose existence can flow from that [intelligible] order, do not exist."[228] God could, that is to say, have eternal knowledge of what the universe would eventually be after it came into existence.[229]

Aquinas recorded two additional arguments for eternity from the eternal attributes of God and he finds that those arguments also fail because the attributes they focus on, although belonging eternally to God, do not imply eternity. One argument went: Since God is a sufficient cause from all eternity, and since whenever a "sufficient cause is given, its effect is given," God's effect must exist from all eternity.[230] In his reply, Aquinas has recourse to the thesis that the creator is a voluntary agent. The "proper effect" of a voluntary cause, he writes, is an effect that comes about as "the will wills," that conforms to the "intention" of the agent. In other words, the proper effect is nothing other than what is willed, and if what a voluntary cause wills should be a noneternal product, a noneternal product and nothing else is the appropriate effect of the cause. An eternal, but voluntary, sufficient cause could produce a noneternal world no less appropriately than an eternal world.[231] A similar response to the same argument for eternity is offered by Bonaventure.[232]

There was, finally, the argument that the world must be eternal because God is eternally the "Lord."[233] Aquinas responds by explaining that God is the Lord in the sense of having the "power of governing" and not necessarily in the sense of actually governing.[234] God therefore can be the Lord eternally although the universe does not exist eternally.

[226]Ibid.

[227]*De Potentia,* q. 3, art. 17 (ad 19). Aquinas refers to Aristotle's characterization of certain things as existing "together by nature"; see *Categories* 7, 7b, 15. Things existing together by nature are things such that if either one of the pair does not exist, the other cannot exist. God's knowledge and the object of his knowledge plainly do not meet the description; for God's knowledge precedes the world and is its cause, whereas the converse is not true. And that is the reason "God can have knowledge even should the object of his knowledge not exist."

[228]*Milḥamot ha-Shem,* VI, i, 4, pp. 303–304.

[229]Cf. 'Abd al-Jabbār's response to the argument that creation would entail change in the creator's knowledge; above, pp. 78–79.

[230]Above, p. 66.

[231]*Summa contra Gentiles,* II, 35 (ad 3). Cf. *Summa Theologiae,* I, 46, art. 1, ad 9; *De Potentia,* q. 3, art. 17 (ad 6).

[232]*Commentary on II Sentences,* d. 1, p. 1, a. 1, q. 2.

[233]Above, p. 64.

[234]*De Potentia,* q. 3, art. 17 (ad 21).

The third form of response to arguments from the eternity of God's attributes consists in showing that it would be utterly impossible for God's attributes to give rise to an eternal universe. Two procedures are in evidence: One rests on the intrinsic impossibility of an eternal universe, without qualification; the other, on the intrinsic impossibility specifically of an eternal universe that has a cause of its existence.

The former procedure is employed by John Philoponus, who builds on his proof of creation. Proclus had argued that the eternity of the world is implied by the deity's "goodness," and Philoponus counters—as he and others had done when answering some of the previous arguments for eternity[235]—with the truism that God can produce only what is possible. Since it can be demonstrated definitively, so Philoponus understands, that an eternal world is absolutely impossible, even God could not produce such a world. The most that God could produce, the most his goodness could express itself in, is a created world. If an eternal world is impossible, the eternity of God's goodness obviously cannot entail the eternity of the world.[236] The same position is taken in the Middle Ages by Āmidī. In the course of an elaborate rejoinder to the argument from God's "beneficence," Āmidī remarks that God's failure to produce the world from eternity betokens no lack of beneficence on his part; for the eternity of the world is demonstrably impossible,[237] and "there is no defect in failing to produce what is impossible."[238]

Ibn Ḥazm employs the other procedure, the procedure resting not on the unqualified impossibility of the world's being eternal, but on the impossibility that the world should be eternal and nevertheless have a cause of its existence. To have a cause of existence and to be eternal are, Ibn Ḥazm explains, mutually exclusive. For "an effect is what passes from nonexistence to existence . . . and thus is tantamount to *that which comes into existence*. The meaning of *that which comes into existence* is *that which does not exist and subsequently exists*, . . . something quite different from *that which is eternal*." To state that the world has a cause and is an effect, yet is eternal, is accordingly to enunciate a "total absurdity," an absolute impossibility. And what is absolutely impossible lies outside even the deity's control. Now the argument from "divine beneficence, wisdom, and power" does acknowledge—in fact presupposes—that the world has a cause, that it is an effect, and hence that it has been brought into existence. Given this acknowledgment, the argument, far from establishing the eternity of the world, falls into a blatant contradiction by concluding that the world is eternal. The argument is therefore wholly invalid.[239] Ibn Ḥazm's response to the argument from divine beneficence, wisdom, and power, as outlined here, would apply, of

[235]Above, pp. 69, 78.
[236]*De Aeternitate*, p. 8.
[237]Cf. below, p. 191.
[238]*Ghāya al-Marām*, p. 270.
[239]*K. al-Faṣl fī al-Milal*, 1, p. 20.

course, to any argument for eternity from God's eternal attributes, and indeed to any argument for eternity which presupposes a cause of the existence of the universe.

Shahrastānī makes virtually the same response to the argument from God's beneficence as Ibn Ḥazm.[240]

3. Summary

The first argument for eternity from the nature of the cause of the universe reasoned that the creator could not have selected a given moment for creation in preference to any other moment. The adherents of creation respond that although all moments before creation were equal, and no explanation can be provided for the creator's having chosen one in preference to another, a particular moment could have been selected arbitrarily by God's eternal will. Subordinate considerations are that there was no time before creation, hence no moment was strictly selected out of an infinite time; that deist adherents of eternity are in the same predicament as the adherents of creation, since they for their part must recognize arbitrary determinations by God in the spatial realm; that adherents of eternity who trace events occurring in the universe back to the divine will must recognize the ability of an eternal will—albeit through intermediaries and within the framework of an already existing universe—to make determinations in the temporal realm; that once creation is demonstrated and the world is known to have come into existence at some moment, it is pointless to ask why creation took place at a certain moment rather than at another.

The second argument for eternity from the cause of the universe reasoned that an ultimate cause must be unchangeable, whereas creation would imply a transition from inaction to action, and hence change, in the cause ultimately responsible for bringing the universe into existence. The adherents of creation usually respond by inverting the thinking; they assert that since the cause of the universe must be unchangeable, creation would not, in his case, imply change. An alternative response is that the creator is eternally and unchangeably in the state of bringing the world into existence, but since an eternal world is impossible, the world could come into existence no sooner than in the finite past.

The third argument was that God's beneficence and other divine attributes express themselves in the existence of the world; and since they are eternal, the existence of the world, which is their expression, must likewise be eternal. The adherents of creation respond either by denying that God eternally has the attributes in question; by denying that the attributes need express themselves in an eternal world; or by showing that it would be intrinsically impossible for the attributes to express themselves in an eternal world.

[240]*K. Nihāya al-Iqdām,* p. 47.

IV

John Philoponus' Proofs of Creation and Their Entry into Medieval Arabic Philosophy

1. Philoponus' proofs of creation

John Philoponus carefully distinguished the negative task of refuting the arguments advanced by proponents of eternity from the positive task of proving creation.[1] To accomplish the latter task, to prove creation, he drew up two sets of proofs, one built around the impossibility of an infinite number, the other based on the principle that a finite body can contain only finite power. Both sets of proofs are mentioned in Philoponus' surviving works, but only in passing.[2] Their systematic development was undertaken by Philoponus in at least two works no longer extant: the *Contra Aristotelem,* and another brief work which was somehow connected to it.[3] The *Contra Aristotelem* and the related work have been partially preserved in Simplicius' commentaries on Aristotle,[4] and either they or Simplicius' excerpts from them were, as will appear, accessible to the medieval Arabic philosophers.

Philoponus' first set of proofs for creation comprises three arguments, each giving a different reason why an infinite series of past events is impossible. The conclusion in each instance is that the world cannot have existed for an infinite time but must have had a beginning.

[1]John Philoponus, *De Aeternitate Mundi Contra Proclum,* ed. H. Rabe (Leipzig, 1899), p. 9; Simplicius, *Commentary on Physics,* ed. H. Diels, *Commentaria in Aristotelem Graeca,* Vol. X (Berlin, 1895), p. 1178, top; S. Pines, "An Arabic Summary of a Lost Work of John Philoponus," *Israel Oriental Studies,* II (1972), 322.

[2]Philoponus, *De Aeternitate,* pp. 8–9, 325.

[3]Regarding that brief work and its connection with the *Contra Aristotelem,* see H. Davidson, "John Philoponus as a Source of Medieval Islamic and Jewish Proofs of Creation," *Journal of the Amerian Oriental Society,* LXXXIX (1969), 358–359.

[4]Simplicius, *Commentary on Physics,* pp. 1171 ff.; *Commentary on De Caelo,* ed. I. Heiberg, *Commentaria in Aristotelem Graeca,* Vol. VII (Berlin, 1894), pp. 28 ff.

(*a.i*) The first of the three arguments pointedly draws from Aristotle. In the *De Generatione et Corruptione,* Philoponus recalls, Aristotle rejected the thesis that the transformation of the physical elements into each other "goes to infinity in a straight line," that is, the thesis that earth, for example, becomes water, water becomes air, air becomes fire, and fire, rather than reverting—circularly, as it were—to air is transformed into a fifth type of element, and so on *ad infinitum*. One of Aristotle's objections to the thesis in question was that at least some of the hypothetical infinite elements could come into existence only after an infinite number of transformations, whereas—as Philoponus spells out Aristotle's meaning[5]—"the infinite cannot be traversed," so that the point never could be reached where those elements actually did come into existence. Consequently, at least some of the elements in the supposed infinite series could never exist, and if part of the series could never exist, the series itself could not be an actuality.[6]

Aristotle's reasoning is subtly turned by Philoponus against the assumption of the eternity of the world. Philoponus does not consider the transformations occurring "in a straight line" through an infinite series of different types of physical elements. Instead he takes up the no less infinite series of transformations that, on the assumption of eternity, necessarily have taken place circularly among the four recognized types of elements, and linearly among individual elemental particles. Philoponus reasons: The existence of a given particle of the element fire must have been preceded by the generation of that fire from a particle of air, the existence of the air must have been preceded by its generation from, let us say, water, the water from either earth or air, and so on. In an eternal world, these transformations would constitute an infinite series. Now it is evident that an infinite number can neither "actually" exist nor be traversed. Therefore, in an eternal world, the infinite series of transformations leading up to the generation of a given particle of fire could never be completed, and the particle known to exist at the present moment could never have come into existence. "The same argument," Philoponus adds, "can be applied to other particular motions" as well, that is, to the various series of transformations leading up to the emergence of whatever individuals exist in the world today.[7] In each instance, an infinite series of transformations would, on the assumption of eternity, have to be traversed in order that what exists today might emerge.

[5]Cf. Philoponus, *Commentary on De Generatione*, ed. H. Vitelli, *Commentaria in Aristotelem Graeca*, Vol. XIV/2 (Berlin, 1897), p. 254.

[6]Cf. Aristotle, *De Generatione et Corruptione* II, 5, 332b, 12—333a, 12. The concept of a *circular* series appears in II, 11, 338a, 4 ff. The principle that an infinite cannot be traversed is not explicitly stated by Aristotle there, but is adduced by him elsewhere in connection with both the spatial and nonspatial realms. Cf. H. Bonitz, *Index Aristotelicus* (Berlin, 1870), p. 74b, lines 30–34, and especially the references to *Posterior Analytics* I, 22, 83b, 6; and *Physics*, 204b, 8–10; 263b, 4; 265a, 19.

[7]Simplicius, *Commentary on Physics*, pp. 1178–1179.

When Philoponus states the argument concisely, he rests it on the general rule that "the infinite cannot be traversed"; and the inference drawn is that if the "number of individuals going up (ἄνω) [into the past]" were infinite, the process of generation could never have been traversed and "come down to each of us."[8] When Philoponus presented the argument more fully, so Simplicius informs us, he employed the same general rule but supplemented it with an additional rule. The additional rule affirms that if something must be preceded by something else in order to be generated, then "the former will not be generated unless the latter is generated prior thereto." In other words, if the prior existence of *y* is a condition for the coming into existence of *x*, then *x* obviously will never come into existence unless *y* has already been in existence. The inference now drawn is that "if for the generation of a given thing there must first exist an infinite number of things that are generated from one another, then the given thing cannot be generated."[9] If *x* cannot come into existence unless an infinite series of things has already come into existence—a condition that cannot conceivably be fulfilled—*x* will never come into existence. Students of Kant will observe the similarity to the proof of the thesis of the first antinomy.

The upshot of Philoponus' first proof of creation is that transformations in the sublunar world cannot be "conceived to precede one another infinitely," but must have a beginning; and since the translunar world is inextricably connected with the sublunar world,[10] it too must have a beginning.[11]

(*a.ii*) Philoponus' second argument for creation from the impossibility of an infinite number employs a principle which although apparently not strictly Aristotelian,[12] does appear in the Peripatetic tradition. The principle affirms that since an infinite cannot conceivably be exceeded, one infinite cannot be greater than another.[13] Philoponus applies the principle thus: "Since motions yet to be generated, when added to those already generated, increase their number, and since, moreover, the infinite cannot be increased, it follows that the motions already accumulated cannot be infinite."[14] That is to say, each new movement in the sublunar world and each new revolution of the celestial spheres add to the number that has gone before. If the number that has gone before should be infinite, each new movement would add to an infinite number. But the infinite cannot be increased.

[8]Philoponus, *De Aeternitate*, pp. 10–11; cf. his *Commentary on Physics*, ed. H. Vitelli, *Commentaria in Aristotelem Graeca*, Vol. XVI (Berlin, 1887), pp. 428–429.

[9]Simplicius, *Commentary on Physics*, p. 1178. On p. 1179, Philoponus is reported to have spoken of "going back up (ἄνοδος)" through the infinite.

[10]Cf. Aristotle, *De Caelo* II, 3; Philoponus, as reported by Simplicius, *Commentary on Physics*, p. 1179.

[11]Simplicius, *Commentary on Physics*, p. 1179.

[12]In fact, *Physics* III, 6, 207a, 1 ff., seems to state the contrary.

[13]Alexander of Aphrodisias, *Aporiai*, ed. I Bruns, *Commentaria in Aristotelem Graeca*, Supplementary Vol. II/2 (Berlin, 1892), III, 12, p. 103. Alexander observes that the assumption of an infinite multiplied infinitely would be the highest absurdity.

[14]Simplicius, *Commentary on Physics*, p. 1179. Cf. Philoponus, *De Aeternitate*, p. 11.

The number of past motions must therefore be finite, and the universe can have existed for only a finite time.

(*a.iii*) Philoponus' third argument is not independent of the previous two. It is a variation of the second, repeating, for good measure, the central thought of the first. Whereas the second argument rested on the principle that an infinite cannot be increased, the third rests on the principle that an infinite cannot be multiplied; multiplication, though, is simply a type of addition. To reinforce the argumentation, Philoponus again refers to the rule that an infinite cannot be traversed.

Philoponus' attention is here restricted to the heavens, and his reasoning is reported thus by Simplicius: "If . . . the motion of the heavens has no beginning, the sphere of the planet Saturn has performed infinite revolutions, the sphere of the planet Jupiter almost three times as many, the revolutions of the sun will be thirty times those of Saturn, the revolutions of the moon will be 360 times as many, and the revolutions of the fixed sphere [which revolves once each twenty-four hours] will be more than 10,000 times as many. Considering that the infinite cannot be traversed even once, is it not beyond all absurdity to suppose the infinite multiplied by 10,000, nay multiplied infinitely? It necessarily follows that the circular motion of the heavens had a beginning."[15] Aristotelians understood circular motion to be an essential expression of the nature of the spheres. As long as the spheres exist they would have to perform such motion and could not exist in a state of rest.[16] Once Philoponus has established that the motions of the heavens had a beginning, he can therefore conclude as well that "the heavens themselves also had a beginning of their existence."[17]

(*b.i*) Philoponus' second set of proofs for creation rests on the Aristotelian principle that a finite body can contain only finite power. The principle has a peculiar history.[18] Its career began in Aristotle's proof of the existence of a first mover, where it served as an essential premise in establishing that the first mover is incorporeal. Aristotle reasoned that infinite power is needed to sustain the motion of the universe for an infinite time; that all bodies are finite; and that a finite body can contain only finite power. The conclusion he drew was that the first cause of the motion of the universe cannot be a body, but must be an incorporeal being.[19] Proclus subsequently took up the Aristotelian principle that a finite body can contain only finite power and arrived at a more far-reaching result than Aristotle. He converted Aristotle's argument that the motion of the universe must depend upon an incorporeal cause into an argument showing that the very existence of the physical universe must depend upon an incorporeal cause. Proclus reasoned that infinite power is needed to sustain not only eternal motion, but

[15] Simplicius, *Commentary on Physics,* p. 1179. Philoponus, *De Aeternitate,* p. 11. Cf. above, n. 13.

[16] Aristotle, *De Caelo* I, 2.

[17] Simplicius, *Commentary on Physics,* p. 1179.

[18] See Appendix A.

[19] Aristotle, *Physics* VIII, 10; below, p. 244.

eternal existence as well; and since the finite physical universe can contain only finite power, the physical universe must depend upon an incorporeal being outside itself for its very continued existence.[20] Philoponus knew of the arguments of Aristotle[21] and of Proclus,[22] and they led him to a proof of creation. Since the corporeal universe contains only finite power, the universe, such will be Philoponus' argument, is not merely incapable of sustaining its own eternal motion or eternal existence; it is incapable of existing eternally no matter what the cause sustaining its existence might be. As will be seen in later chapters, Philoponus' adaptation of the argument from the finite power of finite bodies would elicit a response from Averroes,[23] and Averroes' response would, in turn, elicit from Crescas a critique of the original Aristotelian proof of the incorporeality of the first mover.[24]

Here we are interested in Philoponus' proof. Philoponus is reported by Simplicius to have contended: Inasmuch as "the body of the heavens and of the universe is finite, it contains [only] finite power. And what contains finite power . . . [is] destructible."[25] That is to say, since the corporeal universe contains only finite power, it is, considered in itself, incapable of existing through infinite future time and consequently is liable to destruction.[26] The proposition that the corporeal universe is liable to destruction will not be invalidated even by the assumption that the universe is maintained in existence by a "transcendent" power as, for example, the "will of God."[27] For even should such an assumption be made, the universe will remain incapable of existing infinitely into the future, insofar as it is considered in itself; the universe, considered in itself, will remain liable to destruction. Stated otherwise, the universe will still have the "logos (nature or ground) of destruction."[28] Now a well known and widely employed[29] Aristotelian principle affirms that over an infinite time every possibility must eventually be realized.[30] Accordingly, Philoponus concluded, the logos of destruction in the universe must "sometime come to actuality" and the universe must some day actually be destroyed.[31]

[20]Cf. below, pp. 281–282.

[21]Simplicius, *Commentary on Physics*, p. 1327; Philoponus, *De Aeternitate*, p. 238.

[22]Philoponus, *De Aeternitate*, pp. 238–240.

[23]Below, pp. 323 ff.

[24]Below, pp. 264–265.

[25]Simplicius, *Commentary on Physics*, p. 1327.

[26]Cf. Philoponus, *De Aeternitate*, p. 235.

[27]Simplicius, *Commentary on Physics*, pp. 1330, 1331, 1333.

[28]Ibid., pp. 1331, 1333.

[29]Cf. below, pp. 320, 381.

[30]Aristotle, *Physics* III, 4, 203b, 30; *Metaphysics* IX, 4, 1047b, 4–5; 8, 1050b, 8–15. Cf. J. Hintikka, *Time and Necessity* (Oxford, 1973), pp. 95–96, 103–105, 107.

[31]Simplicius, *Commentary on Physics*, p. 1333. Cf. Aristotle, *De Caelo* I, 12, 281b, 20–25 and 283a, 24–25; *Metaphysics* XIV, 2; Alexander, *Aporiai*, I, 18; and the passage translated by Pines, "An Arabic Summary of a Lost Work of Philoponus," pp. 324–325. In Simplicius, *Commentary on Physics*, pp. 1331, 1333, and Philoponus, *De Aeternitate*, pp. 241–242, Philoponus seems to countenance the possibility of the future eternity of the world.

That is the extent of Simplicius' report, and it falls short of an explicit proof of creation. Another work by Philoponus makes clear, however, how Philoponus passed from the destructibility, and actual future destruction, of the world to creation. He employed yet another well known Aristotelian principle, one that happens to rest on the principle, just cited, according to which every possibility must eventually be realized. The other principle affirms that "whatever is destructible must be generated."[32] If something is destructible, if it has the possibility of not existing, that possibility must have been realized at some point in the infinite past just as it must be realized at some point in the infinite future. Just as a destructible universe must eventually become nonexistent in the future, so too must a destructible universe have been nonexistent in the past. Our universe, therefore, cannot have existed forever, but must have come into existence.[33]

Philoponus' reasoning, then, was this: Since a finite body can contain only finite power, the corporeal universe has the logos of destruction. Since, moreover, every possibility is eventually realized, the universe must eventually undergo destruction. And since whatever is destructible is generated, the universe must have been generated. The Aristotelian principle that a finite body can contain only finite power, taken together with two other Aristotelian principles, leads to the highly un-Aristotelian conclusion that the world can neither exist forever in the future nor have existed forever in the past. Apparently, Philoponus understood that he was demonstrating not merely the creation of the universe in its present form but the creation of the matter of the universe *ex nihilo*; for his contention is that the very matter of the universe, being finite, lacks the power to sustain itself in existence for an infinite time.

Philoponus was not content simply to cite the authority of Aristotle for the principle that a finite body can contain only finite power. He proceeded to offer a set of five or six[34] auxiliary arguments, all ostensibly intended to support the principle. In fact, the auxiliary arguments are not uniform. Some of them do support the proposition that the heavens, or the entire universe, can contain only finite power; the proof for creation is then to be completed through the consideration that what contains finite power is destructible, hence also generated. Inexplicably, however, some of the auxiliary arguments[35] move in the other direction. They start by establishing that the corporeal universe is destructible and infer therefrom that the universe can contain only finite power. The reader is left either again to draw the further inference, now quite redundant, that what

[32]Aristotle, *De Caelo* I, 12, 282b, 2; ibid., 281b, 26—282a, 1.

[33]Philoponus, *De Aeternitate*, pp. 225, 230, 235, 240, 241.

[34]Simplicius, *Commentary on Physics*, enumerates only four auxiliary arguments. But on p. 1332, between the "third" and "fourth," he quotes an additional auxiliary argument not included in the enumeration; and the auxiliary argument that he enumerates, p. 1335, as the "fourth," contains two separate considerations. In the article referred to above, n. 18, I outline all the arguments except the one designated by Simplicius as the "second." That argument is dealt with here, below (b.iii).

[35]See below, (b.iv).

has finite power is destructible, and hence is generated; or else to derive the generation of the universe directly from its finite power. One of Philoponus' auxiliary arguments even drops the issue of finite power altogether.[36]

For our purposes, the first three of Philoponus' auxiliary arguments are of interest.

(*b.ii*) In the "first" of his auxiliary arguments, Philoponus is reported to have reasoned that "the heavens are [composed] of matter and form; what is [composed] of matter requires the matter for its existence; what requires something is not self-sufficient; what is not self-sufficient is not infinitely powerful. From all this he concludes that the heavens, considering their own nature, are not infinitely powerful and therefore are destructible."[37] It remains for us to add that what is destructible must be generated, and consequently the heavens and the rest of the corporeal universe cannot—at least in their present form—have existed for all eternity but must have come into existence.

(*b.iii*) Philoponus' "second" auxiliary argument, as reported by Simplicius, drops the issue of finiteness of power. Philoponus contends: "The essence of matter consists in its being suited to receive all forms. It does not possess that power in vain; the same matter cannot admit several forms at once; nor can matter retain any form eternally insofar as its own nature is considered." Since matter is by nature such that it does not retain any form permanently, "nothing [composed] of matter and form will, considering its matter, be indestructible." The corporeal universe is thus destructible.[38] The reader is left to supply the principle that everything destructible is also generated and to draw the conclusion that the corporeal universe in its present state—clearly not the matter of the universe—must have come into existence.

(*b.iv*) The "third" of the auxiliary arguments opens, like the first, with the contention that the heavens are "composed . . . of substratum . . . and form," the latter here being specified as "solar or lunar form" and the forms of the other heavenly bodies. Anyone who would exclude the distinction of matter and form from the heavens would, Philoponus adds, still have to acknowledge the presence in the heavens of "extension in three dimensions," which too is a mode of composition. Philoponus does not hereupon reason, as he had done in his first auxiliary argument, that composition implies the absence of self-sufficiency, hence the absence of infinite power. Instead, he pursues a line of thought going back to Plato's *Phaedo,* where the proposition had been laid down that the fact of something's being composite implies its being subject to decomposition.[39] What-

[36]Below, (b.iii).

[37]Simplicius, *Commentary on Physics,* p. 1329. The train of reasoning is to a large extent borrowed from Proclus, *Elements of Theology,* ed. and trans. E. Dodds (Oxford, 1963), §127.

[38]Simplicius, *Commentary on Physics,* p. 1329.

[39]*Phaedo,* 78C. Aristotle, *Metaphysics* XIV, 2, 1088b, 14–28, contends that what is composite has the potentiality of not existing, and what has the potentiality of not existing is not eternal.

ever has the "logos (nature or ground) of composition" has, Philoponus explains, the "logos of dissolution" and the "logos of destruction." And what has the "logos of destruction is not infinitely powerful." The heavens, therefore, are not infinitely powerful.[40] Here, as will be observed, is an instance of Philoponus' moving not from the finiteness of the power of the heavens to their destructibility, but vice versa. The argument must be completed either by reasoning, redundantly, back again from the finite power of the heavens to their destructibility and then adducing the principle that everything destructible is generated; or else by inferring the generation of the heavens without returning to their destructibility. The inference might, for example, be that what has finite power cannot maintain itself indefinitely, hence contains the possibility of not existing, and hence cannot have existed forever.

To summarize: Philoponus formulated two sets of proofs of creation, both of which employ Aristotelian principles to draw the un-Aristotelian conclusion that the world is not eternal but had a beginning. Each set of proofs rests on the impossibility of an infinite of one sort or another. The first set argues in three different ways that an infinite number of events cannot conceivably have preceded the present moment. The second set argues that a finite body cannot contain infinite power, that what contains only finite power has the possibility of not existing, and what has the possibility of not existing cannot have existed forever. The second set comprises a general statement of the proof and supporting arguments.

The most important source of the first of the two sets of proofs is Philoponus' *Contra Aristotelem*. The *Contra Aristotelem,* which is no longer extant and which is accessible today solely through Simplicius' excerpts, was available in some form to the medieval Arabs.[41] It is listed by the Arabic bibliographers,[42] a fact that, by itself, does not necessarily mean the book was translated into Arabic. But, in addition, passages from the book, cited in Philoponus' name, have been discovered in the writings of Alfarabi,[43] Sijistānī[44] (912–985), and Avicenna,[45] although unfortunately none of the passages that have been discovered is from the section containing the proofs for creation. Philoponus' second set of arguments, from the finite power of the corporeal universe, was developed in a separate work that might possibly have been written as an appendix to the *Contra*

[40] Simplicius, *Commentary on Physics,* p. 1331.

[41] Simplicius is probably not the source, since his commentaries on the *Physics* and *De Caelo,* which preserve the excerpts from Philoponus, are not mentioned by the medieval Arabic bibliographers. (For the Arabic bibliographers' knowledge of Simplicius, I am relying on M. Steinschneider, *Die Arabischen Uebersetzungen aus dem Griechischen* [repr. Graz, 1960].)

[42] Cf. Ibn al-Nadīm, *K. al-Fihrist* (Leipzig, 1871), I, 254, and, for al-Qifṭī and Ibn Abī Uṣaybi'a, M. Steinschneider, *Al-Farabi* (St. Petersburg, 1869), pp. 162, 220–224.

[43] M. Mahdi, "Alfarabi against Philoponus," *Journal of Near Eastern Studies,* XXVI (1967), 236.

[44] J. Kraemer, "A Lost Passage from Philoponus' *Contra Aristotelem,* in Arabic Translation," *Journal of the American Oriental Society,* LXXXV (1965), p. 319, n. 4, and p. 325.

[45] Ibid., p. 324, n. 27.

Aristotelem.[46] This other work was also available to the medieval Arabs. The Arabic bibliographers list, among the works of Philoponus, *A Single Treatise Showing that Every Body is Finite and has Finite Power.*[47] Again, that is not necessarily evidence that the work was translated into Arabic; but in addition, Yaḥyā ibn ʻAdī (892–973) knew of a treatise in which Philoponus set forth the proof from finite power,[48] and the main thesis of the proof is cited in Philoponus' name by Ibn Suwār[49] in the tenth century, and by Averroes.[50] Further evidence that both sets of proofs were current in the Middle Ages is furnished by a short, recently discovered Arabic text, which styles itself a summary of three "treatises" of Philoponus. In each of the three treatises Philoponus is represented as having given a different proof of creation.[51] One of the three proofs turns out to be a statement of the first of Philoponus' arguments from the impossibility of an infinite number, some of the wording even being identical with the version reported by Simplicius;[52] a second of the three proofs consists in a general statement of the proof from the finite power of finite bodies;[53] whereas the third proof is probably attributed to Philoponus erroneously.[54]

There is thus ample evidence that Philoponus' proofs were accessible to readers of Arabic in the Middle Ages. As will appear in the remainder of the present chapter and in the following chapter, Philoponus became a most important source for medieval proofs of creation.

[46]See Davidson, "John Philoponus as a Source of Medieval Islamic and Jewish Proofs of Creation," pp. 358–359.

[47]References above, n. 42.

[48]Cf. S. Pines, "A Tenth Century Philosophical Correspondence," *Proceedings of the American Academy for Jewish Research,* XXIV (1955), 115.

[49]Cf. the brief treatise by Ibn Suwār published in A. Badawi, ed., *Neoplatonici apud Arabes* (Cairo, 1955), p. 246; French translation in B. Lewin, "La Notion de muhdat dans le kalam et dans la philosophie," *Orientalia Suenica,* III (1954), 91.

[50]Averroes, *Middle Commentary on Physics* (Oxford, Bodleian Library, Hebrew MS. Neubauer 1380 = Hunt. 79), VIII, vi, 2; *Long Commentary on Metaphysics* (*Tafsīr mā baʻd al-Ṭabīʻa,* ed. M. Bouyges [Beirut, 1938–1948]), XII, comm. 41; Steinschneider, *Al-Farabi,* p. 123, citations 2–5. Cf. Davidson, "John Philoponus as a Source of Medieval Islamic and Jewish Proofs of Creation," p. 361, n. 41.

[51]Pines, "An Arabic Summary of a Lost Work of Philoponus," pp. 320–352.

[52]Ibid., pp. 330–336. Compare the three "principles" in Pines' translation of the Arabic text with the three axioms listed by Simplicius, *Commentary on Physics,* p. 1178. In the Arabic text, the key consideration that an infinite cannot be traversed is omitted, and the argument is made to rest on the impossibility of ever counting an actual infinite number.

[53]Pines, "An Arabic Summary of a Lost Work of Philoponus," pp. 323–325.

[54]Ibid., pp. 325–329. The argument is that the lives of individuals mark off the time continuum into segments, each of which is finite; and finite segments of time, no matter how many there might be, could not join together to constitute infinite time. By contrast, Philoponus, *De Aeternitate,* p. 9, takes the common sense position that an infinite number of past time segments, each of which was finite, would indeed constitute an infinite past time.

2. Saadia and Philoponus

The Jewish philosopher Saadia (882–942) never mentions Philoponus by name nor quotes him directly, but Saadia's discussion reveals that he drew from the two sets of proofs of creation formulated by Philoponus.

Saadia's aim was to treat the problem of creation in all its aspects. He begins by offering four proofs of the generation (*ḥadath*) of the world; he demonstrates that the world not only had a beginning but was created *ex nihilo*; and he reviews various alternatives to the doctrine of creation *ex nihilo,* one of which is the theory that the eternal heavens are the cause of the universe. From the entire discussion, seven items are of interest to us: Saadia's four proofs of the "generation" of the world, one of several demonstrations designed to establish that the world was created *ex nihilo,* and two of his refutations—"the third" and "the fourth"—of the thesis that the heavens are eternal. Five of the seven items can be shown unquestionably to be derived from Philoponus, and the other two can be traced to Philoponus with some plausibility.

The comparison with Philoponus will be facilitated by rearranging the seven items from Saadia. The order in which I shall take them up will be: Saadia's fourth proof of creation, the argument for creation *ex nihilo,* the two refutations of the eternity of the heavens, and finally, Saadia's first, second, and third proofs of creation. The following table lists the seven items from Saadia as I have rearranged them, together with the corresponding arguments from Philoponus.

	Saadia		Philoponus
(i)	Fourth proof of creation	(a.i)	The infinite is not traversable
(ii)	An argument for creation *ex nihilo*	(a.i)	The infinite is not traversable
(iii)	A refutation of the eternity of the heavens	(a.ii)	The infinite cannot be added to
(iv)	Another refutation of the eternity of the heavens	(a.iii)	The infinite cannot be multiplied
(v)	First proof of creation	(b.i)	Proof from the finite power of finite bodies
(vi)	Second proof of creation	(b.ii)	Auxiliary argument from composition
(vii)	Third proof of creation	(b.iii)	Auxiliary argument from the succession of forms over matter

(*i*) Saadia's fourth proof of the "generation" of the world carries the title "from time" and runs thus: Assuming that time is eternal, should a person "attempt mentally to ascend in time from the present moment, he would be unable to do so, since . . . thought cannot travel up across the infinite and traverse it. The same reason would prevent existence from traveling down through time and traversing it so as to reach us. But if existence had not reached us, we should not exist. . . . Inasmuch as I find that I do exist, I know that existence has traversed time, . . . and . . . time is finite. . . ." As a sort of appendix to the proof Saadia mentions that his position "on future time" agrees with his position "on past time."[55] He means that future time is, in his opinion, finite, and perhaps also that the finiteness of future time can be demonstrated. If he is indeed alluding to a demonstration of the finiteness of future time, the argument would presumably be that no moment in an infinitely distant future could be reached; for existence could not conceivably travel across and traverse infinite future time so as to reach whatever might be thought to exist at the infinitely distant moment.

The proof of creation just quoted from Saadia unquestionably derives from Philoponus' first proof of creation from the impossibility of an infinite number. The key to Philoponus' proof was the rule that an infinite cannot be traversed.[56] And the same is now the key to Saadia's proof. Characteristic details from Philoponus are also echoed in Saadia, namely the impossibility of things' existing today if they had to be preceded by an infinite past; the impossibility "mentally" (Saadia) to "conceive" (Philoponus)[57] an infinite's extending back into the past; and the comparison between going back "up" through the infinite past, on the one hand, and coming "down to each of us" (Philoponus)[58] or "traveling down . . . to reach us" (Saadia), on the other.

Saadia's proof does differ from Philoponus' in a significant respect. Philoponus considered transformations whereas Saadia considers time, and therefore Philoponus' infinite is an infinite *series,* whereas Saadia's is an infinite *continuum.* The shift from the former infinite to the latter permits a certain simplification in the argument. Philoponus' reasoning established a beginning for the existence of the sublunar region, where everything undergoes transformation, but not for the celestial region, where, in the Aristotelian universe, transformations never occur. In order to extend his proof to cover the celestial region, Philoponus had to add that the sublunar and celestial regions are inextricably connected, so that the beginning of the one implies the beginning of the other.[59] The addition is no longer needed in Saadia's version of the proof, since the contention that existence could not have "traversed" infinite past time to reach the present moment applies equally to the sublunar and celestial regions.[60]

[55]Saadia, *K. al-Amānāt wa-l-I'tiqādāt,* ed. S. Landauer (Leiden, 1880), I, 1, p. 36; English translation with pagination of the Arabic indicated: *Book of Beliefs and Opinions,* trans. S. Rosenblatt (New Haven, 1948).

[56]Above, pp. 87–88.

[57]Above, n. 11. Aristotle, *Posterior Analytics* I, 22, 83b, 6–7, states that it is impossible for an infinite to be "traversed in thought."

[58]Above, n. 8, and Aristotle, ibid.

[59]Above, n. 10.

[60]There is at least one consideration that could make Philoponus' formulation stronger than

Saadia, at any rate, reveals indirectly that he did not recognize the difference between an infinite series and an infinite continuum. He takes up a possible objection to the proof, an objection which is, essentially, Zeno's first paradox,[61] and which is attested to elsewhere in Arabic philosophic literature.[62] It runs: Since distance is infinitely divisible, would not the train of thought embodied in the proof render a person's moving from one place to another impossible, inasmuch as the person would have to traverse an infinite number of parts? Saadia's response is that the objection misleadingly adduces the traversing not of an "actually" existing infinite, but of an infinite existing solely in "imagination"; the proof, by contrast, rests on the fact that "existence does actually traverse [past] time and reach us," and an *actual* infinite cannot be traversed.[63]

The objection faced here by Saadia views distance not as an infinite continuum, as Saadia's proof viewed time, but rather as an infinite series of discrete parts. Since Saadia considers the objection to be pertinent, either he is consciously drawing an analogy between an infinite series of discrete parts and an infinite continuum, or else he did not detect any distinction between the two. In either case the fact that Philoponus' proof addresses the former kind of infinite, whereas Saadia's addresses the latter, can occasion no hesitation about tracing Saadia's proof back to Philoponus; for Saadia, as now seen, does not recognize the difference between the two kinds of infinite.

Saadia's response to the possible objection to his proof incidentally reproduces a further detail from Philoponus: The principle that infinites cannot be traversed applies exclusively to actual infinites.[64]

(*ii*) In a proof intended to establish that creation took place *ex nihilo,* Saadia reasons: If instead of accepting creation *ex nihilo,* "we were to suppose something coming from something, then the second thing in the hypothesis would resemble the first and would have to meet the condition that it can only come from a third thing. The third thing in the hypothesis would resemble the second, and would have to meet the condition that it can only come from a fourth thing. And this would go on to infinity. But since the infinite cannot be completed, . . . we could not exist. Yet behold, we do exist! . . . whereas if the things preceding us were not finite [in number] they would not have been completed so that we might exist."[65] Saadia has, in a more abbreviated fashion, again offered the argument

---

Saadia's, although Philoponus himself does not state it: The assumption of an infinite series of transformations would involve not only the traversal of an infinite; it would also involve an infinite regress of causes, which was rejected by Aristotle on independent grounds. Cf. below, p. 337.

[61]Cf., e.g., W. Ross's edition of Aristotle's *Physics* (Oxford, 1955), p. 72.

[62]Cf. below, p. 118.

[63]Saadia, *K. al-Amānāt*, I, 1, pp. 36–37.

[64]Cf. above, n. 7. On the question whether infinite past time would be an *actual* infinite, see below, p. 128.

[65]Saadia, *K. al-Amānāt*, I, 2, p. 40. The same argument is alluded to in Saadia's *Commentary on Sefer Yeṣira*, published as *Commentaire sur le Séfer Yesira*, ed. M. Lambert (Paris, 1891), Arabic section, p. 3; French translation, p. 16.

that the present could not have been reached if, prior thereto, an infinite past had to be traversed. He has, in other words, again offered a version of Philoponus' first proof of creation from the impossibility of an infinite number. The argument, Saadia now claims, establishes creation *ex nihilo,* although the new version, like Saadia's previous version and like the original proof in Philoponus, does not seem to preclude an eternal matter from which the world might have been formed.[66] What is noteworthy about the new version is that it omits the details of Philoponus' proof which were reproduced in Saadia's previous version, while including two significant details that were absent there. The details appearing here but absent in the previous version are these: Saadia does not consider the time continuum but rather, like Philoponus, the series of transformations leading up to what exists at the present moment. And he appends a statement to the effect that "if the things preceding us were not infinite, they would not have been completed so that we might exist." The statement appears to reflect Philoponus' contention that "if for the generation of a given thing there must first exist an infinite number of things that are generated from each other, then the given thing cannot be generated."[67]

Two items in the list of seven that I wish to examine in Saadia have been shown to derive from Philoponus's first proof from the impossibility of an infinite number, namely, item (i), Saadia's fourth proof for creation, and item (ii), the argument just discussed wherein Saadia claims to have demonstrated creation *ex nihilo*.

(*iii*) After presenting his proofs of creation *ex nihilo* Saadia takes up the unacceptable alternatives, among which is the theory that the heavens are eternal and they bring the world into existence. In the course of refuting the eternity of the heavens he presents an argument entitled "from increase and diminution," which reads: "Every day elapsing from the time of the sphere is an increase over what has passed and a diminution from what is to come. But whatever admits increase and diminution is of finite power; and finiteness implies generation. Should anyone venture to assert that the elapsing of a day does not increase what has passed or diminish what is to come, he would fly in the face of reality and experience."[68] Saadia here has repeated the essential elements in Philoponus' second proof from the impossibility of an infinite number, the proof based by Philoponus on the

[66]The present version could establish creation *ex nihilo* only if one of Saadia's four proofs of creation—which I have designated as items (i), (v), (vi), and (vii)—had already proved that the very matter of the universe is created; but none of those proofs is represented by Saadia as doing so. In fact, the previous version of the proof under consideration—that is, item (i)—could make a stronger argument for creation *ex nihilo* than the present version, although Saadia represents the previous version as a proof merely of creation and the present version as a proof of creation *ex nihilo*. Should it be maintained that formless matter is properly described as existing in time, then it could be argued that since—in accordance with the previous version—past time must be finite, matter cannot have existed forever.

[67]Above, n. 9.

[68]Saadia, *K. al-Amānāt,* I, 3 (eighth theory), p. 60.

principle that an infinite cannot be increased. As in the first item from Saadia which we examined,[69] the argument has been shifted into the realm of time. Saadia reasons that each passing day adds to the time already past; that the infinite cannot be increased; and that time must consequently be finite. Philoponus had made the same points regarding past "motions."[70]

There are two elements in Saadia's proof which do not come from Philoponus. One is the statement that each elapsing day not merely adds to past time but also diminishes future time. The inexorable diminishing of future time contributes nothing to the proof for creation. But the consideration that what is subject to diminution cannot be infinite does appear in Arabic philosophy in a related context, in arguments for the impossibility of an infinite magnitude.[71] Saadia probably intends by his statement to intimate an argument against the future eternity of the world. The thinking would be: Future time is steadily being diminished, the infinite cannot be diminished, therefore future time cannot be infinite.

The other element in Saadia's proof which does not come from Philoponus is the conclusion that past time is "of finite power"; we would have expected instead a conclusion to the effect that past time is of finite extent or of finite duration. Saadia surely has expressed himself poorly; for time cannot properly be described as possessing power, whether finite or infinite. Finite power was, as will be recalled, a central concept in a different proof of creation, the proof from the principle that a finite body can contain only finite power. And that proof, as will appear, was known to Saadia and employed by him.[72] We may conjecture that finite power intruded into the present proof as an echo from the other proof, where it originally belonged. Both proofs concern themselves with finiteness, although of different types, and Saadia—or someone earlier than he in the line of transmission—must have mechanically transferred the expression "finite power" from the proof where it is appropriate to a proof where it is not.

(*iv*) Immediately after the argument just examined, Saadia continues his refutation of the thesis that the heavens are eternal with an argument entitled "from the variation in movements." It reads: "Infinite power does not vary in itself. Since we observe that the movements of the heavens vary to the extent that they are thirty or 365 times one another, and still more, we realize that each [of the movements] is finite. The explanation is as follows: The eastern movement of the highest sphere performs a revolution once in twenty-four hours, whereas the western movement of the fixed stars [i.e., the precession of the equinoxes] proceeds one degree each hundred years, at which rate the complete revolution will be performed in no less than 36,000 years or 13,140,000 days. . . . How can you say that a power whose movement varies so widely is not finite?"[73]

[69] Above, pp. 95–96.
[70] Above, p. 88.
[71] Below, pp. 126–127.
[72] Below, item (v).
[73] Saadia, *K. al-Amānāt,* I, 3 (eighth theory), pp. 60–61.

Partly explicitly and partly implicitly, Saadia has reproduced the key elements in Philoponus' third proof from the impossibility of an infinite number, the proof based on the impossibility of an infinite number's being multiplied. Like Philoponus, Saadia reasons that the heavenly bodies move at varying velocities; their revolutions accumulate in varying multiples of each other; and therefore, "each"—that is to say, the number of revolutions undergone by each of the heavenly bodies—must be "finite."[74] Saadia even remains faithful to the original proof in leaving the discussion within the realm of numbers of motions rather than transferring it—as he did in items (i) and (iii)—to the realm of time. There are small changes in detail: Saadia does not reproduce all the astronomical data given by Philoponus.[75] In one instance, a substitution has been made; the figure "365" (the ratio of the revolutions of the diurnal sphere to the revolutions of the sun, and hence the number of days in a year), if not a scribal error, has been substituted by Saadia or by his immediate source for the less familiar figure "360" (the ratio of the revolutions of the moon to the revolutions of Saturn.[76] And an additional astronomical datum is evinced, with the unmistakable intent of magnifying the effect. The slow movement of the fixed stars, the phenomenon known today as the precession of the equinoxes, had been calculated as one degree each hundred years. Accordingly, to the paradoxes of multiple infinites cited by Philoponus—as reported by Simplicius—Saadia or his immediate source adds that the infinite revolutions of the daily sphere would, over an eternity, have to be no less than thirteen million times as numerous as the infinite revolutions of the fixed stars!

As in the previous item,[77] Saadia speaks of finite power, the references to finite power appearing here not in the body of the argument but in an introductory and a closing statement. The introductory statement, which happens to have a close parallel in Aristotle,[78] affirms that "infinite power does not vary in itself"; and the closing statement affirms that "a power whose movement varies" must be finite. These references to finite power are less incongruous than was the reference to finite power in the previous argument. Movement may after all be described as *due* to a finite or infinite power whereas time can hardly be described as *possessing* either. Nevertheless, the introduction of finite power is again inappropriate. The present argument can be understood solely as reasoning from the nature of number, from the impossibility of one infinite number's being a multiple of another. The fact that the numbers of celestial revolutions accumulate in varying multiples is pertinent when the argument is so understood; it supplies cogent grounds for the conclusion that the number of revolutions of each heavenly body is finite. That fact in no way exhibits, however, the finiteness of the power or powers moving the spheres. The opening reference to finite power and the con-

[74] Cf. above, p. 89.

[75] See ibid.

[76] Ibid.

[77] Above, pp. 98–99.

[78] Aristotle, *De Caelo* I, 7, 275b, 27–29.

cluding statement that "a power whose movement varies" is finite thus merely obscure the issue. The proper conclusion should be that the number of the revolutions of the heavens must be finite, and since the heavens cannot exist without revolving—as Saadia states explicitly elsewhere[79]—they cannot have existed forever. The concept of finite power, it may again be conjectured, has intruded into the present proof as an echo from a separate proof of creation, the proof from the finite power of the body of the universe.

To recapitulate the discussion so far: The four items in Saadia which we have been examining are derived from Philoponus' three proofs for creation from the impossibility of an infinite number. The first of Philoponus' three proofs was the argument that no event could occur if, in order to reach it, an infinite series of transformations had to be traversed. The argument appears in Saadia in two versions, which complement each other in reproducing virtually all the details of Philoponus' proof. Philoponus' second proof was the argument that the number of past events must be finite since an infinite number cannot be added to; and his third proof was the argument that the number of the revolutions of the spheres must be finite since an infinite number cannot be multiplied. Those two proofs reappear in Saadia as part of a refutation of the eternity of the heavens. Saadia's versions reveal departures in detail from the original proofs, but the key elements of Philoponus' reasoning are preserved and Saadia even presents the proofs in the same sequence as Philoponus.

(*v*) I turn to Saadia's first proof of creation. It is entitled "from finitude" and reads: "The heavens and earth clearly are finite inasmuch as the earth is in the center and the revolution of the heavens goes around it. Hence their *power* is finite; for, as is well known, an infinite power cannot be present in a finite body. Since the power that maintains the heavens and earth ceases, they necessarily have a beginning and end." Saadia expands on the evidence for the finiteness of the heavens and earth[80] and concludes: "Since the bodies [of the heavens and earth] are limited, their power is limited, reaching a certain limit where the power stops. They cannot continue after the destruction of that power nor exist before it does."[81]

[79]Saadia, *K. al-Amānāt,* I, 1, p. 33.

[80]Saadia explains that the earth cannot be infinite, since it is circumscribed by the orbits of the heavenly bodies, and what can be circumscribed is not infinite. He explains that the heavens are not infinite, since the heavens revolve around the earth, and an infinite body could never complete a revolution. That argument is derived from Aristotle, *De Caelo* I, 5, 271b, 26 ff. Subsequent to Saadia it can be found in the Kalam thinker Baghdādī, *K. Uṣūl al-Dīn* (Istanbul, 1928), p. 66. Also cf. below, p. 257. Saadia finally explains that the universe cannot be infinite by virtue of being composed of an infinite number of heavens and earths. His grounds for ruling out a multiplicity of heavens and earths are that each of the four elements can have no more than one natural place. That argument too is derived from Aristotle; cf. *De Caelo* I, 8, and below, pp. 274–275.

[81]Saadia, *K. al-Amānāt,* I, 1, pp. 32–34. The proof is alluded to by Saadia in his *Commentary to Sefer Yeṣira,* pp 33 ff.; French translation, pp. 53 ff.

Here Saadia has reproduced Philoponus' proof of creation from the principle that a finite body can contain only finite power. The "well known" proposition cited by Saadia to the effect that "an infinite power cannot be present in a finite body" is exactly the Aristotelian principle upon which Philoponus' proof had been constructed.[82] Saadia establishes that the corporeal universe is in truth finite, and hence can, by the principle in question, contain no more than finite power. And he infers both the generation and the future destruction of the world on commonsense grounds: A finite power must eventually "cease" and "stop," and nothing can continue to exist after its power is exhausted. The inference reflects Philoponus' contention that what contains only finite power is subject to destruction.[83] An indispensable step is, however, omitted by Saadia. Philoponus had explained that since the corporeal universe contains the possibility of not existing, it cannot be maintained in existence eternally even by an infinite power outside itself; for, over an infinite time, the possibility of not existing must, like every possibility, inevitably be realized.[84] Unless the step is included, the corporeal universe is shown by the argument merely to be incapable of maintaining itself eternally in existence through its own power. The universe is not shown to be absolutely incapable of existing eternally. Saadia's proof is, then, a simplified restatement of Philoponus' proof from the finite power of finite bodies, with an indispensable step missing.[85]

(*vi*) Saadia's second proof for creation is entitled "from the joining of parts and the composition of segments." "Bodies," he contends, "consist of combined parts and composed members and thereby reveal . . . signs of generation and of the art of an artisan." As evidence, Saadia cites various types of composition discoverable in the corporeal universe, such as the combination of "earth, stones, and sand" to constitute the lower world, and the combination of the "several layers of spheres" and the "stars" to constitute the celestial region. He concludes: "After establishing joining, conjunction, and composition, which are generated things (*ḥawādith*) in the body of the heavens and the rest [of the universe], I was convinced . . . that the heavens and everything they contain are generated."[86]

What is significant for us in this proof is its location. Saadia presents it immediately after his proof from the finite power of finite bodies. Philoponus, as will be recalled, had supported his own proof from the finite power of finite bodies with a number of auxiliary arguments, intended primarily to show that the corporeal universe can indeed contain only finite power. The first, and also the third, of the auxiliary arguments reasoned from the composite nature of the corporeal

[82]Above, p. 89.

[83]Above, p. 90.

[84]Ibid.

[85]H. Wolfson, "Kalam Arguments for Creation," *Saadia Anniversary Volume* (New York, 1943), p. 203, offers an interpretation of Saadia, the effect of which is to discover in Saadia the technicalities of Philoponus' proof.

[86]Saadia, *K. al-Amānāt,* I, 1, p. 34. Wolfson, "Kalam Arguments for Creation," p. 205, n. 34, notes that the term *parts* at the beginning of the argument might mean *atoms*.

universe.[87] Thus Philoponus' proof from the finite power of finite bodies was followed immediately by an argument from the fact that all bodies in the universe are composite; and Saadia's proof from the finite power of finite bodies is now also followed immediately by an argument from the fact that all bodies in the universe are composite.

In content, Saadia's argument from composition differs from Philoponus'. The evidence cited by Saadia for the composite nature of the corporeal universe is not the evidence cited by Philoponus,[88] and there is no hint in Saadia of a connection between composition and finiteness of power. Yet in whatever light Saadia's proof might be regarded,[89] Saadia undoubtedly has omitted something from his source or left part of his thinking unexpressed, for he fails to explain how the premises lead to the conclusion that the world was created. He merely takes for granted that if things are composite they must be generated. The only intimation of grounds for a conclusion is given by him in the statement: "Conjunction and composition . . . are generated things in the body of the heavens. . . ." The thought behind the statement plays an important role in another proof for creation offered by Saadia, the proof to be examined next. The appearance of the same thought here is perhaps an additional instance of the intrusion of echoes of one proof into another; alternatively, it is a conscious attempt to assimilate the present proof to the other in order to justify the conclusion.[90] But however that thought came to be introduced, it cannot represent the original point of the argument, for then there would be no reason for distinguishing between the present proof from "composition" and the succeeding one "from accidents."

(*vii*) Saadia's third proof for creation is, as just mentioned, entitled "from accidents." It begins: "Bodies cannot avoid accidents which occur in each of them, either from itself or from outside itself."[91] Saadia examines animal life and the earth as a whole, discovers that they are indeed "not free" of change, which

[87]Above, pp. 92–93.

[88]Above, p. 92. Saadia's failure to mention the composition of matter and form might be explained by the fact that the proof came to him after it had been reformulated by Kalam thinkers who rejected the Aristotelian matter and form.

[89]Several arguments for creation from composition were put forward in the Middle Ages; cf. below, pp. 146 ff.

[90]That is suggested by M. Ventura, *La Philosophie de Saadia Gaon* (Paris, 1934), p. 99.

[91]The phrase "either from itself or from outside itself" can be deciphered with the aid of formulations of the proof in other Arabic writers. As a premise, it was common procedure to demonstrate the Kalam doctrine of accidents, the doctrine that none of the characteristics of a physical object flow from, or are dependent upon, an inner essence, but rather all are added to the identical inert atoms that serve as the material base of each object. To establish the existence of accidents it was shown that the characteristics of things cannot be due to the atom or body itself, and therefore must be due to a factor distinct from the atom or body, that is, to an "accident"; cf. below, p. 181. By stating that the accidents he is speaking of do belong to a body "either from itself or from outside itself," Saadia explicitly rejects the proposition that no characteristic of a thing can flow "from itself" and he implicitly rejects the Kalam doctrine of accidents. He would, we may suppose, have been more comfortable with the Aristotelian theory of *form*. But he apparently did not have the philosophic knowledge to allow him its use; cf. Ventura, *La Philosophie de Saadia*, pp. 102–103.

is a "generated thing," and concludes that "what cannot avoid what is generated is known to be of the same character." He hereupon extends his investigation to the heavens, discovers that they too are not free of motion and other accidents, and concludes again: "When I discovered that generated things embrace these bodies, and the latter do not precede the former [in time], I was convinced that whatever does not precede what is generated is of the same character by virtue of its falling under the same class (*ḥaddihi*)."[92]

Saadia's proof addresses accidents and bodies, while a parallel version of the proof known from Islamic writers addresses not accidents and bodies but accidents and atoms.[93] The dichotomy of accident and body, or of accident and atom, is a Kalam analogue of the familiar Aristotelian dichotomy of form and matter: The Aristotelian concept of a *form,* which carries the essential nature of a given object, was rejected by the Kalam, and whatever trace it left was absorbed into the Kalam concept of *accident.*[94] The specific phenomenon that Saadia's proof considers is the continuous and unavoidable presence of generated accidents in each body in the universe; the phenomenon that the parallel version considers is the continuous presence of generated accidents in each atom. That phenomenon, translated back into the Aristotelian framework, would be the continuous presence of generated forms in matter. As will be recalled, the second of the auxiliary arguments with which Philoponus supported his proof from the finite power of finite bodies focused precisely on the succession of forms across matter.[95] Saadia's proof thus corresponds, in a most general way, to Philoponus' second auxiliary argument: Philoponus dealt with the continuous succession of forms over matter, while Saadia deals with the continuous presence of changing generated accidents in bodies. The similarity goes no further. Saadia does not repeat any details from Philoponus; and his argument rests on a problematical rule, not reflecting anything in Philoponus, namely, the rule that "what cannot avoid what is generated is known to be of the same character."[96]

The situation we have discovered is this: Philoponus had offered a proof from the finite power of finite bodies followed directly by a supporting argument from

[92]Saadia, *K. al-Amānāt,* I, 1, p. 35. The proof also appears in Saadia's *Commentary to Sefer Yeṣira,* pp. 33 ff.; French translation, pp. 53 ff. It is alluded to by Saadia in his refutation of Ḥiwi; cf. I. Davidson, *Saadia's Polemic Against Ḥiwi Al-Balkhi* (New York, 1915), p. 75, §65.

The last nine words in the proof are understood by Ventura, Rosenblatt, and Altmann as meaning: "by virtue of its [i.e., that which is generated] entering its definition [i.e., the definition of that which does not precede.]." This would seem to require the following, rather forced, interpretation of the text: Since the concept *accident* is used by the Kalam in defining *body,* and since all accidents are generated, it follows that "what is generated" enters the definition of body, and body is also generated. Definitions of body in terms of accident are in fact to be found in Ash'ari, *Maqālāt al-Islamīyīn,* ed. H. Ritter (Istanbul, 1929–1930), pp. 301 ff., and the first of the definitions given there would be particularly appropriate for our context.

[93]Cf. below, pp. 134 ff.

[94]Cf. Maimonides, *Guide*, I, 73(8); R. Frank, *The Metaphysics of Created Being According to Abū l-Hudhayl* (Istanbul, 1966), p. 42.

[95]Above, p. 92.

[96]Regarding that rule, see below, p. 143.

the composition of all bodies in the universe and thereupon by a supporting argument from the unceasing succession of impermanent forms over matter. Saadia, for his part, offers a proof from the finite power of finite bodies followed directly by a proof from the composition of all bodies in the universe and thereupon by a proof from the continuous presence of impermanent accidents in bodies. Saadia's proof from the finite power of finite bodies is unmistakably derived from Philoponus. The similarity of Saadia's second and third proofs to the corresponding auxiliary arguments in Philoponus is limited. Saadia's proofs consider the general phenomena considered by Philoponus' proofs, and likewise appear directly after the proof from finite power, but they differ from Philoponus' proofs in all particulars.

In each of the three proofs, Saadia's reasoning is incomplete. His restatement of the first proof, the proof from the finite power of finite bodies, omits a step without which the conclusion cannot cogently be drawn.[97] As he formulates his second proof, the inference of creation from composition is not explained. And his third proof, the proof "from accidents," rests on the unproved rule affirming that "what cannot avoid what is generated is known to be of the same character." The proof from accidents must already have been current in Arabic by Saadia's time, since two of his contemporaries, Ash'ari[98] and Alfarabi,[99] expected their readers to be acquainted with it. Ash'ari significantly attributes the proof to the "philosophers" and "those who follow them."[100]

I would suggest that Philoponus' proof from finite power may have circulated in Arabic together with at least two[101] auxiliary arguments as a group; and that group of proofs underlies Saadia's first, second, and third proofs of creation, the items I have designated as (v), (vi), and (vii). If the suggestion is correct, far-reaching changes must have occurred somewhere along the line of transmission: The connection between the auxiliary arguments and the basic proof was forgotten; essential parts of the reasoning were omitted; and in the case of the third proof, the original concepts were translated into their Kalam analogues and provided with a completely new line of argumentation. If the suggestion is not correct, pure chance is the only explanation for what has been shown, namely, that the two proofs following the proof from finite power in Saadia exhibit a correspondence to the two auxiliary arguments following the proof from finite power in Philoponus.

[97] Above, p. 102.

[98] See n. 100.

[99] See below, pp. 134–135.

[100] Ash'ari, *Risāla ilā ahl al-Thaghr, Publications of the Theological Faculty, Istanbul*, II/8 (1928), 89; cf. R. Frank, "Al-Ash'arī's Conception of the Nature and Role of Speculative Reasoning in Theology," *Proceedings of the VIth Congress of Arabic and Islamic Studies* (Stockholm, 1972), p. 139. Ash'ari, more specifically, attributes the argument to "the philosophers, and those who follow them from among the *qadarīya*, the innovators [or: heretics], and the deviators from the prophet(s)." He probably has the Mu'tazilites in mind.

[101] That is to say, the first two or first three auxiliary arguments.

Resumé[102]

Philoponus' three proofs for creation from the impossibility of an infinite number were known to and used by Saadia [items (i)–(iv)].[103] Philoponus had presented the three proofs as a set, but whether Saadia received them as a set is not certain. Philoponus' proof from the finite power of finite bodies was also known to and used by Saadia [item (v)].[104] After his proof from finite power, Saadia offered two other proofs [items (vi) and (vii)] which in their general subject matter and order of presentation correspond to the first two auxiliary arguments whereby Philoponus had supported his own proof from finite power.[105] The resemblance, I have submitted, is not fortuitous.

An interrelation is likely as well between the transmission of Philoponus' proofs from the impossibility of an infinite number and the transmission of his proof from finite power; for in Saadia, two proofs from the former group betray echoes of the latter.[106] Quite possibly, therefore, a list of six—or more—proofs for creation, all of them concerned with finiteness of one sort or another, may have circulated and been available to Saadia. The list would have consisted of Philoponus' three proofs from the impossibility of an infinite number, his proof from finite power, and at least two auxiliary arguments supporting the proof from finite power.

3. Kindi and Philoponus

Several generations before Saadia, Kindi formulated four arguments that also exhibit unmistakable traces of Philoponus. For the purpose of exposition, I shall disregard the order in which the arguments appear in Kindi. The following table of corresponding items may again be helpful.

	Kindi		Philoponus
(i)	Argument for the finiteness of past time	(a.i)	The infinite is not traversable
(ii)	Argument for the finiteness of bodies	(a.ii)	The infinite cannot be added to
(iii)	Another argument for the finiteness of past time	(a.ii)	The infinite cannot be added to
(iv)	Argument from composition	(b.ii, b.iv)	Auxiliary arguments in support of proof from finite power

[102]See table, above, p. 105.
[103]Above, pp. 95–99.
[104]Above, p. 101.
[105]Above, pp. 102–104.
[106]Above, pp. 99, 100.

(*i*) The first item in Kindi to be examined is a proof of the finiteness of time, which reads: "[If past time were infinite], then before every segment of time there would be another segment, *ad infinitum*. But in that case no given time could ever be reached. For [the duration] from the infinite past to the given time would be equal to the duration ascending back from the given time . . . to infinity. . . . If something is infinite, its interval cannot be traversed."[107] After proving that past time cannot be infinite, Kindi explains that time cannot become "actually infinite" in the future either; for no matter what "definite time" might be added to the already accumulated finite past time, the total must remain finite.[108]

Kindi is plainly offering another restatement of Philoponus' first proof for creation from the impossibility of an infinite number. Like Philoponus, he lays down the rule that an infinite cannot be traversed; he draws an analogy between going back up through the infinite past and coming down from the infinite past to the present; and he concludes that the present could never have been reached if an infinite past were to precede it.[109] Kindi departs from Philoponus in transferring the discussion from the realm of past transformations—the realm considered by Philoponus[110]—to the realm of past time, something that Saadia, as will be recalled, also did in his restatement of Philoponus' proof.[111] Kindi's departure from Philoponus is not, however, identical with Saadia's, since Kindi, unlike Saadia, does not treat past time as a continuum but instead as a series of intervals. The conception of time as a series of intervals is quite atypical and would appear to serve a single purpose: It allows Kindi to argue specifically against an infinite past series, as Philoponus had done. Another similarity between Kindi's version and Saadia's, in addition to transferring the discussion to the realm of time, is that both versions supplement Philoponus' proof for the finiteness of past time with a proof, or at least a statement,[112] of the finiteness of future time. The supplement indicates that Kindi and Saadia borrowed from a common line of transmission in which the finiteness of future time—a theme appearing in Philoponus' arguments for creation from finite power—had attached itself to Philoponus' proof for the finiteness of past time. Also to be observed are clear terminological similarities between Kindi and Saadia. Both use the same distinctive terms for *ascending* in time (K: *mutaṣā'idan*;[113] S: *ṣu'ūd*[114]) and *traversing* time (K: *lā tuqṭa*; S: *lam yaqṭa'hu*[115]).

[107]Kindi, *Rasā'il*, ed. M. Abu Rida (Cairo, 1950), I, p. 121. Parallel versions appear, ibid., pp. 197; 205–206. English translation, with pagination of the Arabic indicated: *Al-Kindi's Metaphysics*, trans. A. Ivry (Albany, 1974).

[108]Kindi, *Rasā'il*, I, p. 122. See below, n. 121; Aristotle, *Physics* III, 6, 206a, 25–29.

[109]Above, pp. 87–88.

[110]Above, p. 87.

[111]Above, pp. 95–96.

[112]Above, p. 96.

[113]I have given the reading of a parallel passage, *Rasā'il*, p. 197. The editor of Kindi's *Rasā'il* did not understand the point of the argument and made an incorrect emendation in the present passage, p. 121, l. 10, and note.

[114]Saadia, *K. al-Amānāt*, I, 1, p. 36, and above, p. 96.

[115]Ibid.

Kindi, then, knew and employed Philoponus' first proof for creation from the impossibility of an infinite number and very likely received it through the same line of transmission from which Saadia was subsequently to borrow.[116]

(*ii*) A second argument in Kindi begins by establishing the finiteness of the body of the universe, whereupon it proves the finiteness of time as a corollary. In establishing the finiteness of the body of the universe, Kindi writes: "Supposing that an infinite body does exist, a finite [portion of it] may be assumed to be removed . . . [and subsequently] restored." Now, Kindi continues, the remainder left after a finite portion was removed could be neither finite nor infinite. The remainder could not be finite; for when the subtracted portion is restored, the total, as the sum of two finite magnitudes, would also be finite, whereas the initial supposition was that the total is infinite. The remainder left after a finite portion is removed could just as surely not be infinite. The remainder could not be infinite and yet, on the one hand, fail to be increased when the subtracted portion is restored; for every magnitude undoubtedly becomes greater when another magnitude is added to it. Nor could the remainder be infinite and yet, on the other hand, be increased through the restoration of the finite portion; for "there cannot exist two infinite magnitudes of the same type[117] one of which is greater than the other." Since the remainder left after removing a finite portion could be neither finite nor infinite, an infinite body, Kindi concludes, cannot possibly exist.[118]

The argumentation turns on the proposition that "there cannot exist two infinite magnitudes of the same type one of which is greater than the other." That proposition is equivalent to the principle lying at the heart of Philoponus' second proof of creation from the impossibility of an infinite number, the proof in which Philoponus had reasoned that the number of past motions leading up to the present cannot be infinite, since the number of such motions is subject to increase whereas an infinite cannot conceivably be increased;[119] to affirm, as Kindi does, that one infinite cannot be greater than another is equivalent to affirming that the infinite cannot be increased. Kindi's procedure of mentally removing and restoring a portion of the supposed infinite is designed to bring the principle in question to bear upon an infinite body. Philoponus needed no such device because he was considering the number of motions that have taken place in the universe, and the number of motions is, manifestly, ever growing. Kindi, though, is considering a body. He cannot, for his part, flatly assert that every body is subject to increase,

[116] Saadia does not draw his proof from Kindi, for he has details of Philoponus' proof which Kindi does not have. Neither Kindi nor Saadia drew their proofs from the version published by Pines, "An Arabic Summary of a Lost Work of Philoponus," pp. 330–336 (above, n. 52). For they include key points—the impossibility of traversing an infinite; and the comparison between going back up through the infinite and coming down to the present—which that version omits.

[117] That is to say, two infinite lines, two infinite surfaces, or two infinite bodies.

[118] Kindi, *Rasā'il*, I, pp. 115–116, taken together with pp. 189–191; also cf. pp. 195, 202–203.

[119] Above, p. 88.

and thereupon argue that an infinite body would be subject to increase, which is impossible; for the obvious response would be that infinite bodies are not after all subject to increase inasmuch as nothing exists outside them which might be added to them. The procedure of mentally removing and subsequently restoring a portion serves to reveal that the supposition of an infinite body does in truth involve an infinite's being increased: The remainder left after subtracting a finite portion could not be finite—since the total would then be finite—but would have to be infinite; and when the subtracted portion is restored, an infinite body would have been increased. Once the supposition of an infinite body is known to involve an infinite's being increased, Kindi can reason, as Philoponus did in connection with past events, that an infinite cannot conceivably be increased and consequently an infinite body is impossible.

Kindi's proof of the impossibility of an infinite body is plainly an adaptation of the argument whereby Philoponus had proved the impossibility of an infinite number of past motions.

In the context where the argument appears, Kindi is not interested in the finiteness of bodies for its own sake. He is constructing a proof of creation and, within the economy of the overall proof, the finiteness of bodies is intended to prepare the ground for its more important corollary, the finiteness of time. Kindi has two ways of deriving the corollary.

In several passages he contends that inasmuch as every body is finite, and inasmuch as time is an accident of body, time too must be finite.[120] His reasoning is highly questionable, the sense in which a body is finite being entirely different from that in which time would be finite; time extends in a different dimension from the three dimensions in which bodies extend. It is difficult, therefore, to see how the finiteness of the three dimensions of body can imply the finiteness of the temporal dimension.

(*iii*) Side by side with the foregoing contention, Kindi advances the more apposite contention that the "method" establishing the finiteness of bodies may be employed comprehensively to "show that nothing quantitative can be actually infinite";[121] and inasmuch as "time is quantitative, . . . an actual infinite time is not possible."[122] In other words, the supposed infinite past time would be subject to increase, whereas an infinite cannot conceivably be increased; and hence past time cannot be infinite. Here we have Philoponus' second proof for creation from the impossibility of an infinite number transferred—as in Saadia[123]—from the realm of past motions to the realm of past time.

Both Philoponus' first and second proofs of creation fom the impossibility of an infinite number thus reappear in Kindi, the discussion in both instances being

[120]Kindi, *Rasā'il*, I, p. 116; cf. ibid., pp. 196, 203.

[121]The potentially infinite, in contrast to the actually infinite, consists, as Kindi explains, in the possibility of continually adding to a given quantity. Cf. *Rasā'il*, I, p. 116; Aristotle, *Physics* III, 6.

[122]Kindi, *Rasā'il*, I, p. 116.

[123]Above, pp. 98–99.

transferred to the realm of past time: Philoponus had contended that transformations presently taking place could never have been reached if, prior thereto, an infinite series of transformations had to be traversed; and again that an infinite number of past motions is impossible since the number of past motions is constantly increased whereas the infinite cannot be increased. Kindi contends [item (i)] that the present segment of time could not have been reached if, prior thereto, an infinite series of segments had to be traversed;[124] and again [item (iii)] that an infinite past time is not possible, since past time—like other quantities—is subject to increase whereas the infinite cannot conceivably be increased.

One might have expected Kindi to regard his proofs of the finiteness of past time as self-contained proofs for the creation of the world. In Philoponus, a proof of the impossibility of an infinite number of past transformations or past motions sufficed as a proof of the creation of the world. And in Saadia a proof of the impossibility of infinite past time also sufficed as a proof of creation, since it went without saying for him that if past time is finite, the world cannot have existed forever. Yet Kindi is not satisfied that his proof of the finiteness of past time is tantamount, by itself, to a complete proof of creation. He feels called upon to show explicitly that the finiteness of past time does indeed imply a beginning of the world, or, to use his words, that it implies the finiteness of the body of the universe in respect to "existence" (*innīya*).[125] Kindi's procedure for showing that the finiteness of past time does imply creation consists in demonstrating the coexistence of the physical universe and time. Again, he has various ways of accomplishing his task.

In a certain passage he reasons that if, on the one hand, the body of the universe is not eternal, its very generation was a motion, and therefore the body of the universe and its motion are coexistent; and if, on the other hand, the universe is eternal, it must always have been in motion, for were it ever completely at rest, it could never have started to move.[126] Whether eternal or not, the body of the universe is, then, coexistent with motion.[127] Time is likewise coexistent with motion, since by definition time is the "number of motion."[128] "The body [of the universe], motion, and time" are, consequently, coexistent with each other and "do not ever precede one another."[129] Kindi's previous arguments to the effect that "all time is finite" and his present argument to the effect that "body does not precede time" lead to the desired conclusion that "the body of the universe cannot be infinite in its existence (*innīya*),"[130] but must have a beginning.

[124]Above, p. 107.

[125]Below, n. 130.

[126]The contention that if the universe were eternally at rest it could never begin to move is Aristotelian; cf. *Physics* VIII, 1, 251a, 20 ff. The contention is to be found in the Kalam; cf., e.g., Fakhr al Dīn al Rāzī, *K. al-Arba'īn* (Hyderabad, 1934), p. 16.

[127]Kindi, *Rasā'il,* I, pp. 118–119.

[128]Ibid., p. 117; cf. Aristotle, *Physics* V, 11, 220a, 24–25.

[129]Kindi, *Rasā'il,* I, p. 119.

[130]Ibid., p. 120; cf. pp. 197–198, 204–206.

The reasoning has an Aristotelian cast,[131] and is straightforward enough. Elsewhere, however, Kindi establishes the coexistence of body, motion, and time in a convoluted and much more problematic fashion.[132] Embodied within the convolutions, another of Philoponus' proofs can, I think, be detected.

(*iv*) Kindi writes: "Composition and combination are change, for this [last][133] is a joining and ordering of things. Body is a long-wide-deep, or three dimensional substance, and hence is composed of substance, its genus, and long-wide-deep, its specific difference. It [also] is what is composed of matter and form. Now composition is the change from the state that is noncomposition. Composition is motion. . . . Therefore the body [of the universe] and motion do not precede one another. . . . And accordingly the body [of the universe] does not precede duration numbered by motion [i.e., time]. Therefore the body [of the universe], motion, and time do not precede one another in existence (*innīya*), but are simultaneous in existence. Since time was [already proved to be] finite in actuality, the existence of the body [of the universe] also is necessarily finite in actuality, inasmuch as composition and combination are a certain change. If, on the contrary, composition and combination were not change, the conclusion would not be necessary."[134]

Apparently Kindi is reasoning that the universe is a body; every body is composite, since it consists of matter together with form, or substance together with tridimensionality; composition is a change from the state of noncomposition; change is a type of motion; hence the body of the universe cannot have existed before motion but is coexistent therewith; motion and time are likewise coexistent; hence the universe cannot have existed before time; but time is finite; hence the body of the universe can have existed for only a finite duration.

If such is indeed Kindi's reasoning, it contains a weak link. Kindi lays down that bodies and the universe as a whole are composite (*murakkab*) and infers that the body of the universe has undergone change and motion, since composition (*tarkīb*) is a change. The inference would appear to be accomplished through a blatant equivocation. In Arabic as in English, *composition* can mean both a state and a process; and Kindi appears to be gliding from composition in the universe, in the sense that the universe exhibits the *state* of being composite, to composition in the universe, in the sense that the universe has undergone the *process* of becoming composite. Only thus is he able to infer that the universe has undergone change. To rescue Kindi from the equivocation, we may perhaps understand him to have assumed a hidden premise, the proposition that everything in a state of being composite must have undergone a process rendering it so. The hidden premise does not, however, greatly improve the argument. For the premise is far

[131] Cf. Aristotle, *Physics* IV, 12.

[132] *Rasā'il,* pp. 196–197, offers yet another argument for the coexistence of body and time; the argument is less well developed and is difficult to follow.

[133] I have taken "change" as the antecedent of "this."

[134] Kindi, *Rasā'il,* I, p. 120; cf. ibid., p. 204.

from self-evident and, if employed by Kindi, certainly should have been made explicit and defended.[135]

Moreover, quite apart from the equivocation or hidden premise, Kindi's reasoning is curiously circular. He adduces the fact that the universe has undergone a process of composition, as a single link in an elaborate chain of argumentation leading to the creation of the world. But once the body of the universe, being composed of "matter and form," is acknowledged to have undergone a process of composition—from, as Kindi puts it, "the state . . . [of] noncomposition"[136]—the body of the universe is immediately acknowledged to have come into existence. Kindi is, in effect, contending that given the premise that the body of the universe came into existence and several other premises, the coexistence of the universe and time ensues; and given the coexistence of the universe and time, his earlier proof of the finiteness of past time will imply that the body of the universe—came into existence. The premise that the body of the universe came into existence allows him to conclude, after a number of intermediate steps, that the body of the universe did in truth come into existence.

The most plausible explanation of what has occurred is this: Kindi must have known of an argument proceeding from composition to creation but did not understand or remember exactly how it ran.[137] As a result, he incorporates the argument from composition into the proof of creation from the finiteness of time, employing it merely as a device to help show that the finiteness of past time does imply creation.

The evidence for composition adduced by Kindi indicates the probable source of his argument for composition. Body, Kindi finds, is a "three dimensional substance, hence [it] is composed of substance, its genus, and long-wide-deep, its specific difference"; furthermore, body is "composed of matter and form."[138] Philoponus, as will be recalled, had supported his proof for creation from the finite power of the universe with several auxiliary arguments, the first and third of which took their departure from the presence of *composition* in the universe. The evidence for composition adduced by Philoponus in the first of the auxiliary arguments was the circumstance that "the heavens are [composed] of matter and form."[139] The evidence adduced by him in the third auxiliary argument was, again, that the heavens are "composed of substratum . . . and form." And Philoponus added thereto the observation that even should the distinction of matter

[135]The premise might, conceivably, be alluded to in the first and last sentences of the passage. See also below, p. 148.

[136]Above, n. 134.

[137]Kindi, it should be mentioned, is reported elsewhere to have deduced noneternity directly from composition. According to the report of Ibn 'Adī, he refuted the Christian trinity by contending: "Everything composite is caused, and everything caused is noneternal." Cf. the text published in *Revue de L'Orient Chrétien,* XXII (1920), 4, line 15; French translation in A. Périer, *Petits Traités Apologétiques de Yaḥyā Ben 'Adī* (Paris, 1920), p. 119.

[138]Above, n. 134.

[139]Above, p. 92.

and form in the heavens be disputed, the presence in the heavens of "extension in three dimensions"—which is a mode of composition—must still be acknowledged.[140] The aspects of composition adduced by Kindi turn out to be precisely those that had been adduced by Philoponus in his auxiliary arguments from composition. It seems probable, therefore, that Kindi knew one or both of Philoponus' arguments for creation from composition in the universe, but, at least at the time he was writing, did not recall them exactly. He adduces and utilizes the evidence of composition in the universe not in order to formulate an independent proof for creation nor, like Philoponus, to formulate an auxiliary argument in support of the proof of creation from the finite power of the universe. Instead, he utilizes the evidence of composition for the purpose of showing that the finiteness of time does imply the creation of the world. It is worth mentioning also that although Kindi employs his argument from composition differently, a certain similarity to the overall configuration in Philoponus can be discerned: Kindi too employs the argument from composition in support of another proof for creation, rather than independently.

To recapitulate: Kindi (i) establishes the finiteness of past time through a restatement of Philoponus' first proof of creation from the impossibility of an infinite number. The argument is that an infinite cannot conceivably be traversed, and consequently the present moment could not, on the assumption of eternity, ever have been reached. Kindi (ii) establishes the finiteness of all bodies through an adaptation of Philoponus' second proof of creation from the impossibility of an infinite number; the argument here rests on the principle that an infinite cannot conceivably be increased. Once Kindi proves the finiteness of all bodies, he derives therefrom the corollary that past time must be finite. His reasoning is that since time is an accident of body, and the body of the universe is finite, time too must be finite. And, he adds more appositely, (iii) the principle that an infinite cannot be increased may be applied to time as well as body, thereby establishing that time too is finite. Kindi has thus recast Philoponus' first and second proofs of creation from the impossibility of an infinite number as proofs for the finiteness of past time. For Kindi, a proof of the finiteness of time does not yet, by itself, constitute a proof of creation, and he feels called upon to show that the finiteness of time does imply the finiteness of the "existence" of the body of the universe. One device he employs to that end is (iv) an argument from composition. Examination of his argument from composition discloses, however, that it cannot originally have been intended for the end to which Kindi employs it; and, further, that it cites the very evidence for composition which had been cited by Philoponus in his auxiliary arguments from composition. Philoponus is therefore again indicated as Kindi's likely ultimate source.[141]

[140] Ibid.

[141] A connection between Kindi and Philoponus was also observed by R. Walzer, *Greek into Arabic* (Oxford, 1962), pp. 190–196.

Earlier we saw that Saadia offers adaptations of at least four of Philoponus' proofs for creation, including Philoponus' first and second proofs from the impossibility of an infinite number.[142] Those proofs have now been found in Kindi as well, transferred, as in Saadia, to the realm of time. In addition, Saadia, like Kindi, offers an argument for creation from composition in the universe, and both writers present their arguments from composition in association with other proofs for creation deriving from Philoponus.[143] The reasoning of Saadia's argument from composition differs from the reasoning of Kindi's; but in both instances the reasoning exhibits such obvious gaps that the arguments must be incomplete restatements of a once known, half-forgotten earlier proof.[144] Significantly, Saadia and Kindi use the same three terms—composition (*tarkīb*), combination (K: *i'tilāf;* S: *ajzā' mu'allafa*), and joining (*jam'*)—in formulating their arguments.[145] With regard to Saadia, I suggested that a list or collection of proofs for creation deriving from Philoponus circulated and was utilized by him. I would here suggest that Kindi knew a collection of proofs for creation deriving from Philoponus which resembled that known to and utilized by Saadia. The collection available to Kindi included at least Philoponus' first and second proofs from the impossibility of an infinite number and one or both of his auxiliary arguments from composition. Either the form in which the arguments from composition reached Kindi did not permit him to grasp the original point, or else Kindi forgot the point when he sat down to write. He was consequently left—like Saadia—to employ the argument as best he could.

Two small details in Kindi's discussion of creation may cast added light on his source. The first detail is this: Each time Kindi takes up the issue of creation, he opens with the proof of the finiteness of bodies [item (ii)], whereupon he proceeds to a proof of the finiteness of past time.[146] The proof of the finiteness of bodies contributes only superficially to the proof of the finiteness of past time. The finiteness of bodies prepares the ground for the highly questionable inference that since time is an accident of body, time too must be finite.[147] And the proof of the finiteness of bodies allows Kindi to observe that the same argumentation—an infinite cannot be increased, hence what is subject to increase cannot be infinite—will apply to everything quantitative, so that time too can be proved to be finite.[148] A much simpler procedure would plainly have been to apply the argumentation immediately to time without reference to the finiteness of bodies. But the finiteness of bodies did serve as an essential premise in a different proof of creation, Philoponus' proof from the finite power of finite bodies.[149] In Philoponus' own formulation of that proof the finiteness of the body of the universe

[142]Above, pp. 95–102.
[143]Cf. above, p. 102.
[144]Cf. above, p. 103.
[145]Above, nn. 86 and 134.
[146]Kindi, *Rasā'il*, I, pp. 115–116; 195–197; 203–204.
[147]Above, p. 109.
[148]Ibid.
[149]Above, p. 90.

was presupposed. Aristotle had already demonstrated the finiteness of the body of the universe at considerable length,[150] and Philoponus did not trouble to take the subject up again. Still, completeness would require that the proof should include grounds for the finiteness of the body of the universe; Saadia's version of the proof did, for example, include such grounds.[151] The argument for the finiteness of bodies formulated by Kindi could serve the purpose well. It is an adaptation of one of Philoponus' other proofs for creation[152] and it could fittingly be adduced to support a key premise in Philoponus' proof for creation from the finite power of finite bodies. Kindi's sense that establishing the finiteness of bodies is pertinent to the issue of creation might then echo a version of the proof of creation from finite power in which the finiteness of bodies was explicitly demonstrated.

The other detail that deserves attention is a sentence appearing at the end of Kindi's argument for the finiteness of bodies, in one of the passages in which he advances the argument. Apropos of nothing either preceding or following, Kindi asserts: "Actuality proceeds from power *(al-fi'l khārij min al-qūwa)* inasmuch as the latter is the cause of the former; actuality [therefore] is finite by virtue of the finiteness of power."[153] In its context, the statement is completely incomprehensible. It could, however, fit exactly into a proof of creation from the finite power of finite bodies. After establishing that the body of the universe is finite and finite bodies contain only finite power, such a proof might appropriately explain that "actuality," that is to say, actual existence, must in every instance proceed from a "power"; when actual existence proceeds from "finiteness of power" it must likewise be "finite"; and therefore the physical universe—containing, as it does, only finite power—can have existed for only a finite time.

We have, then, two revealing details: Kindi's feeling that the finiteness of bodies is pertinent to proving creation; and the statement at the end of his argument for the finiteness of bodies to the effect that the finiteness of "actuality" is a consequence of the "finiteness of power." Those two details could well be echoes of a lost version of Philoponus' proof for creation from the finiteness of the body of the universe. The conjectured version would establish the finiteness of the body of the universe by means of the principle that an infinite cannot be increased.[154] It would proceed to establish the principle that a finite body can contain only finite power. It would contend that the "actuality" of a finite body is necessarily finite because of the "finiteness of [the] power" of the body. And it would conclude that finite bodies, including the finite body of the universe, can have existed for only a finite time.

[150] Aristotle, *Physics* III, 5; *De Caelo* I, 5–7. Cf. below, pp. 254–259.
[151] Above, n. 80.
[152] Above, p. 108.
[153] Kindi, *Rasā'il,* I, p. 196.
[154] Cf. above, p. 88; below, pp. 125–126.

4. Summary

Two sets of proofs for creation were formulated by Philoponus. One set rested on the impossibility of an infinite number and argued in three different ways that the number of past transformations and motions cannot be infinite. The other set rested on the principle that a finite body can contain only finite power and reasoned that the finite body of the universe could not contain the power required for existence to have continued for an infinite past time. The latter set consisted of a basic proof supported by auxiliary arguments, among which were to be found two arguments from composition and an argument from the continual succession of forms across matter. Evidence from Arabic philosophers and bibliographers establishes that the works in which Philoponus developed his two sets of proofs were known to the medieval Arabs. Moreover, Arabic texts explicitly cite, in Philoponus' name, the first proof from the former set and the basic proof of the latter.

Kindi and Saadia, who do not mention Philoponus by name, unmistakably employ some of Philoponus' proofs and may have derived other proofs from him. Kindi offers versions of the first and second proofs for creation from the impossibility of an infinite number; and a remnant of the proof from the finite power of finite bodies together with the auxiliary arguments from composition may be reflected in his treatment of the problem of creation. Saadia offers versions of all three of Philoponus' proofs for creation from the impossibility of an infinite number as well as a simplified restatement of the proof for creation from the finite power of finite bodies. The auxiliary arguments from composition and the argument from the succession of forms across matter may also be reflected in Saadia. If my suggestion regarding the argument from the succession of forms across matter is correct, that argument was at some stage converted into a proof from the succession of accidents in bodies.

The next chapter will show that the three proofs of creation from the impossibility of an infinite number were widely employed in the Middle Ages. The proof from the presence of accidents in bodies also enjoyed considerable, although narrower, currency, for it was not taken seriously beyond the boundaries of the Kalam. And several arguments from the composition in the universe were put forward as well.

V

Kalam Proofs for Creation

1. Proofs from the impossibility of an infinite number

As was seen in the previous chapter, John Philoponus drew up a set of three proofs for creation from the impossibility of an infinite number. All three were utilized by Saadia and two were utilized by Kindi, the form they took in Saadia and Kindi indicating that they had circulated in Arabic together with other proofs of Philoponus. The three proofs for creation from the impossibility of an infinite number enjoyed considerable vogue in medieval Islamic and Jewish philosophy, and they underwent several developments.

Philoponus' first proof was the argument that the world must have a beginning since an infinite number of past transformations could not have been traversed.[1] It became, in Kindi and Saadia, the argument that the world must have a beginning since an infinite past time could not have been traversed.[2] In the early period it was also employed by Iskāfī (d. 854) and Naẓẓām (d. 845), two Mu'tazilite thinkers who were approximate contemporaries of Kindi.

Iskāfī is reported to have reasoned: "If there were no first [term] at which things begin, and which is not preceded by anything prior, it would be impossible for anything to come about."[3] The statement is not very precise, but we seem to have, once again, the familiar contention that an infinite series of events or an infinite continuous time—the text does not indicate which of the two Iskāfī might have meant—cannot be traversed, and therefore, the present moment could never have been reached if such an infinite had preceded. Naẓẓām is similarly reported to have reasoned: "The traversal of bodies [*sic*] that has passed must be either finite or infinite. . . . If it were infinite, it would have no first [term], and what has no first [term] cannot be exhausted. The fact that what has passed has been exhausted is a proof of its finiteness."[4] At first glance one might suppose that

[1] Above, pp. 87–88.
[2] Above, pp. 95–96, 107.
[3] Khayyāṭ, *K. al-Intiṣār,* ed. and trans. A. Nader (Beirut, 1957), §5.
[4] Ibid., §20.

Naẓẓām is speaking of bodies' traversing space and that he is undertaking a proof of the spatial finiteness of the universe. But since his argument is represented in our source, Khayyāṭ, as a method for establishing creation,[5] and since it is distinguished there from Naẓẓām's proof of the finiteness of the body of the universe,[6] it must be interpreted in terms of the proof already known to us from Kindi, Saadia, and now—with a bit of hesitation—from Iskāfī: An infinite time, Naẓẓām is contending, could not be traversed by the succession of "bodies" that have existed; therefore, on the assumption of eternity, the past could never be completed, and the present moment could never have been reached.[7]

We are told that a critic, Ibn al-Rawandī, accused Naẓẓām of self-contradiction, since by the side of the argument just cited he held that "what traverses and what is traversed are infinite, and thus by maintaining that the former exhausts the latter Naẓẓām has affirmed the exhaustion of what is infinite."[8] The exact meaning of the objection is clarified by other information concerning Naẓẓām and by recalling an objection that Saadia took up in connection with his fourth proof of creation.[9] Naẓẓām was notorious for having held the unorthodox position that bodies are infinitely divisible, at least in thought, a position implying that space too is infinitely divisible.[10] Accordingly, Ibn al-Rawandī maintains, Naẓẓām himself must assume the traversal of an infinite number of parts whenever a body moves from one place to another; why then should he deny the traversal of an infinite time? This objection to Naẓẓām, which encapsules Zeno's first paradox, is more explicit in a parallel critique of Naẓẓām's proof of the finiteness of the *body* of the universe.[11] No clear solution to the objection is provided by our source, Khayyāṭ, but from other reports Naẓẓām is known to have held that bodies succeed in moving from one place to another by "leaping" over some of the infinite parts of the distance traversed.[12] Saadia must have had knowledge of the exchange over Naẓẓām's proof, for he gives the same proof, raises the same objection, explicitly rejects the solution based on the theory of the "leap," and offers an alternative solution instead.[13]

[5]Ibid.

[6]Ibid., §19.

[7]By "the traversal of bodies," Naẓẓām might also conceivably mean: the total distance traveled by all bodies that have existed in the universe. Still another interpretation is given by H. Wolfson, *The Philosophy of the Kalam* (Cambridge, Mass., 1976), p. 416.

[8]Khayyāṭ, *K. al-Intiṣār,* §20.

[9]Above, p. 118.

[10]References in A. Nader, *Le Système Philosophique des Mu'tazila* (Beirut, 1956), p. 155; H. Wolfson, *The Philosophy of the Kalam,* pp. 495–496, 514.

[11]Khayyāṭ, *K. al-Intiṣār,* §19. For Zeno's paradoxes see Aristotle, *Physics* VI, 9.

[12]References in A. Nader, *Le Système Philosophique des Mu'tazila,* p. 182; H. Wolfson, *The Philosophy of the Kalam,* pp. 515–517.

[13]Saadia, *K. al-Amānāt wa-l-I'tiqādāt,* ed. S. Landauer (Leiden, 1880), I, 1, 36–37; English

The impossibility of eternity was established by Naẓẓām in another way as well, described by Khayyāṭ as one "of the best" the Kalam offered: "The stars either are equal to each other [in velocity] . . . or else some are swifter than others. If they are equal [in velocity], the distance traversed by some is less than what is traversed by all of them together, so that when the distance traversed by some is added to the distance traversed by the rest, the total will be greater than the part." Neither the total distance traveled by all the stars nor the distance traveled by a single star could, then, be infinite, for one infinite cannot conceivably be greater or less than another. "If [the second alternative should be correct and] some stars are swifter than others," the outcome is more obvious. For the distance traveled by the swifter is greater than the distance traveled by the slower, and "whatever is susceptible of *less* and *greater* is [as already seen] finite." That is to say, eternity would involve disparate infinites, whereas one infinite cannot be greater than another. Since the stars have traveled only a finite distance and hence can have performed only a finite number of revolutions, the heavens and the world as a whole must have existed for only a finite time.[14] By reasoning here from the principle that one infinite cannot be greater than another, Naẓẓām is giving a fairly faithful version of Philoponus' second proof of creation, the proof resting on that principle. The illustration employed by Naẓẓām, namely the movements of the stars, is also derived from Philoponus,[15] although Naẓẓām, unlike Philoponus, focuses on the distances that the stars travel rather than on the numbers of revolutions they perform. A new twist, moreover, has been added to the argument by Naẓẓām. Even if the stars happen to have identical velocities, he explains, the assumption of eternity still entails incommensurate infinites, since the distance traveled by each of the stars will be less than the total distance traveled by all.

Philoponus' first proof for creation from the impossibility of an infinite number, the argument that the present could never have been reached if infinite past time or an infinite series of past events had to be traversed, has now been traced in Iskāfī, Naẓẓām, Kindi, and Saadia. It appears with variations in Avicenna (who

translation, with pagination of Arabic indicated: *Book of Beliefs and Opinions*, trans. S. Rosenblatt (New Haven, 1948).

[14]Khayyāṭ, *K. al-Intiṣār,* §20.

[15]In Philoponus' main statement of the proofs from the impossibility of an infinite number, the movements of the stars were the phenomenon upon which the third proof focuses; cf. above, p. 89. In a more casual statement of the proofs, the movements of the spheres, rather than of the stars, were among the phenomena adduced by Philoponus in the course of his second proof; cf. *Commentary on Physics*, ed. H. Vitelli, *Commentaria in Aristotelem Graeca*, Vol. XIV/2 (Berlin, 1897), p. 429. Also cf. Saadia as quoted above, p. 99.

rejects it),[16] Ibn Hazm.[17] Nāṣir-i-Khosraw,[18] Juwaynī,[19] Ghazali,[20] Judah Hallevi,[21] Fakhr al-Dīn al-Rāzī,[22] Averroes (who rejects it),[23] Maimonides (who also rejects it),[24] Āmidī,[25] and Levi Gersonides.[26] It was known to the Christians, being accepted by Bonaventure[27] and rejected by Albertus Magnus[28] and Thomas Aquinas.[29] Spinoza records it,[30] and it filtered down to Kant.[31]

Philoponus' second proof was the contention that the world must have a beginning since the past is continually being increased and an infinite cannot conceivably be increased. The proof has now been discovered in Naẓẓām, Kindi, and Saadia. It appears as well in 'Abd al-Jabbār,[32] Avicenna (who rejects it)[33] Ibn Ḥazm,[34] Ghazali,[35] Baḥya,[36] Judah Hallevi,[37] Shahrastānī,[38] Ibn Ṭufayl (who is

[16]In chapter 4 of an unpublished treatise contained in British Museum, MS. Add. Or. 7473, and summarized by S. Pines, "An Arabic Summary of a Lost Work of John Philoponus," *Israel Oriental Studies,* II (1972), 348.

[17]Ibn Ḥazm, *K. al-Faṣl fī al-Milal* (Cairo, 1964), I, p. 16 (proof 2); Spanish translation: *Abenházam de Córdoba y su Historia Crítica de las Ideas Religiosas,* trans. M. Asín Palacios, Vol. II (Madrid, 1928), p. 102.

[18]Cf. S. Pines, *Beitraege zur Islamischen Atomenlehre* (Berlin, 1936), p. 37, n. 2.

[19]*K. al-Irshād* (Cairo, 1950), p. 26; *K. al-Shāmil* (Alexandria, 1969), p. 216; *Textes apologétiques de Ǧuwainī (Luma')* ed. M. Allard (Beirut, 1968), p. 126.

[20]*al-Iqtiṣād fī al-I'tiqād* (Ankara, 1962), p. 32; *al-Risāla al-Qudsīya,* published as *Al-Ghazālī's Tract on Dogmatic Theology,* ed. and trans. A. Tibawi (London, 1965), pp. 17, 35.

[21]*Kuzari,* V, 18(1).

[22]*K. al-Arba'īn* (Hyderabad, 1934), p. 15. This and Philoponus' other two objections to eternal motion are not given as complete proofs of creation by Rāzī. His complete argument is that if the universe were eternal it would have to be either eternally at rest or eternally in motion, but cannot in fact be either.

[23]*K. al-Kashf,* ed. M. Mueller (Munich, 1859), p. 36; *Tahāfut al-Tahāfut,* ed. M. Bouyges (Beirut, 1930), I, p. 20; English translation, with pagination of the Arabic indicated: *Averroes' Tahafut al-Tahafut,* trans. S. van den Bergh (London, 1954).

[24]*Guide to the Perplexed,* I, 74 (2).

[25]*Ghāya al-Marām* (Cairo, 1971), pp. 13–14.

[26]*Milḥamot ha-Shem* (Leipzig, 1866), VI, i, 11, pp. 344–345.

[27]*Commentary on II Sentences,* in *Opera Omnia,* Vol. II (Quaracchi, 1882), d. 1, p. 1, a. 1, q. 2; cf. E. Gilson, *The Philosophy of St. Bonaventure* (New York, 1938), p. 192.

[28]*Physics,* VIII, i, 12.

[29]*Summa contra Gentiles,* II, chap. 38 (3); *Summa Theologiae,* I, 46, art. 2, obj. 6.

[30]*Cogitata Metaphysica,* II, 10, 11.

[31]*Critique of Pure Reason,* A426/B454. For Mendelssohn's comments on the proof, cf. A. Altmann, "Moses Mendelssohn's Proofs for the Existence of God," *Mendelssohn Studien,* II (1975), 13–14.

[32]*K. al-Majmū' fī al-Muḥīṭ bi-l-Taklīf,* ed. J. Houben (Beirut, 1965), p. 61.

[33]Chapter 8 of the unpublished treatise cited above, n. 16.

[34]*K. al-Faṣl fī al-Milal,* I, pp. 16–17 (proof 3).

[35]*al-Iqtiṣād,* p. 33.

[36]*al-Hidāya* (*Ḥobot ha-Lebabot*), ed. A. Yahuda (Leiden, 1912), I, 5, p. 44.

[37]*Kuzari,* V, 18 (1).

[38]*K. Nihāya al-Iqdām,* ed. A. Guillaume (Oxford and London, 1934), p. 28.

noncommittal regarding its validity),[39] Fakhr al-Dīn al-Rāzī,[40] Maimonides (who rejects it),[41] Ṭūsī,[42] Levi Gersonides,[43] Ījī,[44] and Aaron ben Elijah.[45] It too was known to the Scholastics. Bonaventure accepted it,[46] Aquinas rejected it,[47] and its validity was hotly debated by subsequent Latin writers.[48]

The standard response of the medieval Aristotelians to the proof for creation from the impossibility of an infinite's being increased was that past time or past events are not enumerable objects; consequently, the passage of infinite time would not after all involve the paradox of an actual infinite number's being increased. That response will be examined a little more fully below. A more radical response was also made, however. The more radical response countenances additions to, and subtractions from actual infinite numbers, but maintains that the notions *greater, smaller,* and *equal* are not applicable to the infinite. In other words, an infinite number can indeed be added to or subtracted from; but the resultant new infinite number cannot properly be described as greater than, smaller than, or even equal to the old infinite number. The paradoxes connected with incommensurate infinite numbers thereby fall away: Infinite numbers are neither incommensurate nor commensurate. This response to the proof from the impossibility of increasing the infinite was made by Crescas[49] and certain Scholastic writers;[50] and a similar line of thought was later to be advanced by Bruno and Galileo.[51] As far as is known, only a single medieval Arabic writers, Thābit ibn Qurra, went so far as to countenance the incommensurability of infinites.

[39] *Ḥayy ben Yaqdhān,* ed. and trans. L. Gauthier (Beirut, 1936), Arabic text, p. 81, taken together with p. 76; French translation, p. 61, taken together with p. 58; English translation: *Hayy ben Yaqzān,* trans. L. Goodman (New York, 1972), with pagination of the Arabic indicated.

[40] *K. al-Arba'īn,* p. 15. Cf. above, n. 22.

[41] *Guide,* I, 74 (end).

[42] Glosses to Rāzī's *Muḥaṣṣal* (Cairo, 1905), p. 93.

[43] *Milḥamot ha-Shem,* VI, i, 11, pp. 341, 343, 346.

[44] *K. al-Mawāqif* (Cairo, 1907), VII, 224. Ibn Ṭufayl, Ṭūsī, and Ījī employ the method of "application" (cf. below, pp. 126–127) in formulating the argument.

[45] *'Eṣ Ḥayyim,* ed. F. Delitzsch (Leipzig, 1841), p. 28.

[46] *Commentary on II Sentences,* d. 1, p. 1, a. 1, q. 2; cf. Gilson, *The Philosophy of St. Bonaventure,* p. 190.

[47] *Summa contra Gentiles,* II, 38 (4); cf. *Summa Theologiae,* I, 7, art. 4.

[48] Cf. Maier, "Problem des aktuell Unendlichen," *Ausgehendes Mittelalter,* Vol. I (Rome, 1964), pp. 53–67, 76, 83.

[49] *Or ha-Shem,* III, i, 4. Crescas might have borrowed from the Scholastics; cf. S. Pines, *ha-Skolastika she-aḥare Thomas Aquinas u-mishnato shel Ḥasdai Crescas* (Jerusalem, 1966).

[50] Maier, "Problem des aktuell Unendlichen," as above, n. 48.

[51] G. Bruno, *De la causa,* V (beginning); Galileo, *Discorsi e dimostrazioni matematiche intorno a due nuove scienze,* in *Opere* (Florence, 1898), VIII, p. 78; English translation: *Two New Sciences,* trans. S. Drake (Madison, 1974), p. 40 (dealing with infinite *spatial* magnitude). Also cf. Spinoza, *Ethics,* I, xv, Schol.; H. Wolfson, *Philosophy of Spinoza,* Vol. I (Cambridge, Mass., 1948), pp. 288 ff.

Thābit maintains that one infinite can in truth be "a third or a quarter or a fifth of another infinite."[52]

Philoponus' third proof of creation from the impossibility of an infinite number was the contention that since the planets move at different speeds, eternity would involve the absurdity of one infinite's being the multiple of another. The argument has been shown to appear with minor changes in Saadia.[53] It also appears in Avicenna (who rejects it),[54] in Ibn Ḥazm,[55] Ghazali,[56] Judah Hallevi,[57] Shahrastānī,[58] Abū al-Barakāt,[59] Fakhr al-Dīn al-Rāzī,[60] Gersonides,[61] and a number of Scholastic writers.[62] Saturn, the planet that performs the fewest revolutions and hence gives rise to the greatest paradoxes, caught the fancy of the Muslim writers and is mentioned by most of them who offer the proof.

Because of the similarity between the third proof and the second—one number's being the exact multiple of another is simply a special case of its being larger than the other—the boundary between the two is generally blurred. Even Philoponus had not always kept the two distinct, doing so only in the main presentation of his proofs of creation.[63] With the exception of Saadia and Ghazali, all the instances of the third proof cited here conflate it with the second proof. The reasoning invariably is that since the revolutions of the planets are multiples of one another, eternity would involve the absurdity of one infinite's being larger than another.

The train of thought animating proofs of creation from the impossibility of an infinite number was such that arguments could easily proliferate. Two obvious roads were open: First, additional objects could be discovered which would run to infinity on the assumption of eternity, and secondly, additional grounds could be discovered for the impossibility of an infinite number.

The first road had already been taken by Philoponus. In his main presentation of the proofs from the impossibility of an infinite number, Philoponus had concerned himself with transformations and motions in the universe, explaining that

[52]Cf. S. Pines in *Actes du XIe Congrès International d'Histoire des Sciences* (Warsaw, 1968), III, p. 164.

[53]Above. p. 99.

[54]Chapter 8 of the unpublished treatise cited above, n. 16.

[55]*K. al-Faṣl fī al-Milal,* I, p. 16 (3).

[56]*al-Iqtiṣād,* p. 32; *al-Risāla al-Qudsīya,* pp. 17, 35; *Tahāfut al-Falāsifa,* ed. M. Bouyges (Beirut, 1927), I, §16; English translation in *Averroes' Tahafut al-Tahafut,* trans. S. van den Bergh (London, 1954), p. 9.

[57]*Kuzari,* V, 18 (1).

[58]*K. Nihāya al-Iqdām,* p. 29.

[59]*K. al-Mu'tabar* (Hyderabad, 1939), III, 42.

[60]*K. al-Arba'īn,* p. 15. Cf. above, n. 22.

[61]*Milḥamot ha-Shem,* VI, i, 11, p. 341.

[62]Maier, "Problem des aktuell Unendlichen," as above, n. 48.

[63]Cf. above, pp. 88–89.

on the assumption of eternity, the number of transformations and motions occurring in the universe would run to infinity. But when the proofs appear elsewhere in Philoponus, additional illustrations are given. Philoponus points out that on the assumption of eternity, the members of the human species and indeed the members of each human genealogical line would run to infinity. The number "of horses," "of dogs," "of the other animals and plants as well as the movements of each of the spheres" would be infinite. There would accordingly result a "two-fold," a "threefold," a "many-fold, if not infinitely multiplied . . . actual infinite," all of which is ruled out by the impossibility of an infinite's being traversed, added to, or multiplied.[64]

The Arabic writers discovered a variety of things that would run to infinite on the assumption of eternity. Kindi, as was seen, contended that the segments of past time would run to infinity.[65] Saadia contended that the past time continuum would extend infinitely.[66] Ibn Ḥazm, probably by pure coincidence, adduces some of the examples that Philoponus had offered in his secondary statements of the proofs. Over an eternity, Ibn Ḥazm writes, the number of men, the number of horses, and the total of the two would be infinite, leading once again to the absurdity of one infinite's being larger than another.[67] Gersonides observes that the number of lunar eclipses would be infinite, which would mean that the moon is in an eternal state of eclipse![68] Ghazali, followed by others, argues that an infinite number of immortal souls would accumulate, whereas even the strictest Aristotelian grants that an infinite number of things existing together is impossible.[69]

Ghazali goes a step further. Even if individual human immortality should be denied, he writes, souls or other objects can nonetheless be "supposed" to have come into existence at every moment of past time, to remain in existence, and to accumulate in an infinite number; yet an infinite number of objects is, for the various reasons that have been given, impossible.[70] No more is said by Ghazali, but his intent can be surmised. He undoubtedly was thinking of the Aristotelian definition, or characterization, of the "possible" as what is such that "when it is

[64] *De Aeternitate Mundi contra Proclum,* ed. H. Rabe (Leipzig, 1899), p. 11; *Commentary on Physics,* p. 429.

[65] Above, p. 107.

[66] Above, pp. 95–96; 98–99.

[67] *K. al-Faṣl fī al-Milal,* I, p. 16 (3).

[68] *Milḥamot ha-Shem,* VI, i, 11, pp. 342–343. This argument is descended from Philoponus' third objection to eternal motion (cf. above, p. 89); for Gersonides is contending that it is absurd to suppose that a fraction of an infinite is equal to the whole infinite.

[69] *Tahāfut al-Falāsifa,* I, §22; English translation, p. 13. Also cf. Shahrastānī, *K. Nihāya al-Iqdām,* pp. 24, 28, 50; Maimonides, *Guide,* I, 74 (7); Albertus Magnus, *Physics,* VIII, i, 12, in *Opera Omnia,* ed. A. Borgnet, Vol. III (Paris, 1890); Bonaventure, *Commentary on II Sentences,* d. 1, p. 1, a. 1, q. 2; Gilson, *The Philosophy of St. Bonaventure,* p. 194; Aquinas, *Summa contra Gentiles,* II, chap. 38 (6), and *Summa Theologiae,* I, 46, art. 2, obj. 8; Maier, "Problem des aktuell Unendlichen," pp. 52–53, 56–57, 73, 76. Maimonides, Albertus, Aquinas, and Duns Scotus, all reject the argument; cf. H. Wolfson, *The Philosophy of Kalam,* pp. 455, 457–458, and Maier, ibid.

[70] *Tahāfut al-Falāsifa,* IV, §§8, 19; English translation, pp. 162, 169. Cf. below, p. 369.

assumed to exist, nothing impossible results therefrom."[71] His meaning is that even if objects have not actually come into existence at every moment of past time and remained in existence, for them to have done so is, nonetheless, a logical possibility.[72] Since it is possible, nothing impossible might result from assuming that it occurred. The assumption, however, that objects have at every moment of past time come into existence and remained in existence does result in an impossibility when combined with the additional assumption that the world is eternal; for the two assumptions taken together entail the existence of an infinite number of objects, something known to be impossible. Inasmuch as the former assumption, the "supposition" that an object has come into existence at each moment of past time, is possible, the latter assumption, the eternity of the world, must be rejected.

Ghazali's contention that eternity would imply the actual accumulation of an infinite number of immortal souls or would at least permit "supposing" the accumulation of an infinite number of souls, played a central role in Scholastic speculation regarding the validity of proofs from the impossibility of an infinite number.[73]

The first road to new arguments for creation from the impossibility of an infinite number consisted, then, in the discovery of additional objects that would run to infinity on the assumption of eternity. The second road to new arguments consisted in proposing additional reasons why past events, time, and the like, cannot be infinite. There was the contention that number is finite by its very nature;[74] and consequently, the number of the revolutions of the spheres,[75] of past individuals,[76] or of time itself,[77] must be finite. Alternatively, quantity was held to be finite by its very nature, so that time, which is a species of quantity, must be finite.[78] It was further maintained that whatever has a part must itself constitute a whole, and since a part of time can be marked off, the totality of time must constitute a whole and hence is delimited and finite.[79] Several writers laid down the principle that whatever has an end must have a beginning, whence the inference was drawn that past time, ending as it does at the present moment, must

[71]*Prior Analytics* I, 13, 32a, 18–20.

[72]Ghazali does not take account of the rule that whatever is generated is eventually destroyed. If that rule should be regarded as a law of logic, generated objects would not be able to remain in existence eternally.

[73]Gilson, *The Philosophy of St. Bonaventure*, p. 194; Maier, "Problem des aktuell Unendlichen," pp. 52–53, 56–57, 73, 76. A similar form of reasoning is employed by Aristotle, *Physics* VIII, 1, 243a, 1–2; *Metaphysics* IX, 4.

[74]Cf. Aristotle, *Physics* III, 5, 204b, 7–10.

[75]Shahrastānī, *K. Nihāya al-Iqdām*, p. 28.

[76]Judah Hallevi, *Kuzari*, V, 18 (1); Aaron ben Elijah, *'Eṣ Ḥayyim*, p. 28.

[77]Ibn Ḥazm, *K. al-Faṣl fī al-Milal*, I, p. 18 (4).

[78]Cf. Kindi's argument, above, p. 109; Gersonides, *Milḥamot ha-Shem*, VI, i, 11, pp. 331–332.

[79]Ibn Ḥazm, *K. al-Faṣl fī al-Milal*, I, p. 15 (1); p. 17 (3); Baḥya, *al-Hidāya* (*Ḥobot ha-Lebabot*), I, 5; Judah Hallevi, *Kuzari*, V, 18 (1). S. Pines, *Beitraege zur Islamischen Atomenlehre*, p. 15, n. 1, connects the argument with Alexander of Aphrodisias. Also. cf. Euclid, *Elements*, V, definitions 1 and 2.

have begun at some previous moment.[80] Finally, the consideration was put forward that the number of the revolutions of the spheres must, like all numbers, be either odd or even,[81] whereas an infinite number could be neither.[82] A natural response to the last consideration was elicited from Crescas, who maintained that the distinction of odd and even simply does not apply to infinite numbers.[83]

Thus far we have been examining medieval proofs of creation from the impossibility of an infinite number. At an early date the proofs were adapted by Arabic writers to serve fresh purposes, in particular to establish the finiteness of the body of the universe.

Naẓẓām is reported to have refuted the Manichaeans by contending that the denizens of darkness could never have crossed their own infinite realm to reach the realm of light, which they allegedly attack. For, Naẓẓām explains, "traversing an infinite is impossible, . . . and exhausting something shows it to be finite."[84] Here we have the reasoning of Philoponus' first proof of creation[85] transferred to the subject of the body of the universe. Another consideration is also advanced by Naẓẓām: What is of finite extent in one direction must be so in all directions.[86] That, apparently, is a utilization, vis à vis the body of the universe, of the principle that what has an end must have a beginning.[87]

In the previous chapter, Kindi was seen to have proved the finiteness of every body through a recasting of Philoponus' second proof of creation. Kindi's reasoning was basically that magnitudes are subject to increase whereas the infinite

[80]Job of Edessa, *Book of Treasures,* trans. A. Mingana (Cambridge, England, 1935), I, v, p. 16; Abū al-Hudhayl, according to both Khayyāṭ, *K. al-Intiṣār,* §5, and Shahrastānī, *K. al-Milal wa-l-Niḥal,* ed. W. Cureton (London, 1846), p. 35 (and cf. pp. 326–327); German translation of Shahrastānī, with pagination of the Arabic indicated: *Religionspartheien und Philosophenschulen,* trans. T. Haarbruecker (Halle, 1850–1851); Māturīdī, *K. al-Tawḥīd* (Beirut, 1970), p. 14; Ibn Ḥazm, *K. al-Faṣl fī al-Milal,* I, pp. 18–19 (5); Baḥya, *al-Hidāya* (*Ḥobot ha-Lebabot*), I, 5; Judah Hallevi, *Kuzari,* V, 18 (1); Averroes, *Tahāfut al-Tahāfut,* I, p. 22.

[81]Cf. Aristotle, *Categories* 10, 12a, 7–8.

[82]Abū Bakr ibn Zakarīyā Rāzī, *Opera,* ed. P. Kraus (Cairo, 1939), p. 129 (and cf. p. 130 f.); Ghazali, *al-Iqtiṣād,* p. 32; *al-Risāla al-Qudsīya,* pp. 17, 35; *Tahāfut al-Falāsifa,* I, §16; Gersonides, *Milḥamot ha-Shem,* VI, i, 11, p. 333. The argument seems to have been known to Thābit ibn Qurra; cf. S. Pines, above, n. 52. Averroes agrees that an actual infinite number could be neither odd nor even and hence is impossible, but he does not agree that past events can be assigned an actual number. Cf. *Tahāfut al-Tahāfut,* I, p. 24; *Epitome of Physics,* in *Rasā'il Ibn Rushd* (Hyderabad, 1947), III, pp. 26–27, and *Middle Commentary on Physics* (Oxford, Bodleian Library, Hebrew MS. Neubauer 1380 = Hunt. 79), III, iii, 4, 2, p. 27b.

[83]*Or ha-Shem,* I, ii, 2; H. Wolfson, *Crescas' Critique of Aristotle* (Cambridge, Mass., 1929), pp. 219–221. Cf. Āmidī, *Ghāya al-Marām,* p. 11.

[84]Khayyāṭ, *K. al-Intiṣār,* §19. Cf. passages cited by H. Bonitz, *Index Aristotelicus* (Berlin, 1870), p. 74b, lines 30-34.

[85]Cf. above, p. 87.

[86]*K. al-Intiṣār,* §19. The contention appears in Averroes and in Jewish writers dependent upon him; cf. H. Wolfson, *Crescas,* pp. 429–431.

[87]Cf. above, n. 80.

cannot be increased; hence neither magnitudes nor bodies, which are a species of magnitude, can be infinite.[88] Similar argumentation appears in Baghdādī,[89] 'Abd al-Jabbār,[90] and Shahrastānī.[91] The same line of reasoning, that is, the recasting of Philoponus' second proof of creation to establish the finiteness of bodies, is to be found in Avicenna too, despite his rejection of the original use of the proof to establish creation.[92] One step in Avicenna's procedure is identical with the device, met in Kindi, of assuming that a segment is removed from the purported infinite.[93] Avicenna, however, incorporates added features that deserve separate treatment.

Avicenna is, to be precise, demonstrating the impossibility of an infinite "continuous quantity" of the type whose parts "exist together" and "have [relative] position," that is to say, the impossibility of an infinite line, plane, solid, or place.[94] He proceeds as follows: In order to demonstrate the impossibility of an infinite magnitude of the type specified, account need be taken only of magnitudes supposedly infinite at one end while finite at the other. As for magnitudes that are supposedly infinite at both ends, they can be assumed to be cut in the middle, so that the required finite end is provided. Given, then, a magnitude that is supposedly infinite at one end while finite at the other, the first step is to assume a segment removed from the finite end. The result, in effect, will be two magnitudes, each of which is finite at one end and infinite at the other, namely, the magnitude under consideration before the segment was removed and what remains of it after the segment is removed. The next step is to assume that the smaller of the two magnitudes is superimposed on, or "applied" to, the larger with the finite ends coinciding. The infinite ends could not now coincide; for should they coincide, the smaller magnitude would be equal to the larger, which is absurd. If, however, the infinite ends do not coincide, the smaller magnitude would be finite by virtue of being exceeded by the larger; and the larger would be finite by virtue of consisting of two finite magnitudes, to wit, the smaller magnitude and the segment that had been removed from the larger magnitude. The supposed infinite magnitude turns out to be finite, and an infinite magnitude is consequently impossible.[95]

The procedure just set forth is sometimes called the method of "application," since it involves applying one magnitude to another.[96] The method of application is employed to establish the finiteness of spatial magnitudes by a number of

[88] Above, p. 108.
[89] *K. Uṣūl al-Dīn* (Istanbul, 1928), p. 62.
[90] *K. al-Majmū'*, p. 62.
[91] *K. Nihāya al-Iqdām*, pp. 24–25.
[92] Cf. below, p. 129.
[93] Cf. above, p. 108.
[94] Cf. Aristotle, *Categories* 6, 51, 15 ff.
[95] Avicenna, *Najāt* (Cairo, 1938), p. 124.
[96] Cf. H. Wolfson, *Crescas*, p. 345.

philosophers who were dependent on Avicenna: Ghazali, in his compendium of Avicenna's philosophy;[97] Shahrastānī;[98] Ibn Ṭufayl;[99] Abraham ibn Daud;[100] Fakhr al-Dīn al-Rāzī;[101] and Altabrizi.[102]

According to Avicenna, the method of application rules out not merely an infinite continuous quantity; it also rules out the existence of an "infinite . . . ordered number" of objects existing together.[103] Avicenna does not spell out exactly how the argument is to run when directed against an infinite number of ordered objects. But Fakhr al-Dīn al-Rāzī indicates what Avicenna intended[104] when he, Rāzī, employs the method of application to disprove "an infinite regress of causes and effects [existing together]." Rāzī begins by assuming that several links are removed from the supposed infinite series of causes. Then he imagines the new infinite series "applied" to the original series with the finite ends coinciding. And he contends that the new infinite series could not conceivably extend back as far as the original series, since it has fewer links. Yet it could also not, while remaining infinite, fall short. Consequently neither it nor the original series could, after all, be infinite.[105]

Crescas, as will be recalled, rejected Philoponus' second proof of creation on the grounds that the notions *smaller, greater,* and *equal* do not pertain to the infinite.[106] On the same grounds, he rejects the use of the method of application to demonstrate the finiteness of the body of the universe. The question whether one infinite body applied to another infinite body would be smaller, greater, or equal, is, according to Crescas, illegitimate; for the concepts *smaller, greater,* and *equal,* just do not pertain to the infinite.[107]

2. Responses of the medieval Aristotelians to proofs of creation from the impossibility of an infinite number

The proofs of creation which we have been examining presented a challenge to the medieval Aristotelians. The Aristotelians had to explain how (a) although Aristotle too had denied the possibility of an infinite number,[108] he could nevertheless have advocated the eternity of the world, thereby implying the existence of infinite numbers of past objects and motions; how (b) although Aristotle had

[97] *Maqāṣid al-Falāsifa* (Cairo, n.d.), p. 127.

[98] *K. al-Milal wa-l-Niḥal,* p. 403.

[99] *Ḥayy ben Yaqdhān,* Arabic text, p. 76; French translation, p. 58.

[100] *Emuna Rama* (Frankfort, 1852), I, 4, pp. 15–16.

[101] *K. al-Arba'īn,* p. 28.

[102] As cited by H. Wolfson, *Crescas,* pp. 346–347.

[103] *Najāt,* pp. 124–125.

[104] Maimonides also understood Avicenna's intent; cf. *Guide,* I, 73 (11).

[105] *K. al-Arba'īn,* p. 83. Āmidī, *Ghāya al-Marām,* pp. 9–10, cites, but does not accept, the argument.

[106] Above, p. 121.

[107] *Or ha-Shem,* I, ii, 1(a); Wolfson, *Crescas,* pp. 188–191.

[108] *Physics* III, 5, 204b, 8–10; the reason Aristotle gives is that the infinite cannot be traversed.

denied the possibility of an infinite magnitude,[109] he could have affirmed infinite past time; and how (c) although he had denied the possibility of an infinite series of causes and effects existing together,[110] he could yet have affirmed the existence of an infinite series of causes and effects that succeed one another through time and do not exist together.

From Maimonides and Averroes we learn that the first two of the three difficulties were dealt with by Alfarabi in a work, now lost, entitled *On Changeable Beings* (*Fī al-Mawjūdāt al-Mutaghayyira*). Maimonides' report refers specifically to the problem of infinite past individuals and motions. These, Alfarabi explained, continue to exist only "in imagination"; since they do not exist *together* in actuality, they cannot properly be enumerated and therefore are not affected by the absurdity of an actual infinite number.[111] According to Averroes' report, Alfarabi offered a similar solution to the problem of infinite time, as distinct from the succession of past individuals and motions, which are the subject of Maimonides' account. Some err, Averroes writes, by supposing that time cannot run to infinity just as a straight line cannot. In fact, Alfarabi explained, a straight line cannot extend infinitely only because it "possesses position and exists in actuality"; and since time does not resemble a straight line in either respect,[112] the grounds for rejecting an infinite straight line are absent in the case of time.[113]

The reports of Maimonides' and Averroes', taken in combination, indicate that arguments against infinite numbers of objects or an infinite extension are operative, in Alfarabi's view, only when two conditions are met: The objects in question and the parts of the purported infinite extension must exist together in actuality; and they must possess position. Averroes and Maimonides do not reveal, however, precisely which arguments against the infinite Alfarabi wished to restrict through the two conditions nor what the import of the two conditions is. Subsequently, Avicenna offers an analysis similar to, but more nuanced than, Alfarabi's, and he does explain the import of the conditions. Avicenna connects them with the method of application, which he had employed to rule out certain types of infinite.

The method of application consisted in applying or superimposing one object or one series upon another so that the parts of the one match, and are paired off with, the parts of the other. Avicenna accordingly maintains that the method can be utilized only when the parts of each object or each series, firstly, "exist . . . together," and secondly, either occupy "[relative] position" or are "essentially ordered," essential order being the situation wherein the parts "precede one another

[109]*Physics* III, 5, 204b, 1 ff.

[110]Below, p. 337.

[111]*Guide,* I, 74 (end), supplemented by I, 73 (11), where Alfarabi is not mentioned.

[112]As pointed out by Aristotle, *Categories* 6, 5a, 26–28.

[113]*Epitome of Metaphysics,* ed. and trans. C. Quirós Rodríguez (Madrid, 1919), IV, §4; German translation: *Die Epitome der Metaphysik des Averroes,* trans. S. van den Bergh (Leiden, 1924), pp. 106–107.

naturally." Only when both conditions are met, clearly, can one object or series be "applied" to another. The method of application would therefore rule out an infinite spatial magnitude, inasmuch as the parts of spatial magnitudes all exist together and possess relative position. The method would likewise rule out an infinite number of objects existing together in instances where the objects have an order in nature, for example, when they stand in the relation of cause to effect. But the method would not rule out infinite time and an infinite series of past motions, nor again an infinite number of unarranged objects existing together. The method would fail to rule out infinite time or an infinite number of past motions since neither the parts of time nor the several motions, although arranged in an order, exist together, and one series consequently cannot be applied to another. The method would fail as well to rule out an infinite number of unarranged objects that do exist together, such as an infinite number of "angels," "evil spirits," and the like. For entities of the sort, although existing together, do not satisfy the second condition; they do not possess relative position and one series cannot be applied to another in a manner that will pair the members off.[114]

Such, then, is Avicenna's explanation why an infinite extended object or an infinite series of objects is impossible when the parts of the object and the links in the series satisfy both the condition of existing together and the condition of being arranged in order; and why infinite extension and an infinite series remain possible as long as either of the requisite conditions is not fulfilled. Avicenna's explanation implies a judgment regarding arguments from the impossibility of an infinite number. The argument that an infinite series of past events or infinite past time cannot be traversed is, by implication, dismissed.[115] The sole argument from the impossibility of an infinite number which he takes seriously is the argument from the impossibility of an infinite number's being increased; and he understands it to be valid solely when formulated with the aid of the method of application, solely in situations where sets of actual objects can be superimposed upon one another.

Ghazali, as was seen a little earlier, developed a version of the proof for creation from the impossibility of an infinite number wherein he contended that eternity would entail the accumulation of an infinite number of immortal souls, or at least the possibility of "supposing" the accumulation of an infinite number of souls and other objects.[116] Ghazali proceeds to explain—although without mentioning the method of application—that the infinite number of immortal souls and the like which would have come into existence, or which might be supposed to have come into existence, through infinite past time would satisfy the two conditions we have been considering. Immortal souls would exist together and, since they

[114] *Najāt*, pp. 124–125.

[115] The impossibility of traversing an infinite is dealt with in chap. 9 of the unpublished work of Avicenna's summarized by Pines, "An Arabic Summary of a lost work of Philoponus," p. 349.

[116] Above, p. 123.

come into existence successively through time, would be arranged in order. The doctrine of eternity thus entails an infinite number of objects that meet the two conditions set by Alfarabi and Avicenna for the impossibility of an infinite number. Ghazali concludes therefore that even granting the two conditions, the doctrine of eternity entails an impossibility and is untenable.[117]

The development we have been following merits recapitulation. The Peripatetic principle that an infinite cannot be exceeded underlies one of Philoponus' proofs of creation: Philoponus reasoned that since the past is continually being increased, it cannot be infinite, and consequently the world must have a beginning. In the Middle Ages, Philoponus' proof of creation was readapted to prove the impossibility of an infinite magnitude and even the impossibility of an infinite series of causes. Medieval Aristotelians endorsed the adaptations of Philoponus' proof, while rejecting its original use to rule out an infinite past. The apparent discrepancy was dealt with by Alfarabi and Avicenna. They maintained that infinite objects and infinite series are excluded only when they exist together and possess an order, whereas infinite past time and the infinite series of objects entailed by infinite past time do not meet the two conditions. Ghazali thereupon answers Alfarabi and Avicenna. He points out that the assumption of infinite past time does entail an infinite number of objects satisfying the two conditions determined by them; the doctrine of eternity is therefore refuted from the standpoint of the Aristotelians themselves.

We have seen how a single solution proposed by Alfarabi and Avicenna addressed two difficulties: the discrepancy in Aristotle's rejecting the possibility of an infinite number, while advocating the eternity of the world with its implication of infinite past events and objects; and the discrepancy in his rejecting infinite magnitudes while affirming the infinity of past time. The third difficulty that confronted medieval Aristotelians was the discrepancy between Aristotle's rejection of an infinite regress of causes, on the one hand, and a further implication of the eternity of the world, on the other.

The eternity of the world, at least of an Aristotelian, nonevolutionary world, implies an infinite succession of generations, let us say, of fathers and sons. When the series of fathers and sons is thought of merely as an infinite number of objects, it has to be harmonized with arguments ruling out an infinite number, and we have seen how the harmonization might be accomplished. Quite apart, however, from being thought of merely as an infinite number of objects, past generations can also be thought of as an infinite causal series. Each father is the cause of the existence of his son; and over an eternity, every series of fathers and sons would constitute what might be termed an infinite diachronic causal regress. By advocating the eternity of the world, Aristotle and his followers advocated a diachronic causal regress. Yet he and his followers scrupulously rejected a synchronic causal regress, that is, an infinite regress of causes existing together.

[117] *Tahāfut al-Tahāfut,* IV, §19; English translation, p. 169. The point is further discussed, below, pp. 367–370. Āmidī, *Ghāya al-Marām,* p. 11, repeats Ghazali's criticism of the Aristotelians.

Various reasons were given by various philosophers for the impossibility of an infinite regress of causes existing together.[118] Aristotle's reason was that in a causal series, no link which is activated by something standing behind it can be deemed a true cause. The true cause of whatever occurs in a causal series is the cause standing behind and activating the entire series, that is to say, the first cause. Consequently, if there should be no first cause, there is no true cause, hence "no cause whatsoever," which is absurd.[119] Such being Aristotle's reason for rejecting an infinite causal regress, the problem that poses itself is this: An infinite regress of causes succeeding one another through time would seem to be excluded on the same grounds that Aristotle adduced to rule out an infinite regress of causes existing together: If there were no first cause, there would be no true cause, and hence no cause whatsoever. How then can Aristotle and his school espouse the eternity of the world, with the attendant infinite regress of causes extending back through time?

Averroes explicitly takes up the problem and he solves it with the aid of a distinction—alluded to by Maimonides as well[120]—between an accidental and an essential series. An essential series, as defined by Averroes, is any series in which the prior links are an indispensable "condition" for the existence of the posterior links—*prior* and *posterior* being taken here in a broad sense that comprehends both the temporal and nontemporal.[121] An accidental series is any series in which the prior links are not an indispensable condition for the existence of the posterior. Accordingly, an accidental series will not merely be one in which the links form an obviously noncausal succession. It will also be one in which the "prior is thought to be a cause of the posterior," if in reality the prior link is not a necessary condition of the posterior link.[122] The distinction between essential and accidental series is illustrated by Averroes through the factors in the production of a given man. Although the generations of men seem to be connected causally, in fact "the existence of the prior [man] . . . is not a condition of the existence of the posterior."[123] The series of progenitors standing behind a given man is not a necessary condition for his existing, because the same man could have existed with other progenitors. The necessary and essential conditions of the man's existence, the genuine causes, are the permanent cosmic forces responsible for everything occurring in the universe.[124]

Now, Averroes explains, whereas an essential series, that is to say, a series of genuine causes, cannot regress infinitely, an accidental series may do so, even in instances where the links appear to be connected causally. Averroes might have

[118]Above, p. 127 (the method of application); below, p. 339.

[119]*Metaphysics* II, 2, 994a, 1–19; cf. below, p. 337.

[120]*Guide*, I, 73 (11).

[121]Cf. below, p. 338.

[122]*Tahāfut al-Tahāfut*, I, pp. 20–21. Cf. *Epitome of Physics*, p. 110; *Long Commentary on Physics*, in *Aristotelis Opera cum Averrois Commentariis*, Vol. IV (Venice, 1562), VIII, comm. 15.

[123]*Tahāfut al-Tahāfut*, IV, pp. 268–269. Cf. *Long Commentary on Physics*, VIII, comm. 47.

[124]*Tahāfut al-Tahāfut*, I, p. 59.

put the proposition in the following way, although I did not find that he explicitly does:[125] Aristotle had rested the impossibility of an infinite regress of causes on the consideration that the true cause of a series is the first cause alone, and without a first cause the series would have no true cause. The reasoning clearly pertains solely to an essential series, to causes that are indispensable conditions of what ensues; for only among them is the true cause to be sought. Accidental series, which do not contain indispensable conditions for what ensues, would in no way be prevented from running to infinity.

Having laid down the foregoing distinction between essential series and accidental series, Averroes can maintain that the members of the essential series and only they must "lead upward to an eternal first cause."[126] The first essential cause, which is ultimately responsible for everything in the universe, is "the sphere, or the soul [of the sphere], or the intellect [of the sphere], or all together, or "—the preferable alternative—"the creator."[127] It, rather than any of the progenitors of a given individual, is the true cause of the appearance, as well as the continued existence, of the individual.[128]

As Averroes works out the distinction between accidental and essential series, he adds a further refinement or clarification. Fathers are, after all, observed to play a role in producing their offspring, and, what undoubtedly is at least as important, Aristotle had explicitly stated that a man is begotten by another "man

[125]Cf. *Tahāfut al-Tahāfut,* I, p. 22.

[126]*Tahāfut al-Tahāfut,* IV, pp. 268–269.

[127]Ibid. By "creator," Averroes of course means the eternal cause of the existence of the universe.

[128]In several places Averroes connects the distinction between essential and accidental series with another distinction, that between a rectilinear series and a circular series. Unhappily, he gives differing accounts of the connection. In the *Tahāfut,* he writes that series formed by essential causes are rectilinear, that is to say, the order is irreversible and no link in the chain of causes and effects can ever recur; and series formed by accidental causes are circular. The progenitors of an individual man and the stages in the rain cycle, both of which constitute accidental series, are, in the passage in question, both represented as being circular series. They are circular series because the amount of matter in the universe is finite and the processes of nature can be sustained only if material objects decay and the matter is reused—only if, in contemporary parlance, the matter is 'recycled.' To maintain the processes of life, progenitors must continually die, decay, and be transformed into nourishment for subsequent generations. To maintain the rain cycle, a given particle of water must turn to vapor, precipitate, fall back to earth as rain, and thereupon repeat the process. Cf. *Tahāfut al-Tahāfut,* I, pp. 56–57; IV, pp. 268–269, 274.

In *K. al-Kashf,* p. 37, Averroes characterizes the rain cycle as a "circular" series, but now he characterizes the generations of fathers and sons as a "linear" and "accidental" series. Thus, in contrast to what he writes in the *Tahāfut,* he recognizes series that are accidental as well as linear, the generations of men being represented as such.

A resolution of the discrepancy might be proposed with the aid of Aristotle, *De Generatione* II, 11. There Aristotle writes that series such as the generations of man "appear" to be rectilinear, but in fact are so only "numerically," whereas in respect to "species" they too are circular. It might be tempting to interpret Averroes as follows: In the *Tahāfut,* he means that all accidental series are circular in respect to species. That is to say, the same matter continually recurs in beings that are specifically, although perhaps not individually, the same; for example, the matter of a previous man

. . . together with the sun."[129] In one passage, therefore, Averroes is led to explain that the immediate progenitor is indeed indispensable for a given man's coming into existence, but solely as what Averroes now calls an instrument, and not as a cause in the strict sense: The immediate progenitor is an indispensable instrument for the action of the cosmic powers, which are the genuine causes. According to the present passage, it is the earlier progenitors that are purely accidental factors in the production of a particular man, and they, being accidental, can—or must—regress infinitely.[130]

The question we have been considering is how Aristotle's rejection of an infinite regress of causes existing together could be harmonized with his espousal of the eternity of the world and the attendant infinite regress of causes extending back through time. Averroes' solution is that the former is an essential series whereas the latter is an accidental series, and the impossibility of an infinite regress covers only essential series, not accidental series. The distinction between an essential and an accidental series, together with Averroes' differentiation between the immediate instrument and prior instruments, is repeated by Thomas Aquinas.[131]

Resumé

The present chapter has thus far examined the history, in the Middle Ages, of Philoponus' three proofs of creation from the impossibility of an infinite number. The proofs were employed in their own right. And new versions were developed both through the proposing of additional objects that would run to infinity in an eternal universe and through the proposing of additional reasons, alongside Philoponus' three, for the impossibility of an infinite of one sort or another. Philoponus' proofs of creation, especially the proof turning on the impossibility of an infinite number's being increased, were also recast to serve fresh purposes, specifically, to rule out an infinite spatial magnitude and an infinite series of causes. Even philosophers of the Aristotelian school, for whom the eternity of the universe was a virtual dogma, employed Philoponus' argumentation for those other

---

reappears in a later man. And in *K. al-Kashf,* Averroes means that the rain cycle is a case of a series circular even in respect to its individuals, inasmuch as exactly the same drop of rain can recur; whereas the generations of men are linear in respect to their individuals, since exactly the same individual man cannot ever recur. This proposed interpretation, it must be confessed, does not seem to harmonize with Averroes' remarks on circular series at the end of his *Epitome of the De Generatione*. See the Arabic text of the *Epitome of De Generatione,* in *Rasā'il Ibn Rushd* (Hyderabad, 1947), pp. 32–33; Hebrew text: *Commentarium Medium et Epitome in Aristotelis de Generatione et Corruptione Libros, Textum Hebraicum,* ed. S. Kurland (Cambridge, Mass., 1958), pp. 125–126; English translation: *Averroes' Middle Commentary and Epitome on Aristotle's De Generatione,* trans. S. Kurland (Cambridge, Mass., 1958), p. 136.

[129] *Physics* II, 2, 194b, 13, cited by Averroes, *Tahāfut al-Tahāfut,* IV, p. 268. Cf. Aristotle, *Metaphysics* XII, 5, 1071a, 13–17.

[130] *Tahāfut al-Tahāfut,* IV, p. 269. Cf. *Epitome of De Generatione,* pp. 32–33; *K. al-Kashf,* p. 37. It is not clear how if *c* is essential for *b* and *b* is essental for *a, c* is not essential for *a.*

[131] *Summa contra Gentiles,* II, chap. 38; *Summa Theologiae,* I, 46, art. 2, reply to obj. 7.

purposes. Since the Aristotelians endorsed the utilization of Philoponus' argumentation for certain purposes, and since Aristotle too had ruled out certain types of infinite, the Aristotelians were faced with a problem of consistency. They had to explain why the grounds for ruling out some infinites do not affect other infinites that are implied by the doctrine of eternity. The requisite harmonizations were forthcoming: The grounds for the impossibility of an infinite number were shown to be limited in scope and not to cover the types of infinite implied by the doctrine of eternity. Isolated philosophers, we have also seen, took a more radical position and rejected the various grounds for the impossibility of the infinite. They held that infinite time, infinite past events, and infinite spatial magnitudes, are all equally conceivable.[132]

3. The standard Kalam proof for creation: the proof from accidents

Proofs for creation from the impossibility of an infinite number had a wide currency; but it was another proof, the one Saadia entitled "from accidents," that became the Kalam demonstration par excellence of creation. Briefly, the proof from accidents runs as follows: Since accidents are necessary concomitants of bodies and are subject to generation, bodies too must be subject to generation; the universe, which is a body, must therefore have been generated. In the previous chapter, I suggested that the proof may be an outgrowth of the second of the auxiliary arguments with which Philoponus had supported his overall proof of creation from the finite power of finite bodies. Philoponus had in that second auxiliary argument focused on the continual succession of forms across matter, whereas the proof from accidents focuses on the continued presence of accidents—the Kalam analogue of the Aristotelian forms—in bodies.[133]

The first thinker to offer the proof from accidents is reported to have been the early Mu'tazilite, Abū al-Hudhayl (d. 849).[134] The proof would accordingly go back to the first half of the ninth century. At the turn of the tenth century the proof is found in Saadia, and his version was examined in the previous chapter.[135] The proof was also known to Alfarabi, a contemporary of Saadia's. Alfarabi, a dyed-in-the-wool Aristotelian, could not by any means have accepted the conclusion,[136] but he cites the proof in one of his logical works as an example of a "compound syllogism." In order to add weight to the illustration, Alfarabi spins out the steps, so that the argument runs: "[1] Every body is composite. [2] Everything composite is joined to, and cannot be free of an accident [the accident of composition]. [3] Everything joined to, and not free of an accident is joined to, and not free of what is generated. [4] Everything joined to, and not free of

[132]On the last point, see above, pp. 121–122, 127.

[133]Cf. above, p. 92.

[134]Cf. below, n. 162.

[135]Above, pp. 103–104.

[136]He undoubtedly rejected the argument for the reason that Averroes gives for rejecting it; cf. below, pp. 143–144.

what is generated does not precede what is generated. [5] Everything that does not precede what is generated has its existence together with the existence of what is generated. [6] Everything having its existence together with the existence of what is generated has its existence after nonexistence. [7] And everything having its existence after nonexistence is generated. But the world is a body. Consequently, the world is generated."[137]

The proof from accidents appears in Ibn Suwār, who explicitly labels it a Kalam method of proving creation, points to several flaws in it, and thereupon rejects it in favor of what he identifies as John Philoponus' proof from the finite power of bodies.[138] Ibn Suwār presents the proof in two versions, which he characterizes as reformulations in "a technical arrangement" of the original Kalam argument. What he means is that the original Kalam reasoning had been recast—by him or by someone before him—in syllogistic form. One of Ibn Suwār's syllogisms reads: "Body is not free of accidents nor does it precede them; but whatever is not free of accidents and does not precede them is itself generated; therefore body is generated." Ibn Suwār's other syllogism employs the term "generated things" in place of the term "accidents," and reads: "Body is not free of generated things [etc.]."[139]

The argument from accidents, then, reportedly goes back to Abū al-Hudhayl. It was advanced with approval by Saadia, cited undoubtedly without approval by Alfarabi, and cited with explicit disapproval by Ibn Suwār. Ash'ari also knew of the proof, employing it in one work,[140] but rejecting it in another. In the work where he rejects it, his reason is not that—as Alfarabi and Ibn Suwār thought—the proof is insufficiently philosophical but, on the contrary, that it is too philosophical. He ascribes the proof to "the philosophers, and those who follow them among the *qadarīya* [i.e., the Mu'tazilites], the innovators [or: heretics], and the deviators from the prophet(s)." And he takes the position that the testimony of Scripture is more than adequate and no rational proof of creation is needed.[141]

Despite Ash'ari's reservations, later adherents of his school embraced the proof from accidents. The Ash'arite Bāqillānī reasons: Everything in both the higher

[137]*Epitome of Prior Analytics,* ed. M. Türker, *Revue de la Faculté de Langues, d'Histoire et de Géographie de L'Université d'Ankara,* XVI/3–4 (1958), 263. On p. 262, Alfarabi spells out each of the individual constituent syllogisms in full. Cf. H. Davidson, "John Philoponus as a Source of Medieval Islamic and Jewish Proofs of Creation," *Journal of the American Oriental Society,* LXXXIX (1969), 383.

[138]Cf. above, p. 90.

[139]The text is published in A. Badawi, ed., *Neoplatonici apud Arabes* (Cairo, 1955), pp. 243, 245; French translation in B. Lewin, "La Notion de muḥdaṯ dans le kalām et dans la philosophie," *Orientalia Suenica,* III (1954), 88–93.

[140]*K. al-Luma',* in *The Theology of al-Ash'ari,* ed. and trans. R. McCarthy (Beirut, 1953), §§6, 93.

[141]*Risāla ilā ahl al-Thaghr, Publications of the Theological Faculty, Istanbul,* VIII(1928), 89; cf. R. Frank, "Al-Ash'arī's Conception of the Nature and Role of Speculative Reasoning in Theology," *Proceedings of the VIth Congress of Arabic and Islamic Studies* (Stockholm, 1972), pp. 138–141.

and lower parts of the universe consists of "substances" (i.e., atoms) and "accidents." Accidents are generated; that is shown by "the fact that motion is destroyed with the advent of rest, for it if were not destroyed with the advent of rest, both would be present together in the body, . . . something necessarily known to be impossible." Bāqillānī might be expected hereupon to argue that not merely accidents but atoms too are generated. Yet he does not do so, and proceeds instead to establish that "bodies" are generated. Bodies, he explains, "do not precede generated things nor exist before them," inasmuch as "bodies cannot avoid having their parts touching and joined, or separated." But anything "that does not precede what is generated is likewise generated," since it must either come into existence "together" with or "after" its generated concomitant. Both accidents and bodies are, then, generated, and the reader is left to conclude that since the universe consists only of accidents and of bodies, which are conglomerations of atoms, the entire universe must be generated.[142]

The versions of Saadia, Alfarabi, Ash'ari, Ibn Suwār, and Bāqillānī reveal significant similarities.[143] All undertake to demonstrate the generation of bodies and accidents. All employ the proposition that bodies "cannot avoid" (Saadia, Bāqillānī: *lā yakhlū*) or cannot be "free of" (Alfarabi, Ibn Suwār: *lā yanfakk*; Ash'ari: *lam yanfakk*) their association with accidents. And all conclude that bodies are generated since they "do not precede" (Saadia, Bāqillānī: *lam tasbuq*; Ash'ari: *lam yasbuq;* Alfarabi: *ghayr sābiq;* Ibn Suwār: *lā yataqaddam*) what is generated. In Saadia, Ibn Suwār,[144] and Bāqillānī, the prime illustration of an accident is motion.[145] Alfarabi and Bāqillānī adduce the accident of the *composition* or *joining* of bodies to establish the proposition that bodies are always associated with accidents. Saadia and Bāqillānī seek comprehensiveness by taking into account the higher as well as the lower parts of the universe. They both conclude in similar phrases that what does not precede generated things is "of the same character" (Saadia: *mithluhu*) and "likewise" (Bāqillānī: *kahuwa*) generated.

The Kalam doctrine of accidents is integral to the proof. Yet atomism, which is commonly thought to be a correlate of the Kalam doctrine of accidents, is curiously absent, the preserved early versions establishing the generation of bodies, not atoms. As a result, these versions do not prove creation *ex nihilo:* They establish the creation of the body of the universe, but not the creation of the atoms from which the universe is constituted. The failure to prove the creation

[142] *K. al-Tamhīd,* ed. R. McCarthy (Beirut, 1957), pp. 22–23.

[143] Ash'ari's version, *K. al-Luma',* §§6, 93, is less complete than the others.

[144] As Ibn Suwār records the proof, the principle that no body is free of an accident is established through an analysis of the accidents of motion and rest. Cf. above, n. 139, Arabic text, p. 244; French translation, §3.

[145] Ash'ari, *K. al-Luma',* also has the example of the accident of motion. It should further be recalled that Kindi, in another context, employs the proposition that body does not precede the accident of motion. Cf. above, p. 110.

of atoms is especially conspicuous in Bāqillānī's version. In a preface, Bāqillānī had divided physical beings into bodies, atoms, and accidents;[146] and he opens his argument with the statement that the upper and lower regions of the universe are composed of atoms and accidents.[147] The reader awaits a proof of the generation of accidents and atoms, or of accidents, atoms, and bodies, hence a proof of the creation of all physical existence *ex nihilo*. What Bāqillānī in fact provides is a proof that accidents and *bodies* are generated. Such must have been the object of the ninth-century proof, which underlies the preserved early versions.

Although the Kalam complexion of the proof is unmistakable and although from the beginning the proof was regarded as the property of the Kalam school, Bāqillānī's version alone, among the versions examined thus far, is worked out within a Kalam conceptual frame. Saadia took the existence of accidents for granted, and relied on simple observation as grounds for the propositions that accidents are subject to generation and that bodies are never free of them. For Alfarabi and Ibn Suwār, those points—the existence of accidents, the proposition that accidents are subject to generation, the proposition that bodies are never devoid of accidents—were of minor importance, since Alfarabi and Ibn Suwār rejected the proof in any event. Ash'ari's version is incomplete. But Bāqillānī has a standard analysis of each point.

In his preface to the proof—which I have not quoted yet—Bāqillānī establishes the existence of accidents by considering motion, showing that it cannot be due to the moving body "itself"—were that the case a moving body would never stop moving—but must rather be an added entity, a "something" (*ma'nā*), in the moving body.[148] This is the characteristic Kalam theory that accidents are actual entities inhering in atoms and bodies.[149] And the theory appears in Bāqillānī in association with the related and very peculiar Kalam notion that rest is not merely a privation, that is to say, the absence of the accident of motion, but a positive quality and, no less than motion, a "something."[150]

When Bāqillānī proceeds, within the proof proper, to establish the proposition that accidents are generated, he adduces "the fact that motion is destroyed with the advent of rest."[151] Here he has left an inference implicit which may be reconstructed by consulting later Kalam works.[152] He should have gone on to

[146] *K. al-Tamhīd*, p. 17.

[147] Above, p. 136.

[148] *K. al-Tamhīd*, pp. 18–19. For various views on the origin of the term *ma'nā*, see Wolfson, *The Philosophy of the Kalam*, pp. 147–167.

[149] Cf. below, p. 180.

[150] *K. al-Tamhīd*, p. 22; cf. Maimonides, *Guide*, I, 73 (7).

[151] Above, p. 136.

[152] Cf. 'Abd al-Jabbār, *Sharḥ al-Uṣūl*, (Cairo, 1965), pp. 93–94, 104, and 107; idem, *K. al-Majmū'*, p. 53; Juwaynī, *K. al-Irshād*, p. 20; idem, *K. al-Shāmil*, pp. 186–187. The attribution of the *Sharḥ al-Uṣūl* to 'Abd al-Jabbār is not certain; see the editor's remarks in his introduction, pp. 27–28.

explain that whatever is subject to destruction is also subject to generation—the Aristotelian principle was accepted by the Kalam and is used by Bāqillānī elsewhere[153]—and therefore the accident of motion, if subject to destruction, must be generated. The inference apparently was so familiar to Bāqillānī that he simply takes it for granted. He makes no reference to it and passes at once to another matter, writing: If motion "were not destroyed with the advent of rest, both would be present together in the body, . . . something necessarily known to be impossible." At first glance the addition seems superfluous,[154] since when a body is at rest, motion obviously is not present; and once again we must consult later works to discover the import of the remark. Bāqillānī is forestalling a possible objection to his statement that when the accident of rest is present, the accident of motion ceases to exist. The objection would be that the accident of motion may not completely cease to exist but may rather revert to a state of latency within the body, from which it can be elicited at a future time. In other words, the accidents of motion and rest are, perhaps, not generated after all, but are continually present in each given body, one of them always being in a state of latency, from which and to which it alternately emerges and returns.[155] Bāqillānī forestalls the objection by pointing out, in effect, that the theory of latency is absurd since it transgresses the principle that contraries cannot be present in the same thing at the same time.[156] He is accordingly justified in affirming that when an instance of the accident of motion disappears, it must have been destroyed and hence must also have previously been generated. By an implied generalization, Bāqillānī thereupon assumes that all instances of the accident of motion—not merely instances of motion that are seen to be succeeded by rest—and indeed all instances of all accidents must be generated.

As a further step in his proof, Bāqillānī undertakes to establish the proposition that bodies "do not precede generated things nor exist before them." He reasons that every body has its parts either joined, that is to say, in a state of composition, or separated. The reader is expected to understand that composition and separation are both accidents;[157] that by the earlier implied generalization all accidents are generated; and therefore, since bodies are necessarily associated with either composition or separation, they necessarily are associated with, and do not exist before, "generated things." The full contention, illustrated as here by the accidents of composition and separation, reappears in the same context in later Kalam

[153]Cf. Aristotle, *De Caelo* I, 12, 282b, 2; *K. al-Tamhīd,* p. 20.

[154]A similar statement appears in Philoponus' proof; cf. above, p. 92: "The same matter cannot admit several forms at once."

[155]The theory of latency is reported in the name of Naẓẓam; cf. Khayyāṭ, *K. al-Intiṣār,* §90; Wolfson, *The Philosophy of the Kalam,* pp. 498 ff.

[156]The same objection to the proof is dealt with by Baghdādī, *K. Uṣūl al-Dīn* (Istanbul, 1928), p. 55; 'Abd al-Jabbār, *Sharḥ al-Uṣūl,* p. 105; idem, *K. al-Majmū',* p. 50 (only briefly); Juwaynī, *K. al-Irshād* p. 20; idem, *K. al-Shāmil,* p. 190; Ghazali, *al-Iqtiṣād,* p. 28 (no solution is given).

[157]Cf. Pines, *Atomenlehre,* pp. 6–7.

versions of the proof.[158] In Bāqillānī, at least, the introduction of the accidents of composition and separation is awkward. Since the accidents of motion and rest had been considered earlier in the proof, his presentation would have been tidier had he retained the original pair of examples and, having already shown that motion and consequently also rest are generated, contended now that all bodies necessarily are associated with either the one or the other. Alternatively, he could have used the illustration of the accidents of combination and separation throughout with no change in the argument. Inasmuch as the use of two different pairs of accidents serves no function, it may be presumed to be due to features of the proof's tradition which are no longer available to us. Among the early versions of the proof, Alfarabi's version too, as will be recalled, adduced the composition of bodies when establishing that bodies are always associated with accidents.[159] The appearance of *composition* in the proof from accidents might be explained with the aid of my suggestion that the proof grew out of the second of the auxiliary arguments whereby Philoponus supported his proof from the finite power of finite bodies. Side by side with that auxiliary argument Philoponus had offered an auxiliary argument from composition;[160] and in the course of transmission the reference to composition may have infiltrated from one argument to the other.[161]

Bāqillānī, in sum, employs the standard reasoning whereby the Kalam establishes the existence of accidents. He partially states and partially alludes to standard reasoning for establishing the generation of accidents, the absurdity of the theory of latency, and the necessary association of bodies with "generated things." He rests his proof on the implied generalization from a single sort of accident—motion that is seen to be succeeded by rest—to all instances of all accidents. And he includes one element, the composition of bodies, for what are probably historical rather than intrinsic reasons. The proof must have come to him in a fixed, stylized form, around which a fund of theoretical discussion had already grown up. Either through superficiality or for the sake of brevity, he makes statements that are incomplete in themselves but take on meaning when understood as allusions to familiar discussions.

A short time after Bāqillānī, the Mu'tazilite 'Abd al-Jabbār (d. 1025/26) characterizes the proof we are examining as the most "dependable" one. He, incidentally, is the source of the report tracing the proof to the early Mu'tazilite Abū al-Hudhayl—who, 'Abd al-Jabbār adds, was followed in the use of the proof by

[158]Cf. Baghdādī, *K. Uṣūl al-Dīn,* pp. 49–50 (somewhat different); 'Abd al-Jabbār, *Sharḥ al-Uṣūl,* pp. 111–112; idem, *K. al-Majmū',* p. 50; Juwaynī, *K. al-Irshād* p. 24; idem, *K. al-Shāmil,* pp. 204–205.

[159]Cf. above, p. 134.

[160]Cf. above, p. 92.

[161]In Stoic physics, composition and separation are forces for the production of new objects, and Epicurus also had a doctrine of the mixture of atoms.

"the other sheikhs."[162] 'Abd al-Jabbār's formulation of the proof is more systematized than Bāqillānī's, being explicitly "constructed upon four premises (*da'āwā*)."[163] Each of the four premises turns out to be a parallel of one of the steps in Bāqillānī's argument, but in Bāqillānī neither steps nor premises were designated as such and the structure of the argument was not articulated sharply. The four premises enumerated by 'Abd al-Jabbār are (1) that "things" (*ma'ānin*) are present in bodies, to wit, joining and separation, motion and rest; (2) that these are generated; (3) that body cannot be free (*yanfakk*) of them, nor does it precede (*yataqaddam*) them;[164] and (4) that what cannot be free of, or precede, what is generated is likewise (*mithlahā*) necessarily generated.

The formalization of the four premises upon which the proof is constructed predates 'Abd al-Jabbār. A text entitled *K. al-Majmū' fī al-Muḥīṭ bi-l-Taklīf,* based on 'Abd al-Jabbār's thought and edited sometime after his death,[165] reports that the fully developed form of the proof from accidents had been known to the Mu'tazilite Abū Hāshim (d. 933), the son of al-Jubbā'ī. According to the *Majmū',* Abū Hāshim had insisted that the generation of body, and hence the creation of the world, can be proved only through the "four principles (*uṣūl*)"; and he had declared that anyone denying the existence of accidents will be unable to accomplish the proof.[166] For its part, the *Majmū'*—which, as just mentioned, was not written by 'Abd al-Jabbār but is based on his thought—does not merely use the four "principles" or "premises" to prove creation.[167] It offers elaborate argumentation to support the premises and to remove a variety of possible objections;[168] and it accompanies the proof with a running methodological discussion, by both 'Abd al-Jabbār and his posthumous editor, dealing with the questions whether the present proof is in truth the best way of establishing the creation of the world and whether the best procedure for proving the existence of God is by proving creation first.[169] By the time of the *Majmū'* a theological adept apparently could take the proof for granted, and could busy himself with methodological issues and the dialectic overgrowth that had enveloped the proof.

The proof of creation from accidents is employed by the Ash'arites Baghdādī (d. 1037)[170] and Juwaynī (d. 1085).[171] Juwaynī establishes the creation specifically of "substances," that is, of atoms rather than bodies; his proof therefore concludes with the creation of the world *ex nihilo* and not merely the creation of

[162]*Sharḥ al-Uṣūl,* p. 95.

[163]Ibid. On p. 104, 'Abd al-Jabbār refutes the theory of latency as part of a proof of the generation of accidents.

[164]Regarding these terms in the early version of the proof, cf. above, p. 136.

[165]Cf. the editor's French introduction, p. 8.

[166]*K. al-Majmū',* pp. 30. 63.

[167]The proof appears in *K. al-Majmū',* pp. 30 ff.

[168]Ibid., pp. 32–58.

[169]Ibid., pp. 28–31.

[170]*K. Uṣūl al-Dīn,* II.

[171]*K. al-Irshād,* pp. 17–27; *K. al-Shāmil,* pp. 166 ff.; *Textes apologétiques* (*Luma'*), pp. 120 ff.

the world in its present form. Like 'Abd al-Jabbār, Juwaynī sets down four principles, but he makes a significant change, to be discussed in the next section. The proof is employed as well by Māwardī (d. 1058)[172] and by Bazdawī (d. 1099), a follower of Māturīdī.[173] Juwaynī's version reappears in Ghazali's Kalam writings[174] and in Shahrastānī.[175] The proof from accidents is canonized, as it were, in the creed of Nasafī (d. 1142/43).[176] It is employed by the Isma'ili Nāṣir-i-Khosraw[177] and is cited by Abū al-Barakāt[178] and Ibn Ṭufayl.[179] Averroes, an outspoken opponent of the Kalam,[180] and also Ṭūsī (d. 1273),[181] Āmidī[182] and Ījī[183] describe it as the Kalam proof for creation par excellence.

The proof enjoyed a good deal of popularity among Jewish followers of the Kalam. The Karaite Joseph al-Baṣīr (tenth or eleventh century) explicitly constructs it on four "premises,"[184] and his student Jeshua b. Judah employs it as one of several methods for proving creation.[185] Among the Rabbanites, it appears in Joseph Ibn Ṣaddiq (d. 1149),[186] in a work attributed to Abraham Ibn Ezra (1092–1167),[187] in Judah Hallevi,[188] and in Joseph Ibn Aqnin (Joseph b. Yaḥya) (d. 1226), where it is described as the "current" (or perhaps "preferable" [*meruṣa*]) method of demonstrating creation.[189] The formulations in the works of the two Karaite writers just mentioned and in the work attributed to Ibn Ezra disclose a strong resemblance to 'Abd al-Jabbār's formulation. Finally, Maimonides, like Averroes an outspoken opponent of the Kalam, includes Juwaynī's version in a list of seven Kalam proofs of creation.[190]

172 Māwardī, *A'lām al-Nubūwa* (Cairo, 1971), pp. 13–14.

173 *K. Uṣūl al-Dīn*, ed. H. Linss (Cairo, 1963), p. 15.

174 *al-Iqtiṣād*, pp. 24–32; *al-Risāla al-Qudsīya*, pp. 16–17, 34–35. Cf. *Tahāfut al-Falāsifa*, X, §1.

175 *K. Nihāya al-Iqdām*, p. 11.

176 E. Elder, *A Commentary on the Creed of Islam* (New York, 1950), pp. 28–29 (Nasafī) and p. 33 (Taftazānī's commentary). Cf. A. Wensinck, "Les Preuves de l'Existence de Dieu," *Mededeelingen der Koninklijke Akademie van Wetenschappen te Amsterdam*, LXXXI, Series A, No. 2 (1936), 3.

177 Cf. Pines, *Atomenlehre*, p. 37, n. 2.

178 *K. al-Mu'tabar*, III, p. 31.

179 *Hayy ben Yaqdhān*, Arabic text, p. 81; French translation, p. 62.

180 *K. al-Kashf*, pp. 31–32. Averroes enumerates only three premises, by not counting the existence of accidents as a separate premise.

181 Glosses to Rāzī's *Muḥaṣṣal* p. 89.

182 *Ghāya al-Marām* pp. 261–262.

183 *Mawāqif*, VII, p. 222.

184 P. Frankl, *Ein Mu'tazilitischer Kalām* (Vienna, 1872), pp. 20, 53.

185 M. Schreiner, *Studien ueber Jeschu'a ben Jehuda* (Berlin, 1900), pp. 29–33. On p. 31, n. 2, Schreiner adds further references to the use of the proof.

186 *ha-'Olam ha-Qaṭan*, ed. S. Horovitz (Breslau, 1903), pp. 48–49.

187 *Kerem Ḥemed*, IV (1839), 2–3. Cf. M. Schreiner, *Der Kalām in der juedischen Literatur* (Berlin, 1895), pp. 37–40.

188 *Kuzari*, V, 18 (2).

189 *Treatise as to Necessary Existence*, ed. and trans. J. Magnes (Berlin, 1904), Hebrew section, pp. 17–19; English section, pp. 39–42.

190 *Guide*, I, 74 (4).

In at least two writers, Shahrastānī and Aaron ben Elijah, the Aristotelian matter and form appear in place of body and accident, with the result that the argument approaches or reapproaches the argument of Philoponus from which I have suggested it derives.

Shahrastānī knew of a peculiar tradition according to which Aristotle had espoused the creation of the world.[191] And he attributes to Aristotle an argument that appears simply to be the Kalam proof we have been examining, with the dichotomy of matter and form substituted for the dichotomy of body and accident. Aristotle, Shahrastānī would have us believe, reasoned: "Forms . . . are not *from* one another, but are necessarily *after* one another, so that they succeed one another over matter, and plainly are destroyed and pass away. Whatever passes away must have a beginning; for passing away is an end, and it is one of two terminuses, to which a counterpart must correspond. Now it has been established that the [forms] are something generated out of nothing[192] and that their subject is essentially disposed to receive them and serve as their subject. *They* have a beginning and an end; this shows that their subject has a beginning and an end. . . ."[193]

Another formulation in terms of matter and form is given by the relatively late Jewish Karaite writer Aaron ben Elijah (d. 1369), who undoubtedly wished to free the proof from the burden of Kalam physics and render it more respectable.[194] In the first of four proofs for creation, Aaron begins by establishing the existence of matter and form. To that end he offers a peculiar adaptation of the reasoning used by the Kalam thinkers for establishing the existence of accidents. Kalam reasoning, applied to the accidents of composition and separation,[195] would run: Bodies cannot be in a state of combination or separation by virtue of themselves, since if such were the case, they would either always be in a state of combination or always in a state of separation, whichever their own nature necessitates; consequently, both combination and separation must be "something" (*ma'nā*) added to bodies, that is to say, they are accidents. Aaron alters the meaning and role of combination and separation, reasoning: "Combination and disjunction are contraries, and matter [read: body] receives both. What receives *conjunction* cannot be the cause of *disjunction*; rather we must conclude that combination is due to 'something' (*'inyan* = *ma'nā*) and disjunction is due to 'something' else. Body

[191] *K. al-Milal wa-l-Niḥal,* p. 326. Shahrastānī realizes that the tradition is out of harmony with the position generally attributed to Aristotle, and he too sometimes attributes the doctrine of eternity to Aristotle; cf. ibid., pp. 320–321, 338, 340.

[192] Cf. Philoponus' statements, above, p. 92.

[193] *K. al-Milal wa-l-Niḥal,* pp. 326–327.

[194] One of Maimonides' objections to the proof is that it rests on the theory of atomism, which is not accepted by those who believe in eternity (cf. *Guide,* I, 74 [4]), and Aaron ben Elijah was completely familiar with the *Guide*.

[195] Bāqillānī applied this reasoning to the accident of motion, and 'Abd al-Jabbār applied it to the accidents of composition and separation as well. Cf. above, nn. 148, 163.

thus has two 'things' (*'inyanim*), matter and form, disjunction coming from the former, and conjunction from the latter."[196] Aaron hereupon proceeds in the familiar manner, contending that forms are generated, that matter is never free of forms, and that matter therefore is likewise generated. To justify his conclusion he cites those whom he calls "our scholars," who "have stated that if anything cannot be free of (*yit'areh*) what is generated, it likewise (*kamohu*) is generated."[197]

4. Juwaynī's version of the proof from accidents

At the the heart of the proof from accidents lies the contention that the subject in which generated accidents are ever present must also be generated. The contention is by no means self-evident and unless it is justified in some fashion the conclusion of the proof remains unfounded. A development in the approach to the key contention can be discerned among the philosophers who employ the proof from accidents.

Frequently the bald assertion was made that whatever is unavoidably associated with what is generated is itself generated, and the argument was left at that. Saadia's statement of the argument, for example, concludes: "What cannot avoid what is generated is known to be of the same character."[198] Alfarabi deliberately draws out the argument, with which he himself has no sympathy, into seven steps; but they all merely amount to the assertion that whatever is necessarily joined to what is generated is itself generated.[199] Bāqillānī seems to add something new when he writes that whatever "does not precede what is generated is likewise generated" because it must either come into existence "together" with, or "after" its generated concomitant;[200] but the statement is either a tautology or, as will be seen, an equivocation. 'Abd al-Jabbār declares: "Since body is not free of the aforementioned generated things and does not precede them, its lot (*ḥaẓẓ*) in existence must be like their lot," and he illustrates the assertion by the fact that twins are always of the same age.[201] Baghdādī states: "Inasmuch as bodies do not precede generated accidents, their own generation is entailed; for what does not precede generated things is generated, just as what does not precede any single generated thing is generated."[202]

The weakness in all these statements is, as Averroes points out, the danger of equivocation in such expressions as *whatever is joined to what is generated* and *whatever does not precede what is generated*; for the words *what is generated* can mean either a *single generated thing* or else *generated things*. Obviously, if any body is necessarily joined to a particular accident that is generated, the body

[196]That is to say, form brings things together into a single class, and matter is the principle of individuation.

[197]*'Eṣ Ḥayyim,* chapter 10, p. 28.

[198]Above, p. 104.

[199]Cf. above, pp. 134–135.

[200]Above, p. 136.

[201]*Sharḥ al-Uṣūl,* pp. 113–114; cf. *K. al-Majmū',* p. 58.

[202]*K. Uṣūl al-Dīn,* pp. 59–60.

does not exist before the accident, exists together with it, and also is generated. It is far from obvious that a body must be generated if necessarily joined to one generated accident or another, but not to a particular, given accident; and precisely the latter proposition is at issue.[203] Why—as Ibn Suwār expressed his objection to the proof[204]—can the body of the universe not be joined from all eternity to an infinite series of generated accidents?

Baghdādī may have been cognizant of the weakness in the proof. In his conclusion he writes that "what does not precede generated things is generated, just as what does not precede any single generated thing is generated."[205] The statement seems to be an assurance that the proof does not suffer from the equivocation to be brought out later by Averroes. Baghdādī fails, however, to explain the basis of his assurance. In 'Abd al-Jabbār's *Sharḥ al-Uṣūl* and in the *Majmū'*, an objection identical with Ibn Suwār's is explicitly taken up after the proof is completed. In the *Sharḥ al-Uṣūl,* the objection reads: "Why deny that although body is not free of generated things. . . , one generated thing was generated in it, prior thereto another, prior thereto yet another, *ad infinitum*?"[206] In the *Majmū'* the same objection is introduced by the words: "In order to perfect the proof of the creation of bodies, there is no avoiding . . . showing that generated things have a first [term]."[207] Solutions are given in both works, and finally the *Majmū'* comes to the solution that was to change the character of the proof. 'Abd al-Jabbār, according to the *Majmū'*, "demonstrated the finiteness of these generated things by the presence in them of increase and diminution";[208] that is to say, he contended that the accidents passing over the body of an eternal universe could not form an infinite series, because the argument from "increase and diminution" shows an infinite series to be impossible. 'Abd al-Jabbār thus answered the critical objection to the present proof by adducing what had originally been a separate proof, namely, one of the proofs for creation from the impossibility of an infinite number.[209]

It is Juwaynī who fully and explicitly recognizes that what is at issue here is not merely one of several difficulties to be raised and answered dialectically, but rather the nerve of the entire demonstration. As in the tradition summed up in 'Abd al-Jabbār,[210] Juwaynī bases the proof from accidents on "four principles." First he establishes the existence of accidents, then shows them to be generated, and thirdly shows that no substances—that is, no atom—can be free of them.

[203] *K. al-Kashf,* p. 35. A similar critique is made by Abū al-Barakāt, *K. al-Mu'tabar,* III, p. 31.
[204] See above, n. 139, Arabic text, p. 245; French translation, §5.
[205] Above, n. 202.
[206] *Sharḥ al-Uṣūl,* p. 114.
[207] *K. al-Majmū',* p. 59.
[208] Ibid., p. 61.
[209] Bāqillānī may be alluding to the same type of response to the objection, *K. al-Tamhīd,* p. 49, taken together with p. 25.
[210] Above, p. 140.

But he no longer feels he can assign the status of a *principle* to the proposition that what is unavoidably associated with generated accidents is itself generated.[211] As his fourth premise, he substitutes "the impossibility of generated things without a first [term],"[212] a doctrine that, he goes so far as to aver, is the touchstone separating the believer from the nonbeliever who maintains the eternity of the world.[213] Juwaynī supports his fourth premise as follows: "It is a principle of the heretics that prior to the present revolution [of the sphere] infinite revolutions have been completed. But for something infinite to pass away by having one unit succeed another is impossible. The fact that the revolution[s] [of the sphere] prior to the present revolution(s) have passed away, . . . demonstrates their finiteness."[214] This is simply the familiar and much repeated argument that the world must have a beginning since an infinite series of past events cannot conceivably have been traversed.[215] Juwaynī's complete demonstration accordingly runs: The universe consists of atoms and accidents. Accidents are generated; and atoms are never free of them. But an infinite series of generated accidents is impossible since an infinite cannot be traversed, and hence there must be a first, generated term for every series of accidents. Inasmuch as each atom and the universe as a whole are inextricably associated with finite series of accidents, each atom and the universe itself can have existed for only a finite time, and must have come into existence from nothing. Juwaynī's version rests, as Averroes later was to demand the proof must,[216] on the association of the universe with a *single* generated thing, that single generated thing being the series of generated accidents—or, to be more precise, any single series of generated accidents.[217]—in the universe.

To recapitulate: The key contention in the Kalam proof from accidents cries out for some justification. The lack was eventually felt, and was filled by adducing one or another of the arguments—it makes no difference which—from the impossibility of an infinite number. The classic Kalam proof for creation, which may have derived from Philoponus, was thereby combined with the proofs for creation from the impossibility of an infinite number which undoubtedly did derive from him. Perhaps it would be more accurate—seeing that the proof from accidents failed by itself to demonstrate anything—to say that the entire burden of demonstration fell upon the proofs from the impossibility of an infinite number. The proof from accidents regressed into a mere prelude to those proofs.

The version of the proof from accidents appearing in Ghazali, Shahrastānī, Maimonides, and Ṭūsī,[218] is Juwaynī's version. Averroes, in a critique of the Kalam, first records the earlier version of the proof, discussed in the previous

[211]He does know the proposition; cf. *K. al-Shāmil*, pp. 220–221.

[212]*K. al-Irshād*, p. 18; *K. al-Shāmil*, p. 215; *Textes apologétiques* (*Luma'*), p. 120.

[213]*K. al-Irshād*, p. 25.

[214]Ibid., p. 26; cf. *K. al-Shāmil*, p. 215; *Textes apologétiques* (*Luma'*), p. 126.

[215]Above, pp. 119–120.

[216]Above, p. 144.

[217]Such as the movements of the heavens.

[218]Above, nn. 174, 175, 181, 190.

section. Then he adds: "After the later Kalam thinkers realized the weakness of the proposition [that what is not free of generated things is itself generated], they tried to strengthen it by proving, as they supposed, that an infinite number of accidents cannot pass successively over a single subject."[219] The later Kalam thinkers, Averroes continues, based the impossibility of an infinite series of accidents on the contention that if any event had to be preceded by an infinite number of events, it could never come about; in other words, they employed the argument of proofs from the impossibility of an infinite number. Such, we have seen, is precisely the procedure followed by 'Abd al-Jabbār and more formally by Juwaynī. After Averroes records the version of the proof from accidents attributed by him to the "later Kalam philosophers" he sets forth his own refutation.[220]

The manner in which Shahrastānī and Averroes refer to what I have here called Juwaynī's version of the proof from accidents, indicates that they did not consider him to be its author.[221] Juwaynī, though, does claim credit for the new version[222] and is the earliest writer in whose works I have been able to find it fully articulated.

5. Proofs from composition

Philoponus, as was seen in the previous chapter, offered two arguments from composition in support of his overall proof for creation from the finite power of the physical universe. Both arguments take their departure from the composition of matter and form in the heavens. In one of them, the reasoning was that what is composite is not self-sufficient, hence not infinitely powerful, hence destructible and generated.[223] The reasoning in the other was that what is composite is subject to decomposition, and what is subject to decomposition is not infinitely powerful, the final inference of creation from finiteness of power not being stated.[224] The line of thought of the second of the two arguments goes back to the *Phaedo,* where being composite is held to imply that a thing is subject to decomposition.[225]

The entire enterprise of citing composition in the universe as evidence of creation may well have been prompted by a passage in Aristotle. An argument is found in Aristotle to the effect that anything composite has the possibility of existing in a noncomposite state, and that the possibility of existing in a noncomposite state, like every possibility, must have been realized at some moment in past time. Aristotle inferred herefrom that "the eternal things" cannot be "composed of elements" and consequently do not contain matter and form.[226] Philoponus would plainly proceed to the further conclusion that since eternal things

[219] *K. al-Kashf,* p. 36.

[220] Ibid., pp. 36–37.

[221] They both contrast the present proof with another that they do attribute to Juwaynī. References above, nn. 175, 220.

[222] *K. al-Shāmil* p. 218.

[223] Above, p. 92.

[224] Above, pp. 92–93.

[225] *Phaedo,* 78C; above, p. 92.

[226] *Metaphysics* XII, 2, 1088b, 14–28.

are not composite whereas the heavens are, the heavens cannot be eternal. A different conclusion, more in the spirit of Aristotle's philosophy, might, however, be drawn with equal cogency. The proper conclusion to be drawn from Aristotle's words is, according to Averroes, not that the heavens are generated but that they are free of the composition of matter and form: If eternal beings are not composite and the heavens are known with certainty to be eternal, then—as Averroes spells out Aristotle's intent—the heavens are not composite.[227]

An argument is also to be found in Proclus which concludes that since the physical universe is composite, it must be "generated." But what Proclus means thereby is that the universe is generated eternally, that is to say, eternally dependent on a cause for its existence. He reasons: Anything consisting in the joining together of parts must have a cause responsible for the parts' being joined; nevertheless, the joining together of the parts and the composite product may be eternal, as is the case, so Proclus understands, with respect to the heavens.[228] In arguing from the composition of the physical universe to the generation of the physical universe, Proclus is, in other words, advancing a proof of the existence of God and not a proof of creation.

In medieval Arabic philosophy a number of arguments from composition are in evidence. Some are arguments for the creation of the world, and others, for the existence of God, without reference to creation. The character of the composition from which either creation or the existence of God is derived varies, and at least three strains can be distinguished. (a) In certain instances the general fact of composition in the universe is adduced with no attention given to the specific character of the composition. Here an additional bifurcation can be discerned. Sometimes the inference of creation or the existence of God is put forward without explanation and apparently is regarded as self-explanatory or self-evident; such arguments may stem from Philoponus or, conceivably, even directly from Plato, Aristotle, or Proclus. Sometimes, by contrast, grounds are given for the inference of creation or the existence of God from composition. (b) In other instances, the combination specifically of contrary qualities in the universe is focused on, the contention being that the joining together of contrary qualities could only be effected by an overriding external force. (c) In still other instances, the focus is on the purposefulness of the composition in the universe, and the arguments are of a teleological character. None of the arguments in any of the categories is particularly subtle or profound.

[227]*Middle Commentary on Metaphysics*, Casanatense Library, Hebrew MS. 3083, XII, p. 140 (141)a; *Tahāfut al-Tahāfut*, IV, pp. 280–281.

[228]*Commentary on Timaeus*, ed. E. Diehl, Vol. I (Leipzig, 1903), pp. 290, 297. French translation, with pagination of the Greek indicated: *Commentaire sur le Timée*, trans. A. Festugière, II (Paris, 1967). *Elements of Theology*, ed. and trans. E. Dodds (Oxford, 1936), §47; *Liber De Causis*, ed. and German trans. O. Bardenhewer (Freiburg, 1882), §27.

(*a*) In the previous chapter, Saadia and Kindi were seen to offer proofs for creation from the composition in the universe. Saadia's proof did not explain how composition implies creation,[229] and Kindi's was discovered to be convoluted and circular.[230] Because of the context in which Saadia presented his proof and because of Kindi's terminology and the context in which he presented his, I suggested that both were outgrowths of Philoponus' arguments from composition.[231] A proof of creation from composition is also to be found in the *'Uyūn al-Masā'il,* a work attributed—incorrectly—to Alfarabi. The following syllogism appears there: "The world is composite; everything composite is generated; consequently . . . the world is generated."[232] As in Saadia and Kindi, only creation, not creation *ex nihilo,* is established by the argument; and, as in Saadia, the inference of generation, or creation, from composition is in no way justified. Another argument from composition in which the inference remains unexplained is advanced by the Arabic Christian writer Theodore Abū Qurra (ca. 740–820); now, however, the existence of God is being proved, although with an intimation that creation too might be proved by the argument. Abū Qurra lays down the principle that "whenever something is composite, its parts precede it naturally and usually . . . also temporally."[233] To buttress the principle, he refers to the empirical truth that the existence of building materials "precedes" the construction of a house; but he nowise essays to demonstrate the principle philosophically. His conclusion is that a composite universe must depend for its existence upon an external agent who binds the components together.[234] Similar arguments for both the existence of God and the creation of the world appear among the early Scholastics.[235]

Attempts were sometimes made to explain why the existence of God and the creation of the world do follow from the composition of the universe. Shahrastānī cites an argument in the name of Ash'ari which utilizes Kalam concepts for the purpose. Atoms, the reasoning runs, continually pass from a state of being joined with other atoms to a state of being disjoined from them. Their state of being joined or disjoined cannot flow from their own nature; for if the nature of a given

[229] Above, p. 103.

[230] Above, pp. 111–112.

[231] Above, pp. 102–103, 115.

[232] *'Uyūn al-Masā'il,* §2, in *Alfārābī's philosophische Abhandlungen,* ed. F. Dieterici (Leiden, 1890). German translation: *Alfārābī's philosophische Abhandlungen aus dem Arabischen uebersetzt,* trans. F. Dieterici (Leiden, 1892).

[233] For the different senses in which one thing can be said to precede another, see Aristotle, *Categories* 12.

[234] Arabic text in *al-Mashriq,* XV (1912), 762; German translation: *Des Theodor Abū Ḳurra Traktat Ueber den Schoepfer,* trans. G. Graf (Muenster, 1913), pp. 16–17.

[235] Cf. G. Grunwald, *Geschichte der Gottesbeweise im Mittelalter* (Muenster, 1907), pp. 57–60 (arguments for the existence of God where the proposition that every composite has a cause is treated as self-explanatory); pp. 62–64 (where that proposition is explained by Alan of Lille); pp. 68–69 (an argument for creation where the inference is again treated as self-explanatory).

atom determined that it should be joined, it would always be so, and if its nature determined that it should be disjoined, it would always be disjoined.[236] The passage of atoms from one state to another must therefore depend on an external agent who is responsible for assigning either the one or the other state to each given atom. Thus far, Ash'ari and Shahrastānī have provided a crude proof of the existence of God. To carry the argumentation forward and develop it into a proof of creation, Shahrastānī adds a principle met earlier, the principle affirming that "what does not precede what is generated is likewise generated."[237] The conclusion is not rendered explicit by Shahrastānī, but is clear: Since atoms do not precede the states of conjunction and disjunction, and since every state of conjunction or disjunction is generated, all atoms and therefore the universe as a whole must be generated—generated, be it noted, *ex nihilo*.[238] The argument, which is cited by Shahrastānī in the name of Ash'ari, is recorded by Maimonides in his enumeration of the Kalam proofs of creation. There is a difference, however, in that Maimonides does not venture to explore the argumentation beyond the stage where an external agent responsible for the conjunction and disjunction of atoms is inferred. Maimonides does mention, with a hint of disdain, that the Kalam thinkers thought the argument could establish creation, as well as the existence of God. But the step in which creation was reached was apparently judged by him to be too weak to waste words on.[239]

Once Avicenna appeared on the scene, a new route was made available for completing arguments from composition. Avicenna, as will be seen in a later chapter, devoted considerable effort to the analysis of the concept *possibly existent*. And in the course of analyzing the concept, he established that every compound is possibly existent and that everything possibly existent depends upon something outside itself for its existence.[240] Those propositions, taken together with the empirical fact that the world is compound, furnished Fakhr al-Dīn al-Rāzī with the materials for a proof of the existence of God. The physical universe, Rāzī argues, is compound, everything compound is possibly existent, and everything possibly existent has a cause of its existence; consequently, the physical universe has a cause of its existence.[241] Whereas Rāzī's reasoning concludes with the existence of God, Ījī presented similar reasoning as a proof of both the

[236]The reasoning would have been tighter, had Shahrastānī written: "If the nature of a given atom determined that it should be joined *to a particular other atom*. . . ." In Kalam physics, all atoms in the world are in combination.

[237]Above, p. 136.

[238]*K. Nihāya al-Iqdām*, p. 11; Wolfson, *The Philosophy of the Kalam*, p. 386.

[239]*Guide*, I, 74 (3). The conclusion that atoms are created depends on the questionable principle that "what does not precede what is generated is likewise generated." Cf. above, p. 143. Maimonides himself offers a proof for creation from the composition in the universe. Cf. *Guide*, II, 22; below, p. 208.

[240]Below, p. 296.

[241]*Muḥaṣṣal*, p. 107.

existence of God and creation. The physical universe, Ījī contends, is compound; everything compound is possibly existent; what is possibly existent has a cause that brought it into existence; but whatever is brought into existence is preceded by nonexistence and hence is generated.[242]

(*b*) A second strain of argumentation does not concern itself with composition in general but rather with the composition specifically of contrary qualities. The notion that God or Nature reconciles the contrary forces in the universe goes back at least to the Pseudo-Aristotelian *De Mundo*.[243] A proof of the existence of God using the notion goes back at least to Athanasius (fourth century). Athanasius asks rhetorically how the existence of God can be denied by anyone who "discovers fire mixed with the cold, and the dry mixed with the wet, yet not opposing one another," that is to say, by anyone who beholds the blending together of the elements in nature despite their contrary qualities. The elements, with their opposing qualities, could, Athanasius concludes, only be bound together by an overriding external agent, in other words, by a deity.[244] Similar thinking appears in John of Damascus, Theodore Abū Qurra, and Job of Edessa, each of whom could have served as a bridge to the Islamic world. John of Damascus offers a proof, again of the existence of God, not creation, which reads: "How could such contrary natures as fire and water, earth and air, combine with one another to form one world and remain undissolved unless there were some all-powerful force to bring them together and always keep them so?"[245] Abū Qurra was seen earlier to contend that an external agent must be posited who binds together the components making up the universe.[246] Immediately after having made the general point, Abū Qurra turns his attention to the contrary characters of the four elements. He observes that earth and water naturally descend whereas air and fire ascend, that, further, contrary elements naturally destroy one another; and still the elements are constrained against their contrary natures into combination. The conclusion he draws now is stronger than the conclusion he drew earlier: Not only does composition in the universe indicate a cause; it indicates a cause able to "constrain" and "bind" together contrary qualities "against their nature," hence a cause possessed of "immeasurable . . . and indescribable power."[247] Job of Edessa, who wrote in Syriac, likewise contends that the contrary qualities—"heat," "cold," "wetness," "dryness"—cannot be imagined to combine by "themselves, because if left to themselves they would not have been induced to do

[242]*Mawāqif*, VII, p. 227. Ījī writes that the argument comes from Rāzī. For other arguments to the effect that what is possibly existent must be created, see below, pp. 191, 387.

[243]*De Mundo*, §5. Cf. W. Jaeger, *Nemesius* (Berlin, 1914), p. 112, n. 2; K. Gronau, *Poseidonius und die juedisch-christliche Genesisexegese* (Leipzig, 1914), p. 143, n. 2.

[244]*Contra Gentes*, ed. R. Thomson (Oxford, 1971), §36. Cf. Boethius, *Consolation*, III, prose 12.

[245]*De Fide Orthodoxa*, I, 3.

[246]Above, p. 148.

[247]See above, n. 234, Arabic text, pp. 762–763; German translation, pp. 17–18.

anything opposed to their nature. There must consequently exist a being beyond the elements" which combines them, "and that being is God."[248]

Kalam writers cite the combination of contrary qualities in the universe as evidence both of the creation of the world and the existence of God. The Mu'tazilite Naẓẓām is reported to have proved creation thus: "I find heat and cold joined in a single body despite their contrariety and mutual divergence; and I understand that they cannot be joined by virtue of themselves . . . that the agent who joined them is the agent who created them joined, and who constrained them. . . ."[249] Again: "Heat, I find, is contrary to cold, and contraries cannot be joined through themselves in a single place. Having found that they are joined, I know they have something that has joined and constrained them against their own character. What undergoes constraint . . . is weak. And its weakness and the fact that the constraining agent exercises effective control upon it are a proof of its having been generated and that something has generated and created it."[250]

Māturīdī cites the composition in the universe to prove both the existence of God and the creation of the world. When treating of the existence of God, he notes that the world contains contrary qualities, and he continues: "Any object in which contrary . . . and divergent natures are joined . . . cannot be joined by virtue of itself; consequently, it has a joining agent."[251] When treating of the issue of creation, he writes: "Perceivable objects [including the world as a whole] are unavoidably subject to the joining of differing and contrary natures. Those natures are characterized by mutual repulsion. . . . Consequently, their joining must be due to something other than themselves; and that establishes their generation."[252]

(*c*) In still other instances, attention is focused on the purposefulness of the composition in the universe, be it the purposefulness of composition in general or the purposefulness of the composition of contrary qualities.

Athanasius was just seen to infer the existence of God from the fact that contrary qualities are combined in the universe.[253] But as Athanasius proceeds we find that the joining of contraries implies for him not merely a "superior being and master" who forces them to obey as "slaves obey their lord."[254] The combination of contraries forms an "order" and "harmony," hence implies a "ruler," "director," and "king," who is responsible for the order and harmony.[255] The same configuration recurs in John of Damascus. John of Damascus, as was seen, infers the existence of God from the combination of "contrary natures." But he

[248] *Book of Treasures*, I, iv, p. 15 (cited by A. Ivry, *Al-Kindi's Metaphysics* [Albany, 1974], p. 27).
[249] Khayyāṭ, *K. al-Intiṣār*, §26.
[250] Ibid.
[251] *K. al-Tawḥīd*, p. 18.
[252] Ibid., p. 12.
[253] Above, p. 150.
[254] *Contra Gentes*, §37.
[255] Ibid., §38.

prefaces his proof with the heading: "The very maintenance, preservation, and government of creation teaches us there is a God." And he incorporates into his reasoning the observation that the four elements are "arranged" in "unceasing and unhindered courses."[256] His argument from the combination of contraries has, accordingly, itself entered into combination: It is combined with, or buttressed by, teleological considerations; and Aquinas was indeed to read it as a proof from design.[257] Scholastic literature exhibits additional instances of arguments for the existence of God where the composition of contraries is cited as evidence of design.[258]

The theme that the combination of contraries implies design appears in the writings of the Ikhwān al-Ṣafā'. The Ikhwān recommend the study of plant life in all its vast variety on the grounds that "a well-made product indicates a wise maker." The accompanying explanation is that "the four elements, with their opposing powers and mutually antagonistic natures, could not have been joined and combined [in the plant realm] . . . except through the design of a wise maker"; therefore, the study of plants leads to knowledge of God.[259]

The most complex web is woven by the Jewish writer Baḥya ibn Paquda. Baḥya offers a proof of creation based on the principle that "everything combined is generated," since its parts are "naturally" and "temporarily" prior to the composite.[260] As supporting evidence, he cites three types of composition in the universe. He first notes the construction of the universe from different parts, which are "combined and composed" to produce a fully furnished dwelling place for man, the "householder." Here the theme clearly is teleological. Then he observes that plants and animals are composed of "the four elements, . . . which differ and exclude one another, so that there would be no way for man to combine and order them. . . . Yet their combination . . . is firm and stable. . . . They cannot be mixed by virtue of themselves . . . [and consequently] what does combine them is different from them." Here we have the consideration that contrary qualities can be reconciled only through a constraining agent; and as in Athanasius and John of Damascus, the thought is associated with teleological considerations. As a third piece of evidence for composition in the universe Baḥya observes that the elements themselves are "combined from matter and form, that is to say, from substance [i.e., atom] and accident; their matter is prime matter, . . . whereas their form is the first general form which is the root of every substantial and accidental form."[261] The "first general form," the form common

[256] *De Fide Orthodoxa*, I, 3.

[257] *Summa contra Gentiles*, I, 13.

[258] C. Baeumker, *Witelo* (Muenster, 1908), p. 318.

[259] *Rasā'il, Physics*, vii (Beirut, 1957), Vol. II, p. 152. German translation: F. Dieterici, *Die Naturanschauung und Naturphilosophie der Araber im X Jahrhundert* (Leipzig, 1876), p. 163.

[260] Cf. Abū Qurra, above, p. 148.

[261] *al-Hidāya* (*Ḥobot ha-Lebabot*), ed. A. Yahuda (Leiden, 1912), I, 5 and 6.

to all elements, is nothing other than *corporeal form,* posited by certain commentators as an intermediate stage between Aristotle's prime matter and the proper form of each element. Corporeal form was construed by the Arabic Aristotelians as identical with the tridimensionality of all physical objects, or as the medium for the presence of tridimensionality in prime matter.[262] In elucidating the composition of the elements, Baḥya thus alludes to the composition attendant upon the tridimensionality of physical objects; and he does so although he could have made the more straightforward point that the elements are compounded of matter and the forms of each of the four elements. In his third piece of evidence for composition in the universe, Baḥya refers, then, to both kinds of composition—the composition of matter and form, and the composition of tridimensionality—which had been adduced by Philoponus in the supporting arguments to his proof of creation from finite power[263] and which subsequently appeared in Kindi.[264]

Baḥya's proof of creation from composition has, in fine, woven together the following strands: the argument that combination implies creation since the parts precede the whole; the consideration that combination implies design; the consideration that the combination specifically of contrary qualities implies an external constraining agent; and illustrations of combination which are employed by Philoponus and reappear in Kindi.

[262]Cf. Wolfson, *Crescas,* pp. 582–585.
[263]Above, p. 92.
[264]Above, pp. 111–113.

VI

Arguments from the Concept of Particularization

1. Inferring the existence of God from creation

The standard Kalam procedure for proving the existence of God was to establish the creation of the world and then infer, from creation, the existence of a creator identified as the deity. The procedure has been termed the Platonic mode of proving the existence of God because of a passage in the *Timaeus*. Plato there offered a brief argument for the world's having been "generated and having begun at a first point,"[1] whereupon he deduced the existence of a creator on the grounds that "what comes into existence must perforce come into existence through some cause."[2] Apparently Plato regarded the proposition that nothing can come into existence without "some cause" as self-evident. Galen's *Compendium of the Timaeus,* a text that happens to be known from a medieval Arabic translation, notes as much; Plato, writes Galen, put forward the proposition in question "without . . . demonstration, because it is something manifest to the intellect."[3] A number of Islamic and Jewish thinkers agree, either explicitly or implicitly, that the proposition is self-evident and that the inference of a creator from creation requires no justification.

Ghazali, for example, advances an ostensible argument to show that what comes into existence has a cause of its coming into existence.[4] He adds, however: "In reality, the foregoing . . . is not an argument . . . at all . . . but merely an explication of the terms *generated* and *cause*"; for as soon as anyone "comprehends the meaning of the terms, his intellect will necessarily affirm that whatever is generated has a cause."[5] Fakhr al-Dīn al-Rāzī reports: "The need of what comes

[1] *Timaeus,* 28B. The argument is not related to the medieval Arabic arguments.

[2] *Timaeus,* 28C. See above, Chapter I, n. 6.

[3] Galen, *Compendium Timaei Platonis,* ed. P. Kraus and R. Walzer (London, 1951), Arabic text, p. 4. Cf. also *Corpus Hermeticum,* ed. A. Nock and A. Festugière (Paris, 1945–1954), XIV, 6.

[4] See below, p. 162.

[5] *al-Iqtiṣād fī al-I'tiqād* (Ankara, 1962), pp. 25–26. Cf. also Ghazali, *al-Risāla al-Qudsīya,* ed. A. Tibawi as *Al-Ghazālī's Tract on Dogmatic Theology* (London, 1965), Arabic text, p. 16; English translation, p. 34.

into existence for an agent" was considered by certain Kalam sheikhs, though not by everyone, to be an item of "necessary" and "immediate" knowledge.[6] Ṭūsī has a similar report. He writes: In the view of the "later Kalam thinkers . . . the judgment that whatever comes into existence must inescapably have an agent bringing it into existence is immediate knowedge requiring no proof. . . ."[7] The position taken by Ghazali and recorded by Rāzī and Ṭūsī is endorsed by Ījī and his commentator. "The immediacy of the intellect," they assert, "testifies" that "whatever comes into existence has an agent bringing it into existence."[8]

Several figures who stand on the periphery of the Kalam, while not explicitly characterizing the proposition as self-evident, do treat it as such. Kindi and Joseph ibn Ṣaddiq ask whether the cause bringing a thing into existence might not be the thing itself; but as for the principle that anything coming into existence does have some cause or other, it is treated by them as beyond question.[9] Baḥya too takes for granted that everything coming into existence has a cause. After presenting his proof for creation he does feel called upon to state that it is "impossible for anything to produce itself."[10] But once the statement has been made, he can conclude: "We know by the testimony of healthy intellects" that if the world has come into existence, "something outside the world created and brought the world into existence."[11] In other words, as soon as the impossibility of something's producing itself is recognized, the human intellect testifies that what comes into existence is brought into existence through a cause from without. The more comprehensive principle that things coming into existence do have some cause or other is, Baḥya implies, self-evident.

For each of the authors referred to thus far, the proposition that what comes into existence has a cause is, then, self-evident; and each of them accordingly can—and does—state his proof or proofs of the creation of the world and infer the existence of a creator forthwith. Kalam writers who, on the contrary, did not regard the proposition as self-evident had to justify their inference of a creator from creation. Two lines of reasoning were pursued. Either an analogy was drawn between the coming into existence of objects and events within the world and the coming into existence of the world as a whole; or else recourse was had to an argument turning on the concept of particularization or the kindred concept of *tipping the scales*.

Saadia, in the passage that is germane, does not spell out his position but he seems to pursue the former line. He has an imaginary interlocutor trace the judgment that "there is nothing made without a maker" to what is "testified" in

[6]*K. al-Arba'īn* (Hyderabad, 1934), p. 89.

[7]Gloss to Rāzī, *Muḥaṣṣal* (Cairo, 1905), p. 106.

[8]*Mawāqif* (Cairo, 1907), VIII, p. 3.

[9]Kindi, *Rasā'il* (Cairo, 1950), I, pp. 123, 207; Joseph Ibn Ṣaddiq, *ha-'Olam ha-Qatan*, ed. S. Horovitz (Breslau, 1903), p. 49.

[10]*al-Hidāya* (*Ḥobot ha-Lebabot*), ed. A. Yahuda (Leiden, 1912), I, 6.

[11]Ibid., I, 7(3).

the realm of "the perceptible." The interlocutor's meaning, and probably Saadia's as well, is that in the realm of sense experience things coming into existence are dependent on a cause; and by analogy or induction, things coming into existence outside the realm of human experience, including the world in its entirety, can also be presumed to be dependent on a cause.[12] Māturīdī (d. 944) speaks more directly and clearly. "Building, writing, and ships," he writes, "testify to what we have said. For they can come into existence only through an existent agent, and the present instance," the world's coming into existence in its entirety, "must be similar"; human artifacts are known empirically to come into existence through an agent, and by analogy, the world as a whole—given its creation—must also be the work of an agent.[13] The reference by Saadia and Māturdiī to what is testified to empirically contrasts nicely with the previous references to the testimony of the intellect.[14] Māturīdī gives another version of the line of thought we are considering when he argues: "In the world as perceived (*shāhid*)[15] nothing exists which combines or separates by itself. It follows that such must have been true of the state of the world which cannot be perceived (*ghā'ib*)."[16] That is to say, just as nothing now coming into existence through the combination and separation of atoms can dispense with a cause, so too at the moment when the world as a whole came into existence, the elements constituting the world cannot—whether they themselves came into existence or already existed—have combined and separated spontaneously and without a cause. In Bāqillānī the argument runs: "Writing unquestionably requires a scribe; drawing, an artist; and building, a builder. Should anyone tell us that something written has come about without a scribe, or something molded without a molder, or weaving without a weaver, we would not doubt the speaker's ignorance." But the world is of a "more subtle and wondrous artisanship." Inasmuch as the world is known to "have come into existence," it must, by analogy and with even greater certainty, have a "maker."[17] A similar formulation is found in Baghdādī[18] and, very briefly, in Bazdawī.[19]

'Abd al-Jabbār adds a degree of analysis that raises the discussion from the level of analogy to what may be called the level of induction. "In the realm of what is perceptible (*shāhid*)," he reasons, "our operations . . . stand in need of, and are dependent on, us." So much is obvious from the circumstance that human

[12]*K. al-Amānāt wa-l-I'tiqādāt*, ed. S. Landauer (Leiden, 1880), I, 2, p. 39; English translation, with pagination of the Arabic indicated: *Book of Beliefs and Opinions*, trans. S. Rosenblatt (New Haven, 1948).

[13]*K. al-Tawḥīd* (Beirut, 1970), p. 18.

[14]Above, pp. 154–155.

[15]The root of the word is the same as that in the term *to testify*.

[16]*K. al-Tawḥīd*, p. 17. I have corrected *bi-ghayrihi* to *bi-ghā'ib*.

[17]*K. al-Tamhīd* (Beirut, 1957), p. 23. Bāqillānī may well have in mind the argument from analogy in Ashari, *Luma'*, I, 4. A teleological motif, such as will be discussed in the next chapter, also seems to be present.

[18]*K. Uṣūl al-Dīn* (Istanbul, 1928), p. 69.

[19]*K. Uṣūl al-Dīn*, ed. H. Linss (Cairo, 1963), p. 18.

operations "occur in conformity with our intention." Furthermore, the aspect of human operations rendering them dependent on the human agent can be nothing other than the fact of their "having come into existence." That is obvious from the circumstance that "what occurs in conformity with our intent" is precisely "their coming into existence"; since what occurs in conformity with the agent's intent is precisely the coming into existence of an action or operation, coming into existence and nothing else must be the aspect of actions and operations rendering them dependent on the human agent. Now by the Kalam rule of induction, whenever a primary characteristic or "ground" (*'illa*) is, as far as experience goes, invariably accompanied by another characteristic, the connection between the characteristics can be presumed to obtain where experience does not penetrate.[20] In the instance at hand, the "ground" is the characteristic of having come into existence, and the accompanying characteristic is dependence on an agent. 'Abd al-Jabbār therefore affirms that "whatever has in common with our operations [the characteristic of] having come into existence, must likewise have in common with them [the characteristic of] requiring an agent." Hence once the world is known to have come into existence, an agent must be posited who brought it into existence.[21] The argument that anything coming into existence requires a cause is formulated in almost the same terms by a Jewish Kalam work attributed to Abraham ibn Ezra,[22] and by two Karaite authors, Joseph al-Baṣīr and Jeshua b. Judah, who are known to have stood under the influence of 'Abd al-Jabbār's school. The Karaite authors explain: The "ground" (*'illa*) in human acts rendering them dependent on man for their existence is the fact of their "having come into existence." But when a given ground is invariably observed to be accompanied by another characteristic, whatever possesses the former must—even in areas beyond the scope of observation—possess the latter. In the instance at hand, the world has in common with human actions the characteristic of "having come into existence." It must consequently have in common with human actions the need for "an agent bringing it into existence."[23]

The foregoing quotations from Saadia, Māturīdī, Bāqillānī, 'Abd al-Jabbār, and 'Abd al-Jabbār's Jewish followers, all look like variant formulations of a single theme, though some are expressed more technically than others, and I have termed them a single line of reasoning and contrasted them with the earlier position according to which the dependence of things coming into existence on a cause is self-evident. Fakhr al-Dīn al-Rāzī viewed the matter differently. He drew a sharp distinction between "two approaches" to the inference of a creator

[20]Cf. J. van Ess, *Die Erkenntnislehre des 'Aḍudaddīn al-Īcī* (Wiesbaden, 1966), pp. 361, 381–391. The problem of determining which is consequent and which is ground is obvious.

[21]*Sharḥ al-Uṣūl* (Cairo, 1965), pp. 118–119.

[22]Published in *Kerem Ḥemed,* IV (1839), 3.

[23]Joseph al-Baṣīr, in P. Frankl, *Ein Mu'tazilitischer Kalām aus dem 10. Jahrhundert* (Vienna, 1872), p. 21; and in M. Schreiner, *Studien ueber Jeschu'a ben Jehuda* (Berlin, 1900), p. 39, n. 1. Jeshua, in Schreiner, pp. 38–39.

from creation; and some of the argumentation that has been quoted is subsumed by him under the first approach, while other argumentation is designated as the second.

Rāzī writes: Certain Kalam thinkers who inferred a creator from creation maintained that "the need of what comes into existence for an agent is [a piece of] necessary knowledge." To "prove" their thesis they pointed out that "anyone who sees a building erected or a castle upraised would know necessarily that the structure had a builder and maker; and, indeed, should someone allege that the building might come into existence without a maker and builder, he would be judged mad. We thus recognize the proposition to be [a piece of] immediate [knowledge]."[24] Rāzī is not saying that for the writers in question the coming into existence of the world is analogous to the coming into existence of a building, and just as we perceive the building to be the work of a builder, so by analogy should we judge creation to be the work of a creator. The step from the coming into existence of an effect to the existence of a cause and, more specifically, from creation to a creator is represented, in Rāzī's account, as "immediate" knowledge; and the example of a building and a builder serves not to prove the proposition's truth but rather to "prove" its immediacy and self-evidence. The example serves as a touchstone or mental experiment for eliciting what the intellect apprehends as self-evident truth. Men might not—although it sounds a bit paradoxical—realize at once that creation immediately entails a creator, since the creation of the world is an unfamiliar event. By testing himself with more familiar events like the coming into existence of human artifacts, a person may—such is the burden of Rāzī's account—more easily recognize that the step from something's coming into existence to an agent bringing the thing into existence is self-evident.

That, according to Rāzī, is one approach to the inference of a creator from creation. The proposition that what comes into existence requires a cause is taken as self-evident and the example of human artifacts serves to elicit its self-evidence. The second approach, he continues, is pursued by "most of the Mu'tazilite sheikhs." It treats the proposition that things coming into existence are dependent on a cause not as self-evident but as "something to be proved"; and the proof resorted to turns out to be the argument by induction met in 'Abd al-Jabbār.[25] The reasoning is given by Rāzī as follows: We perceive from our own experience that "the human agent brings his own actions into existence." We discern, moreover, that human "actions stand in need of us precisely because they come into existence after not having existed." The "ground" for their standing in need of an agent is thereby revealed to be nothing other than the fact of their "coming into existence." When a given ground is observed to be invariably accompanied by another characteristic, whatever possesses the former will possess the latter. The "world" as a whole does possess the critical ground of having "come into

[24] *K. al-Arba'īn*, p. 89.

[25] Rāzī refers to other, less familiar Mu'tazilites.

existence." The world must consequently possess the accompanying characteristic of "standing in need of an agent."[26]

Rāzī expects his readers to keep in mind the differing stands taken by Mu'tazilites and Ash'arites on the nature of human actions. Mu'tazilites held the common-sense view that men perform or "create their own actions," while the Ash'arites totally negated natural causation and subscribed to an occasionalistic scheme wherein God, not man, creates every human action. Bāqillānī and Baghdādī were members of the latter school, and Māturīdī belonged to a cognate school. Therefore when they state that things coming into existence require a cause and creation requires a creator, just as a building requires a builder, they cannot, Rāzī surmises, be genuinely thinking of the erection of a building by a human builder; for they believed that buildings and other artifacts are never fashioned by human hands.[27] When these writers compare the dependence of a building on a builder with the dependence of creation on a creator, they should accordingly be read not as drawing an analogy between a human artisan and the divine artisan, but merely as indicating how the self-evidence of the need of what comes into existence for a cause can be elicited. A true analogy or induction from human action to divine action could only be offered by a Mu'tazilite. And 'Abd al-Jabbār, a Mu'tazilite, does unmistakably offer an argument of the sort. He analyzes human acts, undertakes to isolate the characteristic rendering them dependent on the human agent, and generalizes from human actions to everything containing the critical characteristic. Rāzī, an Ash'arite, naturally enough rejects the Mu'tazilite argument by induction. He insists that if the existence of God is to be inferred from creation, the principle that everything coming into existence requires an agent must be accepted as a self-evident truth.[28]

Rāzī, then, demarcates the boundary between those who treat the inference of a creator from creation as self-evident and those who defend the inference by analogy and induction in a different place from where the obvious sense of the texts would locate it. But whatever the merits of his interpretation, one line of reasoning in support of the inference of a creator from creation, should the inference not be regarded as self-evident, did take the form of an analogy or induction. The second line of reasoning in support of the inference brings the concept of particularization into play. Since the two lines are nowise incompatible, several of the writers to be cited—Māturīdī, Bāqillānī, and Baghdādī—could deploy them side by side.

The particularization mode of argument searches for instances in the universe where, it understands, a given alternative has been selected over other, equally possible alternatives; and it submits that the arbitrary selection it discovers implies

[26]*K. al-Arba'īn*, pp. 89–90. Followed by Ījī, *Mawāqif*, VIII, p. 3.

[27]The building, which we are certain has a builder, is in other words fashioned by God.

[28]*K. al-Arba'īn*, p. 90. Juwaynī, *K. al-Shāmil* (Alexandria, 1969), p. 280, struggles with the question how an analogy might be drawn to human actions if men are not truly the authors of their actions.

a particularizing agent or a particularizing factor. In Māturīdī, the term *particularization* is absent, but the thought is unmistakable. Māturīdī builds on his proof of creation and writes: If the world came into existence spontaneously, "no one time would be more appropriate for its coming into existence than another." The world's having come into existence at a given specific time rather than another therefore "proves" that the world came into existence "through something else," through an agent who arbitrarily selected the time for it to appear.[29] Māturīdī might have said, as subsequent writers do, that a particularizing agent must have selected out a particular time for the world to come into existence.

Bāqillānī too has an argument from creation to a particularizing agent; he, however, looks not at the coming into existence of the world as a whole but at the coming into existence of the world's myriad parts, and he is able to draw his inference only by presupposing the Ash'arite denial of natural causation. In his proof for creation, Bāqillānī had established that accidents and bodies—the latter being composed of inert atoms to which accidents are conjoined—must have all come into existence.[30] After completing that proof, Bāqillānī proceeds: Things coming into existence are "similar to one another"; an accident carrying any quality is, in other words, exactly like other accidents carrying the same quality, and a physical object consisting of a number of atoms to which a set of accidents is conjoined is exactly like other physical objects consisting of the same number of atoms and the same set of accidents. Yet "despite . . . being similar to one another," things that come into existence make their appearance at different times, "some earlier . . . than others," and "some later." The cause of a thing's coming into existence at an earlier or a later time cannot be the thing "itself and its genus"; "for should a thing come into existence earlier or later by virtue of itself, everything of the same genus would come into existence at the [same] earlier [or later] time." Bāqillānī ignores the commonsense explanation for some objects' coming into existence earlier and some, later—namely, that natural forces within the universe, which operate in accordance with natural laws, determine the time when things come into existence. He can as a consequence conclude: The fact that a given physical object does come into existence earlier or later than something exactly like it "is proof that it has an agent rendering it early [or late] and assigning it a definite span of existence in conformity with the agent's will." The aggregate of objects making up the physical world must thus depend on an agent—and Bāqillānī will subsequently explain why there can be only one—who arbitrarily assigns each object a time for it to exist.[31] Although Bāqillānī does not expressly speak here of a particularizing agent who brings the world into existence, he employs the terms "particular" and "particularizing" a few lines afterwards in a parallel context.[32] The argument that each physical object

[29]*K. al-Tawḥīd*, p. 17.
[30]Cf. above, p. 136.
[31]*K. al-Tamhīd*, p. 23.
[32]Ibid., and below, p. 178.

coming into existence must have the time for its emergence assigned by an agent is advanced by Baghdādī as well; and Baghdādī does expressly call the latter a "particularizing agent" (*mukhaṣṣiṣ*).[33]

Juwaynī also has a particularization argument, but he, like Māturīdī, applies it to the world as a whole and thereby frees it again of the occasionalistic burden with which Bāqillānī and Baghdādī loaded it; if the world has come into existence as a whole there are—even on a naturalistic picture of the universe—no natural forces operating in accordance with natural laws which might have determined the moment for it to come into existence. Juwaynī's reasoning goes: The creation of the world has been demonstrated, and whatever comes into existence has an equal possibility of "existing . . . and not existing." The world, moreover, might have come into existence at different "possible" times. From the world's having come into existence at a given time rather than "continuing in a state of nonexistence," the "intellect immediately judges that the world requires a particularizing agent who selected out existence for it" at the time when it came into existence.[34] Juwaynī's use of the phrase "the intellect immediately judges" is revealing. The particularization argument was called into play by Kalam thinkers in order to furnish underpinning for the inference of a creator from creation and not leave it an item of self-evident knowledge. But, Juwaynī recognizes, the step from the selection between equal possibilities to a particularizing agent who makes the selection must itself ultimately be taken as self-evident.

The thought that creation points to a particularizing agent who selected out the particular moment for the world to come into existence is employed by Bazdawī,[35] a follower of Māturīdī, and by Ghazali;[36] and it is recorded by Judah Hallevi in his summary of Kalam doctrines.[37]

The passage quoted from Juwaynī merits further comment, since it reveals, in fact, not one, but two intertwined motifs. Juwaynī is saying both that a particularizing agent must have selected out a moment for the world to emerge in preference to other times when the world might have emerged; and that a particularizing agent must have selected out existence for the world in preference to nonexistence.[38] The former is the motif met already in Māturīdī, but the latter is a new motif, which must have been suggested to Juwaynī by Avicenna's analysis of the concepts *possibly existent* and *necessarily existent*.[39] Juwaynī realized that the two motifs are distinct.[40] And in a composition that was designed as a mere outline of his thought and not as a full-fledged theological work he restricts himself to the second of the two. Given creation, he explains there, the world

[33] *K. Uṣūl al-Dīn,* p. 69.

[34] *K. al-Irshād* (Cairo, 1950), p. 28; cf. *K. al-Shāmil,* pp. 262–265, 267.

[35] *K. Uṣūl al-Dīn,* p. 18.

[36] *al-Risāla al-Qudsīya,* Arabic text, p. 16; English translation, p. 34.

[37] *Kuzari,* V, 18(3).

[38] *K. al-Shāmil,* pp. 263–265, shows how Juwaynī came to combine the motifs.

[39] Cf. below, p. 290. [40] Below, p. 177.

must have a creator. For "things coming into existence" are "possibly existent (*jā'iz al-wujūd*), inasmuch as either their existence or—in preference to their existence—their continued nonexistence can be supposed. When they are particularized in [or: by] actual possible existence (*wujūd mumkin*) they stand in need of a particularizing agent."[41]

Echoes of Avicenna are more audible in a variation of the particularization argument offered by Ghazali, Fakhr al-Dīn al-Rāzī, and Ījī. Avicenna had insisted that when something possibly existent becomes actual, actual existence must be "differentiated out" for the object over nonexistence.[42] It was not Avicenna's intent that the world is anything but eternal, and he certainly did not have in mind a temporal differentiation out of actual existence for the world as a whole. Ghazali and the others, however, turn Avicenna's analysis to their own purpose. They are once more justifying the inference of a creator from creation. They focus exclusively on the selection of existence for the created world in preference to nonexistence; and they now employ not the language of *particularization,* but kindred language, that of *tipping the scales*. Everything coming into existence, they write, plainly is "possibly existent" prior to actually existing; for if a thing were "impossible," it would never exist, whereas if it were "necessarily existent by virtue of itself," it would always exist. But anything whose existence is possible is, in itself, equally capable of existing and not existing. Something of the sort can, therefore, enter the domain of actual existence only through an agent that "tips the scales (*murajjiḥ*) in favor of its existing." The world is known, thanks to Kalam proofs of creation, to be an object that came into existence, and hence an object that was possibly existent before actually existing. When the world entered the domain of actual existence an agent must, then, have tipped the scales in favor of its existence.[43]

Such were the Kalam arguments supporting the inference of a creator from creation. A supplementary detail was sometimes provided both by those who treated the inference of a creator from creation as self-evident and those who offered arguments to support the inference.

The rhetorical question is posed: Granted that the world must have had a cause bringing it into existence, why might not the world itself have been the cause? Why cannot an object bring itself into existence and select out the particular moment for its own emergence? The most common response goes back at least to Proclus' commentary on Plato's *Timaeus*. When expatiating on Plato's statement that "what comes into existence must perforce come into existence through

[41] *Textes apologétiques de Ǧuwainī (Luma')*, ed. M. Allard (Beirut, 1968), pp. 128–129.

[42] *Najāt* (Cairo, 1938), p. 226; *Shifā': Ilāhīyāt*, ed. G. Anawati and S. Zayed (Cairo, 1960), p. 39; French translation, with pagination of the Arabic indicated: *La Métaphysique du Shifā'*, trans. G. Anawati (Paris, 1978).

[43] Ghazali, *al-Iqtiṣād*, pp. 25–26; Rāzī, *K. al-Arba'īn*, p. 86, together with p. 71; cf. idem, *Muḥaṣṣal*, p. 106, together with pp. 53–54; Ījī, *Mawāqif*, VIII, p. 3.

some cause,"[44] Proclus wrote: "Nothing can possibly bring itself into existence; for if it did, it would exist before coming into existence" which is absurd.[45] Saadia, Māturīdī, Baghdādī, Baḥya, and Joseph ibn Ṣaddiq argue in the same vein. Before something has come into existence, they point out, it surely cannot be described as the agent bringing it into existence; for it does not yet exist, and what does not exist cannot function as an agent. But after a thing has already come into existence it again cannot be described as the agent bringing it into existence, since it already exists and no longer needs to be brought into existence. There is thus no time at which a thing can be described as bringing itself into existence; and the world cannot have been brought into existence by itself.[46] 'Abd al-Jabbār says more or less the same in different words. Nothing, he contends, can bring itself into existence, since only what has power can bring things into existence, and "what has power over anything must precede its effect. If the agent bringing a body into existence were the body itself, it would have to have power while still nonexistent," which is impossible.[47]

Kindi and Ibn Ḥazm introduce the following consideration: To state that an agent brings something into existence is to affirm, in effect, that two entities exist, one of which acts upon the other. The supposition that something brings itself into existence would, hence, be tantamount to supposing that a single thing is two distinct things. Since the supposition is self-contradictory and absurd, nothing, including the world, can bring itself into existence.[48]

Additional considerations were advanced which, we should probably agree, fall somewhat below the threshold of philosophy. Bāqillānī reasons: The world contains death; but what contains death cannot be a creator; therefore the world cannot be a creator.[49] Saadia and Māturīdī reason: If the world had the capability of creating itself, it would likewise have had the capability of desisting from creating itself. To speak, however, of the world's desisting is to imply that the world existed, whereas to speak of its desisting from creating itself would be to imply that the world had never been created and never existed. The notion of the world's desisting from creating itself is thus self-contradictory and impossible. By definition, the possible is such that when it is assumed to exist, nothing

[44]Above, n. 2.

[45]*Commentary on Timaeus*, ed. E. Diehl (Leipzig, 1903–1906), I, p. 260, lines 8–9; French translation, with pagination of the Greek indicated: *Commentaire sur le Timée*, trans. A. Festugière (Paris, 1966–1967).

[46]Saadia, *K. al-Amānāt*, I, 2, p. 38; Māturīdī, *K. al-Tawḥīd*, p. 18; Baghdādī, *K. Uṣūl al-Dīn*, p. 69; Bahya, *al-Hidāya* (*Ḥobot ha-Lebabot*), I, 5; Ibn Ṣaddiq, *ha-'Olam ha-Qatan*, p. 49. Cf. below, n. 48. The argument was known to Hume, *Treatise of Human Nature*, I, iii, 3.

[47]*Sharḥ al-Uṣūl*, p. 119.

[48]Kindi, *Rasā'il*, I, pp. 123–124; Ibn Ḥazm, *K. al-Faṣl fī al-Milal* (Cairo, 1964), I, p. 18. At first Kindi seems to be arguing as Saadia, Baghdādī, and the others did, but then he slides into this argument.

[49]*K. al-Tamhīd*, p. 24. Cf. Māturīdī, *K. al-Tawḥīd*, p. 17.

impossible results therefrom.[50] Since the hypothesis that the world had the capability of creating itself leads to the impossible notion that the world could have desisted from creating itself, since the hypothesis does result in an impossibility, it fails to satisfy the definition of the possible. That the world might have created itself is therefore not possible.[51]

Saadia, to give a final instance, writes: When a thing exists, it is stronger than when it does not exist. The world in its state of existence is too weak to create a world. *A fortiori*, the world would not have been strong enough to create a world before existing.[52]

To recapitulate: The standard Kalam procedure for proving the existence of God was to infer a creator from creation on the grounds that what comes into existence must have a cause. Some Kalam authors treated the proposition that things coming into existence must have a cause bringing them into existence as self-evident. Of those who did not treat the proposition as self-evident, some supported the inference of a creator from creation through an argument by analogy or induction. And some supported it through an argument turning on the concept of particularization or the kindred concept of tipping the scales. Once satisfied that the coming into existence of the world does require a cause, a number of Kalam writers appended a supplementary detail. They explained why nothing can bring itself into existence and why the creator of the world cannot have been the world itself.

Adherents of the Kalam do not stop here. They proceed to argue that the cause of the existence of the world must be eternal; several undertake to prove that what brought the world into existence cannot be a *nature* or *necessary cause*; invariably they argue that the cause of the world must be one and incorporeal and that it must possess power, knowledge, life, and will. Unfortunately, Kalam thinkers do not make clear which of the foregoing they consider to be part of the concept of God, indispensable conditions such that nothing can merit the designation *deity* without them; and which are ancillary attributes, not integral to the very concept of deity. If an attribute is deemed integral to the concept of the deity, a proof of the existence of God will not be complete until the cause of the universe is shown to possess it. If, for instance, 'God' is taken to mean the 'single incorporeal cause of the world,' the existence of God will not have been established until the world is shown to have a cause that is one and incorporeal, whereas if 'God' means merely the 'cause of the world,' a proof of the existence of God can be achieved without arguments for unity and incorporeality.

Of the attributes that have been mentioned, unity, especially, might be expected to be integral to the Kalam concept of the deity; the unity of God was so central a Kalam doctrine that 'establishing the unity [of God]' (*tawḥīd*) was the Kalam

[50] Aristotle, *Prior Analytics* I, 13, 32a, 18–20.

[51] Saadia, *K. al-Amānāt,* p. 38; Māturīdī, *K. al-Tawḥīd,* p. 18.

[52] *K. al-Amānāt,* p. 37.

term for 'natural theology.' The texts dealing with the unity of God are, as happens, ambiguous. Writers propose to demonstrate that "God is one,"[53] and by expressing themselves in that way may appear to grant that God can be known to exist before he is known to be one. And yet in the most popular proof for unity, the pivotal thought is that the hypothesis of two deities cannot be squared with what is meant by 'God'; and there the presupposition would appear to be that unity is after all an indispensable condition for the deity. The writers who follow their proof for the existence of a creator with an argument establishing that the creator cannot be a nature or necessary cause undoubtedly consider this characteristic at least, the characteristic of not being a nature or necessary cause, to be an indispensable specification for the deity. And a few other attributes also are represented as indispensable specifications.[54] But as for the remainder, including unity, it is hard to determine which the generality of Kalam writers did, and which they did not, understand to be specifications for the deity, and where, consequently, their proofs of the existence of God end and their arguments for the ancillary attributes begin.

At any rate, virtually all who infer a creator from creation add that not merely does the existence of the world depend upon a cause; it depends upon a first eternal cause. The stock explanation is that an infinite regress would otherwise ensue. As Baghdādī puts it: "If the creator himself came into existence, he would stand in need of an agent to bring him into existence. If the latter too came into existence, he in turn would stand in need of a third agent. And the series would regress (*yatasalsal*) infinitely, which is absurd." To avoid an infinite regress, the "creator" of the world—or, to be more precise, the ultimate creator—must be judged "eternal."[55] The argument for the eternity of the creator from the impossibility of an infinite regress is given by Bāqillānī,[56] 'Abd al-Jabbār,[57] Juwaynī,[58] Ghazali,[59] Baḥya,[60] Joseph al-Baṣīr,[61] Judah Hallevi,[62] Abraham ibn Ezra,[63] Fakhr al-Dīn al-Rāzī,[64] and Ījī.[65] It contrasts slightly with an argument employed by Aristotle and his medieval disciples. Aristotle and the medieval Aristotelians reached a first cause, not a first creator, of the world with the aid of a parallel

[53]'Abd al-Jabbār, *Sharḥ al-Uṣūl,* p. 277; Juwaynī, *K. al-Irshād,* p. 53; Ghazali, *al-Iqtiṣād,* p. 73; Ījī, *Mawāqif,* VIII, p. 39.

[54]See below, p. 167.

[55]*K. Uṣūl al-Dīn,* p. 72.

[56]*K. al-Tamhīd,* p. 25.

[57]*Sharḥ al-Uṣūl,* p. 181.

[58]*K. al-Irshād,* p. 32; *K. al-Shāmil,* pp. 617–618.

[59]*al-Iqtiṣād,* p. 35.

[60]*al-Hidāya* (*Ḥobot ha-Lebabot*), I, 6; 10.

[61]Frankl, *Ein Mu'tazilitischer Kalām,* p. 24.

[62]*Kuzari,* V, 18(4).

[63]*Kerem Ḥemed,* IV, p. 4.

[64]*K. al-Arba'īn,* p. 92 (in the name of the "sheikhs" of the Kalam).

[65]*Mawāqif,* VIII, pp. 4–15.

consideration, the consideration that an infinite regress of causes is impossible. But their grounds for ruling out an infinite regress of causes were that the nature of causation precludes causes' running to infinity;[66] whereas the adherents of the Kalam rule out an infinite series of agents bringing the world into existence at a particular moment on the grounds that an infinite regress of any sort is impossible. The difference may sound insignificant. Yet Ghazali made a good deal of it and undertook to show that the Aristotelian approach to the infinite regress embodies an inconsistency and as a consequence is, unlike the Kalam approach, untenable.[67]

Kalam argumentation has arrived at a first eternal cause of the creation of the world. Bāqillānī, 'Abd al-Jabbār, Juwaynī, and Jeshua b. Judah, furnish an argument establishing that the first cause cannot be a "nature" or a "[necessary] cause" (*'illa*), but must be a "voluntary agent." The reasoning goes: The cause of the world is known to be eternal. Both a nature and a necessary cause operate in an unvarying manner; and if an eternal nature or eternal necessary cause had acted to produce the world, it would have acted as long as it existed, and the world would likewise be eternal. But the Kalam proofs of creation have demonstrated that the world was created. The world therefore cannot have been produced by a nature or a necessary cause, and must instead be the work of a "voluntary agent."[68]

A variety of grounds, all highly dialectical, were adduced to establish the unity of God. Most popular was (a) the argument from mutual interference (*tamānu'*), an argument whose provenance apparently was Greek. The author of the Hermetic Corpus contended that if more than one deity existed, "rivalry" would beset them.[69] And John of Damascus defended the unity of God with the argument that if "several gods" existed, "conflict" between them could not be avoided, and the world would be "broken up and utterly destroyed."[70] For their part, the Islamic writers traced the argument from mutual interference not to Greek sources but to Scripture.

Quran 21:22 reads: "If gods other than God were in them [i.e., in the heavens and earth], both [the heavens and earth] would fall into ruin." And Quran 23:91 reads: "There is no god along with him; else each god would assuredly have championed what he created, and one would have overcome the other." Ash'ari spells out the intent of the Quranic verses: On the hypothesis of two creators, one of the two might will something while the other willed the contrary. For example, one might "will to have a man live, while the other willed to have him die." Should such occur, it would not be possible for the will of both to be accomplished; nor for the will of neither to be accomplished; nor again for the

[66] See below, p. 337.

[67] See Appendix A.

[68] Bāqillānī, *K. al-Tamhīd*, pp. 34–35; 'Abd al-Jabbār, *Sharḥ al-Uṣūl*, p. 120; Juwaynī, *K. al-Irshād*, pp. 28–29; Schreiner, *Studien ueber Jeschu'a ben Jehuda*, p. 38. Baghdādī, *K. Uṣūl al-Dīn*, p. 69, has a different argument showing that the creator is not a "nature."

[69] *Corpus Hermeticum*, XI, 9.

[70] *De Fide Orthodoxa*, I, 5.

will of one to be accomplished, while the will of the other was frustrated. Clearly, "what both will could not conceivably be accomplished; for a body cannot conceivably be alive and dead at the same time." Nor might "what neither wills be accomplished." The reason therefor is that if the will of neither creator were accomplished, both assumed creators would "perforce be powerless, whereas," Ash'ari postulates, "what is powerless cannot be God or eternal." Later writers would here interpose an additional, logical reason why it would be impossible that the will of neither supposed deity should be accomplished; if the will of neither were accomplished, the body would be neither dead nor alive; and the law of the excluded middle would be violated.[71] Finally, writes Ash'ari, it is impossible that the will of one of the two assumed divine creators should be accomplished while the will of the other is frustrated. For if "what one of the two assumed creators wills is accomplished to the exclusion of the other," the former alone would be God, seeing that the other would be powerless, and "anything powerless cannot be God or eternal." Thus, no more than one divine creator can, Ash'ari concludes, exist.[72] The argument from mutual interference is employed by Saadia[73] and Māturīdī,[74] who like Ash'ari were active in the early tenth century; and it undoubtedly was already a stock argument by the end of the previous century.[75]

When Bāqillānī subsequently gave his version, he put Ash'ari's remark concerning power and eternity a trifle more explicitly. "Powerlessness," Bāqillānī comments in the course of restating Ash'ari's argument, "is a mark of having come into existence," and for that reason "the eternal cannot be powerless."[76] The presupposition in Ash'ari and, more explicitly, in Bāqillānī—a presupposition other Kalam thinkers employing the argument concur in—is that eternity and power are part of the irreducible concept of the deity, and nothing can be designated a deity without being eternal and powerful. Inasmuch as the unity of God is being demonstrated through the presupposition that eternity and power are integral to the concept of God, unity too is perhaps being construed as integral to the concept of God.

The argument from mutual interference invites an objection, and the objection is taken up by 'Abd al-Jabbār. He presents the argument from mutual interference much as Ash'ari did, although the imagined conflict between the deities which he outlines is slightly different; he assumes not that one of the supposed deities wills life for a given body while the other wills death, but rather that one wills

[71]Baghdādī, Juwaynī, Shahrastānī, and Ījī, in passages cited below, nn. 80, 81, 83, 84.

[72]*K. al-Luma'*, in *Theology of al-Ash'arī*, ed. and trans. R. McCarthy, (Beirut, 1953), §8.

[73]*K. al-Amānāt*, II, 3, p. 82.

[74]*K. al-Tawḥīd*, p. 20.

[75]The argument seems to be present in al-Muqammiṣ (ninth century). See G. Vajda, "Le Problème de l'Unité de Dieu d'après al-Muqammiṣ," *Jewish Medieval and Renaissance Studies*, ed. A. Altmann (Cambridge, Mass. 1967), p. 57.

[76]*K. al-Tamhīd*, p. 25.

motion while the other wills rest. Should conflict of the sort occur, it would, 'Abd al-Jabbār shows, be impossible for the will of both supposed deities to be accomplished; it would be impossible for the will of neither to be accomplished; and if the will of only one were accomplished, the one whose will was accomplished would alone be the deity.[77] 'Abd al-Jabbār concludes that only one deity can exist, whereupon he turns to the objection.

The objection runs: Granted that two deities could not exist who will contrary effects, might not two deities "be wise and never interfere with each other?" Might there not exist two deities who always will identical events and always see their will accomplished? 'Abd al-Jabbār's response is that the argument from mutual interference requires not the "reality" of a conflict of wills, but merely the "supposition" (*taqdīr*) or "possibility" (*ṣiḥḥa*) of a conflict. Since on the hypothesis of two deities the two would have the possibility, at least, of willing contrary effects, the question remains which of them would, in the event of a conflict, be able to execute his will. And the one possessing the ability to execute his will could alone be called the deity.[78] 'Abd al-Jabbār may well have in mind the definition of the possible as what is such that "when it is assumed to exist, nothing impossible results therefrom."[79] His thinking would be that the hypothesis of two deities entails the possibility, at least, of their willing anything in their power, and hence their willing contrary events. The existence of two deities who will contrary events has, however, been shown to be an impossibility. The hypothesis of two deities is thus found to be an assumption from which something impossible does result, and the hypothesis fails to satisfy the definition of the possible. No more than one deity can, consequently, exist.

The argument from mutual interference, together with the response to the objection that two deities might perhaps always will identical effects, is put forward by Baghdādī,[80] Juwaynī,[81] Joseph al-Baṣīr,[82] Shahrastānī,[83] and Ījī.[84] The argument without the objection and response to it is given by Taftāzānī,[85] and is alluded to by Baḥya[86] and the Karaite Judah Hadassi.[87]

[77]*Sharḥ al-Uṣūl,* p. 278.

[78]*Sharḥ al-Uṣūl,* p. 283, with further elaborations. Cf. *K. al-Majmū' fī al-Muḥīṭ bi-l-Taklīf,* ed. J. Houben (Beirut, 1965), pp. 215–217, where there are even more elaborations.

[79]Cf. *Prior Analytics* I, 13, 32a, 18–20. Ījī, below, n. 84, quotes the definition of possible in his version of the argument.

[80]*K. Uṣūl al-Dīn,* p. 85.

[81]*K. al-Irshād,* pp. 53–54, 57; *K. al-Shāmil,* pp. 352–382, with an elaborate discussion of the eight "principles" underlying the argument, and with an excursus showing that Mu'tazilites, who recognize the efficacy of human will, are guilty of inconsistency in employing the argument.

[82]Frankl, *Ein Mu'tazilitischer Kalām,* p. 27.

[83]*K. Nihāya al-Iqdām,* ed. A. Guillaume (Oxford and London, 1934), pp. 91–92; 94–96.

[84]*Mawāqif,* VIII, p. 42.

[85]Taftāzānī, *A Commentary on the Creed of Islam,* trans. E. Elder (New York, 1950), p. 37.

[86]*al-Hidāya* (*Ḥobot ha-Lebabot*), I, 7 (end).

[87]*Eshkol ha-Kofer* (Eupatoria, 1836), §26.

Averroes and Maimonides reject the argument on the grounds that two deities might exist who have separate responsibilities, each creating a segment of the world and not interfering with the other.[88] Āmidī rejects the argument because he cannot accept the response made by 'Abd al-Jabbār to the objection that two deities might always agree. As far as the argument from mutual interference goes, Āmidī surrejoins, two deities might indeed exist who always agree; for the assumption of two deities does not entail even the possibility of conflict. The fact that when the two are considered in isolation from each other,[89] one of them would be able to will an effect and his compeer would be able to will the contrary does not mean that they would have the possibility of doing so at the same time. The situation, Āmidī proceeds, may be clarified by considering a single divine agent. A single divine agent can, at any moment, will an event or its contrary, but he cannot conceivably will both the event and its contrary at the same moment. Similarly, each of two divine agents would be able to will an effect or its contrary when each and his exercise of will is taken in isolation; but it does not follow that one of the two could will the effect while his compeer was willing the contrary. Two deities might then exist who do not have even the possibility of willing contrary events; and the argument from mutual interference collapses.[90]

Further arguments for the unity of the creator were current in Kalam circles. In (b) an extension or attenuation of the argument from mutual interference, it was contended that the hypothesis of two divine creators embodies a contradiction. Joseph ibn Ṣaddiq explains: Should two agents produce the world cooperatively, they would "stand in need of each other"; they would be weak; they would not be "eternal, inasmuch as the eternal cannot be weak"; and yet the world has been shown to have an eternal cause. "If, by contrast, one of the two agents produced the world by himself, . . . the other would be weak and not . . . eternal." The existence of two divine creators is therefore impossible.[91] Ījī puts the thought differently: If two deities were equally "powerful" (*qādir*), any possible "object of activity" (*maqdūr*) for them would have an "identical . . . relation . . . to each"; and as a consequence, they could perform no action whatsoever. No action could be performed by them cooperatively; for a given single action is, by definition, performed by one and not by two agents.[92] Nor could an action be performed by one to the exclusion of the other. For since every object of

[88] Averroes, *K. al-Kashf*, ed. M. Mueller (Munich, 1859), p. 49; German translation, with pagination of Arabic indicated: *Philosophie und Theologie von Averroes aus dem Arabischen uebersetzt*, trans. M. Mueller (Munich, 1875); Maimonides, *Guide to the Perplexed*, I, 75(1).

[89] Āmidī apparently has in mind Juwaynī, *K. al-Irshād*, p. 54.

[90] *Ghāya al-Marām* (Cairo, 1971), pp. 151–152.

[91] *ha-'Olam ha-Qatan*, p. 50. Similar reasoning appears in al-Muqammiṣ, in Vajda, "Le Problème de l'Unité de Dieu," pp. 57–58; and Baḥya, *al-Hidāya* (*Ḥobot ha-Lebabot*), I, 7(7). Cf. also Shahrastānī, *K. Nihāya al-Iqdām*, p. 92; Maimonides, *Guide*, I, 75(5).

[92] I understand that Ījī is defining a single action as the action done by a single agent. Should two agents perform an action, we would in fact have two actions.

activity would have an identical relation to each of the two agents, neither of them could, to the exclusion of the other, undertake the activity—unless some wholly inexplicable factor were to materialize and "tip the scales," thereby determining which of the two equally powerful agents should act. Inasmuch as equally powerful divine agents would be incapable of performing any action whatsoever, the world must be the handiwork of a single powerful divine agent.[93]

Saadia and Baḥya advance (c) an argument that is recorded as well by Juwaynī, where it is attributed to the Mu'tazilites and "many of our [i.e., Ash'arite] scholars."[94] It goes: Creation requires that a creator be posited; the requirement is satisfied as soon as a single creator is acknowledged; hence a single creator is all that should be posited.[95] A specialized version is offered by Shahrastānī and recorded by Āmidī. As was seen, one line of reasoning in support of the inference of a creator from creation had been that an agent would be needed to "tip the scales" in favor of the world's coming into existence. The act of tipping the scales, Shahrastānī, now contends, "points" to an agent tipping the scales, but not to "two agents." Consequently only a single agent who tipped the scales in favor of the world's coming into existence should be posited.[96] Juwaynī, who refers to the general form of the present argument, and Āmidī, who refers to the specialized form, reject the argument. They remark, cogently, that the "nonexistence of a proof" is not tantamount to "proving the nonexistence of a thing." Accordingly, the failure of the proof from creation to demonstrate more than a single creator is nowise tantamount to its demonstrating that no more than a single creator exists.[97] Maimonides was familiar with all sides of the discussion. He records the Kalam argument to the effect that no more than one divine creator should be posited because no more than one had been demonstrated. He repeats the objection raised by Juwaynī and Āmidī. He even reports a rejoinder to the objection, and submits a surrejoinder of his own.[98]

The foregoing have been arguments for the unity of God the burden of which is, in each instance, that the world could not have more than one creator. In addition, Kalam writers put forward considerations that do not build on the prior inference of a creator from creation. A number of writers maintain (d) that only one deity can existence because two entities possessing the nature of the deity could not be differentiated. The explanation takes divers forms.[99] A common form (d.i), which has something of an Aristotelian cast,[100] goes: The deity is

[93]Ījī, *Mawāqif,* VIII, p. 41. Similar arguments appear in Juwaynī, *K. al-Shāmil,* p. 384 (attributed to the Mu'tazilites); and Āmidī, *Ghāya al-Marām,* p. 154.

[94]Juwaynī, *K. al-Shāmil,* p. 387.

[95]Saadia, *K. al-Amānāt,* p. 80; Baḥya, *al-Hidāya* (*Ḥobot ha-Lebabot*), I, 7(3).

[96]*K. Nihāya al-Iqdām,* p. 93, intertwined with other considerations.

[97]Juwaynī, *K. al-Shāmil,* p. 387; Āmidī, *Ghāya al-Marām,* p. 153.

[98]*Guide,* I, 75(4).

[99]The motif appears as well in John of Damascus, *De Fide Orthodoxa,* I, 5.

[100]*Metaphysics* XII, 8, 1074a, 33–37.

incorporeal. Inasmuch as two identical incorporeal beings could neither occupy different places nor exist at different times, they could in no way be distinguished from each other. Consequently, no more than one entity possessing the nature of the deity can exist.[101] An alternate form (d.ii) has a purely Kalam cast. Two entities with the nature of the deity, it is argued here, could be differentiated from each other only if one or both contained an added element setting it apart. The added element would have the status of an accident; everything affected by an accident is known—through the proof for creation from accidents[102]—to belong to the class of beings that come into existence; and the ultimate cause of the world cannot have come into existence. Hence, no more than one entity possessing the nature of the deity can exist.[103] A closely related form (d.iii) was: If two entities possessing the nature of the deity were to be differentiated, one or both would have to contain an added element setting it apart. The entity containing the added element would be composite; what is composite is generated;[104] what is generated cannot be the eternal first cause. Only one entity with the nature of the deity can, therefore, exist.[105] Still another closely related form (d.iv) incorporates concepts from Avicenna's philosophy and runs: If two deities could be differentiated, one or both would, as before, have to be composite. But the cause of the universe is known to be necessarily existent by virtue of itself, and what is necessarily existent by virtue of itself cannot be composite.[106] Consequently, two deities could not be distinguished, and no more than one can exist.[107]

Several writers, finally, utilize the teleogical mode of thought and maintain (e) that the unity of the plan of creation discloses the unity of the creator responsible for the plan.[108]

Such were the Kalam arguments for the unity of God. A separate repertoire of dialectical arguments was utilized for establishing that God is incorporeal.

(*a*) David al-Muqammiṣ, Māturīdī, Bāqillānī, ʻAbd al-Jabbār, Juwaynī, Ghazali, Joseph al-Baṣīr, Jeshua b. Judah, Judah Hallevi, Fakhr al-Dīn al-Rāzī, and Ījī, give the following argument: The Kalam proofs of creation have established that atoms, accidents, and bodies all come into existence. But the first cause of

[101] Juwaynī, *K. al-Shāmil,* pp. 384–385 (attributed to Muʻtazilites); Ghazali, *al-Iqtiṣād,* pp. 74–75; Joseph al-Baṣīr, in Frankl, *Ein Muʻtazilitischer Kalām,* p. 27; Shahrastānī, *K. Nihāya al-Iqdām,* p. 93; Ibn Ezra, in *Kerem Ḥemed,* IV, p. 4; Maimonides, *Guide,* I, 75(2); Āmidī, *Ghāya al-Marām,* p. 153 (a peculiar version).

[102] Above, p. 134.

[103] Baḥya, *al-Hidāya* (*Ḥobot ha-Lebabot*), I 7(6) (with a slight difference); Ibn Ezra, in *Kerem Ḥemed,* IV, p. 4; al-Muqammiṣ, in Vajda, "Le Problème de l'Unité de Dieu," p. 53.

[104] Above, p. 147.

[105] Ibn Ḥazm, *K. al-Faṣl fī al-Milal,* , pp. 36–37; Baḥya, *al-Hidāya* (*Ḥobot ha-Lebabot*), I, 7(4).

[106] Below, p. 296.

[107] Ījī, *Mawāqif,* VIII, p. 39.

[108] Below, pp. 218, 221, 224, 230.

the universe is known to be eternal.[109] The first cause of the universe cannot, therefore, be a body or a constituent of a body.[110]

(*b*) An argument that enjoyed considerable currency went: Every body has *particular* dimensions to the exclusion of other dimensions that it might have;[111] every body occupies a *particular* place to the exclusion of other places it might occupy;[112] and every body possesses *particular* qualities to the exclusion of other qualities it might equally possess.[113] A "particularizing agent" must assign the body its dimensions, its place, and its qualities. Inasmuch as every body is thus dependent on an agent outside it, the ultimate cause, which is dependent on nothing whatsoever, cannot be a body.

(*c*) A number of writers laid down the premise that bodies are incapable of creating other bodies; their justification of the premise was either that induction reveals bodies to be incapable of creating other bodies;[114] or that nothing can, in principle, create its like.[115] If a body cannot create another body, the being that created the physical world obviously cannot be a body.

(*d*) Finally, Fakhr al-Dīn al-Rāzī and Ījī borrow again from Avicenna's philosophy and offer a proof of the incorporeality of God which does not rest on the premise of creation. They argue that every body is composite; the ultimate cause of the universe is known by demonstration to be necessarily existent by virtue of itself; and what is necessarily existent by virtue of itself cannot be composite. The ultimate cause of the universe therefore cannot be a body.[116]

A first cause of the physical universe has now been reached which is eternal, which is not a mere nature or necessary cause, and which is one and incorporeal. The Kalam writers go on to argue that God possesses certain key attributes, attributes that the Ash'arite school construed as, in some sense, real distinct things within the deity, but that the Mu'tazilite school construed as having no real distinct existence.

[109]Above, p. 165.

[110]Vajda, "Le Problème de l'Unité de Dieu," p. 50; Māturīdī, *K. al-Tawḥīd,* p. 38; Bāqillānī, *K. al-Tamhīd,* p. 25; 'Abd al-Jabbār, *Sharḥ al-Uṣūl,* p. 218; Juwaynī, *K. al-Irshād,* p. 43; idem, *K. al-Shāmil,* p. 411; Ghazali, *al-Iqtiṣād,* pp. 38–40; idem, *al-Risāla al-Qudsīya,* Arabic, p. 18, English, p. 36; Joseph al-Baṣīr, in Frankl, *Ein Mu'tazilitscher Kalām,* p. 25; Schreiner, *Studien ueber Jeschu'a ben Jehuda,* p. 39; Judah Hallevi, *Kuzari,* V, 18(6); Rāzī, *K. al-Arba'īn,* p. 104; Ījī, *Mawāqif,* pp. 21, 26.

[111]Baghdādī, *K. Uṣūl al-Dīn,* pp. 73, 77; Juwaynī, *K. al-Shāmil,* p. 412; Shahrastānī, *K. Nihāya al-Iqdām,* pp. 105–106; Ghazali, *al-Iqtiṣād,* p. 39; Ījī, *Mawāqif,* VIII, p. 26; Āmidī, *Ghāya al-Marām,* p. 181 (with critique); Maimonides, *Guide,* I, 76(3).

[112]Juwaynī, *K. al-Shāmil,* p. 413; Jeshua b. Judah, in Schreiner, *Studien ueber Jeschu'a ben Jehuda,* p. 40(3); Ījī, *Mawāqif,* VIII, p. 20.

[113]Rāzī, *K. al-Arba'īn,* pp. 104–105 (4;7); Ījī, *Mawāqif,* VIII, p. 26.

[114]'Abd al-Jabbār, *Sharḥ al-Uṣūl,* pp. 221–222; Joseph al-Baṣīr and Jeshua b. Judah, in Schreiner, *Studien ueber Jeschu'a ben Jehuda,* p. 39(2) and n. 3. The argument is rather complicated.

[115]Juwaynī, *K. al-Shāmil,* p. 413; Ibn Ṣaddiq, *ha-'Olam ha-Qatan,* p. 51. Also see Saadia, *K. al-Amānāt,* II, introd., p. 78; Āmidī, *Ghāya al-Marām,* pp. 184–185.

[116]Rāzī, *K. al-Arba'īn,* p.104(1); Ījī, *Mawāqif,* VIII, p. 21. See a similar argument for the unity of God, above, p. 171.

To establish that God is "powerful,"[117] it was argued that only a powerful agent would be able to create a world;[118] or, more specifically, that only a powerful agent would be to create a world as well-designed as ours discloses itself to be.[119]

To establish that God is "knowing," Kalam writers submit that an agent exercising power must know what he is doing, and hence a powerful agent possesses knowledge;[120] or else that the creation of a well-designed world, such as ours is discovered to be, implies knowledge.[121]

To prove that God is "alive," they reason that whatever has power together with knowledge must likewise have life;[122] or that only something alive would have the ability to create.[123]

To show that God possesses will, they argue that power entails will, and therefore every being possessed of power is possessed of will.[124] Or they contend that the decision to bring a world into existence after none had existed before involves the exercise of will.[125] Or else they argue more specifically that the decision to create the world at a particular moment or with a particular set of characteristics to the exclusion of equally possible moments and equally possible characteristics involves will.[126] Of all the divine attributes the present attribute,

[117]See above, pp. 167, 169, where some writers connect power to eternity.

[118]Ash'ari, *K. al-Luma'*, §14; Māturīdī, *K. al-Tawḥīd,* p. 45; Bāqillānī, *K. al-Tamhīd,* p. 24; 'Abd al-Jabbār, *Sharḥ al-Uṣūl,* p. 151; *al-Majmū'*, p. 103; Juwaynī, *K. al-Irshād,* pp. 61–62; *K. al-Shāmil,* p. 621; Ghazali, *al-Iqtiṣād,* p. 81; Ibn Ṣaddiq, *ha-'Olam ha-Qatan,* p. 57; Shahrastānī, *K. Nihāya al-Iqdām,* p. 170; Rāzī, *K. al-Arba'īn,* p. 129 (power excludes necessity); Āmidī, *Ghāya al-Marām,* p. 45 (with reservations); Ījī, *Mawāqif,* VIII, p. 49 (power excludes necessity).

[119]Ghazali, *al-Iqtiṣād,* p. 80; idem, *al-Risāla al-Qudsīya,* Arabic, p. 20, English p. 40.

[120]'Abd al-Jabbār, *K. al-Majmū'*, p. 107; Ījī, *Mawāqif,* VIII, pp. 66–67.

[121]Ash'ari, *K. al-Luma'*, §13: Māturīdī, *K. al-Tawḥīd*, p. 45; Bāqillānī, *K. al-Tamhīd*, p. 26; 'Abd al-Jabbār, *Sharḥ al-Uṣūl*, p. 156; *al-Majmū'*, p. 113; Juwaynī, *K. al-Irshād*, p. 61; idem, *K. al-Shāmil*, p. 621; Ghazali, *al-Iqtiṣād*, pp. 99–100; idem, *al-Risāla al-Qudsīya*, Arabic, p. 20, English, p. 41; Judah Hallevi, *Kuzari*, V, 18(7); Ibn Ṣaddiq, *ha-'Olam ha-Qatan*, p. 57; Shahrastānī, *K. Nihāya al-Iqdām*, p. 171; Rāzī, *K. al-Arba'īn*, p. 133; Āmidī, *Ghāya al-Marām*, p. 45 (with reservations); Ījī, *Mawāqif*, VIII, p. 65.

[122]Bāqillānī, *K. al-Tamhīd,* p. 26; 'Abd al-Jabbār, *Sharḥ al-Uṣūl,* p. 161; Juwaynī, *K. al-Shāmil,* p. 622; Ghazali, *al-Iqtiṣād,* pp. 100–101; idem, *al-Risāla al-Qudsīya,* Arabic, p. 21, English, p. 41; Shahrastānī, *K. Nihāya al-Iqdām,* p. 171 (not quite clear); Rāzī, *K. al-Arba'īn,* pp. 154–155; Ījī, *Mawāqif,* VIII, p. 80.

[123]Ash'ari, *K. al-Luma'*, §14; Juwaynī, *K. al-Shāmil,* p. 622.

[124]'Abd al-Jabbār, *Sharḥ al-Uṣūl,* pp. 107; 147; cf. Judah Hallevi, *Kuzari,* II, 6.

[125]Māturīdī, *K. al-Tawḥīd,* p. 45; 'Abd al-Jabbār, *Sharḥ al-Uṣūl,* p. 120; Rāzī, *K. al-Arba'īn,* p. 129; and see above, p. 166, the argument that the cause of the world cannot be a nature or necessary cause.

[126]Bāqillānī, *K. al-Tamhīd,* p. 27 (in connection with the decision to create all things, and not just the world); Juwaynī, *K. al-Irshād,* p. 64; Ghazali, *al-Iqtiṣād,* p. 101 (not specifically in regard to the time the world as a whole came into existence); idem, *al-Risāla al-Qudsīya,* Arabic, p. 21, English p. 41; Judah Hallevi, *Kuzari,* V, 18(9); Shahrastānī, *K. Nihāya al-Iqdām,* p. 171; Rāzī, *K. al-Arba'īn,* p. 147; Āmidī, *Ghāya al-Marām,* p. 45 (with reservations); Ījī, *Mawāqif,* VIII, p. 82. The contention survives in Muḥammad 'Abduh, *K. al-Tawḥīd* (Cairo, 1966), p. 33; English translation: *The Theology of Unity,* trans. I. Musa'ad and K. Cragg (London, 1966), p. 50.

will, is what distinguishes the Kalam concept of the deity from, for example, the Aristotelian concept. The intrinsic status of will is underlined by Ghazali, who insists that the existence of God has not been demonstrated until the cause of the universe is shown to possess will;[127] and it is likewise made explicit by those who follow their proof of a cause of the world with an argument showing that the cause of the universe could not possibly be a nature or necessary cause.[128] But the centrality of will for the Kalam concept of the deity is undoubtedly recognized as well by adherents of the Kalam who do not dwell upon the point and who treat will routinely together with the other divine attributes.

The upshot of Kalam natural theology, in fine, is that a first cause exists who brought the world into existence; that the cause is eternal, one, and incorporeal; that he is possessed of power, knowledge, life, and will.[129] Precisely which attributes are, and which are not, integral to the concept of the deity—and hence precisely where the proof of the existence of God ends and the arguments establishing ancillary attributes begin—is, as has been mentioned, difficult to determine. Yet it would seem that every adherent of the Kalam expected a proof of the existence of God to establish, at the minimum, the existence of an eternal, powerful, first cause of the universe.[130] At least some Kalam thinkers plainly regard will as integral to the concept of the deity. And some, perhaps, regard unity too as integral to the concept.[131]

2. Arguments from the concept of particularization

At the heart of the arguments to be examined in the present section is the notion that when an object has a given characteristic but could have alternative characteristics, something must *particularize* the object in its characteristic or—the phraseology can also go—particularize the characteristic for the object. Something, that is to say, must choose the particular characteristic the object does have from among the totality of possible characteristics it might have. The distinctiveness of the particularization mode of thought can be brought out by contrasting it with the Aristotelian mode. The former, in its classic versions, supposes that all characteristics of physical objects are equally in need of an explanation; and the explanation provided is that each characteristic is the outcome of an arbitrary choice. The latter considers only certain characteristics of objects to be in need of an explanation; and its explanations are formulated in accordance with what it understands to be the laws of nature. To take an example, the particularization approach is as eager to ask why an object should be at rest and not in motion as

[127] Above, p. 3.

[128] Above, p. 166.

[129] Other divine attributes, of a theological character, are also established by the Kalam authorities, e.g., that God is "hearing," "seeing," and "speaking."

[130] See above, p. 167.

[131] Ibid.

to ask why an object should be in motion and not at rest. On the Aristotelian approach only the motion, and not the rest, of a physical object has to be explained, for empirical and analytic reasons—just as in a later physics, the single question requiring an answer is why an object has undergone a change of state; and the Aristotelian explanation of an object's motion is made in conformity with uniform laws of nature. Light is likewise cast on the particularization approach to the world by contrasting it with the teleological approach. Whereas the teleological approach seeks out characteristics in objects which are so well designed that no human observer can fail to detect the imprint of a designing agent who chose intelligently, the particularization approach views the characteristics of objects indifferently, and the choice it detects is an arbitrary one.

Arguments from the concept of particularization are usually associated with an occasionalistic picture of the universe. When Kalam thinkers ask why a given individual object came into existence at a given time or why the object has given characteristics to the exclusion of others, they ignore the commonsense answer that the time when the object came into existence or the characteristics possessed by the object are determined by the situation obtaining within the world prior to the object's coming into existence and by circumstances surrounding the object. The association of the particularization concept with an occasionalistic picture of the universe is not, however, absolute. The concept could be applied as well to general features of the world which would not be amenable to any natural explanation. General features that might be other than they are and that do not permit a natural or rational explanation must—it could be argued—surely reflect an act of deliberate choice.

The particularization notion has been traced to a detail of Stoic speculation. Chrysippus and certain adversaries debated whether human actions are wholly determined by natural causes or whether, on the contrary, man has an autonomous power to opt for one course of behavior over another. The adversaries held that the human mind does have the ability to "incline" in favor of one course of behavior as against another, an ability "especially manifest" in instances where man is faced with "indistinguishable alternatives." Chrysippus responded that neither the "fall" of the "dice," nor the "inclination" of the "scales," nor any other event in the universe ever occurs spontaneously and "without a cause."[132] The remark of Chrysippus regarding an inclination of the scales of an actual physical weighing apparatus may prefigure the metaphor of *tipping the scales* in Islamic literature. And the notion of tipping the scales is akin to the notion of particularization; to say that the scales were tipped in favor of a given characteristic is equivalent to saying that the characteristic was particularized for its subject.

[132]Plutarch, *Moralia: De Storicorum Repugnantiis,* 23; S. Horovitz, "Ueber den Einfluss des Stoicismus auf die Entwickelung der Philosophie bei den Arabern," *Zeitschrift der Deutschen Morgenlaendischen Gesellschaft,* LVII (1903), 190. See also N. Rescher, "Choice without Preference," *Kant-Studien,* LI (1959–1960), 143–166.

A nearer and more likely route through which the particularization notion may have been introduced into Arabic thought was the thrust and parry in connection with a recurring argument for eternity. The argument in question reasoned that creation is impossible because no moment in empty infinite time could have lent itself to the world's coming into existence in preference to the identical earlier and later moments when the world might have come into existence. A recurring response going back to Augustine and Philoponus was that God, through his will, would be capable of arbitrarily selecting one moment for creation in preference to the other equivalent moments at which the world might have emerged.[133] Here we have a given moment's being selected from among equally possible moments through a sheer act of will; and this, if not the sole source of inspiration for Islamic particularization arguments, undoubtedly fostered them.

The particularization style of thought was long-lived. A vestige survives in Leibniz's "great principle . . . that nothing happens without a sufficient reason," a principle that compels one to ask "why there is something rather than nothing" and why, moreover, things "exist so and not otherwise."[134] Leibniz's principle of sufficient reason diverges, though, from the classic particularization principle in that the former is understood by Leibniz to operate with the highest rationality and not arbitrarily.[135] The philosophic problem of how a particular moment might have been selected for creation remained a topic for discussion in even later centuries. Hume and Kant[136] can still be discovered wrestling with it.

The concept of particularization found many applications in Arabic philosophic literature. As has already been shown, it furnished Kalam thinkers with a rationale for defending their inference of a creator from creation. In addition, it supplied the nerve of a new proof of the existence of God when Kalam thinkers contended that, quite apart from creation, the characteristics exhibited by the world must have been selected by an agent outside the world. It supplied the nerve of a new method of proving the creation of the world when adherents of the Kalam and one nonadherent contended that the selecting out of particular characteristics for the world could not have been effected from eternity. And it served as a dialectical tool in miscellaneous contexts both connected with, and independent of, the issues of creation and the existence of God.

Particularization arguments, it was seen in the previous section, constituted one of two general lines of reasoning whereby the inference of a creator from creation could be supported. Within that single line of reasoning, three or four

[133] Above, pp. 68–69.

[134] Leibniz, *Principes de la Nature et de la Grâce*, §7; cf. *Théodicée*, §14.

[135] Cf. *A Collection of Papers which Passed between the Late Learned Mr. Leibnitz and Dr. Clarke* (London, 1717), IV, §1: "In things absolutely indifferent there is no choice and consequently no election or will, since choice must have some reason or principle."

[136] Hume, *Treatise of Human Nature*, I, iii, 3; Kant, *Critique of Pure Reason*, A427/B455 (antithesis of first antinomy).

subordinate strands can be distinguished. Three subordinate strands were recognized by Juwaynī, who makes the following perceptive observation: In "establishing the need of what comes into existence for an agent bringing it into existence," Kalam thinkers employed varying "formulations" and "terminologies." All the formulations presume that when an object contains a "possible characteristic," that is to say, a given characteristic to the exclusion of others it might possibly have, the characteristic it does have "depends on a particularizing agent." But each formulation focuses on a different type of characteristic. Attention may be directed, first, to "existence and nonexistence"; or secondly, to "the earlier appearance of some objects coming into existence and the later appearance of others"; or thirdly, to "the particularization of bodies in [or:by]" sundry "attributes, . . . shapes, or manners of composition,"[137] In the first instance, the contention would be that what comes into existence might equally have remained nonexistent, and a particularizing agent must have selected existence for it in preference to nonexistence. In the second instance, the contention would be that what comes into existence at a given moment might have come into existence at an earlier or later moment, and a particularizing agent must have chosen the moment for the object to come into existence. In the third, the contention would be that a physical object might have come into existence with other attributes, another shape, and another composition than those it has; and a particularizing agent must have selected the features with which the object did come into existence. These three strands of argumentation distinguished by Juwaynī increase to four if we further differentiate between the contention which was put forward to the effect that a particularizing agent must have selected a time for the emergence of the world as a whole, and the contention which was put forward to the effect that a particularizing agent must select a time for the emergence of every object within the world.

Four separate strands may, then, be differentiated in the deployment of the argument from particularization to support the inference of a creator from creation. Three of the four strands have already been examined. Juwaynī and others, it was seen, brought the particularization line of argumentation to bear on the selecting out of existence for the world in preference to nonexistence; they inferred the existence of the creator as the particularizating agent or the agent tipping the scales who made the selection.[138] Māturīdī, without use of the term particularization, and Juwaynī, who did use the term, brought the particularization line of argumentation to bear on the time when the world as a whole came into existence; building on their prior proof of the creation of the world, they inferred the existence of a creator as the agent who chose a particular moment for the world's appearance.[139] Bāqillānī, without using the term particularization, and Baghdādī,

[137] *K. al-Shāmil*, p. 272.
[138] Above, pp. 161–162.
[139] Above, pp. 160–161.

who has the term, brought the particularization notion to bear not on the time when the world as a whole came into existence, but on the time when each of the individual physical objects comprising the world came into existence. They built on their proof of the creation of every accident and body. And they presupposed the Ash'arite denial of natural causation. Within a physical universe where natural causation reigns, the moment when individual objects emerge is determined by forces operating in accordance with natural law; but Bāqillānī and Baghdādī supposed that the state of physical objects is nowise determined by what takes place within the universe. They could therefore reason: Each object coming into existence might have come into existence at a different time from the time when it did. Each object consequently had to have a time for its emergence selected out by an agent apart from it. And since all bodies and accidents are known to have been created, an agent must be posited who selected out the time for each to come into existence.[140]

Side by side with his application of the particularization notion to the times when objects come into existence, Bāqillānī—in a passage that has not been taken up yet—brings the notion to bear on the shapes and configurations possessed by objects coming into existence. He is again presupposing that the state of a body is not determined by forces within the physical universe. He does now expressly use the terms *particular* and *particularizing*; and he writes: "Each body in the world" has the "possibility (*ṣiḥḥa*) of receiving" a different "composition," that is to say, a different configuration of atoms from the configuration it has. What is "square" could be "round," and vice versa; and what has the "form of one animal" could have "the form of another." Bodies, moreover, are constantly "transformed" from one "shape to another shape." Plainly, a body that "is particularized in [or:by] a specific, particular shape" cannot have been "particularized in its shape by virtue of itself or merely by virtue of [its having] the possibility of receiving the shape." For then a body would "have to receive, at the same time, every shape it has the possibility of receiving," which of course does not, and cannot, occur. The conclusion drawn by Bāqillānī is that "whatever possesses a shape" received its shape through a "combining agent who combined it and an intending agent who intended that it should be as it is."[141]

Bāqillānī's ostensible aim here is to support the inference of a creator from creation; he is ostensibly reasoning that since all bodies have come into existence, and since they might have come into existence with a different set of characteristics from the set they have, a particularizing agent must have brought them into existence with the characteristics they do have. The premise of creation is not, however, required for the purpose. Bāqillānī might have set the issue of creation aside and still argued for a voluntary agent who assigns to bodies one out of all

[140] Above, pp. 160–161.
[141] *K. al-Tamhīd*, pp. 23–24.

the conceivable sets of characteristics they might have. He is, in other words, on the verge of a proof for a voluntary cause of the universe which dispenses with the premise of creation, a proof wherein a voluntary cause is inferred directly from the presence in things of particular characteristics.

To repeat, Kalam writers adducing a particularization argument to support their inference of a creator from creation, contended either that the existence of the world would have to be selected in preference to the world's nonexistence; that a time would have to be selected for the coming into existence of the world as a whole; that a time would have to be selected for the coming into existence of every body in the world; or that a set of characteristics would have to be selected out for every body coming into existence within the world.

Besides being deployed in support of the inference of a creator from creation, particularization arguments were utilized by the Kalam in other contexts. When establishing that the creator of the world cannot be a "necessary cause" or "nature," Juwaynī argued that inasmuch as a necessary cause and a nature act in an unvarying manner as long as they exist, neither an eternal nor a noneternal necessary cause or nature could be the cause of the world. An eternal necessary cause or nature could not be the cause of the world because an eternal necessary cause or nature would produce the world from eternity, whereas the world is known to have been created. But a noneternal necessary cause or nature—or, for that matter, any noneternal being—could also not be the cause of the world. For a noneternal necessary cause or nature would need a "particularizing agent" to assign existence to it at the time when it came into existence; if that particularizing agent were noneternal, it would require another particularizing agent to assign existence to it; and so on *ad infinitum*. The creator therefore cannot, Juwaynī found, be anything except a "voluntary agent."[142]

Ījī established the unity of God by arguing that if two deities existed, neither could undertake any action unless a factor should inexplicably materialize and "tip the scales," thereby determining which of the two equally divine agents was to perform the action.[143] A number of Kalam writers proved the incorporeality of God by arguing that every body has particular dimensions, a particular location, and other particular qualities; every body accordingly depends on a "particularizing agent" that assigns it its dimensions, place, and qualities; hence the first cause, being dependent on no agent, cannot be a body.[144] Kalam writers established the presence of will in the deity by arguing that the selection of a particular moment for the world to come into existence in preference to the infinite alternative moments when the world might have come into existence entails will.[145]

[142]*K. al-Irshād,* pp. 28–29.
[143]Above, pp. 169–170.
[144]Above, p. 172.
[145]Above, p. 173.

The foregoing have been arguments pertaining to the issues of creation and the existence of God, but particularization arguments were resorted to elsewhere as well. They crop up constantly in the dialectical give and take typical of the Kalam, notably in the heavy dialectical web of real and artificial objections, rejoinders, surrejoinders, and sur–surrejoinders which later writers such as Juwaynī, Rāzī, and Āmidī wove around every topic they touched. The language of particularization could even be employed by philosophers of the Aristotelian school. They, however, use that language with no connotation of arbitrariness, but simply as synonymous with the language of natural causation.[146]

An application of the particularization concept which has not been mentioned yet but which is of interest for us is its use to establish the existence of accidents. The accident as an actual entity that carries a characteristic and, when conjoined to an inert atom or to a collection of atoms constituting a body, imparts the characteristic to them was an idiosyncratic feature of theoretical Kalam physics. The proposition that accidents do exist served as a key premise in the most distinctive Kalam proof of creation.[147] Yet the existence of accidents as actual entities is hardly obvious, and proofs had to be furnished.

An argument for the existence of accidents which reportedly goes back to the ninth century ran as follows: If "motion," to take an example, were not an actual "thing" (*ma'nā*) in the moving body, there could be no reason why one body is "in motion in preference to another" body, and no reason why a body "moves at the time it does move in preference to moving at an earlier time." Similarly, if "blackness" and "whiteness" were not "things" in bodies there would be no reason why a certain body is black or white "to the exclusion of another body's" being black or white. The circumstance that motion and rest, blackness and whiteness, and the "remaining" qualities, occur in some bodies but not in others, and the further circumstance that they occur sometimes but not at other times, demonstrates—the argument went—that they occur by virtue of "things," by the inherence in bodies of accidents construed as real entities.[148]

Bāqillānī split the argument into two versions, the first of which considers why a body should be in motion at one time to the exclusion of another time, while the second considers why one body should be in motion to the exclusion of another body. The first version reads: A body is observed to be in "motion . . . subsequent to its being at rest" and to be at "rest subsequent to its moving." A body's being

[146]See Alfarabi (?), *Ta'līqāt,* in *Rasā'il al-Fārābī* (Hyderabad, 1931), pp. 10, 14; Avicenna, *Shifā': Ilāhīyāt,* p. 411; *De Anima,* ed. F. Rahman (London, 1959), p. 229; Averroes, *Tahāfut al-Tahāfut,* ed. M. Bouyges (Beirut, 1930), p. 412; English translation, with pagination of the Arabic indicated: *Averroes' Tahafut al-Tahafut,* trans. S. van den Bergh (London, 1954); Maimonides, *Guide,* I, 73(10); 74(5).

[147]Above, p. 137.

[148]Ash'ari, *Maqālāt al-Islāmīyīn,* ed. H. Ritter (Istanbul, 1929–1933), pp. 372–373, in name of Mu'ammar. Cf. Khayyāṭ, *K. al-Intiṣār* (Beirut, 1957), §34; Wolfson, *Philosophy of the Kalam* (Cambridge, Mass., 1976), pp. 149 ff.

at rest or in motion must come about "either by virtue of [the body] itself or by virtue of a cause (*'illa*). But if the body were in motion by virtue of itself, it could not [as long as it exists] ever possibly be at rest. The possibility of its being at rest after being in motion proves that it moves by virtue of a cause," by the presence in it of the accident of "motion." A similar analysis, Bāqillānī adds, will cover "colors, tastes, odors, combination, . . . and the like," and show that each of them occurs through something actual, through an accident construed as an entity.[149]

In his second version of the argument for the existence of accidents, Bāqillānī considers why one body should be in motion to the exclusion of another. A body in motion, he begins, must "move either by virtue of itself or by virtue of a thing (*ma'nā*)." But a body cannot "possibly be in motion by virtue of itself; for if such were the case, everything belonging to its genus"—everything of exactly the same nature—"which exists at the specific moment would have to be in motion. . . . The fact that members of the genus of mobile atoms and bodies are sometimes seen not to be in motion proves that atoms and bodies in motion do not move by virtue of themselves . . . but by virtue" of a *thing*, by the presence of the accident of "motion."[150]

The grounds adduced by Bāqillānī for the existence of accidents are, it is to be noted, almost identical with the grounds he adduced for the existence of a particularizing agent who selects out the characteristics of bodies coming into existence. When inferring a particularizing agent, he reasoned that a body coming into existence with a certain shape or configuration cannot have been "particularized in its shape by virtue of itself or merely by virtue of [its having] the possibility of receiving the shape"; for then a body would "receive, at the same time, every shape it has the possibility of receiving." Inasmuch as shape and configuration cannot flow from the physical substratum of the body, the body must, he concluded, be particularized in its shape and configuration by a "combining . . . and intending agent" who determines in what form it should exist.[151] When establishing the existence of accidents, Bāqillānī reasons again that a body with a certain characteristic cannot have the characteristic "by virtue of itself"; for were such the case, the body would always have the characteristic, and everything capable of having the characteristic would also always have it. His conclusion here, however, is that inasmuch as the characteristic cannot flow from the physical substratum of the body, the body has the characteristic thanks to a *thing*, thanks to an accident construed as an actual entity. Virtually identical argumentation leads him in one context to an external agent who selects out characteristics for bodies coming into existence and in another context to the existence of accidents conceived as entities that inhere in bodies and impart characteristics to them.

[149] *K. al-Tamhīd*, p. 18.
[150] Ibid., p. 19.
[151] Above, p. 178.

Other writers offer an argument for the existence of accidents which sounds like an argument for the existence of a particularizing agent. Baghdādī, who employed a particularization argument to support his inference of a creator from creation,[152] points out in his discussion of accidents that "a body moves after being at rest" and that a body is "black after being white." The body, he proceeds, surely does not move and is not black "by virtue of itself"; for the body "itself" is in existence both at the time when it does and the time when it does not move, both at the time when it is black and the time when it is not black. Inasmuch as motion and blackness cannot be due to the body itself, they must come about "by virtue of a thing (*ma'nā*) that inheres in the body."[153] 'Abd al-Jabbār, who did not employ a particularization argument to support the inference of a creator from creation, writes in regard to accidents: A body acquires a certain characteristic—'Abd al-Jabbār's example is the body's having its atoms in a state of "aggregation"—when it could as well retain the characteristic it previously had. "The situation is the same, and the conditions are the same. There is, hence, no avoiding something, a particularizing factor (*mukhaṣṣiṣ*), by the presence of which" it acquired the new characteristic; for "otherwise" the body "would not be [characterized] in the given fashion in preference to the contrary fashion." The Arabic term translated here as "particularizing factor" is the term I translated earlier as "particularizing agent," the Arabic simply meaning "particularizer." The particularizer that 'Abd al-Jabbār is positing in order to account for the appearance of new characteristics in bodies is not an agent; it is "nothing . . . other than the existence of a thing (*ma'nā*)," an actually existent accident.[154] Juwaynī affirms the existence of accidents as actual entities on the grounds that an atom is observed to be "stationary" and subsequently is observed to be "in motion in a particular spot . . . distinct from the spot from which it started. . . . Its being particularized in its spot is possible, not necessary, . . . and . . . stands in need of a determining factor (*muqtaḍīn*) that determines the positive particularization for it." The "determining factor" cannot be "the atom itself"; "for if it were, the atom would have been particularized in the given spot as long as it, the atom, was in existence." The determining factor must be a "thing," which is "added to the atom,"[155] that is to say, an accident construed as an actual entity.

Bāqillānī thus employs virtually identical argumentation when establishing the existence of accidents as physical entities that are present in bodies and impart characteristics to them, and when inferring the existence of an external agent who selects out characteristics for bodies. Baghdādī and Juwaynī offer an argument for the existence of accidents which resembles their arguments for the existence of a particularizing agent who selected a time for individual objects to

[152] Above, p. 161.

[153] *K. Uṣūl al-Dīn*, p. 37.

[154] *Sharḥ al-Uṣūl*, p. 96. The argument, to be precise, is proving the existence of the most basic and general sort of accidents, the *akwān*.

[155] *K. al-Irshād*, pp. 18–19.

come into existence, or who selected existence for the world when it came into existence, or who selected one time for the world to come into existence in preference to alternative times.[156] ‘Abd al-Jabbār does not use the particularization mode of reasoning when inferring a creator from creation, but does have a particularization argument for the existence of accidents; and he goes as far as to call the accident a "particularizing factor" or, more literally, a "particularizer," the very term whereby others designated the particularizing agent. It is hard to avoid asking why both the creator as an external particularizing agent and the accident as an inhering particularizing factor are required. Why, to be precise, must an external agent be posited if accidents account for the characteristics of atoms and bodies, and why must accidents as actual inhering entities be posited if the external particularizing agent can account for those characteristics?

An answer to the first half of the question is supplied by the Kalam proofs of the generation of accidents.[157] If the accidents that impart characteristics to atoms and bodies come into existence, an added external cause has to be posited who brings the accidents into existence. The step from the coming into existence of accidents to a cause bringing them into existence might either be treated as self-evident, supported by analogy, or supported through a particularization argument.[158] Should a particularization argument be used, the reasoning could be either that a particularizing agent must select out existence for the accident, conceived as a particularizing factor, in preference to nonexistence; that a particularizing agent must select out a moment for the accident to come into existence; or that a particularizing agent must decide which accident is to be joined to which atom.[159]

An answer to the second half of the question is indicated by Bāqillānī. Besides giving the two versions of the argument for the existence of accidents already quoted, Bāqillānī advances a totally different argument. He now approaches the subject of accidents not from the side of characteristics observable in a body, but from the side of the external agent producing the characteristics. When an agent "exercises power," he asserts, the agent's power must be "attached to some object of power" which possesses actual existence, just as "knowledge . . . is attached to an object of knowledge" which possesses actual existence, and "memory, . . . to an object of memory" which possesses actual existence. When, for instance, the agent "exercises his power . . . in moving a body," the exercise of power must have an actually existing "object of power," it must give rise to some actually existing thing in the body moved. And the object of power, the actually existing thing produced by the agent, can be nothing other than the concrete accident of motion in the moving body.[160] Bāqillānī has hereby explained in effect why the existence of accidents as actual entities has to be accepted in addition to the agent

[156] Above, pp. 161–162.

[157] Above, pp. 137–138.

[158] Above, pp. 155, 159.

[159] Above, pp. 160–162, 178.

[160] *K. al-Tamhīd*, p. 19.

who selects out characteristics for bodies: When an agent acts, his action must bring about something concrete in the body acted upon.

Juwaynī does not, as far as I could detect, expressly address the question why both a particularizing agent and an accident construed as a particularizing factor must be posited. But a subtle answer to both halves of the question can be extracted from his discussion of accidents. Juwaynī was troubled by an objection to the theory of accidents which was of the 'third man' type.[161] An accident, the challenge went, is supposedly the factor through which an atom is, for example, "particularized" in a certain "spot." But if an atom occupying a given spot has to be particularized in that spot in preference to others, the accident responsible for particularizing it there also has to be "particularized" for the atom with which it is connected in preference to all the other atoms with which it might be connected. And if what particularizes the atom in the spot it occupies is a "determining factor," to wit an accident, which has the status of a real thing, the accident too should have to be particularized for its atom by a further determining factor, which would likewise have the status of a "thing." The further determining factor particularizing the accident in its atom would, in turn, have to be selected out and particularized for its accident to the exclusion of other accidents by still another "thing"; and "an infinite regress would ensue."[162]

Mu'ammar, the earliest Kalam figure known to have formulated the argument for the existence of accidents from the concept of particularization, realized that the argument, if allowed to run its logical course, ends in an infinite regress. But Mu'ammar, instead of recoiling from paradoxes, could embrace them with relish; and an infinite regress of accidents did not discomfit him in the least. He is reported to have maintained that motion occurs in a body by virtue of a "thing" (*ma'nā*), and that the thing—the accident of motion—is connected to the body by an infinite series of other "things," all of which come into existence "at the same moment."[163] Juwaynī, by contrast, was incapable of admitting an infinite regress of accidents in every atom or body; and he faced the objection that if each atom receives its characteristics through an accident conceived as a particularizing factor, the accident would have to be particularized for its atom by a further factor, which is particularized by still another, *ad infinitum*.

The objection, writes Juwaynī, can be handled in two ways. It can, in the first place, be sidestepped by understanding that each accident is unique and suitable only for its own atom. The accident would accordingly "be particularized in its subject by virtue of itself," and no further "thing" would be required in order to determine which accident is to inhere in which body.[164] In the second place, the

[161]Cf. Plato, *Parmenides*, 134; W. D. Ross's edition of Aristotle's *Metaphysics*, Vol. I (Oxford, 1924), p. 195.

[162]*K. al-Shāmil*, p. 174.

[163]Ash'ari, *Maqālāt al-Islāmīyīn*, p. 372; Khayyāṭ, *K. al-Intiṣār*, §34; cf. Wolfson, *Philosophy of the Kalam*, pp. 149–162.

[164]*K. al-Shāmil*, p. 174. The determination of a particular time for the accident to appear is not explained.

objection can be resolved as follows: Each "accident" can be understood to be "particularized in its subject by the intention of an intending agent" who decided to "particularize it" specifically there.[165] The accident would thus not need a further immanent factor to particularize it in its atom, and no infinite regress ensues. This second way of explaining how a given accident comes to be connected to a given subject indirectly answers the question why accidents are insufficient in themselves to account for the occurrence of characteristics in objects and why a particularizing agent must be posited as well. The answer indicated is not very different from the answer implied by the Kalam proof of the generation of accidents: In order to avoid an infinite regress, a particularizing agent must be posited who determines that a given accident should inhere in a given atom to the exclusion of the other atoms in which the accident might inhere.

Juwaynī has another point to make. The imaginary or real opponent who is challenging the theory of accidents might counter that the foregoing exposition permits accidents to be dispensed with altogether. If the principal accident can be assigned to an atom through an external agent with no help from further accidents, why, the opponent may riposte, might not a characteristic be assigned directly to its atom by the external agent with no help from any accident whatsoever. Juwaynī fends off the riposte. The real or imaginary opponent challenging the theory of accidents has, he avers, forgotten the diverse natures of atoms and accidents. Accidents, in the dominant Kalam physical scheme, remain in existence for no more than a moment of atomic time, whereas atoms enjoy an existence extending over a number of moments.[166] An agent, even the divine agent, can exercise his "power" solely at the moment when he brings the object of his power into existence. Since the divine particularizing agent brings accidents into existence anew at each moment, he can at each renewal assign a given accident to a given atom; and no supplementary factors are needed to link the accident to its atom. But the divine agent allows atoms to exist over a stretch of moments, and he could by his own act assign a characteristic to an atom—locating the atom, for example, in a particular spot—for no more than the first moment. An actual real accident must therefore constantly inhere in the atom—or, to be more precise, must constantly be recreated in the atom—in order to locate the atom in its spot for all subsequent moments.[167] Juwaynī's position is perhaps problematical,[168] but it does, after a fashion, explain why the particularizing agent is insufficient and why concrete accidents have to be assumed as well. If the particularizing agent acted directly on an atom, he could do so for the duration of

[165] *K. al-Shāmil,* p. 175. Mu'ammar reportedly held that God does not create accidents; Ash'ari, *Maqālāt,* p. 199.

[166] Cf. Wolfson, *Philosophy of the Kalam,* pp. 522 ff., where no less than eight theories of the duration of accidents are differentiated and discussed.

[167] *K. al-Shāmil,* p. 175. Juwaynī adds that since the accident is needed for subsequent moments, it would be needed for the first moment as well.

[168] If an accident is required to locate an atom in a particular spot and the accident exists for only a moment, it is hard to see how the atom can be described as existing for more than a moment.

no more than a moment, and there would be nothing to impart qualities to the atom during the remaining moments.

We have this picture: Bāqillānī, Baghdādī, 'Abd al-Jabbār, and Juwaynī, establish the existence of accidents as real *things* through the consideration that some factor must be present in a body or atom to particularize the body or atom in each of its characteristics. In a separate context, Bāqillānī supported the inference of a creator from creation through the consideration that a particularizing agent must select out the characteristics possessed by bodies coming into existence; Bāqillānī and Baghdādī supported the inference through the consideration that a particularizing agent must select out the time when individual bodies come into existence; and Juwaynī supported the inference through the consideration that a particularizing agent must select out existence for the world in preference to nonexistence, and a moment for the world to come into existence in preference to alternative moments when the world might have come into existence. As to why both a particularizing agent and a particularizing factor are needed, half the question is answered by Kalam proofs of the generation of accidents. Granting the existence of accidents, a particularizing agent must be posited who selects out existence for the accident, or a time for the accident to come into existence, or an atom for the accident to inhere in. Bāqillānī indicates an answer to the other half of the question. He explains why, assuming the existence of a particularizing agent, the existence of accidents as the carriers for characteristics has to be recognized: When an agent acts it must produce something concrete in the body acted upon. Juwaynī indicates answers to both halves of the question. The particularizing agent is needed so that one accident to the exclusion of another can be assigned to a given atom without an infinite regress' ensuing; and the accident is needed so that an atom can retain characteristics, such as location in a particular spot, over a stretch of moments. The discussion, as cannot have been missed, is entirely rooted in the peculiar Kalam physical universe, with its inert atoms, accidents construed as actual entities, atomic time, and denial of natural causality.

And the dialectic again approaches a proof for the existence of God in which the issue of creation is set aside. One of the contentions whereby Bāqillānī supported the inference of a creator from creation was already seen to border on an argument for the existence of a particularizing agent who assigns characteristics to all bodies, whether or not the world has come into existence.[169] The dialectic of the theory of accidents has now maneuvered Juwaynī into stating that to avoid an infinite regress, a particularizing agent must be posited who assigns accidents to atoms. Having been led to that statement, Juwaynī might have based the existence of the cause of the universe on the need to posit a particularizing agent who assigns accidents to atoms and bodies whether or not atoms and bodies are created. As it turns out, Juwaynī and others do advance such an argument.

[169] Above, p. 178.

3. Particularization arguments for the existence of God without the premise of creation; particularization arguments for creation

A proof of the existence of God from the concept of particularization without reference to creation is perhaps to be detected in the Ikhwān al-Ṣafā', the so-called "Brothers of Purity." The Ikhwān remark in one passage that a "body cannot move in every direction [or: to every spot] at the same time," and they conclude that a body's "movement in a certain direction to the exclusion of another must be due to a cause."[170] The meaning seems to be that a particularizing agent must exist who arbitrarily chooses the direction in which bodies and especially—as the text indicates—the heavenly bodies move. The passage happens, however, to be interlaced with teleological motifs and may be animated solely by the teleological, and not the particularization, outlook. The Ikhwān may, in other words, be concerned only with evidence of design in the world, not with evidence of arbitrariness. The meaning of the passage would then be that since the movements of the celestial spheres in their several directions disclose design, they are undoubtedly the work of a divine designer.[171]

A fully conscious argument for a voluntary particularizing cause is given in a composition of Juwaynī's from which I have not quoted yet. The composition, a later work, undertakes to establish the existence of God by "methods" Juwaynī had "hitherto not pursued," methods that he pronounces the "most useful and finest" he had ever met.[172] In the new procedure, Juwaynī notes that everything in the physical universe which is observable and, by analogy, everything not observable as well, has "possible" characteristics. Any given body might, for example, "conceivably" have a "different shape" from the one it has. "What is at rest" could "conceivably" be "in motion," and vice versa. Physical objects that move upwards might move downwards. Objects that perform a circular motion, such as the heavenly bodies, might move in other orbits. The stars might be arranged in the heavens differently. And the world as a whole might have an alternative location in space. Since the parts of the world and the world as a whole could be different from what they are, "a determining agent" (*muqtaḍī*)[173] must have selected out for the parts of the world, and for the world as a whole, the characteristics that they do have.[174] Juwaynī goes on to show, with the aid of typical Kalam considerations, that an agent who selects out characteristics and assigns them to the world would possess the familiar properties of the deity, namely unity, incorporeality, power, knowledge, life, and will.[175] He thus arrives

[170]Ikhwān al-Ṣafā', *Rasā'il* (Beirut, 1957), III, p. 336.

[171]See below, p. 225.

[172]*al-'Aqīda al-Niẓāmīya* (Cairo, 1948), pp. 8, 13; German translation, with pagination of the Arabic indicated: *Das Dogma des Imām al-Ḥaramain al-Djuwaynî*, trans. H. Klopfer (Cairo, 1958).

[173]In other contexts I translated this term as "determining factor."

[174]*al-'Aqīda al-Niẓāmīya*, pp. 11–12.

[175]Ibid., pp. 17, 29; the arguments are similar to those examined above pp. 166 ff.

at a single incorporeal cause of the world, a cause possessed of power, knowledge, life, and will; and he has not had to start with creation. From the Kalam standpoint, his argument has the virtue of at once establishing a specifically voluntary cause of the world, it being evidence of the arbitrary exercise of will that leads him to a cause of the world. His argument likewise has virtues from a non-Kalam standpoint. It looks at the characteristics of things without insisting on the existence of accidents as actual entities. It looks, moreover, not merely at individual characteristics of individual objects within the world, characteristics that commonsense and Aristotelian philosophers ascribe to the workings of natural forces, but also to structural features of the world which cannot easily be traced to immanent natural forces.

Shahrastānī cites the argument in Juwaynī's name, although the evidence of arbitrary choice in the world which he adduces diverges somewhat from the evidence offered by Juwaynī; and he agrees with Juwaynī's assessment of the argument, finding it to be a "superlatively fine and perfect . . . method" for proving the existence of God. Shahrastānī's version runs: The constituents of the physical universe, to wit "earth, . . . water, . . . air, . . . fire, . . . and . . . the spheres," might possibly occupy alternative places, might have an alternative "shape and magnitude," and might be "larger or smaller" than they are. But "whatever is particularized in a certain way . . . to the exclusion of other . . . equally possible ways" is judged "by the necessity of intellect."[176] to stand in "need of a particularizing agent." All the parts of the world depend, therefore, on a particularizing agent.[177] An agent capable of selecting between equal possibilities plainly acts by "power and choice," not by "nature." The agent upon which the universe depends must accordingly be possessed of power and will.[178] Later chapters of Shahrastānī's book explain why a cause of the world would possess the remaining divine attributes; and the explanation given there is intended to supplement all his arguments for a cause of the world, including the present argument, and raise them all to the level of complete proofs of the existence of God.

Fakhr al-Dīn al-Rāzī too has an argument for a particularizing agent who is the cause of the world, but Rāzī's argument returns to a wholly occasionalistic framework by focusing exclusively on the characteristics of individual bodies. "Bodies," he contends, "are similar in their quiddity and essence," and as a consequence, "any attribute" that is connected to a given body might equally be connected to "other bodies." Any body with a "particular attribute" hence "stands in need . . . of a particularizing agent and an agent tipping the scales" who selects out the attribute for it. And since the world consists of bodies that might

[176]See above, p. 161.
[177]*K. Nihāya al-Iqdām*, pp. 12, 14.
[178]Ibid., p. 14.

have alternative attributes, the world in its totality stands in need of a particularizing agent.[179] Rāzī's formulation of the argument is copied by Ījī.[180] Both Rāzī and Ījī bracket it with other arguments for a cause of the world, and the considerations whereby they establish that the cause of the world is eternal, one, incorporeal, and so on, are designed to supplement all their arguments and render them all complete proofs of the existence of God.

The difference between the standard Kalam procedure for proving the existence of God and the procedure just examined merits comment. The standard procedure began by laying down a set of premises, among which is the existence of accidents; and a particularization argument provided one rationale for affirming that accidents indeed exist. After the premises had been laid down, the conclusion was drawn that the world was created. Then a creator was inferred from creation; and a particularization argument could again be called upon to support the inference. In contrast to the standard procedure, the new procedure is more direct. Without the customary premises, with no insistence on the existence of accidents, and without first establishing the creation of the world, Juwaynī and the others take a thread from the standard procedure, the thought that the world and its parts might have alternative characteristics and that something must arbitrarily select out the characteristics the world and its parts do have. A voluntary particularizing agent is inferred directly from the arbitrary selection of characteristics for the parts of the world and for the world as a whole.

Together with its role in a new, more straightforward proof of the existence of a cause of the universe, the particularization concept supplied the nerve of a new argument for creation. ʿAbd al-Jabbār and his anonymous editor record such an argument, having as its pivotal premise the assertion that nothing can "particularize" an eternal body "in one spot to the exclusion of another." Every body, the reasoning goes, obviously has the possibility of occupying alternative locations in space. Since nothing might have particularized an eternal body in a single spot to the exclusion of others, an eternal body would have to occupy either "every spot" simultaneously or no "spot whatsoever." Both suggestions are preposterous. Hence no eternal body can exist; and the body of the universe must have been created.[181] The premise affirming that nothing could particularize an eternal body in one spot to the exclusion of another is not elucidated in ʿAbd al-Jabbār's report, but Kalam literature reveals how it might have been defended. Several Kalam writers, as will appear, argue that the act of particularization involves will and that the exercise of will is incompatible with eternity. An underlying and unstated step in the argument recorded by ʿAbd al-Jabbār may accordingly be that inasmuch as the particularization of a body in a specific spot would involve will, it cannot have occurred from eternity.

[179] *K. al-Arbaʿīn,* p. 84; cf. *Muḥaṣṣal,* pp. 107-108.
[180] *Mawāqif,* VIII, p. 4.
[181] *K. al-Majmūʿ*, pp. 63–64.

An argument for creation from the concept of particularization is advanced by Juwaynī in connection with the argument already quoted wherein he employed the particularization concept to establish a cause of the world. Juwaynī's object was to provide a combined proof of the existence of God and creation, although I have disentangled the components and discussed them separately.[182] In the passage that has been quoted, Juwaynī arrived at a particularizing agent who arbitrarily selects a set of particular characteristics for the world and its parts. He thereupon proceeds: A "necessary cause" is incapable of selecting between equivalent possibilities, and if faced with equivalent possibilities must embrace them indiscriminately. For example, a purging medicine is incapable of working on the right side of the body to the exclusion of the left and perforce affects both sides equally. When an arbitrary choice has been made between equivalent possibilities, a voluntary, and not a necessary, cause is therefore responsible.[183] Now whereas a necessary cause acts unvaryingly as long as it exists and, if eternal, would act eternally, a voluntary cause cannot act from eternity. And inasmuch as the arbitrary selection of characteristics for the world does show the world to be the handiwork of a voluntary cause, the world—whether only its form, or its matter as well, is not made explicit by Juwaynī[184]—cannot have been brought into existence from eternity but must have been created.[185] The keystone of the argument, the rule that a voluntary cause cannot act from eternity, is not so much explained by Juwaynī as postulated. An "object of will," he asserts, is something "particularized that did not exist and subsequently existed . . . while, on the contrary, the existence of what exists eternally cannot possibly be dependent on will. . . . In general, . . . what exists through will is an effect produced by a voluntary agent who brings it about in conformity with his will, whereas what exists eternally is not effected [by an agent at all]."[186] Juwaynī, it is to be noted, is saying two things here: that eternity is incompatible with the exercise of will, and also that eternity is incompatible with a cause's having brought an effect into existence. The two points will reappear in writers to be discussed presently; the second point, moreover, is a perennial motif with a long history in medieval philosophy.[187]

Fakhr al-Dīn al-Rāzī has an argument from the particularization concept which, although ostensibly a proof merely of creation, could, like Juwaynī's argument, stand as a combined proof for both creation and the existence of a cause of the

[182]Shahrastānī, *K. Nihāya al-Iqdām*, p. 14, does the same.

[183]I am using language that Juwaynī would reject. He would insist that a being who acts voluntarily should be called an *agent*, not a *cause*.

[184]Perhaps Juwaynī was thinking that the particularization of atoms in a given spot entails their creation.

[185]*al-'Aqīda al-Niẓāmīya*, p. 12.

[186]Ibid., p. 12.

[187]See above, p. 3; below, pp. 210, 387.

world. Every body and the world as a whole, Rāzī begins, are of "finite magnitude," and what is of finite magnitude could "conceivably" be larger or smaller. A "particularizing agent" must, as a consequence, have "tipped the scales" and selected out the exact magnitude every body has from among the alternative magnitudes it might have. A particularizing agent selecting between equal alternatives is a "voluntary agent,"[188] an agent operating by "choice and intent." But the "intent to produce something" is operative during the period of the thing's "nonexistence" or, if one prefers, at the "moment of its coming into existence; and in either event, anything coming about through a voluntary agent" is preceded by nonexistence and "is created." Since every body and the world as a whole have their magnitudes assigned by a voluntary agent, the world and everything contained therein have been created.[189] An abbreviated version of the same train of thought is found in Āmidī[190] and Ījī.[191]

Rāzī has an additional argument for creation from the concept of particularization, and there he proposes a separate reason for the rule that a particularizing agent must precede its effect. The reason in the argument of Rāzī's just examined was that will excludes eternity; the reason now is to be that the eternity of an object is excluded by its having been brought into existence. Both reasons were already to be discovered in Juwaynī.[192]

The world, Rāzī submits, is "possibly existent" and what is possibly existent has need of an external agent that "tips the scales" in favor of its existence.[193] To maintain that something needs an "agent" to tip the scales in favor of its existence when it already exists would be absurd. The world must consequently have had the scales tipped in favor of its existence when it did not yet exist; and the world was created.[194]

The contrast between these arguments for creation and the standard Kalam procedure for proving creation and the existence of God again merits comment. The standard procedure set forth an elaborate proof of creation. It thereupon inferred a creator, often on the grounds that a particularizing agent must have selected a moment for the world's emergence or other characteristics for the world and the parts of the world when they came into existence. And it added that a particularizing agent must possess will. The new arguments borrow a thread, the particularization line of reasoning, from the standard procedure; and they invert the sequence of thought, concluding, rather than commencing, with the creation of the world. They observe that sundry characteristics of the world disclose an arbitrary choice between equivalent alternatives. From the evidence of arbitrary

[188] *K. al-Arba'īn*, pp. 27–29. Very similar reasoning appears in Shahrastānī, *K. Nihāya al-Iqdām*, p. 13.

[189] Ibid., pp. 17, 29.

[190] *Ghāya al-Marām*, pp. 250–251.

[191] *Mawāqif*, VII, p. 227.

[192] Above, p. 190.

[193] See above, p. 162.

[194] *K. al-Arba'īn*, pp. 30–31; see below, p. 387, n. 54.

choice they infer an agent who exercises will. And they conclude that an agent exercising will cannot have acted from eternity but must have brought the world into existence after not having done so.

Averroes was familiar with Juwaynī's argument for creation from the concept of particularization; he records it with explicit reference to the work of Juwaynī's in which it appeared. The argument, Averroes finds, rests on "two premises." The first states that "the world and everything therein" might have different characteristics from those they have; and the second states that whatever might have alternative characteristics must be brought into existence because it depends on an "agent," more specifically a "voluntary . . . particularizing agent," who "fashioned it in one possible mold in preference to another"[195] As might be expected, Averroes admits neither premise.

In connection with the first premise, he recognizes a distinction between the individual characteristics of individual objects within the world and characteristics touching the structure of the world. The supposition that the characteristics of individual objects might be other than they are is dismissed by him as "patently false," on the grounds that such characteristics are, unquestionably, determined by natural forces. As to characteristics touching the structure of the world, the possibility of their being other than they are might, Averroes concedes, seem plausible; for the "causes" of general phenomena—as, for example, the movements of one celestial sphere to the "west" rather than "east," and the remainder to the "east" rather than "west"—are often "hidden from man." Nevertheless, the scientific presumption should in every instance be that structural features of the universe are unqualifiedly "necessary" or at least represent the "best" and "most perfect" adaptation of natural objects for the functions they fulfill. It can thus not be taken for granted either that individual or general characteristics in the world might be different from what they are, and that certain characteristics have hence been selected out arbitrarily in preference to others.[196]

That is Averroes' refutation of Juwaynī's first premise. Averroes likewise rejects the premise stating that what is dependent for its characteristics on a voluntary agent must have been brought into existence after not having existed. Juwaynī had explained, or postulated, that an "object of will" cannot be eternal. Averroes rejoins that an eternal agent possessed of an eternal will not only would be able to exercise his will eternally, but could not help doing so. Consequently, Averroes clinches his refutation, even if the world were to reveal evidence of arbitrary selection, the agent making the selection would be capable of acting eternally, and the world need not have been created.[197]

Maimonides records a Kalam argument for creation from the concept of particularization which echoes both Juwaynī's argument and Shahrastānī's refor-

[195] *K. al-Kashf,* pp. 37–38, 40.
[196] Ibid., p. 38. Averroes undoubtedly has in mind the passage of Ghazali, quoted below, p. 195.
[197] Ibid., p. 40. Cf. above, p. 76.

mulation of it.[198] Like Averroes, Maimonides rejects the supposition that the characteristics of individual objects within the world might be other than they are; he too traces them to natural forces operating in conformity with natural law.[199] Maimonides does not, however, rule out the pertinence of the particularization notion to structural features of the world. He leaves open the possibility that these might be other than they are, and that a particularizing agent can be inferred who selected out structural features the world does have over features the world might have. In addition, Maimonides differs from Averroes on the subject of will; he agrees with Kalam thinkers that the exercise of will is incompatible with an eternal product. Because he can accept much of the present argument, Maimonides deems it the "best" of the Kalam arguments for creation, and he subsequently will rework it in what he believes to be an unexceptionable form.[200]

Maimonides also records an argument for creation, attributed by him to one of the "later" Kalam thinkers and described by him as a variation of the "preceding," which turns on the world's being "possibly existent." The argument is identical with one in Fakhr al-Dīn al-Rāzī.[201] What is "possibly existent," the reasoning goes, "has the possibility of existing as well as . . . not existing" and must, if it does exist, have had the scales "tipped in favor" of its existence. But an agent tipping the scales in favor of existence would not have acted from eternity. Consequently, the world, which is possible existent, cannot have existed from eternity. This train of reasoning is entirely rejected by Maimonides. The term "possibly existent," he explains, has two connotations, being applicable to what is, as well as to what is not, eternal.[202] To be possibly existent is therefore by no means tantamount to having the possibility of not existing. An object can be known to have a possibility of not existing only if it is known to be both possibly existent and noneternal, whereas anything possibly existent and eternal—as "our adversary," that is to say, Avicenna and his school, judged the world to be—would not have the possibility.[203] By taking for granted that everything possibly existent does have the possibility of not existing and hence does need to have the scales tipped in favor of its existence, the argument under consideration

[198]*Guide,* I, 74(5). Some of the features that, in Maimonides' version, could be other than they are, are the size and shape of the world and its parts, the location of the world, the natural place of the elements. Maimonides also enumerates features not mentioned by Juwaynī or Shahrastānī.

[199]*Guide,* I, 73(10).

[200]*Guide,* I, 74(5), in conjunction with I, 73(10).

[201]Above, p. 191. Maimonides and Rāzī were contemporaries and were separated, geographically, by half a world. The argument very likely was circulating in the schools, and Maimonides may have learnt it there.

[202]That is to say, something might in theory be possibly existent and eternal, although the advocate of creation would not believe that anything in reality is such.

[203]See Avicenna's discussion of the "possibly existent by virtue of itself, necessarily existent by virtue of another," below, p. 292.

assumes from the start that everything possibly existent is noneternal and has been created. The argument thereby begs the question and is invalid.[204]

4. Ghazali and Maimonides

Ghazali and Maimonides develop particularization arguments that are purged of Kalam elements. Ghazali's formulation is put forward in his *Tahāfut al-Falāsifa,* a work confronting Aristotelian philosophy—as recast by Avicenna—in its own terms. Maimonides, for his part, had no sympathy with the Kalam view of the universe and subscribed to the Aristotelian view, at least as far as the sublunar realm is concerned. Both philosophers accordingly sever the particularization mode of thought from its Kalam seedbed and transplant it to an Aristotelian framework by seeking out features of the world which are, even from an Aristotelian standpoint, unamenable to natural or rational explanation.

Ghazali allows himself to be led to his formulation by a dialectical exchange of a kind he enjoyed. The *mise en scène* is the refutation of an old proof for eternity, the proof maintaining that no given moment in an empty infinite time could, to the exclusion of other moments, have lent itself to the creation of the world and recommended itself to the creator as the proper moment for bringing the world into existence.[205] As Ghazali sums up the contention to be refuted, nothing "could differentiate one specific time [for the world's coming into existence] from earlier and later times, seeing that it would not have been impossible" for the world to come into existence "earlier or later."[206] The core of Ghazali's refutation goes back to Augustine and Philoponus; the creator, he responds, would have been able to choose a moment for creation by the exercise of will.[207] But a proponent of eternity might, Ghazali recognizes, surrejoin that the exercise of will is never a matter of pure arbitrariness, that will invariably opts for what it sees as the preferable alternative; and when the alternatives are alike in every respect, neither the faculty of will nor any other "attribute" of an agent can "differentiate" one alternative from another. The creator's will would thus not be able to fix upon a given moment to the exclusion of wholly similar alternative moments, and the world must after all be eternal. To counter the surrejoinder, Ghazali attempts to convince the advocates of eternity that they cannot help acknowledging instances in the universe of the "particularization of one thing over what is similar."[208]

He begins by remarking that God's will is seen to decide upon "whiteness to the exclusion of blackness" and assign the former to a subject although "the subject is as receptive of blackness as of whiteness"; and that God's will is seen

[204] *Guide,* I, 74(6).

[205] Above, p. 53.

[206] *Tahāfut al-Falāsifa* (Beirut, 1927), I, §28; English translation in *Averroes' Tahafut al-Tahafut,* trans S. van den Bergh (London, 1954), p. 18.

[207] Above, p. 69.

[208] *Tahāfut al-Falāsifa,* §34; English translation, p. 21.

to decide that a body should be in "motion" or "rest" although the body is no less receptive of the alternative. By the same token, God's will would have been able to decide upon one moment for creation to the exclusion of the identical alternative moments.[209] At this stage Ghazali is apparently working from the old Kalam presupposition that when an individual object in the world is black or white, in motion or rest, its color and state are not determined by natural forces within the universe. The illustrations he adduces, "whiteness and blackness, motion and rest," are moreover redolent of the classic Kalam arguments for the existence of accidents as actual entities.[210] No adversary, Ghazali knew, was likely to be convinced by a rebuttal that presupposes an occasionalistic picture of the universe and that provocatively employs illustrations of so dubious a pedigree. He therefore quickly passes on to more satisfactory evidence of the "particularization" of a given characteristic to the exclusion of equivalent alternatives.

The world, he observes, has a "particular shape" although it might equally have alternative shapes. Hence the shape it does have must be selected out arbitrarily by the cause of the world from among the alternatives it might have; and by the same token, a moment might arbitrarily be chosen by the cause of the world for creation.[211] Yet an obstinate adversary, Ghazali realizes, might still not be convinced. The adversary might retort here that if the world had a different shape, if it were "smaller or larger" than it is, or if it had, for example, a different "number of spheres and . . . stars," the "universal order" of the world would be impaired.[212] In other words, although the overall structure of the world, unlike the characteristics of individual objects, is not traceable to immanent natural forces, it is perhaps explicable on rational grounds as the optimum structure. And whereas the cause of the universe can, the adversary will hold, make a rational choice and opt for a superior alternative in preference to an inferior, no agent can choose arbitrarily between wholly identical alternatives. The upshot of the antagonist's retort to Ghazali's rebuttal would again be that a creator could not, through the exercise of will, have selected a moment for creation from among infinite identical and indifferent alternatives.

Ghazali at last is driven to the stage he planned to reach from the outset. He discovers two features of the celestial region where no natural explanation is feasible and where, in addition, there can be no basis for rational, as distinct from arbitrary, choice. The two features are the directions in which the celestial spheres move, and the location of a pair of points in the outermost sphere which function as poles and around which the spheres rotate. Analogous items had been adduced by Juwaynī and perhaps by the Ikhwān al-Ṣafā' as evidence of arbitrary

[209] Ibid., §28. I am altering the order of some of Ghazali's statements.
[210] Above, pp. 180, 182.
[211] Ibid., §35. I have followed the reading given by *Tahāfut al-Tahāfut,* I, p. 41.
[212] Ibid., §36.

particularization.[213] But Ghazali, unlike his predecessors, endeavors to show how these characteristics of the world do evince arbitrariness.

As regards the second of the two features, he writes that since all parts of the celestial sphere are, on the astronomical theory of the day, alike in nature, nothing could recommend any one pair of opposite points to the exclusion of another pair as the location of the poles. Therefore, an act of "particularization of one thing over what is similar" must be acknowledged, that is, a decision nowise dictated by the merits of the alternatives, but completely arbitrary. As regards the first of the two features, Ghazali concedes that the rotation of the outermost celestial sphere in a direction contrary to that in which the remaining spheres rotate is not yet evidence of arbitrariness; the rotation of the spheres in contrary directions may be an optimum arrangement for having the influences of the heavens intermesh and act on the earth in the most productive manner. Instead, his contention is that the selfsame result could be achieved in a universe that would be the inversion of ours, in a universe where the highest sphere rotated to the east rather than to the west, as it does in the Ptolemaic and medieval astronomical systems, while the remaining spheres rotated west rather than east. Since the movement of the spheres in the directions they do move has nothing to recommend it over the reverse arrangement, it surely is evidence of arbitrariness and of an act of particularization between identical, indifferent alternatives. Ghazali's conclusion is that even the proponents of eternity must admit instances in the universe of particularization between indifferent alternatives. They cannot, accordingly, object to creation on the grounds that no moment might have lent itself to creation to the exclusion of another; for the particularizing agent who arbitrarily selected the location of the poles and the direction of the movements of the spheres could have selected a moment for creation.[214]

Ghazali, in sum, tacitly concedes that the individual characteristics of individual objects within the sublunar region do not furnish convincing evidence of arbitrary particularization. Such characteristics can be ascribed to the workings of immanent natural forces. After restricting his attention to structural features of the world, Ghazali explicitly concedes that features rendering ours the best of all possible worlds do not constitute convincing evidence of arbitrary particularization. They may reflect rational, not arbitrary, choice by the cause of the world. In the end Ghazali allows his attention to be narrowed to two features of the celestial realm which he views as wholly indifferent and as containing nothing whatsoever to recommend them over the alternatives. These, he insists, can only be explained on the thesis that the cause of the universe selected them by an exercise of sheer will. Ghazali's aim throughout is the modest one of refuting a familiar proof for eternity by showing that the creator would be able to select a moment for creation. He might, however, have framed an independent argument,

[213] Above, p. 187.
[214] *Tahāfut al-Falāsifa*, I, §§37–41.

akin to the argument of Juwaynī's,[215] wherein evidence of arbitrary choice proves that the world is dependent on a voluntary agent, and the dependence of the world on a voluntary agent is found to entail creation. Such an argument would go beyond Juwaynī in completely divorcing itself from Kalam physics. Maimonides rethinks Ghazali's reasoning and does formulate a particularization argument for creation wholly within an Aristotelian, as distinct from a Kalam, framework.

Maimonides' argument is offered for the sole purpose of proving creation and not the existence of God. His rationale in not utilizing the argument for the latter purpose apparently is that the particularization procedure takes its departure from nothing positive and plays instead upon the inexplicability of various features of the physical universe. There could be no way of absolutely precluding an eventual evolution of human science to a level where the features in question might be explained. As a consequence, an argument from the concept of particularization lacked, for Maimonides, the probative weight of the proofs he does offer for the existence of God, they being "demonstrations" (*burhān*). A demonstration takes its departure from true and certain premises and proceeds to an indisputable conclusion. Since Maimonides was confident that a demonstration of the existence of God was available to him, he would not have knowledge of the existence of God rest on anything less. As regards creation, however, where no apodictic demonstration was available, he advances the strongest argumentation at his disposal.[216]

Maimonides studiously opposes his use of the particularization concept to the use made by the Kalam. He expressly rejects the Kalam doctrine of the "atom and the continual generation of accidents"; and he refuses to draw any inference from the characteristics of things in the sublunar realm, since they can be construed as "particularized through the powers of the sphere . . . just as Aristotle taught us." Characteristics of objects within the world are, that is to say, "particularized" in the naturalistic sense of having been determined by physical forces, and not in the more significant sense of having been selected through an act of will. In contradistinction to Kalam writers, Maimonides undertakes to establish the presence of "particularization where it should be established," and to do that through "philosophical premises derived from the nature of what exists."[217] The domain where he believes particularization, in the significant sense, can be established is the celestial region; and two aspects of the celestial region attract his attention. The first is the circumstance that the movements of the spheres do not fall into a regular pattern. The second is the location of the stars and planets in their several spheres.

Whereas Ghazali concerned himself exclusively with the nine main celestial spheres, Maimonides looks at all the spheres, primary as well as subordinate,

[215] Above, p. 190.
[216] *Guide*, I, 71; II, 16; 19.
[217] Ibid., II, 19.

which were recognized by ancient and medieval astronomy. As many as fifty-odd spheres were sometimes hypothesized in order to accomplish the task ancient and medieval astronomy set for itself, the task of reducing to motion in circles around the earth what modern astronomy represents as an elliptical motion of the planets around the sun. Each planet, it was theorized, is embedded in a sphere that rotates at the surface of another, and it, at the surface of still another. Each of the primary and subordinate spheres in the system was assumed to rotate with a constant velocity of its own, but some had to be assumed to rotate with a greater and some with a lesser velocity than others; and some had to be assumed to move in one direction, while others move in another. What, Maimonides asks, might account for the diverse velocities and directions in which the spheres rotate? He knows that the diversity cannot be explained through a diversity of material substance. The spheres, Aristotelian science taught, have a common material substance, which expresses itself—or, viewed from the opposite angle, is symptomized—by something common to them all, not by anything diverse; the common material substance of the spheres expresses itself in, and is symptomized by, the general circularity of motion common to all the spheres. The diversity in velocity and direction must then be due to a factor apart from the substance of the spheres. If a regular pattern were detectable in the sequence of spherical motions, a uniform natural explanation would be feasible whereby each celestial sphere determines the motion of the succeeding sphere. The outermost sphere could, for instance, be understood to communicate a proportion of its motion to the next sphere, the latter, to communicate an analogous proportion of its motion to the next, and so forth. But no pattern was detectable which would permit such a hypothesis; for in the supposed sequence of the spheres, westward and eastward movements, rapid and slow movements, appeared to succeed each other haphazardly. Inasmuch as the directions and velocities of the motions of the spheres fail to disclose any regular pattern, they can, Maimonides insists, be subsumed under no necessary natural law. The sole tenable thesis is that a voluntary agent "particularized each sphere with whatever direction and velocity of motion he wished."[218]

The second feature of the structure of the heavens engaging Maimonides' attention is the location of the stars and planets in the spheres. The spheres, on the enlightened medieval consensus, rotate constantly, whereas the stars and planets imbedded in the spheres undergo no motion of their own. Maimonides deduces herefrom that the substance of the spheres must be radically unlike the substance of the stars and planets. In addition, the supposed fact that the spheres are transparent whereas the stars are luminous also proves to Maimonides that the substance of the spheres is radically different from the substance of the stars

[218] *Guide*, II, 19. Elsewhere in the chapter, Maimonides brings in the forms of the spheres. He writes: A particularizing agent must be posited "who particularized the substrata [of the spheres] and prepared them to receive the diverse forms [of which the several motions of the spheres are an expression]."

and planets. No law of nature and no principle of regularity can, Maimonides argues, explain how stars of one substance come to be imbedded in spheres of a completely different substance. And no law of nature or principle of regularity can account for something still "more extraordinary," namely, the distribution of stars in the sphere assumed to contain all the fixed, or true stars, as distinct from the wandering stars or planets. In some areas of the sphere of the fixed stars, stars of varying magnitudes were seen to be clustered in constellations; in some areas, stars were seen to be scattered at random; and still other areas were devoid of stars. Since the presence and distribution of the stars in their several spheres—like the motions of the spheres—exhibit no regular pattern and can be subsumed under no necessary law of nature, they too, Maimonides maintains, are explicable only as an act of choice by a "particularizing agent" who "intended" that the stars should be located where they are.[219]

The evidence of particularization adduced by Maimonides is unmistakably related to the evidence that Ghazali eventually settled upon. As did Ghazali, Maimonides cites the movements of the spheres and the location of objects—planets and stars in his case, poles in Ghazali's—within the spheres. Maimonides, though, has introduced a change in that he deems not the arbitrariness, but the irregularity of the phenomena to be crucial. The reason for the change can be gathered from a totally unrelated context, a theological discussion where Maimonides incidentally deals with the selection by the deity of one from among several indifferent possible alternatives. In situations of the sort, he writes, it is meaningless to ask "why one possibility and not another came to pass; for an identical question would occur if the other possibility had come to pass instead of the one that did."[220] A choice in any event being called for, no weight can be attached to the circumstance that the deity happens to have selected one alternative over another. The consideration that when alternatives are wholly similar and a choice has to be made, it is senseless to ask why a given alternative was preferred to another, had much earlier been brought to bear in the debate over eternity and creation. Philoponus had put forward that consideration in connection with the very proof for eternity which Ghazali was rebutting. Faced with the question how a given moment could have recommended itself to the creator as the moment for creation in preference to other similar moments, Philoponus explained that once it is known that the world was created the question why a certain moment was chosen becomes meaningless; for an identical question might be posed if another moment had been chosen.[221] The same consideration presumably prevented Maimonides from inferring a "particularizing" agent who acts through "intention" either from the circumstance that the movement of the heavens in

[219] Ibid.

[220] *Guide*, III, 26. Maimonides is dealing with the question why Scripture prescribes the sacrifice of a certain number of animals rather than a larger or smaller number.

[221] Above, p. 69. Ghazali also makes the point, *Tahāfut al-Falāsifa*, I, §34; English translation, p. 21.

their present directions has nothing to prefer it over their moving in the reverse directions, or from the circumstance that the location of the poles in their present positions has nothing to prefer it over their location at any other pair of opposite points. Seeing that one of the two indifferent sets of directions and one from among numberless indifferent pairs of points had to be selected anyway for the world to exist, it is, in Maimonides' view, meaningless to ask why a given alternative happens to have been selected over another.

The noteworthy aspect of the movements of the spheres and the locations of the stars within them is, for Maimonides, not their indifference but their irregularity. If the movements and dispositions of the heavenly bodies formed a necessary scheme, the scientific observer would expect a certain ratio between the velocities of the inner spheres and their distances from the rapid outermost sphere, he would expect some affinity between the substance of the planets and stars and the substance of the spheres in which they are imbedded, and he would further expect that the fixed stars be evenly spaced throughout the heavens. Inasmuch as these phenomena lack regularity, they cannot be subsumed under any natural law and can be accounted for only on the thesis of a particularizing agent who selected them. The selection, it should be mentioned, is not thought by Maimonides to be ultimately "purposeless" or, despite the appearance, "haphazard." Believing as he does that the universe is the work of an intelligent cause, he is sure that a hidden rationality underlies the irregularity exhibited by the celestial realm.[222]

Having come this far, Maimonides adds that "particularization"—or "intention," or "choice," or "will" in the proper sense[223]—implies the ability of the particularizing agent to control the outcome of his choice. A man cannot, for example, be described as "particularizing" himself in, or as "intending" to possess, the shape of a being with "two eyes" and "two hands," since he cannot help possessing that shape. "Particularization" and "intention" are, consequently, conceivable solely in connection with "something nonexistent that has the possibility of both existing and not existing as intended and particularized"; they thus go hand in hand with a product's coming into existence after not existing. Maimonides has not forgotten that the Arabic Aristotelians spoke of an eternal emanation of the universe through an eternal exercise of will on the part of the deity,[224] but he dismisses statements of the sort as verbal legerdemain. If the world exists eternally, it exists necessarily, necessity and eternity being mutually implicative by virtue of an Aristotelian principle that Maimonides does not explicitly refer to but undoubtedly had in mind.[225] And to suppose that "existence through necessary emanation" can be combined with a thing's existing by "intention and

[222] *Guide,* II, 19. Cf. above, p. 176.

[223] At the end of *Guide,* II, 20, Maimonides writes that an agent may be described as *willing* his effect merely in the sense that he has pleasure in it.

[224] Cf. above, pp. 60, 75–76.

[225] Aristotle, *De Generatione et Corruptione* II, 11, 337b, 35 ff.; *Metaphysics* VI, 2, 1026b, 27.

will" would be very "close . . . to combining two contraries," thereby violating the law of contradiction. The world either exists eternally and necessarily, or it exists by particularization, intention, and will, and has come into existence after not existing. Given the evidence of "particularization," "choice," "will," and "intention" in the celestial region, Maimonides concludes that the celestial region and the rest of the world, which is dependent on the celestial region, cannot be eternal but must have come into existence.[226] To be precise, the form of the physical universe is what must have come into existence. Maimonides' argument says nothing about the matter of the universe and does not pretend to be a proof of creation *ex nihilo*.

Resumé

Kalam writers employed the particularization concept to support their inference of a creator from creation. They argued that a particularizing agent must be posited who selected out a time for the world to come into existence, who selects a time for individual objects to come into existence, who selected existence for the world in preference to nonexistence, or who selects out the characteristics of each object coming into existence. The contention that a particularizing agent must select a full complement of characteristics for each object coming into existence borders on a proof for a cause of the world wherein the premise of creation is set aside; for it could be argued that whether the world is created or not, a particularizing agent must be posited who selects out a complement of characteristics for the world and its parts.

Besides serving to support the inference of a creator from creation, the particularization concept found a variety of applications, most notably in arguments for the existence of accidents as real entities. A particularizing factor, it was reasoned here, must be present in an atom or body to tie each characteristic to the given atom or body to the exclusion of the alternative atoms and bodies that might have the characteristic, and to tie the characteristic to the atom or body at a given time to the exclusion of the alternative times when the characteristic might be there. In the course of resolving a difficulty in the theory of accidents, Juwaynī indicated why the accident as a particularizing factor is not sufficient and a particularizing agent must be assumed as well; the agent must be posited to explain how the accident, the particularizing factor, comes to be assigned to one atom to the exclusion of others. Juwaynī's explanation again borders on a proof for the existence of a cause of the world which sets aside the premise of creation. The world, it could be argued, stands in need of an external agent who assigns accidents to atoms and bodies.

Juwaynī, in one of his later works, did formulate a proof of the existence of God from the concept of particularization which dispenses with the premise of

[226] *Guide,* II, 20.

creation. Since the parts of the world and the world as a whole might, he reasoned, have alternative characteristics, an agent must exist who selects out the particular characteristics that the world and its parts do have from among all those they might have; and a particularizing agent capable of selecting out characteristics for the world would have to possess the properties of a deity. Similar arguments for the existence of God were offered by other Kalam writers. Juwaynī also formulated a new argument for creation. The agent who arbitrarily selects out a set of characteristics for the world must, he argued, act through will; but anything produced through the exercise of will exists only after not having existed; the world, therefore, being the product of an exercise of will, must have come into existence after not having existed. Similar arguments for creation likewise appear in other Kalam writers. The new arguments for the existence of God and creation contrast with the standard Kalam procedure for proving the same doctrines. The standard procedure inferred a creator from creation. The new procedure for proving the existence of God derives a cause of the world directly from the presence in the world of characteristics that might be different from what they are. And the new procedure for proving creation derives creation from the dependence of the world on a voluntary, particularizing agent; it concludes with, rather than commences with, creation.

Ghazali had a further application for the particularization concept, employing it in his refutation of one of the familiar proofs for eternity. He endeavored to show that even the advocate of eternity must acknowledge the ability of the cause of the world to make arbitrary decisions and, hence, arbitrarily to have chosen a particular moment for the world to come into existence. Maimonides, finally, developed a particularization argument for creation in a wholly Aristotelian framework. He sought out features of the celestial region which cannot be subsumed under any necessary natural law and which are therefore explicable solely as the result of particularization and intention. Whatever is produced by particularization and intention cannot, he agreed with the Kalam thinkers, be eternal, whence it follows that the celestial region together with the rest of the world must have come into existence.

Such were the principal arguments turning on the particularization concept. A progression is to be discerned as well in the phenomena adduced as evidence of particularization. The Kalam writers envisaged a world in which the characteristics of individual objects are not determined by natural forces and they could discover arbitrary selection in the individual characteristics of individual objects. Juwaynī's combined proof of the existence of God and creation cited, no doubt deliberately, both the characteristics of individual objects within the world and structural features of the world as evidence of arbitrary selection. Ghazali tacitly conceded that the individual characteristics of individual objects do not constitute convincing evidence of particularization, since they can be explained through the workings of natural forces; and he explicitly conceded that structural features of the universe rendering ours the best of all possible worlds are also not convincing evidence of arbitrary selection. He discovered the evidence of sheer arbitrariness which he required in features of the celestial region having nothing at all to recommend them over the alternatives. Maimonides, finally, understood that where a choice must be made in any event, no especial significance can be attached to one option's happening to be preferred over the equivalent alternatives. What was crucial for Maimonides was not arbitrariness but irregularity in the structure of the world. Irregularity would be incompatible with necessary natural law and would consequently imply an exercise of choice.

5. Additional arguments for creation in Maimonides and Gersonides

Maimonides advances two more arguments for creation, without however enumerating them as two distinct and coordinate arguments; what I call the first of the two appears as a brief appendix to the second. The additional arguments are bracketed by Maimonides with the argument of his examined in the previous section for the reason that the new arguments again find the structure of the physical universe to be explicable not by any necessary law, but solely as the work of a voluntary agent. The first of the additional arguments has, moreover, an inner resemblance to the argument already examined. That argument looked at the movements of the several celestial spheres and contended that the movements of the spheres cannot be subsumed under a necessary, natural law; the new argument looks at the spheres' forms and contends that these cannot be subsumed under any necessary law. Since the movements of the spheres are related to their forms—movement being an expression of form—the new argument can be regarded as an extension of the other.

The first of the additional arguments is concerned with the matter as well as the form of the celestial spheres, and with the interaction between celestial matter and celestial form. A central Aristotelian thesis had held that objects in the celestial region have a different material substratum from sublunar objects. Evidence therefor was drawn from the supposed fact that the natural motion of sublunar objects is rectilinear whereas the natural motion of the celestial spheres is circular.[227] According to Maimonides' reading of Aristotle and according to what he understood to be the best scientific description of the universe, the world contains no less than three separate types of matter: Sublunar objects have their common matter, as is indicated by the fact that their natural motion is rectilinear; the celestial spheres have another matter, as indicated by the fact that their natural motion is circular; and the stars have still a third matter, as indicated by the fact that, unlike both the celestial spheres and sublunar objects, they perform no motion of their own.[228] Celestial objects, Maimonides further understood, do resemble sublunar objects in a certain respect, insofar as they do contain the

[227] *De Caelo* I, 2.
[228] *Guide*, II, 19.

distinction of matter and form.[229] The celestial spheres are thus understood by him to have a matter common to them and different from the matter of the other objects in the universe; and each individual sphere consists in the union of the common matter of the spheres with the unique form of the particular sphere. The stars similarly have a matter common to them and different from the matter of both spheres and sublunar objects; and each star consists in the union of the common matter with the unique form of the particular star.

John Philoponus had long since called attention to an anomaly in Aristotle's thinking regarding the material substratum of the heavens. In the Aristotelian scheme, the material substratum of objects in the sublunar world constantly exchanges one form for another, whereas the substance of the heavens is unchanging. But if, Philoponus remarked, matter as conceived in the Aristotelian system, is "adapted to receive all [possible] forms," the matter of the celestial region should be so adapted; it should be "adapted to receive every one of the forms of the celestial [bodies]." Just as matter in the sublunar world can, and does, exchange one sublunar form for another, each portion of matter in the heavens should be capable of exchanging its form for any of the forms suitable to celestial matter.[230] Philoponus' comment was made as part of a proof for creation which proceeded in an entirely different fashion from the proof Maimonides is to offer. His contention was that since neither the matter of the sublunar world nor the matter of the celestial region can by nature retain a form permanently, "nothing [composed] of matter and form" can be "indestructible," and the physical universe cannot have existed from eternity.[231]

A nearer and more probable source of inspiration for Maimonides' proof is a passage in Ghazali's compendium of Avicenna's philosophy. The composition of the celestial region is once more the subject of discussion. Ghazali makes the unusual statement that the spheres do not after all have a common material substratum, that each sphere has a matter peculiar to itself.[232] Very likely at the back of his mind was the thought that if each sphere has a matter peculiar to itself, an answer is at hand to the question why the substratum of celestial objects is not seen to exchange one form for another: Perhaps no portion of celestial matter ever assumes a new form because only a single form is suitable for each unique portion of matter.[233] The intimation would be that should celestial objects indeed

[229]This was Avicenna's reading of Aristotle; Averroes, by contrast, rejected the notion that the spheres contain the distinction of matter and form. See citations in H. Wolfson, *Crescas' Critique of Aristotle* (Cambridge, Mass., 1929), pp. 594–598.

[230]As cited by Simplicius, *Commentary on the Physics*, ed. H. Diels, *Commentaria in Aristotelem Graeca*, Vol. X (Berlin, 1895), pp. 1329–1330.

[231]Above, p. 92.

[232]*Maqāṣid al-Falāsifa* (Cairo, n.d.), p. 247. I could not find the theory Ghazali is presenting in Avicenna's Arabic works, but did find it in his *Dānesh Nāmeh*; see *Le Livre de Science*, trans. M. Achena and H. Massé (Paris, 1955), p. 193.

[233]See an analogous notion in Juwaynī, above, p. 184.

possess a single common matter, as Aristotle had maintained, the substratum of each celestial object might exchange its form for any of the other possible forms known, by observation of the heavens, to be appropriate for celestial matter.

The problematic character of the substratum of the heavens and, it may be ventured, the intimation in Ghazali's statement[234] suggest a new proof of creation to Maimonides. The essence of matter, Maimonides argues, is to receive, in succession, every form appropriate to it. If the spheres are all constituted of a common matter, the material substratum of each given sphere is as adapted to receive the form of any other sphere as it is to possess the form of its own sphere. By the same token, if the stars are constituted of a common matter, the material substratum of each given star is as adapted to receive the form of any other star as it is to possess the form of its own star. Should events take their natural course, the matter of the spheres and of the stars would behave as the matter of the sublunar region does, repeatedly shedding one form and adopting another. The matter of each sphere would shed its form and successively adopt all the possible forms of spheres, with the changes manifesting themselves through repeated changes in the direction and velocity of the sphere's motion. The matter of each star would likewise shed its form and successively adopt all the possible stellar forms, the changes here manifesting themselves by changes in the quality of the light radiated by the star. But events plainly do not take their natural course. Neither the matter of the spheres nor the matter of the stars behaves as matter should behave, successively receiving all the forms it is adapted for.

The situation perceived by Maimonides is, then, as follows: The spheres have the common characteristic of moving circularly, and that should betoken a common matter. The stars have the common characteristic of not undergoing any motion of their own, and that should also betoken a common matter. Yet neither the common matter of the spheres nor the common matter of the stars behaves as a common matter should, by nature, behave. In neither case does the common matter receive all the forms it is adapted to receive. Since the situation does not lend itself to a natural explanation, Maimonides submits his nonnatural explanation. A conscious voluntary agent must permanently assign a specific form to the material substratum of each individual sphere, a form expressing itself in a distinctive circular motion; and it must assign a specific form to the material substratum of each individual star, a form expressing itself in a distinctive radiated light.[235] Maimonides could as well have used the language of particularization and written that a particularizing agent must have assigned a particular form to the material substratum of each sphere and star.

[234]In *Guide*, II, 22, Maimonides mentions, as something preposterous, the possibility that "someone should assert that the matter of each sphere is different from the matter of the others."

[235]Maimonides apparently means that God produces each sphere or star by permanently assigning—through an act of will—a unique form to a portion of the matter common to all the spheres or all the stars.

Having again come this far, Maimonides concludes as before: Eternity implies necessity and is incompatible with the action of a voluntary agent. The "intending" agent who "particularized" each sphere and star in its fixed form cannot therefore have acted from eternity; and the body of the heavens together with the sublunar region, which is dependent on the heavens, must have been created.[236] Nothing has been said regarding the creation of the matter of the universe, and the argument is not a proof of creation *ex nihilo*.

Maimonides' remaining argument for creation is set forth against the background of a theory of emanation which had been fathered upon Aristotle. Aristotle, as most medieval Arabic philosophers read him, bore a Neoplatonic guise. That is, he was understood to have recognized a first cause not merely of the motion, but also of the existence, of the universe; and he was understood to have maintained that the first cause emanates the universe continually and eternally. Aristotelian philosophy in its Neoplatonic guise had, accordingly, to face a perennial Neoplatonic problem. It had to explain how a highly complex universe can flow out of an absolutely simple first cause, considering that, as the formula went, "from one, only one can proceed."[237]

Alfarabi and Avicenna propounded a solution to the problem, a solution that probably was rooted in late Greek Neoplatonism. Alfarabi explained: The first cause, which consists in pure thought,[238] eternally emanates its effect by the mere act of thinking. Since it has a single object of thought, namely itself, what it emanates is a single being; and the latter, flowing from an incorporeal being that consists in pure thought, is likewise incorporeal and consists in pure thought. There is nevertheless a respect in which the incorporeal being that is emanated differs from the incorporeal first cause that emanates it. Whereas the first cause has one object of thought, the emanated being has two, namely itself and its cause. And inasmuch as the second being has two thoughts, it, through the mere act of thinking, emanates two things. It eternally emanates a celestial sphere and an additional incorporeal being consisting, again, in pure thought. The additional incorporeal being similarly has two objects of thought, itself and the first cause; and it too emanates two things, a second celestial sphere and yet another incorporeal being consisting in pure thought. The eternal process is repeated over and over, each incorporeal being emanating two effects, until the emanation of the active intellect, which is the last of the beings consisting in pure thought, and the sublunar corporeal region, which is the last stage of corporeal existence.[239]

[236] *Guide,* II, 22. There is a certain irony in the argument: It is just the unchanging character of the stars and spheres that proves, according to Maimonides, their noneternity.

[237] Avicenna, *Shifā': Ilāhīyāt,* p. 405; Ghazali, *Maqāṣid,* p. 218; idem, *Tahāfut al-Falāsifa,* III, §29, English translation, I, p. 104. Cf. Plotinus, *Enneads,* V, 1, 6; V, 2, 1; V, 3, 15. In Avicenna and Ghazali, the term for "proceed" is *yūjad* or *yaṣdur.*

[238] Cf. Aristotle, *Metaphysics* XII, 9. Plotinus maintained that the first cause, the One, is above thought, but exactly what he meant thereby is open to interpretation.

[239] Alfarabi, *K. Arā' Ahl al-Madīna al-Fāḍila,* ed. F. Dieterici (Leiden, 1895), p. 19; German

Such was the scheme of emanation promulgated in the name of true Aristotelianism by Alfarabi. Avicenna added a nuance and was able to distinguish not two, but three thoughts in each of the incorporeal emanated beings. One of the three thoughts, in Avicenna's version, gives rise to the next incorporeal being in the series; the second thought gives rise to the soul of a celestial sphere; and the third gives rise to the body of the sphere.[240]

The theory was subjected to a harsh critique by Ghazali and, following him, by Judah Hallevi,[241] the aim of both being to expose the depths of absurdity into which philosophy can fall. Ghazali registers a number of considerations, two of which are germane here. If the first cause is completely free of composition and has no more than a single thought, its thought of itself, the being emanated from it should, Ghazali objected, be equally free of composition and have no more than a single thought, the thought of itself. The first emanated being could in its turn give rise to no more than a single additional being, a being that for its part would still have just a single thought. The emanation theory of Alfarabi and Avicenna might therefore account at most for a series of unitary incorporeal entities, but not for the actual physical universe in all its complexity.[242] Furthermore, Ghazali objected,, even granting that the first emanated being thinks as many as three thoughts, each of which gives rise to a new entity, the three thoughts could not explain the complexity of the actual physical universe. One of the thoughts in the first emanated being is supposedly the source of the outermost sphere. But a single thought could not be sufficient for the task, since the sphere is complex, containing "form and matter" as well as secondary aspects.[243] And the thought in the second incorporeal being which supposedly produces the second celestial sphere would certainly not suffice, seeing that the second sphere contains not merely aspects of its own but a thousand-odd stars to boot.[244]

Ghazali's objections, the aim of which was negative, are transformed by Maimonides into an argument for creation. The principle at the heart of the issue is ascribed by Maimonides to "Aristotle and everyone who has philosophized," and the pregnant formulation Maimonides gives the principle reads: "From a simple

---

translation, with pagination of the Arabic indicated: *Der Musterstaat,* trans. F. Dieterici (Leiden, 1900). Cf. H. Davidson, "The Active Intellect in the *Cuzari* and Hallevi's Theory of Causality," *Revue des études juives,* CXXXI (1972), 356.

[240]*Shifā': Ilāhīyāt,* p. 406. In Avicenna's version, each emanated incorporeal being, or intelligence, is possibly existent by virtue of itself, necessarily existent by virtue of its cause, and accordingly has the following objects of thought: itself insofar as it is a possible being; itself insofar as it is a necessary being; the first cause.

[241]Hallevi, *Kuzari,* IV, 25; Davidson, "The Active Intellect in the *Cuzari,*" p. 358.

[242]*Tahāfut al-Falāsifa,* III, §§37–40, 49; English translation, pp. 109–110, 139.

[243]The secondary aspects are the size of the sphere, and the location of the poles around which the sphere rotates; cf. above, pp. 195–196.

[244]*Tahāfut al-Falāsifa,* III, §§54–59; 65–66; English translation, pp. 142–145, 149.

thing only one simple thing can *necessarily proceed* (*lazima*)."[245] Maimonides' argument is to be that on the hypothesis of necessary emanation no satisfactory account of the actual universe is feasible. For the process of necessary emanation from a simple incorporeal cause could never give rise to a composite effect; and even granting that it could do so inasmuch as the emanated being has two distinct thoughts, duality of thought is far removed from the manifold composition exhibited by the universe.

Maimonides writes: Though the hierarchy of emanated beings flowing out of the simple first cause should descend "through thousands of stages," no stage would possess a greater degree of composition than the preceding one, and the "last [emanated being] . . . would be simple," exactly like the first cause. The process of necessary emanation from an absolutely simple being could, consequently, never produce a composite effect. But further, conceding for the sake of argument that the second incorporeal being does contain a duality of thought, the full complexity of the universe is unexplained. Maimonides makes the point in several ways. He contends: (a) In necessary emanation "a correspondence always obtains between the cause and the effect" so that "a form cannot proceed necessarily from matter, nor a matter from form." Hence an incorporeal being, conceding the duality of its thought, could produce only things of the same kind, in other words, incorporeal beings and not a corporeal celestial sphere; and corporeality in the universe remains unaccounted for. Moreover, granting once more for the sake of argument that an incorporeal being could emanate something corporeal, the emanation scheme of Arabic Aristotelianism is still inadequate. For (b) a single one of the two thoughts of the supernal incorporeal beings could not emanate a full-blown celestial sphere. A celestial sphere consists of no less than four distinct factors, the matter and form of the sphere and the matter and form of the star imbedded in the sphere. On the assumption of necessary emanation, the thought that is the source of the celestial sphere would itself have to contain four distinct aspects to serve as the source of the four factors in the sphere.[246] Finally (c) the complexity of the sphere of the fixed stars would remain totally unaccounted for. That sphere contains stars of various types, and each type is comprised of two factors, its matter and its form. The single thought from which the sphere is assumed to emanate could not give rise to all those factors.

The actual complexity and corporeality of the universe are thus completely at odds with the principle affirming that "from a simple thing only one simple thing can necessarily proceed." The principle, as comprehended and formulated by Maimonides, appertains, however, exclusively to necessary emanation. An incorporeal cause acting, by contrast, not through necessity but voluntarily could

[245] *Guide*, II, 22.

[246] As it happens, one of Hallevi's criticisms of the emanation theory was that each succeeding incorporeal being would have more objects of thought than the preceding incorporeal being, inasmuch as it has more entities above it; and as a consequence, each should have a larger number of emanated product than the previous one. See *Kuzari*, IV, 25.

produce composite as well as simple, and corporeal as well as incorporeal, effects. The corporeality and the composition of the universe, which are inexplicable on the assumption that the first cause produced the universe through a process of necessary emanation, can therefore be satisfactorily explained on the contrary assumption that the first cause acted voluntarily. And since a voluntary agent does not act eternally, since it acts after not having acted, the world must, Maimonides concludes, be created.[247] Elsewhere Maimonides stresses that the sole conceivable kind of causation attributable to incorporeal beings is the process of emanation.[248] Here, then, he is advocating a theory of voluntary, noneternal emanation in which God initiated the process at a given moment through the exercise of his will; God switched on the emanation process, as it were, and brought the entire universe into existence.

The argument, as will be noted, is designed to establish that both the matter and the form of the world were created; it is an argument not merely for creation, but for creation *ex nihilo*.

Gersonides framed an argument for creation which, although thought out differently from Maimonides' arguments borrows from them. Repeatedly in the history of medieval philosophy, one encounters the contention that anything having a cause for its existence cannot exist from eternity; and that the world, since it does have a cause of its existence, cannot be eternal, but must have been created.[249] The argument Gersonides offers is a refinement of the contention. He does not maintain unqualifiedly that anything with a cause of its existence cannot be eternal. What he espouses is the narrower thesis that a subsistent entity with a cause of its existence cannot be eternal.

Like some of the writers already met, notably Juwaynī and Maimonides, Gersonides works from a prior proof of the existence of God, his being a teleological argument—to be examined more fully in the next chapter—which establishes that the physical universe has "come into existence" and is the "product of an agent."[250] Proving that the world was produced by an agent is something less, Gersonides makes clear, than proving that the world was created; for, he understands, there are two distinct ways in which an agent might bring its effect into existence. The existence of the effect may, in the first place, be the fruit of "an absolute coming into existence"; that is to say, the effect may be endowed with existence "solely at the moment of its coming into existence," as when a "house" is produced by a "builder."[251] But existence may, in the second place, be "emanated continually" from the cause, as when "motion" is produced continuously—

[247] *Guide*, II, 22.
[248] *Guide*, II, 12.
[249] Above, p. 190; below, p. 387.
[250] Below, p. 231.
[251] *Milḥamot ha-Shem* (Leipzig, 1866), VI, i, 6, p. 309; 7, p. 312.

in the pre-Newtonian universe—by a cause of motion. Should something be brought into existence in the first way, should it all at once acquire an existence that it did not previously have, it obviously cannot have existed from eternity. Should, however, something be emanated in the second way, it may well—although it not necessarily will—have existed from eternity, since the cause of its continual emanation may always have emanated it. An illustration of this possibility is the motion of the spheres, which is "emanated" from the "thinking of the [incorporeal] movers of the celestial spheres"; inasmuch as the motion of the spheres is emanated continually, the process might be eternal, the movers of the spheres continually and eternally bringing about the spheres' motion.[252]

Having laid down the distinction between things that are brought into existence once and for all and things that are emanated continually and perhaps eternally, Gersonides reduces the question of creation to another question. His teleological proof for the existence of God had established that the celestial region and, through it, the rest of the world were brought into existence by a cause. He now asks whether the cause bestowing existence on the celestial region did so once for all or whether it does so through a continual and perhaps eternal process. Ruling out the latter hypothesis will prove that the world was created.

Consider, writes Gersonides, the hypothesis that the celestial region is emanated from the deity continually. The hypothesis is certainly not viable if its purport is that the deity sustains the heavens in existence by, at each moment, converting already existent celestial bodies into new, exactly identical celestial bodies. On such a construction the celestial region would exist unchangingly and unchanged, "no act at all" would ever occur, and to speak of a cause of the existence of the heavens would be nonsensical. The hypothesis that the celestial spheres are emanated by their cause continually can have meaning only on the supposition that "immediately upon the spheres' being emanated from God, they are destroyed [and return] to nothingness, whereupon they are immediately re-emanated . . . from God out of nothing." But the supposition that the heavens are constantly being destroyed and reemanated is on its face ludicrous; and analysis, further, shows that it embodies a logical absurdity. The heavens would, on that supposition, not just come into being and forthwith be destroyed. If gaps are not incessantly to interrupt their existence, the heavens would have to come into existence, be destroyed, and come into existence again at the very same moment. The heavens would thus have to be both existent and nonexistent at every moment in time, a situation precluded by the law of contradiction.[253]

The supposition that the spheres are constantly emanated, destroyed, and re-emanated has, moreover, bizarre corollaries. Time would not be continuous, but would consist of the contiguous discrete moments at which the heavens exist. The heavenly region would have potential, and not actual, existence, because

[252]Ibid., 6, p. 308; 7, p. 314.
[253]Ibid., 7, p. 313.

each discrete episode in the existence of the spheres would last no more than a moment, and "as is proven in [Averroes' commentary on] the *Physics,* anything existing for merely a moment exists potentially, not actually."[254] And the heavens would be unable to undergo continuous motion, or for that matter any motion whatsoever. For a new set of heavens would be reemanated at each moment, and no set of heavens would exist long enough to move.[255]

It is to be noted that Gersonides' reasoning, although spelled out in conjunction with a distinctive picture of the celestial region, is not wedded to any physical scheme. Gersonides is attempting to refute the entire notion of an entity's being produced through continuous emanation. An entity produced through continuous emanation would, he reasons, either have to exist forever unchanged; and to speak of its being caused would then be nonsensical. Or else the entity would have to be constantly destroyed and recreated, which is ludicrous and ultimately illogical. As a "general rule," he summarizes, "things that endure cannot possibly" be emanated continually and, if caused, must be brought into existence once for all; "only accidents that do not endure, such as motion" can be emanated continually, and they alone might be emanated from eternity.[256] As regards the heavenly bodies, since they are subsistent entities and do have a cause of their existence, their cause cannot be of the sort that emanates its effect continually, but must be of the sort that brings its effect into existence at a single moment. Neither the celestial region nor the sublunar region, which is dependent on it, can therefore have existed from eternity.[257]

Gersonides buttresses his conclusion through an auxiliary argument, obviously inspired by Maimonides, that details the "things found in the heavens" which are incompatible with the "nature of the heavens." From the "physical sciences" he knows that the substratum of all the heavenly bodies is of a "single . . . nature . . . containing no diversity whatsoever."[258] If the heavenly bodies are all constituted of exactly the same homogeneous substratum, they should be completely alike; and in fact no more than a single celestial body should exist. Yet not merely does more than one heavenly body exist, the heavenly bodies differ from each other in a number of respects. They are, Gersonides finds, endowed with different forms, as can be inferred from the differing influxes descending from them into the sublunar realm. The spheres are of different sizes, even though size is, by

[254]See Averroes, *Middle Commentary on Physics,* (Oxford, Bodleian Library, Hebrew MS., Neubauer 1380 = Hunt. 79), VIII, v, 3, pp. 94a–b, referring to Aristotle, *Physics* VIII, 8, 263a, 23 ff.

[255]*Milḥamot ha-Shem,* VI, i, 7, pp. 312–313.

[256]Ibid., p. 314. Gersonides' thinking presumably is: Since an accident such as motion is not subsistent, new motion neither is produced out of the previous motion, nor does it require the previous motion to be destroyed.

[257]Ibid.

[258]*De Caelo* I, 2–4.

nature, a function of the "mixture" of matter (*mezeg*) in an object,[259] and the celestial bodies share the same material substratum. The spheres occupy different places in space, one encompassing the other, even though objects of the same substance should have an identical natural place.[260] The spheres differ in the number of stars they contain, the sphere of the fixed stars containing a thousand-odd stars, other spheres[261] containing one each, and still others containing none. The heavenly bodies differ insofar as some, the spheres, are transparent, whereas others, the stars and planets, are luminous and nontransparent. The stars and planets, for their part, differ from each other in their colors and in the diverse influences they exercise upon the earth. These differences, Gersonides submits, although wholly inexplicable from the standpoint of the nature of the celestial substratum can be accounted for on the thesis that the celestial region was brought into existence through "will and choice." The upshot of his proof is that the celestial region as well as the sublunar realm, which is dependent on it, must have been brought into existence after not having existed, by a transcendent agent who exercised will and choice.[262]

[259]As Gersonides explains, the mixture of matter in an ant determines a certain size and that in a camel determines another size. The minor variations from the norm in the size of an ant or a camel are due to their being "compounded from contraries," and even such minor variations should not be exhibited by the spheres, since their matter does not contain contraries.

[260]In other words, they should not stand one above the other, but all should occupy exactly the same place, at the same distance from the center of the earth.

[261]Those containing the wandering stars, or planets.

[262]*Milḥamot ha-Shem,* VI, i, 8, pp. 316–320.

VII
Arguments from Design

1. Cosmological, teleological, and ontological proofs of the existence of God

The preceding chapters have dealt primarily with arguments for eternity and creation, but several, relatively simple proofs of the existence of God have also been met. The latter are all associated with the Kalam tradition, though not in every instance thought out in a specifically Kalam framework. The most popular of the proofs of the existence of God met thus far is the one resting on a prior proof of creation: Given the creation of the world, the conclusion is drawn that an agent must have been responsible for the world's coming into existence. The weight in that procedure falls upon the preliminary proof for creation. Once the creation of the world is established, the existence of a creator is inferred, either as something self-evident[1] or by analogy with objects observed to come into existence within the world,[2] or through the subtler thought, turning on the concept of *particularization,* that an agent would be required to select out the *particular* moment for the world to come into existence as well as other particular characteristics for the world when coming into existence.[3] Besides the proof from creation we have met a proof of the existence of God from the concept of *particularization* in which the premise of creation is dispensed with; and proofs of the existence of God from composition in the universe. The proof from the concept of particularization dispensing with the premise of creation takes its departure from features of the physical universe which might conceivably be other than they are; and it reasons that since the universe could as well have had alternative characteristics, an agent must be posited who selected out the characteristics the universe does have.[4] Proofs of the existence of God from composition in the universe reason either in general that the composition of the universe indicates an intelligent and powerful cause,[5] more narrowly that the

[1]Above, pp. 154–155.
[2]Above, pp. 156–157.
[3]Above, pp. 161, 178.
[4]Above, p. 187.
[5]Above, pp. 148–149.

composition of contrary qualities in the universe indicates such a cause,[6] or else that the design evidenced in the composition of the universe must be due to such a cause.[7] Arguments establishing a cause of the universe are not in themselves complete proofs of the existence of God, although the Kalam thinkers unfortunately do not state what, precisely, a complete proof must comprise. After arriving at a cause of the universe, adherents of the Kalam proceed to argue that the ultimate cause of the universe is eternal,[8] one[9] and incorporeal,[10] possessed of power, knowledge, life, and will[11]—but they do not tell us precisely which of these attributes are being established as part of their proofs of the existence of God and which are ancillary attributes.

The distinction between ontological, cosmological, and teleological[12] arguments for the existence of God is always illuminating. Of the proofs met thus far, all but one are cosmological, insofar as they fit Kant's paradigm exactly,[13] but in the broader sense that they reason from the existence of the world or from the existence of something in the world, to a cause of the world. One proof, that from the design evidenced by the composition in the universe, is teleological. None is ontological. None, that is, arrives at the existence of God purely through an analysis of concepts.

The chapters to follow will have as their subject medieval Islamic and Jewish proofs of the existence of God which are associated with the Aristotelian tradition. These are: the Aristotelian proof from motion; an offshoot of the proof from motion which has been called the proof from 'logical symmetry'; Avicenna's proof of the existence of a being necessarily existent by virtue of itself; and a family of arguments from the impossibility of an infinite regress of causes. All of them take their departure from the existence of something, be it motion in the world, an individual object in the world, or, less frequently, the world as a whole. And they reason therefrom to the existence of a first cause. They are, accordingly, like the majority of proofs associated with the Kalam tradition, cosmological.

One of the proofs associated with the Aristotelian tradition might, if read carelessly, be misinterpreted as an ontological rather than a cosmological argument. Central to Avicenna's proof of the existence of a being necessarily existent by virtue of itself is the analysis of the critical concept, the concept *necessarily existent by virtue of itself*. And a superficial reading might lead to the misapprehension that the existence of a being corresponding to the concept is derived by

[6]Above, p. 151.

[7]Above, p. 152.

[8]Ghazali, *Tahāfut al-Falāsifa*, ed. M. Bouyges (Beirut, 1927), III, §§3, 16; English translation in *Averroes' Tahafut al-Tahafut*, trans. S. van den Bergh (London, 1954), pp. 89, 96; above, p. 165.

[9]Above, pp. 166–171.

[10]Above, pp. 171–172.

[11]Above, p. 173.

[12]I employ the expressions *argument from design* and *teleological argument* as synonyms, although by etymology an argument showing order without purpose and functionality should not, perhaps, be called teleological. Kant's name for the argument from design was the "physico–theological proof."

[13]See Kant, *Critique of Pure Reason*, A604–606/B632–634.

Avicenna solely from an analysis of the concept.[14] The error might, moreover, be abetted by the presence in European philosophy of ontological arguments for the existence of God which do consist exclusively in the analysis of a similar concept, that of *necessary being*.[15] Avicenna's proof, it turns out, does not arrive at the existence of a being necessarily existent by virtue of itself solely through analysing a concept, and his proof is unambiguously cosmological.[16]

Were a desperate need felt to unearth a specimen of ontological argumentation in medieval Arabic philosophy, a more promising area to search would be not the writings of Avicenna himself, but the minor philosophic works that his writings inspired. It might, for example, be imagined that a passage in the *'Uyūn al-Masā'il* offers an ontological argument. The *'Uyūn al-Masā'il,* a text sometimes wrongly ascribed to Alfarabi, is a philosophic florilegium, a catena of excerpts and paraphrases from philosophic writings, especially from Avicenna. One paragraph in the text reads: "To assume that the necessarily existent does not exist is impossible. Its existence has no cause, and cannot be (?) from another."[17] Should the statement—which is based on Avicenna's definition of the necessarily existent by virtue of itself[18]—be taken as a self contained proof of the existence of God, the intent could only be that the mere inspection of the concept *necessarily existent* renders the denial of the existence of a being corresponding to the concept absurd. The passage would then constitute an ontological argument. But that construction is militated against by the failure of the author, or compiler, of the *'Uyūn* to intimate that his statement is in fact intended to stand as a self-contained proof of the existence of God. Immediately prior to the passage, a brief cosmological proof had been given, and now the author apparently is saying: Once a necessarily existent being is known to exist through the preceding argument, such a being cannot fail to exist at all times; for it has no cause and is dependent on nothing else. Placing an ontological construction on the passage in question is also militated against by the absence of any suggestion that the author or compiler had a notion of what ontological argumentation might be. The *'Uyūn al-Masā'il* may be safely judged free of the ontological argument of the existence of God, and, more comprehensively, medieval Islamic and Jewish philosophy may as a whole be judged free of ontological argumentation.

[14]See Jak. Guttmann, *Die Religionsphilosophie des Abraham ibn Daud* (Goettingen, 1879), p. 121; H. Wolfson, "Notes on Proofs of the Existence of God in Jewish Philosophy," reprinted in his *Studies in the History of Philosophy and Religion,* Vol. I (Cambridge, Mass., 1973) p. 569; P. Morewedge, "Ibn Sina and Malcolm and the Ontological Argument," *The Monist,* LIV (1970), 238 and 242.

[15]See below, pp. 392–393.

[16]See below, pp. 298, 303–304, 403–404.

[17]In *Alfarabi's philosophische Abhandlungen,* ed. F. Dieterici (Leiden, 1890), p. 57; German translation, with pagination of the Arabic indicated: *Alfarabi's philosophische Abhandlungen aus dem Arabischen uebersetzt,* trans. F. Dieterici (Leiden 1892). Other examples are *R. Zaynūn,* in *Rasā'il al-Fārābī* (Hyderabad, 1931) pp. 3–4, and *al-Da'āwā al-Qalbīya,* pp. 2–3, in the same volume.

[18]Cf. below, p. 291.

Most proofs of the existence of God associated with both the Kalam and Aristotelian traditions are, to repeat, cosmological. The sole teleological argument encountered until now is the proof from the design evidenced by the composition of the universe. Additional arguments from design can, however, be gleaned from the writings of the Islamic and Jewish philosophers. The arguments have differing purposes; they seek to prove the existence of God or to establish certain attributes in God. Teleological arguments are more common among writers standing in the Kalam tradition, but they appear as well among those who stand in the Aristotelian tradition and those who would eschew labels. In the medieval, as in the ancient, period, teleological arguments are not closely reasoned. An effort was virtually never made to determine precisely what constitutes evidence of design, the supposition being that anyone who sees design will intuitively recognize it. Furthermore, arguments that are ostensibly proofs of the existence of God generally fall short of establishing the existence of a being possessed of all the specifications of the deity and therefore remain something less than complete proofs of the existence of God. A critique of teleological argumentation has been preserved from the ancient period, but it is rudimentary,[19] nowise approaching the critique of Hume, for example.[20] I was able to discover no more than a soupçon of a critique in medieval Islamic and Jewish literature.[21] It may be of the nature of teleological arguments that they are the least sophisticated proofs of the existence of God and yet, when all has been said, remain the most plausible.

2. Teleological arguments

The teleological arguments preserved from ancient philosophy pursue one of two lines, discovering evidence of design either on a minor, or a cosmic scale. The evidence of design which is discovered is either the functionality of nature or an aesthetic quality in nature, its orderliness and beauty; and there is a partial correlation between the two lines of argumentation and the two kinds of evidence which are adduced. Arguments on the minor scale pass in review individual details from various realms of nature and usually focus, specifically, on the functionality of each detail. Arguments conducted on the cosmic scale may similarly adduce the overall functionality disclosed by nature as evidence of design, but to a greater extent they concern themselves with the aesthetic quality, the orderliness and beauty, of the cosmic panorama.

Xenophon relates that the first of the two lines of argumentation constituted a part of Socrates' positive teaching. Socrates, the report goes, would point out the functionality of divers aspects of man's physiological and psychological

[19] See Cicero, *De Natura Deorum*, III, ix-x, 24–25; xi, 28

[20] Cf. Hume, *Dialogues concerning Natural Religion*, II.

[21] Judah Hallevi, *Kuzari*, I, 19–20. Hallevi himself employs teleological arguments; see below, pp. 228–229.

endowments and also phenomena in the world surrounding man which contribute to man's well-being. He postulated that what "clearly is for a purpose" must be the result of "forethought" (γνώμη), not "chance." And he concluded that man must have been made by a "wise demiurge"; and again that the "gods take care to furnish men with the things they need," that nature discloses divine "providence" and "love of man."[22] The same procedure was employed, in an expanded form, by the spokesman for Stoicism in Cicero's *De Natura Deorum*. More data are offered there; the inquiry is extended to phenomena of botany, zoology meteorology, and geology, which have nothing to do with man; and the wealth of evidence discloses that functionality and hence design are ubiquitous in nature. The conclusion of the spokesman for Stoicism is that "the world is governed by the providence of the gods"—that "everything in the world is governed, for the welfare and preservation of all, by divine intelligence and deliberation," and not "by chance."[23] The style of reasoning employed by Socrates and by the Stoic was likewise used by Galen, who uncovers the intelligent operation of "nature" or the "demiurge" in numerous details of human and animal physiology.[24]

The second line of teleological argumentation in the ancient period, the one that searched for design on a cosmic scale, directed its attention in some instances exclusively to the heavens; and what was cited was the utility of the heavens, their immense beauty, and the regularity of celestial motion. The motif goes back to Plato's *Laws*[25]—and indeed to the Book of Isaiah.[26] It was expressed eloquently by the younger Aristotle in his dialogues.[27] And it was enunciated several times by the spokesman for Stoicism in Cicero's *De Natura Deorum*. The "uniformity" of celestial motion and the "utility, beauty, and order" of the heavenly bodies prove to the spokesman for Stoicism that a "mind . . . governs" the universe.[28] In other instances the aim of the argument was even more ambitious and a cosmic plan was discovered not merely in the heavens, but in the universe as whole. An argument of the sort was employed, once more, by the spokesman

[22]Xenophon, *Memorabilia,* I, iv; IV, iii. Scholars have debated whether the sections in question are correctly attributed to Xenophon and whether the account that is given of Socrates' teachings is accurate.

[23]Cicero, *De Natura Deorum,* II, xxix, 73; xlvii, 120—lxvi, 167; and more briefly in *Tusculan Disputations,* I, xxviii, 68.

[24]Galen, *De Usu Partium,* X, 9; XVII. The theme had also appeared in Aristotle's *De Partibus Animalium,* but there design in nature is clearly regarded as immanent and nonconscious. See *De Partibus Animalium* I, 1, 641b, 12; III, 1, 661b, 24; III, 2, 663b, 23–24; IV, x, 687b.

[25]Plato, *Laws,* 966–967; cf. 897–899.

[26]Isaiah 40:26. Cf. also Quran 71:15–16.

[27]Cicero, *De Natura Deorum,* II, xxxvii, 95 (designated as Aristotelian fragment #12); Sextus, *Adversus Physicos,* I, 20–22 (fragment #10). These passages are assigned by scholars to the lost dialogue *De Philosophia*.

[28]*De Natura Deorum,* II, v, 15; see notes of J. Mayor and A. Pease in their editions, *ad locum*. The theme also appears in *De Natura Deorum,* II, ii, 4; xxxviii, 97; xliv, 115; and in Sextus, *Adversus Physicos,* I, 26.

for Stoicism in Cicero,[29] and Philo.[30] As expressed in the language of Stoic philosophy, the "sympathetic, harmonious, all-pervading affinity of things . . . forces" one to recognize "a single divine, all-pervading spirit."[31] Whether it is in the heavens alone or in the universe as a whole that an all-embracing design was sought, the ancient philosophers had a repertoire of analogies at their disposal. They compare the heavens or the entire universe to a house,[32] a wrestling school,[33] a deliberative assembly,[34] a city,[35] a ship,[36] a book,[37] or a piece of machinery such as a clock.[38] No impartial observer encountering any of these objects and perceiving how well ordered it is, how well its parts mesh, and how well they function, would—so the contention went—imagine that the object before him had come about spontaneously. No one would imagine that it could have arisen by pure chance, that it is not the handiwork of an intelligent designer. By the same token, no impartial observer gazing upon the heavens or the universe in its entirety can imagine that they came about through pure chance and are not the handiwork of an intelligent being who designed them. A certain advantage is to be noted in the cosmic line of teleological argumentation, especially from the standpoint of medieval thinkers, for whom the unity of God was more fundamental than for the ancients. Teleological argumentation on a cosmic scale lends itself to a demonstration of the unity, as well as of the existence, of a cause of the universe. If the universe exhibits a single overall design, the conclusion is more easily drawn—although it is not necessarily drawn—that a single designer is responsible.

Both lines of teleological argumentation, that from the details of nature and that from the overall design exhibited by nature, reappear in medieval Arabic philosophy. Usually, though not always, arguments pursuing the first line adduce functionality as evidence for design, while arguments pursuing the second, adduce the orderliness of nature. That is to say, Arabic arguments from design stress the functionality, but sometimes the orderliness or also the beauty of the details of nature; or else they stress the orderliness or perhaps the beauty, but sometimes the functionality too, of nature as a whole. A number of teleological considerations deployed in medieval Arabic literature are akin to those deployed in classical

[29] *De Natura Deorum,* II, vii, 19; cf. xlv, 115.

[30] *Legum Allegoria,* III, xxxii, 99.

[31] *De Natura Deorum,* II, vii, 19; cf. notes of Mayor and Pease.

[32] Cicero, *De Natura Deorum,* II, v, 15; Philo, *Legum Allegoria,* III, xxxii, 98. See Pease's note to Cicero, *ad locum,* for further examples.

[33] Cicero, ibid.

[34] Ibid.

[35] Philo, *Legum Allegoria,* III, xxxii, 98.

[36] Cicero, *De Natura Deorum,* II, xxxiv, 87; 89: "the course of a ship"; Philo, *Legum Allegoria,* III, xxxii, 98: "a ship"; Sextus, *Adversus Physicos,* I, 27: a ship seen travelling across the oceans. Further examples in Pease's note to Cicero, *ad locum.*

[37] Cicero, *De Natura Deorum,* II, xxxvii, 93.

[38] Ibid., xxxiv, 87–88; xxxviii, 97.

literature, and the resemblance is such that a historical continuity is unmistakable. At least one route whereby the transmission was effected can be traced.

Both lines of teleological argumentation passed into patristic literature, where writers use them to prove the existence of God and the pervasiveness of divine providence in the universe.[39] As a rule, literary links between the Church Fathers and Islamic thought are not detectable, but in the case of the argument from design the stages in the transition can be observed. The links are a Greek work by the fifth-century Christian cleric Theodoret, entitled *On Providence*; and an Arabic work, extant in at least three slightly different recensions, which is sometimes attributed to the well known litterateur Jāḥiẓ and sometimes to one Anbārī, apparently a Christian.[40] The Arabic work appears under varying names, each of which contains the term *reflection*. In the printed edition, where the book is attributed to Jāḥiẓ, it is called: *The Book of Proofs and Reflection regarding Creation and Divine Governance* (*K. al-Dalā'il wa-l-I'ti'bār 'alā al-Khalq wa-l-Tadbīr*). Theodoret's treatise *On Providence* undertakes, as the title promises, to set forth the operation of divine providence in the universe. A large amount of evidence for design is presented, much of it strikingly similar to evidence adduced by the spokesman for Stoicism in Cicero. Since Cicero utilized Stoic sources in Greek and Theodoret probably was restricted to Greek materials, Cicero and Theodoret very possibly borrowed from the same Stoic work—perhaps a lost work of Posidonius'[41] which was still available to the Church Fathers;

[39]Lactantius, *De Opificio Dei; Institutiones,* I, 2; VII, 3; Ps. Clement, *Recognitiones,* VIII, chaps. 20, 22, 23, 44; Basil, *Hexameron,* I, 7; VI, 1 and 10; *Sermon on Deuteronomy* 15:9, §§7–8; Gregory of Nyssa, *De Anima et Resurrectione,* III; Augustine, *De Civitate Dei,* XI, 4; Nemesius, *De Natura Hominis,* chap. 42; K. Gronau, *Poseidonius und die jeudisch-christliche Genesisexegese* (Leipzig, 1914); M. Pohlenz, *Die Stoa* (Goettingen, 1955–1959), note to p. 431.

[40](1) *K. al-Dalā'il wa-l-I'tibār 'alā al-Khalq wa-l-Tadbīr,* attributed to Jāḥiz (Aleppo, 1928). (2) *K. al-'Ibar wa-l-I'tibār,* attributed to Jāḥiẓ (British Museum, MS. Or. 3886); described by H. Gibb, "The Argument from Design," *Ignace Goldziher Memorial Volume,* Vol. I (Budapest, 1948), pp. 150–162. (3) *K. al-Fikr wa-l-I'tibār,* attributed to Jibrīl b. Nūḥ b. Abī Nūḥ al-Anbārī (Aya Sofia, MS. 4836/2, pp. 160–187). A photograph of the Aya Sofia manuscript was made available to me by Josef van Ess, and he called to my attention that a certain Abū Nūḥ al-Anbārī, who is presumably identical with the grandfather of our Jibrīl, is known as a Christian; see P. Krauss, "Zu ibn al-Muqaffa'," *Rivista degli Studi Orientali,* XIV (1933), 10–11. Militating against the attribution of the text to Jibrīl al-Anbārī is a statement in the introduction to the British Museum manuscript to the effect that one of four sources used by the author of the text was a poorly written and ill-organized book by the same Jibrīl b. Nūḥ al-Anbārī. See below, n. 42.

The Aya Sofia manuscript is exactly the same as the printed Aleppo text except for occasional changes or omissions of a word or two, and except for occasional missing sentences, which are possibly just marginal glosses that were introduced into the printed text. The body of the British Museum manuscript is identical with the body of the printed text. But the British Museum manuscript has an introduction and conclusion of its own, each running several pages; and the printed text has a long concluding section of eighteen pages which does not appear in the British Museum manuscript.

[41]See J. Mayor's edition of *De Natura Deorum,* Vol. II (Cambridge, Mass., 1883), pp. xx-xxiii; Gronau, *Poseidonius und die juedisch-christliche Genesisexegese.*

and at any rate they draw from a common Stoic pool of thought. The *K. al-Dalā'il,* for its part, also puts forward evidence of design in the universe, the aim here being to establish the existence of a "creator," to establish more precisely that the entire universe is the handiwork of a single creator, and to establish that faultless providence and wisdom pervade the world. In the introduction to one of the recensions of the *K. al-Dalā'il,* Jāḥiẓ, or pseudo-Jāḥiẓ, names the sources he used, and included among them is Theodoret.[42] The body of the Arabic work, moreover, evinces parallels with Theodoret's *On Providence*. The *K. al-Dalā'il* is known to have been in circulation in the Middle Ages; for adaptations and excerpts have been identified in a number of Arabic writers, both Islamic and Jewish.[43] A complete itinerary is thus traceable in which the teleological argumentation of the ancients can be seen traveling through Theodoret to the *K. al-Dalā'il,* whence it diffused into Islamic and Jewish literature. A few illustrations may be instructive.

(*a*) Cicero reports that Cleanthes and Chrysippus of the old Stoa argued for the existence of the "gods" with the aid of an analogy between the design apparent in a house and the design apparent in the universe. The evidence they cited is of an aesthetic character. Cleanthes contended: "When someone comes into a house" and finds everything regulated and arranged systematically, he cannot "suppose that these [things] came about without a cause"; with far more reason, the "motions . . . and order" of the world, which over an infinite past time have "never . . . played false," must convince every observer that nature is "directed by a mind."[44] Chrysippus compared the "beauty of the heavenly bodies . . . and the great magnitude of the ocean and lands" to "a giant and beautiful house"; and he inferred that the world must have been constructed by the "immortal gods" as a "domicile" for themselves.[45] Philo too was familiar with the analogy and he credited it to members of the school "whose philosophy is reputed the best." The philosophers in question—presumably the Stoics—maintained: "Should a man see a carefully built house . . . he will get a notion of the craftsman. . . . Just so, anyone entering, as it were, the great house of the world . . . and beholding the heavens, . . . planets, stars, . . . which move rhythmically, harmoniously, and for the benefit of the whole, . . . beholding as well the [arrangement and variety of] earth, . . . water, . . . air, . . . and fruits, . . . will surely conclude that these [things]" are the work of "a creator, God."[46] The analogy recurs in the patristic writers,[47] and among them is Theodoret, his intent being not so much

[42]The passage is published by C. Rieu, *Supplement to the Catalogue of the Arabic Manuscripts in the British Museum* (London, 1894), pp. 466–467; translated by Gibb, "The Argument from Design," pp. 153–154.

[43]See below, pp. 223–224.

[44]*De Natura Deorum,* II, v, 15.

[45]Ibid., vi, 17.

[46]*Legum Allegoria,* III, xxxii, 98–99. Cf. *De Specialibus Legibus,* I, vi, 33–35.

[47]Lactantius, *Institutiones,* II, 9; Ps. Clement, *Recognitiones,* VIII, 20; Pease's note to *De Natura Deorum,* II, v, 15.

to demonstrate the existence of God as to determine God's nature. "At the sight of a very skillfully made house," Theodoret writes, "we forthwith admire the craftsman . . . [although] not present, . . . and we attribute the entire beautiful form . . . to his craftsmanship. Similarly, when we see the heavens and the beneficial procession of lights across them, the magnitude and beauty of the creatures give us some conception of the creator."[48] And the motif reappears in the *K. al-Dalā'il*. Jāḥiẓ, or pseudo-Jāḥiẓ, states at the beginning of his treatise: The "early writers . . . did well" when they compared the "world" to "a house that has been constructed and made ready, a house containing every appointment," while man is "like the householder placed in charge of everything therein." As construed by the author of the *K. al-Dalā'il,* the analogy suffices not only to establish the existence of an intelligent cause of the universe; on the grounds that the world reveals a single unified design, the Arabic text concludes that a single creator must have made the world and provide for it. The resemblance of the world to a house is a "clear proof that the world is created with governance (*tadbīr*), planning (*taqdīr*) . . . and order, . . . and that the creator of the world is one, he being [the agent] who fitted the parts together and arranged them."[49]

(*b*) Xenophon relates that Socrates cited the functionality of the human speech organs as a piece of physiological evidence for divine providence. The "gods," Socrates taught, rendered the human tongue "capable of producing articulate speech through touching different parts of the mouth at different times," all with the purpose of letting men "communicate whatever they wish to each other."[50] The mechanics of human speech are subsequently cited as evidence of divine providence by Cicero's Stoic. The Stoic speaker sees the wisdom of "God" or of "nature" in the tongue, which "renders the sounds of the voice distinct and clear by striking the teeth and other parts of the mouth." And here the design inherent in the speech organs is brought out through an analogy with a musical instrument. "Our school," the speaker remarks, "is wont to compare the tongue to the plectrum [whereby the lyre is plucked]; the teeth to the strings; and the nostrils to the horns of the lyre [i.e., to the hollow arms extending from the body of the instrument]. . . ."[51] The comparison of the tongue to the plectrum of a lyre and the teeth to the strings became commonplace,[52] and Theodoret was one of the writers adopting it. But Theodoret was careful to add a reservation. He recalls that in truth art imitates nature, and not nature, art;[53] hence, the art of fashioning musical instruments must have imitated the operations of providence and not vice versa.

[48]Theodoret, *De Providentia,* in *Patrologia Graeco-Latina,* ed. J.-P. Migne, Vol. LXXXIII (Paris, 1864), col. 608.

[49]*K. al-Dalā'il,* p. 3; cf. p. 63.

[50]*Memorabilia,* I, 4.

[51]*De Natura Deorum,* II, lix, 147; 149. On II, vi, 18, Cicero refers to Xenophon's reports regarding Socrates' teleological teachings.

[52]See notes of Mayor and Pease to *De Natura Deorum,* II, lix, 149.

[53]See Aristotle, *Physics* II, 2, 194a, 21–22.

"In place of the teeth, art has stretched the strings. In place of the lips, art has inserted the bronze [sounding board]. And the plectrum serves as a tongue for [plucking] the strings."[54] The workings of divine providence in "the human . . . voice organs" also engage the attention of the *K. al-Dalā'il*. Its author too knows of the analogy with a musical instrument, and once again cites the analogy in the name of the "early writers." The comparison now drawn, however, is different from Theodoret's, the organs of speech being likened—more appropriately—to a wind, rather than to a stringed instrument. The Arabic passage cannot, as a consequence, be directly, or solely, dependent on Theodoret. Yet since the thought is unchanged, and the remark regarding the relationship between art and nature is repeated, the Arabic passage is unmistakably continuing the same tradition. According to the *K. al-Dalā'il*: "The [human] throat is like a pipe for producing the voice, and the tongue, lips, and teeth [serve] to shape the letters and melodies." The "early writers therefore . . . did well" when they "compared the throat to the tube of a wind instrument (*mizmār*) . . . and drew an analogy between—on the one hand—the lips and teeth that shape the voice into letters and melodies, and—on the other hand—the fingers that travel over the mouth of the wind instrument and shape its piping sounds into tunes." As for the question which is imitating which, the analogy with a musical instrument is, writes the *K. al-Dalā'il,* legitimate enough "for the sake of demonstration and instruction," that is to say, as a pedagogical device for exhibiting the presence of design in the world. It must nonetheless be remembered that "in truth the wind instrument copies the . . . voice. The instrument is made by art, whereas the voice is natural; and art copies nature."[55]

(*c*) To take an additional example, Socrates is reported to have insisted upon the functionality for man not only of the light of day but also of the darkness of night. Inasmuch as man "stands in need of rest," Socrates explained, the "gods provide [him] with a most beauteous time for resting."[56] The notion is repeated by the spokesman for Stoicism in Cicero's *De Natura Deorum*. Among the gifts of divine providence, the Stoic counts "the alternation of day and night," which "affords a time for acting and another time for resting."[57] This motif as well became a commonplace.[58] In Theodoret's formulation, divine providence is disclosed in the movement of the sun. "For in rising, the sun produces the day, while by setting and, as it were, hiding itself, the sun yields to the night." During the day, especially during the long days of summer, men attend to their affairs. Yet "night provides men with no less benefit than day," particularly[59] insofar as

[54]*De Providentia,* col. 592.

[55]*K. al-Dalā'il,* pp. 50–51.

[56]Xenophon, *Memorabilia,* IV, 3.

[57]*De Natura Deorum,* II, liii, 132. The notion also appears in Philo, *De Specialibus Legibus,* II, xx, 100–103.

[58]See Pease's note to Cicero, *ad locum*.

[59]Another function of night, according to Theodoret, is that it prevents men from becoming sated with daylight.

night prompts men to "rest [their] weary bodies awhile," insofar as it compels "even exceedingly industrious men to desist from labor."[60] In the same vein, the *K. al-Dalā'il* detects the hand of providence in the "rising and setting of the sun." The rising of the sun permits men "to busy themselves in their affairs." The "usefulness" of the setting of the sun consists, primarily,[61] in furnishing men with the opportunity of "repose for the recovery of their bodies," and indeed in forcing them, when necessary, to rest; for human greed is so great that "many would never rest except for . . . the darkness of . . . night."[62]

In addition to the preceding examples, the *K. al-Dalā'il* accumulates hundreds of details from plant biology, zoology, human physiology, and human psychology, and the treatise sets forth the functionality of each detail; some, though by no means all, of the details are akin to those in the classical sources.[63] The "order," "permanence," and "purposefulness," evinced by the world as a whole and the functionality of the several parts of the world demonstrate to the *K. al-Dalā'il* that the world cannot have come about "by accident" and "neglect" (*ihmāl*), but must have been made in accordance with a "plan" by a "creator . . . [who] is one."[64]

As already mentioned, material from the *K. al-Dalā'il* can be identified in other Arabic texts. A book attributed to Ghazali and entitled *The Wisdom in God's Creatures* (*al-Ḥikma fī Makhlūqāt Allāh*) consists largely of excerpts from the *K. al-Dalā'il*.[65] The entire *K. al-Dalā'il* was recast as a dialogue by a Shiite author.[66]

[60] *De Providentia,* cols. 565, 568.

[61] Night, according to the Arabic text, also has the function of preventing the world from becoming overheated.

[62] *K. al-Dalā'il,* p. 4. The Quran too had cited the alternation of night and day as evidence of divine design; see below, p. 227.

[63] The following are a few other examples of continuing motifs:

The human eye is providentially protected by eyelids, lashes ("a palisade"), and eyebrows ("a roof"). Cf. Xenophon, *Memorabilia,* IV, 3; Aristotle, *De Partibus Animalium* II, 15; Cicero, *De Natura Deorum,* II, lvii, 143; Theodoret, *De Providentia,* col. 601; Jāḥiẓ, *K. al-Dalā'il,* p. 52.

The digestive and excretory functions show clear evidence of design. Cf. Xenophon, *Memorabilia,* I, 4; Cicero, *De Natura Deorum,* II, lv, 137–138; Theodoret, *De Providentia,* col. 612; Jāḥiẓ, *K. al-Dalā'il,* pp. 46–47, 57.

Man's erect stature is evidence of divine providence. Cf. Xenophon, *Memorabilia,* I, 4; Cicero, *De Natura Deorum,* II, lvi, 140; notes of Mayor and Pease, ibid.; Jāḥiẓ, *K. al-Dalā'il,* p. 47. The purpose served by man's erect posture was explained in various ways; see notes of Mayor and Pease.

The willingness of animals such as the ox, the horse, the dog, and the ass, camel, or elephant to submit to man, although they are stronger than he, is evidence of divine concern for man. Cf. Xenophon, *Memorabilia,* IV, 3; Cicero, *De Natura Deorum,* II, lx, 151, and lxiii, 158–159; Theodoret, *De Providentia Deorum,* cols. 633–637; Jāḥiẓ, *K. al-Dalā'il,* pp. 28–29.

[64] *K. al-Dalā'il,* pp. 2, 3, 47, 66, 67.

[65] *al-Ḥikma fī Makhlūqāt Allāh,* (Cairo, 1908). See D. Baneth, "The Common Teleological Source of Bahye ibn Paqoda and Ghazzali" (in Hebrew with English Summary), *Magnes Anniversary Volume* (Jerusalem, 1938), pp. 23, 28–29.

[66] *K. al-Tawḥīd,* attributed to Mufaḍḍal b 'Umar and incorporated into al-Majlisī's encyclopedic work, *Biḥār al-Anwār.* See Baneth, "The Common Teleological Source," p. 27, n. 33; J. van Ess, *Die Gedankenwelt des Ḥāriṯ al-Muḥāsibī* (Bonn, 1961), p. 172.

Sections were reworked in a treatise written by one Aḥmad b. Sulaymān, also a Shiite.[67] And excerpts are incorporated by the Arabic Jewish writer Baḥya into his *Duties of the Hearts*.[68] The teleological theme is fairly frequent in medieval Arabic and Hebrew literature, as the examples to follow will show. The theme happens, however, to be prominent in the Quran.[69] Consequently, in most of the examples to be given, it can only be conjectured how far the teleological mode of thought is an outgrowth of the classical tradition, mediated through the *K. al-Dalā'il* or another channel, and embraced because of the congeniality of the theme; and how far that mode of thought may have sprouted directly from the seedbed of the Quran.

Muḥāsibī (d. 857) has a teleological argument for the unity of the cause of the universe which might serve equally well as an argument for the existence of such a cause. Throughout the universe, in inanimate nature, plant life, animal life, and human life, Muḥāsibī discovers that "each part fits together with the others"; and a number of details he adduces seem distinctly to echo details in the *K. al-Dalā'il*. The interconnections reveal that the universe, from its lowest to its highest level, forms "one whole," and the unity of "governance" evidenced by the universe leads Muḥāsibī to infer the unity of the cause of the universe.[70]

A brief teleological argument for the existence of God is given by the Shiite author Qāsim b. Ibrāhīm (785–860). The "imprints" of perfect wisdom and the "signs" of good governance manifest in the universe prove to Qāsim that a wise and good deity must be responsible.[71] Māturīdī (d. 944) offers a brief argument wherein he derives the existence of a "creator" (*muḥdith*) from what he calls the "principle" that everything in the universe exhibits "wondrous wisdom."[72] The unity of the creator is subsequently inferred from the fact that each process in nature—the seasons of the year, the paths of the heavenly bodies, the life cycles of plants and animals—observes its own unvarying and uninterrupted course.[73] Māturīdī does not state, but perhaps he meant, that these several processes of nature are interdependent and mesh, and hence stand in need of a single, all-envisaging architect. In a separate teleological argument for the unity of the cause of the universe Māturīdī points to the circumstance that diverse species and widely scattered individuals belonging to the same species have their needs provided for throughout the universe. The interdependence of the workings of providence, he concludes, shows a "single . . . provider (*mudabbir*)" to be responsible.[74]

The Ikhwān al-Ṣafā', the "Brothers of Purity,"—who are thought to belong to one of the Shiite branches of Islam—offer a teleological argument for the existence of God in which the heavens supply the evidence of design. The celestial

[67]Gibb, "The Argument from Design," p. 152, n. 5.

[68]Baneth, "The Common Teleological Source," pp. 28–29.

[69]See Quran 2:164; 3:190; 30:20–25; 45:3–5; 55.

[70]See van Ess, *Die Gedankenwelt des Ḥāriṯ al-Muḥāsibī*, pp. 163–166, 170–171.

[71]See W. Madelung, *Der Imām al-Qāsim ibn Ibrāhim* (Berlin, 1965), p. 106.

[72]Māturīdī, *K. al-Tawḥīd* (Beirut, 1970), p. 18.

[73]Ibid., p. 21.

[74]Ibid., pp. 21–22.

"spheres" (*falak*) and "stars," the Ikhwān observe, differ in respect to size; and the motions of the heavenly bodies differ, moreover, in respect to velocity as well as direction, some moving "east, [some] west, [some] south, [some] north." Those data could, as was seen in the previous chapter, furnish the underpinning for an argument from the concept of particularization, an argument that detects arbitrariness or irregularity in celestial phenomena and proposes to explain the arbitrariness or irregularity.[75] But what the Ikhwān al-Ṣafā' detect is design and wisdom. The Ikhwān conclude, with no ado, that the ingenious arrangement of the celestial spheres cannot have come about by chance but must have "occurred through the intention of an intending agent . . . [who is] wise and powerful."[76]

A more circumstantial teleological argument for the existence of God is advanced by Ibn Ḥazm. Ibn Ḥazm had formulated an elaborate proof of the existence of God from creation, and his teleological argument is appended almost as an afterthought.[77] He uncovers the requisite evidence of design at two levels, on a cosmic scale, in the arrangement of the celestial region, and on a lesser scale, in miscellaneous details of biology. The celestial phenomena attracting Ibn Ḥazm's attention are the phenomena that had constituted evidence of arbitrariness or irregularity in the aforementioned argument from particularization, but that had furnished evidence of design in the passage just quoted from the Ikhwān al-Ṣafā'. In working out his argument, Ibn Ḥazm enters into astronomical technicalities to a greater extent than did the Ikhwān.

The primary theoretical task of ancient and medieval astronomy had been to construe the motion of the planets in conformity with the principle that celestial motion is circular and the further principle that the earth stands at the center of the universe. What modern astronomy sees as an elliptical motion of the planets around the sun had, accordingly, to be reduced to motion in circles around the earth. Some astronomers explained the motion of the planets through the hypothesis of eccentric spheres. An eccentric sphere is a sphere whose center is different from the earth's, and the hypothesis of eccentric spheres assumes each planet and star to be embedded in the surface of such a sphere: The eccentric sphere rotates circularly at a constant velocity; the center around which the sphere rotates is itself in circular motion; and the interaction of the circular motions was understood to give rise to the noncircular motion of the planets.[78] Other astronomers explained the motion of the planets through the hypothesis of epicycles. Epicyclical spheres are secondary spheres rotating at fixed points on the surface of other spheres—which may themselves rotate at the surface of still others—the underlying, major spheres in the system being concentric to, and in constant circular motion around, the earth.[79]

[75]Above, pp. 187, 197–199.

[76]Ikhwān al-Ṣafā, *Rasā'il* (Beirut, 1957), III, pp. 335–336.

[77]Cf. above pp. 120, 122.

[78]Cf. J. Dreyer, *A History of Astronomy from Thales to Kepler* (New York, 1953), pp. 144–146.

[79]Ibid., pp. 152–157.

Ibn Ḥazm does not distinguish between the two hypotheses, and in fact they are not mutually exclusive.[80] With the characteristics of eccentric spheres in mind, he marvels at the circumstance that the celestial spheres "have different centers" around which they rotate, yet despite their having different centers, they "fit together tightly" and are able to "maintain their circular motion" and unvarying velocities.[81] With the characteristics of the epicyclical spheres in mind, be marvels at the "variance" between the "motion of the epicyclical spheres" and the "motion . . . of the [main] supporting spheres"; and at the additional circumstance that the "revolutions of all the [main] spheres [with the exception of the ninth are] from west to east" whereas "the revolution of the ninth [and outermost] sphere," which is responsible for the daily movement of the heavens, is "from east to west." The interaction of the motions of the secondary, epicyclical spheres with the motions of the underlying, main spheres, and the interaction of the eastward motions of all but one of the main spheres with the "conflicting," westward motion of the outermost sphere give rise to the intricate courses of the planets; and the entire epicyclical arrangement is so ingenious that it must be the fruit of conscious planning. The operations of the eccentric spheres and of the epicycles thus lead "us . . . necessarily" to recognize the hand of a "mover."[82]

When Ibn Ḥazm turns from the celestial to the terrestrial region, he assembles data from animal and plant biology. Unlike most medieval proponents of teleological argumentation, who, when treating the details of nature, saw only their functionality, he underlines the aesthetic side of the details of nature. He admires the skill by which the limbs of the human body are fitted together, a motif touched on in the *K. al-Dalā'il*;[83] the uniform color patterns of sundry "animals, . . . birds, tortoises, reptiles (*ḥasharāt*), and fish"; the variegated plumages of other species of bird, also touched on in the *K. al-Dalā'il*;[84] the fact that palm tree fiber has a texture as skillfully woven as fabric from a loom, again a motif present in the *K. al-Dalā'il*.[85] Ibn Ḥazm concludes: It is incontrovertibly "known through the necessity of intellect" that the celestial and terrestrial regions must have come about by the "deliberation of a maker" who "exercises choice and invention." And the evidence of design on both the macrocosmic and microcosmic levels is, he avers, sufficient not merely for concluding that the universe has a maker, but for concluding as well that it has a "single" maker.[86]

A simple teleological argument for the existence of God is put forward by Ghazali in one of his less technical works. Like Ibn Ḥazm, Ghazali offers the

[80]Ibid., pp. 155–156.

[81]Maimonides touches upon problems of this sort in *Guide to the Perplexed,* II, 24.

[82]*K. al-Faṣl fī al-Milal,* Vol. I (Cairo, 1964), pp. 18–19; Spanish translation: *Abenházam de Córdoba y su Historia Crítica de las Ideas Religiosas,* trans. M. Asín Palacios, Vol. II (Madrid, 1928), pp. 111–113. Ibn Ḥazm does not draw the conclusion until after giving his examples from the terrestrial region.

[83]*K. al-Dalā'il,* p. 47.

[84]Ibid., pp. 38–39, also in Nemesius, *De Natura Hominis,* chap. 42.

[85]Ibid., p. 23.

[86]*K. al-Faṣl fī al-Milal,* p. 19.

teleological argument side by side with the proof from creation, the proof that first establishes the creation of the world and then infers the existence of a creator. But unlike Ibn Ḥazm, Ghazali indicates a preference for the teleological argument—which he describes as "inborn" in man and as so evident that "setting up a demonstration" is, in reality, superfluous. Ghazali begins by quoting passages from the Quran which contain the teleological theme, for example Quran 2:164: "In the creation of the heavens and the earth, in the alternation of night and day, in the ship (*fulk*!) that runs on the sea, . . . in the driving forth of winds and clouds, . . . there are surely signs for people who understand." After quoting several such passages, Ghazali declares: Nobody "possessing the least intelligence who reflects upon . . . these verses, who gazes upon the wonder of God's creation on earth and in heaven, who gazes upon the marvelous formation of animals and plants," can doubt that "the well adapted arrangement" depends on a "maker who governs . . . and adapts it."[87]

The teleological argument for the existence of God is also employed by Baḥya ibn Paquda, a writer who in addition used teleological material borrowed from the *K. al-Dalā'il* in contexts apart from the one to be examined now.[88] Once again, in Baḥya, the teleological argument for the existence of God appears in the company of an argument for the existence of God from creation. Here, however, the two arguments are not juxtaposed and coordinate. Rather, the argument for the existence of God from creation is a more comprehensive train of reasoning into which teleological considerations are incorporated at several stages. In the more comprehensive argument, Baḥya considers composition in the universe. He points out, as was seen in an earlier chapter,[89] that the world is composite; from composition he concludes that the world was created; and from creation he infers a creator. But in the first stage of the argumentation, when ostensibly adducing evidence of composition, Baḥya makes a comment that can stand by itself as a self-contained proof of the existence of God. The thinking is that the world as a whole has been designed to accommodate man: and Baḥya's source is clearly the *K. al-Dalā'il*. He writes: "Looking at the world, we find it to be joined and composite . . . like a house that has been constructed and made ready in every appointment," while man is "like the freeholder of the house, who makes use of everything therein"; for "the intent in each part [of the world] is the benefit and well-being of rational beings."[90] At a subsequent stage of the argumentation, after having completed the body of the wider proof, after having deduced creation from the composition in the universe and inferred the existence of a creator from creation, Baḥya confronts cavillers who may retort that the world has no cause and exists "by chance." His rebuttal turns out to be wholly

[87] *al-Risāla al-Qudsīya*, published as *Al-Ghazali's Tract on Dogmatic Theology*, ed. and trans. A. Tibawi (London, 1965), Arabic text, p. 16; English translation, pp. 33–34.

[88] See Baneth, "The Common Teleological Source," pp. 28–29; below, p. 235.

[89] Above, pp. 152–153.

[90] *al-Hidāya* (*Ḥobot ha-Lebabot*), ed. A. Yahuda (Leiden, 1912), I, 6. Cf. above, p. 221.

independent of the evidence he had adduced for composition in the universe and equally independent of the inference of a creator from creation. No one of "sound" mind, he submits, would suppose that a device such as the waterwheel whereby farmland is irrigated can come about "without the design of a maker."[91] How then could the "great wheel" of the heavens, which serves the "well-being of the entire earth and everything in it," have come about "without the intent of an intending agent and the planning of a wise and powerful agent?" No one, moreover, would imagine that a book might have been produced merely by "spilling ink on blank paper";[92] how then might the world, which is of "an infinitely . . . finer craftsmanship," have come about "without the intent of an intending agent, the wisdom of a wise agent, and the power of a powerful agent?"[93] The world must be the handiwork of an intelligent maker.

When Baḥya takes up the question of the unity of God, he bases that tenet too on evidence of design. He writes: The "traces of wisdom in God's creatures are similar and alike," the entire world being governed by "a single order and a single motion, which comprehends each and every part." These "traces of the creator's wisdom" in "the least and most exalted parts of creation testify that they derive from a single wise creator."[94]

A teleological proof diverging from the arguments discussed so far is offered by Judah Hallevi. Hallevi's proof has its roots in a strain of medieval speculation regarding the source of the formal, as distinct from the material, side of natural objects. And his is a proof that goes a bit beyond mere intuition, beyond the presupposition that no one of sound mind can fail to recognize design in the details of the world and in the world as a whole. *Form* in the Aristotelian sense is the inner principle that makes any object what it is, that gives the object its essence. The contention of certain medieval philosophers, notably Avicenna, had been that the formal side of a natural object is radically different from the inert, material side, and that a cause can never give rise to what is totally different from itself. The conclusion drawn was that the formal aspect of any natural object cannot emerge from within, from the material aspect, but must enter from without, from a cause consisting, like the effect, in form. The formal aspect of each natural object must come to the material substratum from a transcendent *giver of forms*.[95] Against the background of that line of speculation, Hallevi formulates a teleological argument. He reasons that the only qualitative traits of physical

[91] See above, n. 38.

[92] See above, n. 37.

[93] *al-Hidāya* (*Ḥobot ha-Lebabot*), I, 6.

[94] Ibid., I, 7 (2, 3, 7). Baḥya has no less than seven proofs of the unity of God. The second is wholly teleological, and the teleological motif is woven together with other strands in the third and seventh proofs.

[95] See H. Davidson, "Alfarabi and Avicenna on the Active Intellect," *Viator,* III (1972), 149–150, 156–157; idem, "The Active Intellect in the *Cuzari* and Hallevi's Theory of Causality," *Revue des études juives,* CXXXI (1972), 368–377.

objects attributable to the blind forces of nature are the elemental and wholly physical qualities, that is to say, heat, cold, wetness, and dryness. By contrast, "the instilling of form, the evaluation [of the appropriateness of a given portion of matter for a specific form], . . . and whatever involves wisdom [acting] towards a goal, can be attributed solely to a wise and powerful agent" who is beyond nature.[96] Whenever a natural form appears in a material substratum, the form cannot, consequently, emerge from the matter itself but must be the handiwork of an external source, operating consciously and towards a goal. Hallevi does not go beyond the affirmation that the world is the product of a "wise and powerful agent." His argument therefore falls short of establishing either a single cause or a first cause and, like most teleological arguments, cannot pretend to be a complete proof of the existence of God.

Another writer who makes use of teleological reasoning is Averroes, and the context in which he does so is illuminating. Averroes makes clear in several works that from a scientific and philosophic standpoint, the only fully adequate proof of the existence of God is Aristotle's proof of the existence of a first mover.[97] In the treatise where he expresses approval of the teleological argument the tone and subject matter are not scientific and philosophic, the subject being, instead, the proper method of teaching fundamental truths, particularly the existence and unity of God, to nonphilosophers. Methods whereby different theological schools and especially the Kalam sought to establish the existence and unity of God are passed in review; and the various methods are rejected by Averroes on the grounds that they are both invalid and at variance with the spirit of the Quran.[98] As Averroes interprets the Quran, two arguments for the existence of God are recommended there. First, the Quran recommends what is in effect a simplified cosmological argument, an argument concluding in a nontechnical fashion that some entity must be responsible for the occurrence of events in the world. Secondly, it recommends an argument running as follows: "Everything in the world is adapted" to the needs of the human species and reveals "providence." "Day and night, sun and moon," the earth and everything therein, the organs of the human body—all serve the needs of man. The functionality exhibited throughout the world cannot conceivably be due to "chance." It must "perforce" be the doing of "an agent . . . who intends . . . and wills it"; and the "existence of a creator" is thereby established.[99] When Averroes, in the present treatise, turns to the unity

[96]*Kuzari,* I, 77. See the article in the *Revue des études juives,* cited in the previous note.

[97]*Tahāfut al-Tahāfut,* ed. M. Bouyges (Beirut, 1930), pp. 393–394; English translation, with pagination of the Arabic indicated: *Averroes' Tahafut al-Tahafut; Epitome of Metaphysics,* ed. and Spanish trans. C. Quirós Rodríguez (Madrid, 1919), IV, §3; German translation: *Die Epitome der Metaphysik des Averroes,* trans. S. van den Bergh (Leiden, 1924), pp. 105–106.

[98]*K. al-Kashf,* ed. M. Mueller (Munich, 1859), pp. 27–28; German translation, with pagination of the Arabic indicated: *Philosophie und Theologie von Averroes aus dem Arabischen uebersetzt,* trans. M. Mueller (Munich, 1875).

[99]Ibid., pp. 43–44.

of God, the argument which he finds to be recommended by the Quran is only marginally teleological. The Quran, as read by Averroes, observes that the universe is a single unified effect and from the unity of the effect it infers the unity of the cause.[100] As for the incorporeality of God, no argument for it is derived by Averroes from the Quran. In his opinion, the incorporeality of God is not a doctrine that should be imparted to common folk and hence not a doctrine with which Scripture would deal.[101]

Averroes further states that the two proofs for the existence of God which can be extracted from the Quran—the simplified cosmological argument and the teleological argument—are exactly the "procedure" philosophers employ when proving the existence of God. The philosophic formulation of the proofs and the popular formulation differ merely "in degree," the philosophic formulation being of greater comprehensiveness and greater "profundity."[102] What Averroes means when referring to a philosophic formulation of the scriptural cosmological argument is plain. The precise, philosophic formulation of the cosmological argument would be nothing other than Aristotle's proof from motion. What he means when referring to a philosophic formulation of the scriptural teleological argument is evidently to be taken in the same vein. He must be permitting himself a certain liberty; and his meaning must be that in a loose sense the proof from motion subsumes the teleological argument, and the latter can be thought of as a popular version of the former.[103]

Averroes' position, then, is that purposefulness is to be discovered in the workings of nature,[104] and the teleological argument contains sufficient grains of truth, and is sufficiently plausible, to answer the requirements of nonphilosophers. The Quran, which addresses an audience of nonphilosophers, accordingly reasons teleologically. The teleological argument is not, however, a fully adequate demonstration of the existence of God, the chief reason presumably being that argumentation from the functionality of nature views the universe anthropocentrically. The universe is represented as if it existed exclusively for

[100]Ibid., pp. 47–48.

[101]Ibid., p. 67.

[102]Ibid., pp. 46, 48.

[103]Averroes can sometimes be seen to change his mind from one work to another; but there is no reason to suppose that we have an instance of that in his characterizing the Aristotelian proof from motion as the only adequate proof for the existence of God in certain of his works, while apparently referring in the present work to a philosophic version of the teleological argument. Averroes' changes of mind seem always to concern the proper interpretation of a specific passage in Aristotle and never to be conscious departures from Aristotle.

It is curious that Aristotle offered a teleological argument for the existence of God in one of his early popular dialogues (which was lost by the Middle Ages and was unknown to Averroes), but omitted that form of argument in his later more technical works (see above, n. 27); and now Averroes characterizes the teleological argument as appropriate for nonphilosophers, whereas in his own technical works he recommends only the Aristotelian proof from motion.

[104]See Aristotle, *Physics* II, 8; *De Partibus Animalium*.

the sake of man, whereas in truth the celestial realm cannot exist "primarily" to serve man, since the "superior" never exists "for the sake of the inferior."[105] The sole proof of the existence of God which meets the standards of serious philosophers would be Aristotle's proof from motion.

Gersonides was highly dependent on Averroes and yet he strained constantly for originality and independence. As it happens, despite Averroes' patronizing assessment of the teleological proof for the existence of God, Gersonides adopts the proof. Gersonides' version stands in the tradition of teleological arguments that search for a single overall purpose in the heavens. But his version is unique in its broad astrological assumptions.

Gersonides begins by laying down "the most distinctive property" of things whose existence has a cause, that is to say, the characteristic found in all such things and found exclusively in them. "The most distinctive property" of things whose existence has a cause, whether they exist by "nature" or "art," is, he writes, that "they exist for a certain purpose."[106] In the terminology of the Aristotelian theory of four causes distinguishable in objects—a material cause, formal cause, final cause, and efficient cause—Gersonides' principle means that anything disclosing a final cause must also have an efficient cause.[107] Examination of the celestial region, Gersonides continues, reveals that "everything" in the heavens is adapted "in the highest possible degree, to bring terrestrial beings to perfection"; and the functionality is so extensive that it cannot conceivably be "accidental." The number of stars and planets, their diverse magnitudes, their varying distances from earth, their distribution through the heavens, the diversity in their radiation, their several movements, are adjusted and attuned in a way that allows the influences descending from the heavens to work the maximum benefit for sublunar existence. Since the heavens unmistakably exist for a purpose, and since they show evidence of a "final cause," they must belong to the class of beings that have an "efficient cause" of their existence.[108] In a separate passage,

[105] *Tahāfut al-Tahāfut,* pp. 484–485.

[106] *Milḥamot ha-Shem* (Leipzig, 1866), VI, i, 6, p. 308. For the definition of *property,* see Aristotle, *Topics* I, 5, 102a, 18–19; Averroes, *Middle Commentary on Porphyry's Isagoge and on Aristotle's Categoriae,* trans. H. Davidson (Cambridge and Berkeley, 1969), pp. 16–17.

[107] See Aristotle, *Physics* II, 7; E. Zeller, *Die Philosophie der Griechen,* Vol. II, Part 2, (4th ed.; Leipzig, 1921), p. 328.

[108] *Milḥamot ha-Shem,* VI, i, 7, pp. 310–312. Gersonides later finds evidence of design and of a designer in an aspect of the sublunar world, in the fact that dry land exists although the natural place of the element earth is beneath the natural place of the element water. He reasons: The existence of dry land obviously serves a purpose; it makes possible the existence of land animals. Given the rule that anything existing for a purpose has an efficient cause of its existence, an efficient cause of the existence of dry land must be posited. And since the phenomenon is contrary to nature, that cause must not be a natural, but a transcendent agent. See *Milḥamot ha-Shem,* VI, i, 13, p. 350. The motif goes back to Stobaeus, *Geography,* XVIII, i, 36, and it is found in Arabic literature. Cf. Ikhwān al-Ṣafā, *Rasā'il,* II, pp. 56–57; G. Vajda, "Commentaire Kairouanais," *Revue des études juives,* CXII (1953), 7, 17, and 18, where reference is also made to the same motif in Fakhr al-Dīn al-Rāzī; idem, CXIII, (1954), p. 49, line 4.

Gersonides argues that the ultimate cause of the existence of the heavens is one and incorporeal. He takes it as an empirical fact that the course of events on earth is directed by the cumulative influence of the stars; and as an astronomical fact that every star is imbedded in a celestial sphere, which is sustained in circular motion by an "intelligence," a being consisting in pure thought. Now, Gersonides argues, each intelligence is responsible for the motion of a given sphere and star or stars,[109] and each is aware only of the influences emanating from its star. The influence of the various stars interact and dovetail, thereby bringing about the total harmonious "order" (*nimus*) of events on earth. It is well known that whenever artisans belonging to "different crafts" cooperate to produce "one single artifact," their cooperation is a result of their being "subservient to," and directed by, "a single [master] craft."[110] By the same token, the presence of a single overall order in the universe, unless "an accident of some sort," must be due to a single "intelligence that has knowledge of the [overall] order of existence" and that assigns the appropriate operation to each intelligence and, thence, to each star.[111] The celestial intelligences, the heavens, and, through them, the rest of the universe must be dependent on a single ultimate cause. Furthermore, the cause that governs the intelligences, the cause whose unitary, all encompassing thought embraces and subsumes their partial thoughts, must stand to them in the relation of "form" to "matter." It must therefore be judged an incorporeal being even more certainly than they are judged incorporeal.[112] The design evinced by the universe thus establishes the existence of a single incorporeal cause of the universe. And Gersonides has carried teleological argumentation farther than others by formulating what can stand as a complete teleological proof of the existence of God.[113] In addition, Gersonides expands his teleological proof of the existence of God into a proof of creation; but the reasoning he employs is not specifically teleological, and hence not pertinent here.[114]

Gersonides' teleological argument for the existence of God was copied by a later Jewish writer, Simon Duran (1361–1444).[115] Besides repeating Gersonides' argument, Duran advances another, much simpler teleological argument based on two propositions, the proposition that "the world together with its parts is adapted (*na'ot*) for the existence of man and for the existence of other beings"; and the proposition that "everything adapted . . . and ordered to a single end

[109] The eighth sphere contains all the fixed stars; by contrast, each of the wandering stars, or planets, has its own sphere.

[110] The notion is both Platonic and Aristotelian; cf. J. Burnet, *The Ethics of Aristotle* (London, 1900), p. xxiv.

[111] *Milḥamot ha-Shem,* V, iii, 8, pp. 272–273. Averroes had offered a similar argument; see *Epitome of Metaphysics,* IV, §§37–38.

[112] Ibid., 8, 272–273; 12, 280–283.

[113] It should be noted that Gersonides does not advance all the parts of the argument in the same place, and that I have brought them together.

[114] See above, pp. 209–211.

[115] *Magen Abot* (Livorno, 1785), p. 95a.

necessarily has a maker." The conclusion Duran immediately draws is that "the world has a maker."[116]

The foregoing have been teleological arguments for the existence of a cause of the universe and, in several instances, for the unity of the cause of the universe. In one instance, in Gersonides, the teleological argument for existence and unity supplies grounds for the incorporeality of the cause of the universe as well, and a complete proof of the existence of God is offered. Arabic philosophic literature also knows of teleological arguments that take the existence of God as given and attempt merely to establish the presence in the deity of certain attributes, namely wisdom or knowledge, power, and providence. Teleological arguments for divine attributes usually focus their attention on the details of nature, but they sometimes look at nature as a whole. The evidence of design which they cite is either the functionality or the marvelous order in nature. The arguments are especially frequent in Kalam works, both in works of Ash'arite writers, who maintained that attributes somehow exist as distinct real things in God, and of Mu'tazilite writers, who denied that divine attributes have distinct objective existence. The *K. al-Dalā'il*—and by means of it, the tradition going back to the Church Fathers, the Stoics, and Socrates—could here too have served as the model. Readers of the *K. al-Dalā'il* are exhorted to "cogitate upon" and "contemplate" numerous features of the celestial and terrestrial regions; for through contemplating the functionality of the details of nature, they will be led to recognize God's "governance," or providence, and his "wisdom."[117]

As was mentioned earlier, the *K. al-Dalā'il* is in some manuscripts ascribed to the well-known litterateur Jāḥiẓ, but the ascription is probably incorrect. The genuine Jāḥiẓ, a Mu'tazilite, happens to be one of the authors who employed the teleological motif. Jāḥiẓ assures readers of his work on the animals that what prompted him to investigate the seemingly lowly subject was not any intrinsic interest the brute animals might have, but instead the evidence they exhibit of the "wondrousness of God's governance and the subtlety of his wisdom."[118] The thought that design in nature reveals God's wisdom or knowledge plays a role in Bāqillānī. After having proved the creation of the world and the existence of God, Bāqillānī asserts that in the domain of human activity, "acts" can have an "arrangement and order" only if they come from a "knowing" agent. The products of the arts of "goldsmithery, carpentry, writing, and weaving" are examples; for in each case, the product undoubtedly entails a craftsman endowed with knowledge. Since "God's acts" are far "more subtle and better adapted (*aḥkam*)" than any human act, they "prove still more plainly" than human acts that the agent responsible for them—that is to say, God—is "knowing"; and God must

[116]Ibid., p. 4a.
[117]*K. al-Dalā'il*, pp. 2, 5, *et passim*.
[118]*K. al-Ḥayawān* (Cairo, 1938), II, p. 109.

consequently be understood to possess the attribute of "knowledge."[119] Similar reasoning in the same context recurs in the Mu'tazilite 'Abd al-Jabbār, who, however, unlike Bāqillānī denies that divine *knowledge* has any sort of distinct objective existence. After completing his demonstration of the existence of God from creation, 'Abd al-Jabbār writes: "The capability of [executing] a well adapted act" is "proof" that the agent is "knowing." But God's "creation of the animals, with the marvels they contain, his producing the circular motion of the spheres, his fitting the spheres within one another,[120] his subjugation of the winds,[121] his measuring off of winter and summer—these exhibit finely adapted action to a more patent and ample degree than fine writing does." The agent capable of them must therefore be deemed "knowing" with more reason than the human agent.[122] Juwaynī, an Ash'arite, advances a teleological argument for the divine attribute of "knowledge," again as a corollary to the proof for the existence of God from creation. Given the creation of the world and the existence of a creator, Juwaynī submits, "the subtleties of creation, . . . the harmony, arrangement, perfection, and consummate execution . . . of [everything in] the heavens and earth," reveal that the agent responsible "is knowing" and in possession of the attribute "knowledge."[123] Ghazali's Kalam writings adhere to the same pattern. As a corollary to his proof of the existence of a creator, Ghazali finds that the world is "well adapted and ordered, perfected and well arranged, encompassing divers marvels." It must accordingly be the handiwork of an agent who is "powerful" and "knowing"; and the creator must possess the attribute "knowledge" and the attribute "power." To support his conclusion, Ghazali cites the analogy of the art of writing: "Anyone who saw regular lines of writing proceeding in an orderly fashion from a scribe, yet who doubted that the scribe has knowledge of the art of writing, would be a fool." Surely, anyone who doubts that the maker of the world has power and knowledge is no less a fool. Ghazali goes further. He calls the conclusion that the maker of the world is powerful and knowing a "necessary" inference, by which he means an inference requiring no demonstration because the "intellect confirms it without proof."[124]

In another work, the *Iḥyā',* where the stress and tone are devotional, not doctrinal, Ghazali adduces numerous details from the higher and lower realms, very much in the style of the *K. al-Dalā'il.*[125] His subject is the religious virtue of "cogitation" (*tafakkur*), a term that reverberated through the *K. al-Dalā'il.* He

[119] *K. al-Tamhīd* (Beirut, 1957), pp. 26, 197. The arguments of Bāqillānī and the others which establish creation and a creator, and to which the arguments for divine knowledge and power are attached as corollaries, were discussed earlier in Chapters V and VI.

[120] Cf. Ibn Ḥazm's argument, above, p. 226. [121] Cf. Quran 2:164; 45:5.

[122] *Sharḥ al-Uṣūl* (Cairo, 1965), pp. 156–157. For the nonreal status of divine attributes, see ibid., pp. 182–183.

[123] *K. al-Irshād* (Cairo, 1950), pp. 61–62; *K. al-Shāmil* (Alexandria, 1969), p. 621.

[124] *al-Iqtiṣād fī al-I'tiqād* (Ankara, 1962), pp. 80–81, 99–100; *al-Risāla al-Qudsīya,* Arabic section, p. 20; English translation, pp. 40–41. Cf. above, pp. 154–155, 226.

[125] *Iḥyā' 'Ulūm al-Dīn* (Cairo, 1937), XV, pp. 78–100.

bids his reader to "cogitate" over the functionality, order, and "wonders" of the details of nature as well as the splendid panorama presented by nature as a whole. The reader will thereby be brought to an understanding of God's "knowledge and wisdom, the efficacy of his will, his power," "his majesty, and his glory," all with the ultimate goal of serving God properly.[126]

Additional illustrations can be given. There is a book entitled *al-Ḥikma fī Makhlūqāt Allāh, The Wisdom in God's Creatures,* which is ascribed, probably in error, to Ghazali. The book is composed of excerpts from the *K. al-Dalā'il* and from the devotional work of Ghazali which was just quoted.[127] In it, the unknown author underlines the functionality of numerous details of nature. And each detail furnishes him with evidence of God's "knowledge," his "governance" or providence, his "glory and power, . . . the efficacy of his will, and . . . his wisdom."[128] Baḥya, in a devotional context, borrows material from the *K. al-Dalā'il*[129] and adds related material of his own, his topic being an exposition of the virtue of "reflection" (*i'tibār*). The term *reflection,* as will be recalled, occurs in the title of the various recensions of the *K. al-Dalā'il*;[130] and Baḥya's virtue of reflection turns out to be very close to the virtue of cogitation which had been treated by Ghazali, as reflection is treated by Baḥya, in a devotional context. "Reflection," writes Baḥya, consists in a "contemplation of the signs of divine wisdom" in the world.[131] Its object is to bring man to an understanding of God's essence and, ultimately, to inculcate proper worship of God.[132] Joseph al-Baṣīr, a Jewish Karaite author who usually echoes Mutazilite thinking, supplements his proof of the existence of God with a teleological argument showing God to be "knowing."[133] Shahrastānī, in a Kalam work, records teleological arguments for the divine attributes of knowledge, wisdom, and power.[134] Judah Hallevi offers a teleological argument for God's "wisdom" and "providence."[135] Ibn Ṭufayl, in no sense a member of the Kalam school, supplements his proof of the existence of God with a teleological argument for God's "perfection," "power," "wisdom," and "knowledge."[136] Fakhr al-Dīn al-Rāzī[137] and Ījī[138] follow their Kalam forebears and append to their proofs of the existence of God a teleological argument showing the creator to be "knowing," and hence possessed of the attribute of knowledge.

[126]Ibid., pp. 77–78.

[127]Cf. Baneth, "The Common Teleological Source," pp. 28–29.

[128]*al-Ḥikma fī Makhlūqāt Allāh,* pp. 2, 3, 4, 5, 6, 8, 62, 64.

[129]Cf. Baneth, "The Common Teleological Source," pp. 28–29.

[130]Above, p. 219.

[131]*al-Hidāya* (*Ḥobot ha-Lebabot*), II, 1.

[132]Ibid., II, introduction.

[133]P. Frankl, *Ein Mu'tazilitischer Kalām aus dem 10. Jahrhundert* (Vienna, 1872), p. 22.

[134]*K. Nihāya al-Iqdām,* ed. A. Guillaume (Oxford and London, 1934), pp. 171, 174, 400–401.

[135]*Kuzari,* III, 11; 17; V, 20(1).

[136]*Ḥayy ben Yaqdhān,* ed. and trans. L. Gauthier (Beirut, 1936), pp. 88–89; French translation, pp. 66–67.

[137]*K. al-Arba'īn* (Hyderabad, 1934), p. 133.

[138]*Mawāqif* (Cairo, 1907), VIII, p. 65.

3. Summary

Much, though probably not all, Islamic and Jewish argumentation from design is a direct outgrowth of a Greek tradition running from Socrates to the Stoics and Church Fathers, and thence into Arabic. Islamic and Jewish arguments from design, like the ancient arguments, search for design on either a minor or a cosmic scale, and as evidence they cite either the functionality or the orderliness and beauty of nature. From evidence of design they reason to an intelligent cause of the universe; and in some instances they reason from what they discern as a single overall plan to a single intelligent planner. A concomitant series of arguments from design takes the existence of God as given and contends that the deity, however he be known to exist, plainly exercises providence and possesses such attributes as wisdom and power.

Medieval Islamic and Jewish teleological arguments, like the ancient arguments, generally do not inquire into what constitutes evidence of design; they do not justify the step from design to a designer; and they do not explain why a single overall plan cannot be a cooperative enterprise by a group—we would say a committee—of architects. The stated or unstated presupposition is that no intelligent person encountering design will fail to recognize it and that no person of sound mind can doubt that design entails an intelligent designer. The best thought-out teleological argument would appear to be that of Gersonides', a philosopher who, perhaps significantly, stands not in the Kalam, but in the Aristotelian, tradition. Gersonides furnishes an Aristotelian explanation of the step from design to a designer, and with the aid—unhappily—of broad astrological assumptions he is able to establish that the cause of the overall design in the universe must be a single agent and incorporeal. The remaining medieval Islamic and Jewish arguments seem to be aptly covered by Averroes' evaluation of the teleological argument for the existence of God. Averroes characterizes the argument as appropriate for instructing nonphilosophers, but as unable to meet the standards of formal philosophy.

VIII

The Proof from Motion

1. Aristotle's proof from motion

Aristotle's proof for the existence of a first mover is not the most widely adduced proof of the existence of God in medieval Islamic and Jewish philosophy,[1] but it does appear frequently. It is to be found in one form or another in a work attributed to Alexander of Aphrodisias but known only in the Arabic,[2] in the Jābir corpus,[3] in Miskawayh,[4] in a brief work mistakenly attributed to Alfarabi,[5] in Avicenna,[6] Shahrastānī,[7] Ibn Ṭufayl,[8] Abraham ibn Daud,[9] Averroes,[10] Maimonides,[11] and Aaron ben Elijah.[12] Averroes considered it to be the only completely cogent proof of the existence of God.[13] In the same spirit, Maimonides

[1]The most widely adduced proofs are discussed in Chapters VI and IX.

[2]Alexander of Aphrodisias, *Mabādi' al-Kull,* in *Arisṭū 'ind al-'Arab,* ed. A. Badawi (Cairo, 1947), pp. 259–263.

[3]Jabīr ibn Ḥayyān, *Textes Choisis,* ed. P. Kraus (Cairo, 1935), pp. 518–521.

[4]Miskawayh, *al-Fawz al-Aṣghar,* I, §§3 and 4. The book was not available to me and I am relying on Kh. Abdul Hamid, *Ibn Miskawaih, A Study of His al-Fauz al-Aṣghar,* (Lahore, 1946), pp. 14–21.

[5]*'Uyūn al-Masā'il,* §13, in *Alfarabi's philosophische Abhandlungen,* ed. F. Dieterici (Leiden, 1890); German translation: *Alfarabi's philosophische Abhandlungen aus dem Arabischen uebersetzt,* trans. F. Dieterici (Leiden, 1892), p. 100.

[6]Avicenna, *Najāt* (Cairo, 1938), pp. 128–130.

[7]Shahrastānī, *K. al-Milal wa-l-Niḥal,* ed. W. Cureton (London, 1846), pp. 385–386 (Shahrastānī's account of Avicenna's philosophy). German translation with pagination of Arabic indicated: *Religionspartheien und Philosophenschulen,* trans. T. Haarbruecker (Halle, 1850–1851).

[8]Ibn Ṭufayl, *Ḥayy ben Yaqdhān,* ed. and trans. L. Gauthier (Beirut, 1936), Arabic text, pp. 84–85; French translation, p. 64; English translation: *Hayy ibn Yaqzān,* trans. L. Goodman (New York, 1972), p. 132. Ibn Ṭufayl maintains that a first cause of motion is *ipso facto* a first cause of existence.

[9]Abraham ibn Daud, *Emuna Rama* (Frankfurt, 1852), I, 5; II, i.

[10]*Long Commentary on Metaphysics,* XII, comm. 41; Arabic text: *Tafsīr mā ba'd al-Ṭabī'a,* ed. M. Bouyges (Beirut, 1938–1948), p. 1632; *Epitome of Metaphysics,* ed. and trans. C. Quirós Rodríguez (Madrid, 1919), IV, §3; German translation: *Die Epitome der Metaphysik des Averroes,* trans. S. van den Bergh (Leiden, 1924), pp. 105–106.

[11]Maimonides, *Guide to the Perplexed,* II, 1 (1).

[12]Aaron ben Elijah, *'Eṣ Ḥayyim,* chap. 5.

[13]Averroes, *Epitome of Metaphysics,* IV, §3; German translation, pp. 105–106.

called it the "strongest argument" and Aquinas called it the "more manifest way" of proving the existence of God.[14]

Philosophers and historians have pointed out that two philosophic principles lie at the heart of the proof from motion: the principle of causality, and the impossibility of an infinite regress of causes.[15] In Aristotle's theory of motion the principle of causality expresses itself as the rule that an object can move only as long as something sustains its motion.[16] The two philosophic principles just mentioned accordingly give rise to an argument to the effect that each moving object[17] has a cause sustaining it in motion; the series of such causes cannot regress indefinitely; therefore, the motion of each moving object must be sustained, ultimately, by a first cause. Here and in other cosmological proofs, *first cause* does not of course mean *temporally first cause*. Should causes and effects happen to precede each other back through time, a first cause would indeed be temporally first, as well as first in a more significant sense. But when the links in a causal series do not precede each other temporally, the first cause—and this is the more significant sense of the term, the sense intended in cosmological proofs—is that which, although existing together with the other links, stands behind them all and is responsible for the causation running through the series.[18] The necessity of reaching a first cause in the nontemporal sense is what is affirmed by the principle of the impossibility of an infinite causal regress.

In its skeletal formulation, the proof from motion approaches another demonstration of the existence of God, which I shall call the proof from the impossibility of an infinite regress of efficient causes.[19] That too rests on the two key principles, the principle of causality and the impossibility of an infinite regress of causes, whence the existence of a first uncaused efficient cause is now inferred. The proof from motion and the proof from the impossibility of an infinite regress of efficient causes differ most obviously and most fundamentally in that the former focuses on causes of motion, whereas the latter abstracts from various categories of causation and considers efficient causation in general. The former, consequently, arrives at a first mover, which it identifies as the deity, whereas the latter arrives at a first efficient cause, which it too identifies as the deity. The differing focuses of the proofs will be taken up more fully in a later chapter.[20]

Apart from their differing focuses, the proof from motion and the proof from the impossibility of an infinite regress of efficient causes also differ in the ways

[14]Maimonides, *Guide,* I, 70 (end); Aquinas, *Summa Theologiae,* I, 2, art. 3, resp. (1); cf. Aquinas, *Commentary on Physics,* VIII, §970.

[15]Cf. Aquinas, *Summa contra Gentiles,* I, 13; C. Baeumker, *Witelo* (Muenster, 1908), p. 324; H. Wolfson, *Philosophy of Spinoza* (Cambridge, Mass., 1934), I, p. 193.

[16]Cf. Aristotle, *Physics* VII, 2; VIII, 10.

[17]Motion, here, means not merely locomotion, but change in general. Cf. *Physics* VIII, 7, and below, n. 33.

[18]This is the second sense of *priority* defined by Aristotle in *Categories,* 12.

[19]Cf. below, Chapter XI.

[20]Below, pp. 337, 344.

they elaborate the central, skeletal argument. The reason why the skeletal argument must in each instance be elaborated is that a first cause is not necessarily a deity. The first cause reached by tracing back a series of moving causes or efficient causes might be an inanimate object; it might be an animate physical object; or there might exist a multitude of first causes. Until the incorporeality and unity of the first cause are established, the minimum specifications for a deity, at least in the view of the medieval Aristotelian philosophers, are not met.[21] Both the proof from motion and the proof from the impossibility of an infinite regress of efficient causes expend more effort on establishing the incorporeality and unity of the first cause than they do on establishing the existence of such a cause.

Aristotle, the originator of the proof from motion, provided arguments for both the incorporeality and the unity of the first mover. His argument for the incorporeality of the first mover took its departure from the eternity of motion; he reasoned that since no corporeal object could contain power sufficient to sustain eternal motion, the first mover cannot be a corporeal object.[22] Aristotle provided two arguments for the unity of the first mover. One inferred the unity of the mover from the unity of the fundamental underlying motion of the universe;[23] the other, which appears in a different work, deduced the attribute of unity from the attribute of incorporeality. Here the reasoning was: "All things that are many in number have matter"; and since the first mover has no matter, it cannot be many in number.[24] Recent scholars have questioned whether Aristotle's most considered position was indeed that only a single first mover exists,[25] but medieval writers had no such doubts.

Whereas the skeletal argument is simple and direct, Aristotle's complete proof of the existence of God from motion is quite complex. The skeletal argument is supplemented not only by arguments for incorporeality and unity, but also by an argument designed to show that the first cause of motion is, more comprehensively, the first cause of all types of change occurring in the universe. The complete proof winds tortuously through Books VII and VIII of Aristotle's *Physics*,[26]

[21]Cf. Maimonides, *Guide*, I, 71 (end).

[22]Aristotle, *Physics* VIII, 10.

[23]Aristotle, *Physics* VIII, 6, 259a, 13–20.

[24]Aristotle, *Metaphysics* XII, 8, 1074a, 33–37.

[25]Cf. H. Wolfson, "The Plurality of Immovable Movers in Aristotle, Averroes, and St. Thomas," reprinted in his *Studies in the History of Philosophy and Religion*, Vol. I (Cambridge, Mass., 1973), pp. 1–21; D. Ross, "The Development of Aristotle's Thought," in *Aristotle and Plato in the Mid-Fourth Century*, ed. I. Duering and G. Owen (Goeteborg, 1960), pp. 12–14; and, for further bibliography, D. Frede, "Theophrasts Kritik am umbewegten Beweger," *Phronesis*, XVI (1971), 65.

[26]The relationship of *Physics* VII to *Physics* VIII is problematical. Cf. Averroes, *Long Commentary on Physics*, in *Aristotelis Opera cum Averrois Commentariis*, Vol. IV (Venice, 1562), VIII, comm. 9; idem, *Derushim Tib'iyim*, ed. H. Tunik, Ph.D. dissertation, Radcliffe College (1956), §7; Aquinas, *Commentary on Physics*, VII, introduction; H. Wolfson, "Notes on Proofs of the Existence of God in Jewish Philosophy," in his *Studies in the History of Philosophy and Religion*, Vol. I, p. 580; F. Solmsen, *Aristotle's System of the Physical World* (Ithaca, 1960), p. 228, n. 19. All of these find that the discussion in *Physics* VIII relies on the conclusions of *Physics* VII. D. Ross in the introduction to his edition of the *Physics* (Oxford, 1936), pp. 15–19, finds that Book VII is an interpolation into the core of the *Physics*.

and many more than the two central philosophic principles come into play. In the Middle Ages the full argumentation of Aristotle's proof was unraveled, reassembled, and presented afresh by Averroes,[27] Maimonides, and Aquinas.[28] Most systematic is Maimonides' version. It painstakingly spells out the principles and steps, implied or explicit, the full proof being shown to rest on no less than nineteen formal principles of Aristotelian philosophy. Maimonides' articulation of the premises and steps in the proof facilitated and invited critical scrutiny; and his version, did, in fact, elicit a most searching critique by Ḥasdai Crescas. Crescas does not transcend his age. He does not reject, in totality, the overall Aristotelian framework within which the proof is worked out. And he does not question the proposition in the proof which would be most questionable for the modern reader, to wit, the principle that motion continues only as long as sustained by a cause.[29] Crescas' critique is, however, as radical a critique as could be drawn up in Aristotelian terms by a student of Aristotelian physics.[30]

The present chapter will examine Maimonides' version of the proof from motion, that version being considered as a systematic articulation of what is explicit and implied in the proof. Then Crescas' critique will be examined, it being considered as a capital illustration of the extent to which the proof can be subverted from within the framework of Aristotelian physics. Finally, an offshoot of the proof from motion will also be examined together, again, with Crescas' critique.

2. Maimonides' version of the proof from motion

The proof from motion as reformulated by Maimonides carefully establishes each of the theses that are, in his view and in the view of other medieval philosophers, required for a complete demonstration of the existence of God. That is to say, he undertakes to establish (a) the existence of a first cause, (b) the incorporeality of that cause, and (c) its unity. In the course of establishing the three theses, Maimonides, as already mentioned, cites nineteen principles of Aristotelian philosophy. Each of those principles has its philosophic justification in Aristotle, but Maimonides states them flatly, without supporting reasoning, because he was wary of overwhelming his reader with technical detail.[31] Crescas in his critique will, before setting forth his own objections, lay bare the presuppositions and reconstruct the reasoning underlying each of the principles.

(a) The existence of a first mover

The object of the Aristotelian proof from motion is to establish a first cause of locomotion, or motion in place. In addition, though, the proof proposes to trace

[27] See above, n. 10.

[28] *Summa contra Gentiles,* I, 13.

[29] That principle stands in opposition to Newton's first law of motion, the law of inertia. Cf. A. Kenny, *The Five Ways* (London, 1969), pp. 28–31.

[30] Cf. H. Wolfson, *Crescas' Critique of Aristotle* (Cambridge, Mass., 1929), pp. 114–127, especially pp. 125–127.

[31] Maimonides, *Guide,* I, 71.

all other changes in the universe to motion in place, so that the first cause of motion in place will emerge, more comprehensively, as the first cause of all change and motion in the universe.[32] Maimonides in his systematization of the proof accordingly cites principle (1) affirming that there are "four categories" in which change occurs, namely "the category of substance, . . . of quantity, . . . of quality, . . . and of place."[33] In reasoning back to a first cause of motion in place, Maimonides starts by considering a single given instance of change in the category of substance: He inspects the coming into existence of an actual physical object. Principle (2) states that every "individual compound substance" has "matter and form" as its components, and also that a substance of the sort comes into existence only after some factor "prepares" the given matter to receive the appropriate form.[34] Here, as will be observed, is the principle of causality applied to the most radical type of change, the coming into existence of a new object.

Whenever a given portion of matter receives a new form, some factor, then, prepares the matter. If the factor preparing a portion of matter to receive a new form should itself be brought to the state wherein it performs its causal function, a further causal factor must be responsible for the change; here the principle of causation is applied to changes of state. If that factor, in turn, is also brought to the state wherein it performs its causal function, there must lie behind it yet another factor. According to principle (3), however, an infinite regress of causes is impossible.[35] The series of causal factors lying behind the generation of a given physical object cannot, therefore, regress indefinitely, and a first term in the series must be posited.[36] Now, whereas the end product of the series that Maimonides is considering is a change in substance, the change directly brought about by the first cause in the series must be motion in place. For principle (4) affirms that "motion in place" is the "most primary" of the four categories of change, inasmuch as a change of place ultimately stands behind all the other kinds of change.[37] The series of causes leading up to each change in the universe has thus been traced back to a first cause whose immediate effect is motion in place.

The generation of a single given substance is not an isolated event; it is part of a continuing process. And for the processes of nature to continue uniformly, without interruption, the underlying motion in place from which other motions

[32] Cf. *Physics* VIII, 7.

[33] *Guide,* II, introduction, prop. 4. Cf. *Physics* III, 1, 201a, 4–9; V, 1, 225a, 34 ff. In the former passage Aristotle states that "motion" and "change" take place in the four categories. In the latter passage, he observes that motion and change take place in the categories of quantity, quality, and place; but in the category of substance only change, not motion, occurs.

[34] *Guide,* II, introduction, prop. 25.

[35] Ibid., prop. 3. Cf. Aristotle, *Physics* VIII, 5, 256a, 4—256b, 3; *Metaphysics* II, 2, 994a, 5–8; 11–19; below, p. 337.

[36] The first cause is *first* in the sense defined above, p. 238. It is also temporally prior to at least some effects in the series, namely those effects that come into existence after not having existed.

[37] *Guide,* II, introduction, prop. 14; Aristotle, *Physics* VIII, 7.

and changes derive must be continuous.[38] Principle (5) states that circular motion in place is the sole type of change capable of continuing indefinitely and without interruption.[39] The effect directly produced by the first cause of a given event—that is to say, by the first cause of the continuing process of which the given event is a part—must hence specifically be circular motion in place. Side by side with these abstract considerations, the analysis of actual events in the sublunar world revealed to the Aristotelian observer that all sublunar events are traceable to the circular motion of the heavens.[40] Maimonides consequently affirms that every motion and change in the sublunar world is traceable to the continual circular movement of the heavens, and that the cause of the motion of the heavens is the ultimate cause of all motion and change in the sublunar world.

To complete the argument and raise it to a genuine demonstration of the existence of God, the first cause of the movement of the heavens must be shown to be an incorporeal being beyond the heavens, and there must be shown to exist only one first incorporeal cause. Restating the same in the terms of Aristotelian astronomy, the cause of the motion of the celestial spheres must be shown to be a single incorporeal being beyond the spheres. Once that has been proven, a single incorporeal first cause of all motion and change in the universe will have been established.

(b) The incorporeality of the first mover

Maimonides hereupon cites principle (6), which is, again, the principle of causality in one of its guises. Principle (6) states that "whatever undergoes motion, necessarily has a mover," that is to say, a cause sustaining motion as long as the object continues to move;[41] and, the same principle continues, the cause of motion obviously exists either "outside" the moving object or "within it."[42] The system of celestial spheres must hence be sustained in motion either by something outside, or else by something within, the system of spheres. Each of those two possibilities, the possibility that the mover of the spheres is to be found outside the spheres and the possibility that the mover is within, subdivides in two; and a total of four alternatives results. For should, on the one hand, the mover of the spheres lie outside them, the mover must be either—the first alternative—another body, or—a second alternative—an incorporeal entity. And should, on the other hand, the mover lie within, principle (7) comes into play. Principle (7) states that whatever is in a body is a power of one of two conceivable sorts. It is either a power that strictly does "exist within the body," or else it is a power "through" which "the body exists, such as the natural form."[43] Accordingly, should the

[38]This is not stated explicitly by Maimonides, but cf. Aristotle, *Physics* VIII, 6. On the problematic character of the proposition, see Ross's introduction to his edition of the *Physics*, pp. 91–92.

[39]*Guide*, II, introduction, prop. 13; cf. Aristotle, *Physics* VIII, 8.

[40]*Guide*, II, 1 (1). Cf. Aristotle, *Metaphysics* XII, 6, 1072a, 10–18; *De Generatione et Corruptione* II, 10.

[41]Cf. Aristotle, *Physics* VII, 1; VIII, 10.

[42]*Guide*, II, introduction, prop. 17.

[43]*Guide*, II, introduction, prop. 10.

mover of the spheres be present within the body of the spheres, it would be either—a third alternative—strictly within the body of the spheres, in other words, a physical "power distributed through the entire sphere [or spheres]" and hence "divisible";[44] or else—a fourth alternative—the mover would be a natural form that is "not divisible," in other words, a "soul" or "intellect." *Intellect* in this fourth alternative is to be distinguished from *incorporeal entity* in the second alternative. Intellect here has the sense of a soul, possessing the faculty of reason, which is attached to a body. The incorporeal entity in the second alternative is an entity consisting in pure thought, that is to say, an intellect, which is not attached to a body.

The conceivable ways of construing the cause of celestial motion are thus exhausted by four alternatives. For the purpose of his exposition, Maimonides arranges the four alternatives in the following order. The cause of the motion of the spheres and, thereby, the ultimate cause of motion and change in the universe must be either: (i) a body beyond the spheres; (ii) a power distributed through the body of the spheres—or, put more generally, a power distributed through the body of the universe; (iii) a natural form or "power" present in, but not distributed through, the body of the spheres, in effect, a "soul" or "intellect" attached to the spheres; or (iv) a purely incorporeal entity, distinct from the spheres. Maimonides undertakes to establish the correctness of alternative (iv) by eliminating the other three alternatives. In formulating four exhaustive alternatives and eliminating three of them in order to establish the correctness of the fourth, Maimonides is introducing something not found in Aristotle. There he is systematizing the proof. However, the overall contention that the unceasing movement of the spheres implies an incorporeal mover is genuinely Aristotelian.[45] And all of the argumentation whereby Maimonides eliminates the three unacceptable alternatives is drawn from Aristotle's statement of the proof. Maimonides seems even to have taken pains to find a role, a niche, for strands of argumentation in Aristotle which he could well have regarded as superfluous or redundant.

(i) The alternative that the mover of the spheres is a body beyond the spheres.

If the system of spheres should be moved by a body beyond the spheres, Maimonides reasons, the body in question must itself undergo motion because of principle (8); principle (8) states that one body can move another only when it is itself in motion.[46] The body moving the spheres would likewise have to have a cause of its motion. And if it were moved by yet another body, the latter too would have to undergo motion and would require a cause of its motion. To assume an infinite regress of moving bodies would entail the simultaneous existence of an infinite number of bodies. But the simultaneous existence of an infinite number of bodies is ruled out by principle (9), which affirms, in general, the impossibility

[44] Cf. ibid., prop. 11.

[45] Cf. *Physics* VIII, 10.

[46] *Guide,* II, introduction, prop. 9; cf. Aristotle, *Physics* VII, 1.

of the simultaneous existence of an infinite number of magnitudes.[47] Consequently, a cause of the motion of the spheres must eventually be reached which is not a body.

It should be noted that principle (9) and the inference drawn therefrom are superfluous. Maimonides had already established a first cause of motion with the aid of principle (3), the impossibility of an infinite regress of causes. He could, therefore, now more appropriately and more simply have contended that the first cause of motion already established cannot be a body; for it would then be in motion, hence have a cause of its motion, and not after all be the first cause. Something undoubtedly led Maimonides to rule out an infinite regress specifically of moving bodies, in so curiously oblique a fashion—by arguing that such a regress would entail the assumption of an infinite number of bodies, whereas the existence of an infinite number of bodies is impossible. What did lead Maimonides to argue as he does can be surmised. A similar argument appears in Aristotle's *Physics* VII,[48] and Maimonides presumably wished to find a niche for that argument of Aristotle's in his own restatement of the proof from motion.

In any event, the system of celestial spheres and, in fact, every series of moving bodies, has been shown to owe its motion to a cause which is not a body. The question recurs: What is the nature of the cause?

Maimonides proceeds to adduce an argument showing that the first cause of motion cannot be a power distributed through a body, and a separate argument showing that the cause cannot be a power of the type that, although not distributed through a body, is present in one.

(ii) The alternative that the mover of the spheres is a power distributed through a body

Aristotle had established that the first mover cannot "have magnitude" through an argument running thus: Every entity having magnitude, including the corporeal universe as a whole, must be finite, since nothing of infinite magnitude can possibly exist. Whatever is of finite magnitude can contain only finite power. Finite power can produce motion for only a finite time. But motion, specifically the motion of the celestial spheres, is eternal and therefore dependent upon an infinite power. From these premises, Aristotle concluded that the first mover must be "indivisible, without parts, and without magnitude."[49] It must, in other words, be an incorporeal being. Without spelling out the several alternatives to an incorporeal first mover, Aristotle excluded them all through a single argument.

[47] *Guide*, II, introduction, prop. 2.

[48] *Physics* VII, 1; here Aristotle does not establish that an infinite number of bodies cannot exist, but rather than an infinite number of bodies would be unable to move.

[49] *Physics* VIII, 10, summed up in 267b, 17–26. The argument has four premises: every body is finite; a finite body can contain only finite power; a finite power can produce motion for only a finite time; the motion of the heavens is eternal. In Maimonides' reformulation, the principle that a finite power can produce motion for only a finite time will not have the status of a separate premise.

As Maimonides read Aristotle, he apparently found the argument just outlined to be inappropriate for eliminating alternative (i), the alternative that the mover of the spheres is another body. That alternative, as just seen, is eliminated by Maimonides through a different argument.[50] And Maimonides apparently found Aristotle's argument inadequate for eliminating alternative (iii), the alternative that the mover of the spheres is a power of the type which is not distributed through a body. While one can only guess why Maimonides considered the argument inappropriate for eliminating alternative (i),[51] a reason why he might find it inadequate for eliminating alternative (iii) is easily discovered: A power not distributed through a body is not strictly *contained* in a magnitude and consequently is not covered by the consideration that a finite magnitude can contain only finite power.[52] The single purpose for which Maimonides does use Aristotle's argument is to eliminate alternative (ii), the alternative that the spheres are moved by a power of the type which *is* distributed through a body.

In reformulating Aristotle's argument, Maimonides cites principle (10), according to which every power distributed through a body is divisible in the same manner that its body is divisible.[53] Principle (11) adds that no body can be "of infinite magnitude."[54] Together, (10) and (11) imply (12), the principle that every power distributed through a body is finite.[55] A finite power could not, however, produce

[50]Above, pp. 243–244.

[51]The argument strictly eliminates the thesis that the mover of the spheres is a power distributed through a body, not the thesis that the mover of the spheres is a body; but I am not sure that Maimonides could conceive of a body's producing motion by itself, i.e., by its mass as distinct from its power. In any event, though, Maimonides would be led to use a separate argument for eliminating alternative (i) by his desire to utilize every strand of argumentation provided by Aristotle; cf. above, p. 244.

[52]Cf. S. Munk's note to his translation of Maimonides, *Le Guide des Égarés,* Vol. II (Paris, 1861), p. 32, n. 2.

[53]*Guide,* II, introduction, prop. 11.

[54]Ibid., prop. 1; cf. *Physics* III, 5.

[55]*Guide,* II, introduction, prop. 12. The reasoning whereby (10) and (11) imply (12) would be as follows: The power distributed through a body, being divisible, may be assumed actually to be divided. Since the body is finite, it can only be divided into a finite number of parts, and the power distributed through it, which is similarly divided, will also have a finite number of parts. Each partial power will, moreover, clearly be finite; for each part will be exceeded by the larger whole, and nothing that is exceeded can be infinite (cf. above, p. 88). Inasmuch as each partial power is finite and the number of partial powers is finite, the total power must be finite.

This train of reasoning—which is similar to the argument whereby Aristotle proves that a finite power can produce motion for only a finite time, *Physics* VIII, 10, 266a, 12 ff.—is spelled out by Alexander, cited in Simplicius, *Commentary on Physics,* ed. H. Diels, *Commentaria in Aristotelem Graeca,* Vol. X (Berlin, 1895), p. 1326; Avicenna, *Najāt* (Cairo, 1938), p. 130; Ghazali, *Maqāṣid al-Falāsifa* (Cairo, n.d.), p. 209; Ibn Ṭufayl, *Ḥayy ben Yaqdhān,* Arabic text, p. 84; French translation: p. 64; English translation: p. 84; Shem Ṭob's commentary to Maimonides' *Guide,* II, introduction, prop. 12. Also cf. Munk's translation of the *Guide,* II, p. 32, n. 2. Aristotle proves principle (12) differently; cf. *Physics* VIII, 10, and below, p. 261. In an article entitled "The Principle that a Finite Body Can Contain Only Finite Power," *Studies in Jewish Religious and Intellectual History Presented to Alexander Altmann* (University, Alabama, 1979), p. 77, I suggest a reason why Alexander and the others might have wished to substitute their reasoning for Aristotle's.

motion for an infinite time;[56] this proposition is so obvious to Maimonides that he does not even register it in his list of formal principles. Yet principle (13), which Maimonides accepts merely for the sake of argument, asserts that the universe and its motion are eternal.[57] Since eternal motion has to be sustained by infinite power, and since no power distributed through a body can be infinite, the eternal motion of the spheres, so Maimonides concludes, could not be due to a power distributed through a body.

The alternative that the motion of the spheres is due to a power distributed through a body is thus ruled out. Earlier the alternative that the motion of the spheres is due to a body beyond the spheres was also ruled out. Two conceivable explanations still remain: The cause of the motion of the spheres must be either a power present within, but not distributed through, the body of the spheres; or else it is an incorporeal being.

(iii) The alternative that the mover of the spheres is a power present in, but not distributed through, the body of the spheres

In the course of his proof of the existence of a first incorporeal mover, Aristotle had at one point made the statement that the first cause of eternal motion cannot itself undergo motion accidentally, "since what is accidental . . . has the possibility of not being."[58] Maimonides, characteristically, utilizes the thought embodied in Aristotle's statement for a single specific purpose, to show that the ultimate mover of the spheres cannot be a power present within, although not distributed throughout, the body of the spheres, in other words, to show, that the motion of the spheres is not produced by a soul attached to the spheres.

Maimonides' formulation of the argument begins with principle (14), which affirms that physical bodies and nothing else can undergo motion "essentially." By contrast, the same principle continues, powers in a body, whether or not they are distributed through their body, do undergo motion "accidentally" as a consequence of their presence "in" something that undergoes motion essentially.[59] Principle (15) adds that nothing undergoing motion accidentally can continue to move indefinitely;[60] for the accidental, having the possibility of not existing, cannot be eternal.[61] It follows that every power in a body must eventually come

[56]*Guide*, II, 1 (1); cf. Aristotle, *Physics* VIII, 10.

[57]*Guide*, II, introduction, prop. 26; cf. *Physics* VIII, 1, and above, Chapters II and III. Maimonides' position is that the proof from motion is needed only on the hypothesis of eternity, whereas on the opposite hypothesis the existence, incorporeality, and unity of the creator can be inferred from the world's having come into existence. Cf. *Guide*, I, 71; II, 2; above, pp. 4–5.

[58]*Physics* VIII, 5, 256b, 10; cf. ibid., 6, 259b, 28–31.

[59]*Guide*, II, introduction, prop. 6; cf. Aristotle, *Physics* VIII, 4, 254b, 7 ff.

[60]*Guide*, II, introduction, prop. 8; cf. Aristotle, *Physics* VIII, 5, 256b, 8–10.

[61]The principle that every possibility must eventually be realized was met above, p. 90, and will be met again below, pp. 320, 381.

to rest. When a power in a body would come to rest, the body inextricably associated with it and moved by it would also necessarily come to rest.[62] As already seen in principle (13), our discussion is proceeding on the hypothesis that the motion of the celestial spheres is eternal. The primary motion in the universe cannot, consequently, be due to a power in a body, even a power of the type that is attached to, without being distributed throughout, its body; for such a power could not sustain eternal motion.

The reasoning put forward here, as will be observed, applies equally to powers that are, and powers that are not, distributed through a body, as long as those powers are properly described as *in* a body. Powers of both types would, by the reasoning given, undergo motion accidentally, and hence not be able to undergo motion indefinitely; and whatever is inextricably attached to them and moved by them would likewise be unable to undergo motion indefinitely. Maimonides could thus have utilized the present reasoning earlier when he was eliminating the previous possible explanation of celestial movement, the alternative that (ii) the cause of the motion of the spheres is a power distributed through the spheres. He could have dispensed with the argument to the effect that a finite body can contain only finite power, and finite power can sustain motion for only a finite time; and he could have reasoned instead that a power distributed through a body would itself undergo accidental motion and therefore be unable to produce motion eternally. Maimonides presumably employed the other train of reasoning when ruling out alternative (ii) because of his desire to find a role for each strand of argumentation in Aristotle's proof from motion.

(iv) The remaining alternative: The cause of the motion of the spheres is an incorporeal being

The cause of the motion of the celestial spheres has now been shown to be neither a body, nor a power distributed through a body, nor a power present in, although not distributed through, a body. The remaining alternative is that the first cause of the motion of the spheres and, thereby, the ultimate cause of motion and change in the sublunar world is (iv) an incorporeal being. Principle (16)

[62]This step raises difficulties. Maimonides has argued not merely that a power undergoing accidental motion cannot produce motion eternally; he has argued that such a power or anything else undergoing accidental motion cannot itself be in motion eternally. That would mean that the spheres cannot move eternally—cannot even be maintained eternally in motion by a transcendent incorporeal cause—should anything attached to them undergo accidental motion. For example, the spheres would be unable to move eternally if a soul is inextricably attached to them; or, as Narboni observed in his commentary, if such qualities as *transparence* and *radiance* are attached to them; or indeed, as Crescas observed, if their own surface is attached to them! For the soul, the qualities *transparence* and *radiance,* and the surface of the spheres, are not bodies. Hence they move not essentially but accidentally and, by Maimonides' reasoning, would eventually have to cease moving. Cf. Narboni's commentary on *Guide,* II, introduction, prop. 8; Crescas, *Or ha-Shem,* I, i, 8; Wolfson, *Crescas,* pp. 250–251, 551–561; and Aristotle, *Physics* VIII, 6, 259b, 28–31. Below, p. 271, I shall suggest that there is an equivocation in Maimonides' use of the term *accidental*.

states that whatever is capable of undergoing motion is divisible (and, conversely, whatever is not capable of undergoing motion is indivisible).[63] Principle (17) states that motion is a kind of change (and change is a kind of motion).[64] Since the first cause of motion cannot, as already seen, undergo motion either essentially[65] or accidentally,[66] it must by principles (16) and (17) be indivisible and exempt from all change. Principle (18) affirms that "time is an accident consequent upon motion. . . . and whatever is exempt from motion does not fall under time."[67] The unmoved first mover hence exists outside the domain of time.

The first mover must then be an indivisible, unchanging, timeless, incorporeal substance.

(c) The unity of the first mover

In Aristotle's fullest statement of the proof from motion, the unity of the first mover is inferred from the unity of the underlying primary motion of the universe, a single continuing motion implying a single mover.[68] In a condensed statement of the proof from motion, in the *Metaphysics,* Aristotle derived the unity of the first mover from the attribute of incorporeality.[69] The latter method is employed by Maimonides.

Principle (19) is a version of the rule that matter is the ground whereby objects are individuated. According to principle (19), "what is not a body . . . or a power in a body . . . cannot conceivably by enumerated" except in one instance. The sole instance in which incorporeal beings can be differentiated from each other and enumerated is when they stand in the relation of "causes and effects."[70] Within the class of causeless beings the differentiation and enumeration of separate entities as causes and effects would of course be impossible; one causeless being obviously cannot be the cause or the effect of another causeless being. Consequently, no more than a single incorporeal uncaused mover can exist. Yet should there also exist other incorporeal movers of the spheres, in addition to the first unmoved mover, those other incorporeal movers would have to stand in relation to each other as causes and effects; for only then would they be distinguishable from each other and numerable. Thus principle (19) arranges the incorporeal movers in the Aristotelian scheme of the universe into a causal series with a single uncaused cause at the apex.

[63] *Guide,* II, introduction, prop. 7; cf. Aristotle, *Physics* VI, 4, 234b, 10–20. In this and the following proposition I have added the clauses in brackets. Without the additions, the logic of the inference Maimonides is about to draw would be faulty.

[64] *Guide,* II, introduction, prop. 5.

[65] Alternative (i).

[66] Alternatives (ii) and (iii).

[67] *Guide,* II, introduction, prop. 15; cf. Aristotle, *Physics* IV, 11.

[68] *Physics* VIII, 6, 259a, 13–20.

[69] Above, n. 24.

[70] *Guide,* II, introduction, prop. 16. The principle is never stated in just this way by Aristotle.

With the unity of the first mover, the proof from motion is complete. The proof has established the existence of a single incorporeal first mover of the celestial spheres and, through the mediacy of the spheres, of the sublunar world.

3. Ḥasdai Crescas' critique of the proof from motion

Crescas' critique of the proof from motion contains at least ten separate objections arranged in an artificial scheme.[71] In the present section, I am rearranging Crescas' objections to conform to the three main moments in the proof, that is, the arguments for the existence, for the incorporeality, and for the unity of the first mover. Crescas' critique of the arguments for incorporeality and unity are of greater interest than his critique of the argument for the existence of the first mover. And his critique of the argument for incorporeality is so far ranging that the train of thought is difficult to follow. Crescas' intent throughout is not free-thinking, but conservative. His aim is to dispute not the existence, incorporeality, or unity of God, but the ability of philosophy to demonstrate those three theses.

(a) The existence of a first mover

Principle (3), the impossibility of an infinite regress of causes, is essential to the proof from motion. It alone allows the causes behind the appearance of a given physical object to be traced back to a first term; and unless a first term in the series of causes is reached, the proof accomplishes nothing. Crescas ostensibly refutes principle (3), the impossibility of an infinite regress of causes, thereby ostensibly invalidating the proof entirely.[72] In fact, the principle that causes cannot regress infinitely contains two separate points. The principle asserts both that every chain of causes must terminate at a first cause and that the number of links in a causal chain must be finite. Of the two points, only the former is essential for demonstrating the existence of God, whereas, once a first cause is posited, a demonstration of the existence of God can dispense with the latter point, the proposition that the number of causes and effects must be finite. And Crescas' refutation is addressed exclusively to the latter point. He acknowledges that a first cause must indeed be posited for every series of causes and effects, and merely contends that, granted a first cause, the number of links between it and the final effect might conceivably run to infinity.[73] Crescas' refutation of the principle that causes cannot regress infinitely thus in no way goes to the heart of the proof from motion. His reservations on the principle are of some interest, but they will best be examined in a later chapter; the grounds for the impossibility of an infinite regress which were known to him grow out of a certain proof for

[71]Crescas lists seven principles employed in the proof which, he contends, are unsubstantiated. Then he gives three reasons why the proof is invalid even if all the principles are granted. *Or ha-Shem,* I, ii, 15.

[72]*Or ha-Shem,* I, ii, 3; 15 (following the reading of the Vienna edition); Wolfson, *Crescas,* pp. 224–229.

[73]Cf. below, p. 365.

the existence of God advanced by Avicenna, and his refutation will be best taken up after examining Avicenna's proof.[74] At the same time I shall also take up a more trenchant critique of the principle that causes cannot regress infinitely, a critique developed by Ghazali.[75]

(b) The incorporeality of the first mover

Maimonides established the incorporeality of the first mover by laying down the Aristotelian principle (6), according to which objects move only as long as they are sustained in their motion by a cause, and by eliminating all but one of the conceivable ways of construing the cause of celestial motion. The mover of the spheres cannot, Maimonides argued, be (i) a body beyond the spheres, (ii) a power distributed through a body, or (iii) a power present in, although not distributed through, a body, in effect, a soul of the spheres; therefore the mover of the spheres must be the sole alternative remaining, (iv) an incorporeal being independent of the spheres.[76] Crescas challenges the conclusion. But whereas a modern critic might question principle (6), which affirms that motion continues only as long as sustained by a cause,[77] Crescas has no thought of taking so radical a course. His procedure is instead to refute the several arguments whereby each of the three alternatives to an incorporeal mover had been ruled out.

(i) The alternative that the mover of the spheres is a body beyond the spheres

To rule out the alternative that the mover of the spheres is a body beyond the spheres, Maimonides had reasoned: One body can produce motion in another only if it is itself in motion. To assume that every body producing motion is moved by another amounts to assuming an infinite series of moving bodies. And that is excluded by principle (9), the impossibility of the simultaneous existence of an infinite number of magnitudes. The mover of the spheres cannot therefore be a body beyond the spheres; or, to be more precise, if the mover of the spheres is a body beyond the spheres, a mover must ultimately be reached which is not a body.[78]

Crescas' critique consists in rejecting the various grounds that had been adduced to support principle (9), the impossibility of an infinite number of magnitudes. Ostensibly, he thereby defends the thesis that motion in the universe may be explained through an infinite regress of moving bodies all existing together.[79] In fact, Crescas' refutation of principle (9) does not affect the proof from motion,

[74]Below, ibid.

[75]Below, pp. 366–372.

[76]Cf. above, pp. 243–247.

[77]Crescas will propose the thesis that the elements are moved to their proper places, and the spheres are moved circularly, by their natures; cf. below, p. 269. But that falls far short of Newton's first law of motion.

[78]Above, pp. 243–244.

[79]*Or ha-Shem,* I, ii, 2; 15; Wolfson, Crescas, pp. 218–221.

and for that reason need not be explored here.[80] For even without principle (9), an infinite regress of moving bodies is excluded by the more fundamental principle—principle (3)—which affirms the impossibility of an infinite regress of causes. Although Crescas rejected part of principle (3) as well, he accepted its more significant part, namely, the proposition that every series of causes and effects has a first cause;[81] and he thus cannot consistently defend a picture of the universe in which each body is moved by another body and a first cause is never reached. Crescas was too deeply under the sway of Aristotelian physics to reject the proposition that every motion must be sustained by a cause, or the proposition that bodies produce motion only when themselves in motion, or again the proposition that causes cannot regress indefinitely without reaching a first cause. And unless one of those propositions is rejected, the Aristotelian conclusion is unavoidable that motion is ultimately due to a first cause which is not a body.

It is in what follows that Crescas genuinely comes to grips with the proof from motion.

(ii) The alternative that the mover of the spheres is a power distributed through a body

Even granting that the Aristotelian proof from motion can establish a first cause which is not a body, the proof still has to show that the first cause is not a power in a body. Should the first cause of motion be the kind of power which is distributed through a body, it would be inanimate and clearly not an appropriate candidate for a deity. Should the first cause be the other kind of power, the kind that is attached to but not distributed through a body—in other words, a soul—it would still not deserve the title of deity in the view of the medieval philosophers. In their view, a proof of the existence of God has to establish the existence of a first cause that is incorporeal and that transcends the universe; an immanent soul of the universe would not be the deity.

The possibility that the mover of the spheres is a power distributed through a body had been ruled out by Aristotle and Maimonides through an argument that ran: No body can be of infinite magnitude [principle (11)]. Hence, no power contained within a body can be infinite [principle (12)]. Finite power is, however, insufficient to sustain eternal motion. And the universe and its motion are eternal, or at least may be assumed to be so for the sake of the argument [principle (13)]. The comprehensive conclusion originally drawn herefrom by Aristotle was that the cause of the motion of the spheres and of the universe as a whole must be "indivisible, without parts, and without magnitude."[82] The conclusion drawn in

[80]Crescas explains that principle (9) depends on principle (11), the impossibility of a single infinite magnitude. And he refutes principle (11); cf. below, pp. 254 ff. Alternatively, he explains, principle (9) might be defended through the proposition that an infinite number is impossible. And he refutes that proposition too; cf. above, p. 121.

[81]Above, p. 249; below, p. 365.

[82]Above, n. 49.

Maimonides' reformulation of the proof from motion is, more specifically, that the first cause of motion cannot be a power contained within, and distributed through, the body of the spheres.[83]

Crescas subjects the argument to a most minute critique. He contends that three of the four premises in the argument are unfounded. Those premises are the assumption of eternity, the principle that no body can be of infinite magnitude, and the principle that a body of finite magnitude cannot contain infinite power. Then he contends that even if all the premises are granted, the argument still fails since it confuses two distinct senses of *finiteness of power.* And finally he contends that the argument can in any event be circumvented by construing the motion of the spheres as due not to a power distributed through the body of the spheres or the body of the universe, but rather as due to the *nature* of the spheres. Crescas thus finds the argument wanting on five separate scores. Keeping the particulars of Crescas' discussion in place demands something of the juggler's skill. He is refuting an argument designed not to prove anything directly, but rather to eliminate one of three conceivable alternatives to an incorporeal first mover. And his attention is focused mainly on the presuppositions underlying the argument's premises. He is, that is to say, examining the presuppositions of the premises of the argument whereby a single alternative to an incorporeal first mover had been ruled out. While Crescas' discussion is annoyingly labyrinthine, the circumstance that he focuses on ultimate presuppositions enhances the significance of his critique. For he successfully exposes both the extent to which the proof of incorporeality, and hence the entire proof from motion, is bound to the presuppositions of Aristotelian physics, and the extent to which those presuppositions can be subverted from within the Aristotelian framework.

Crescas' critique of principle (13), the assumption of eternity. The eternity of the world is, Crescas insists, an indispensable premise in the argument showing that the mover of the spheres cannot be a power distributed through a body. Infinite power has to be posited solely in order to explain motion that is sustained over an infinite time. Should, on the contrary, the spheres not exist and move for an infinite time, a "power which is distributed through a body and therefore finite" would suffice to sustain their motion for as long as they do exist. But the doctrine of eternity is "undoubtedly false." Consequently, Crescas concludes, the argument showing that the first mover cannot be a power distributed through a body loses an indispensable premise and collapses.[84]

Crescas' thinking on the issue of the eternity of the world was touched upon in an earlier chapter[85] and need not be taken up again here for the following

[83] Above, pp. 245–246.

[84] *Or ha-Shem,* I, ii, 14 (end); 15.

[85] Above, p. 67. Crescas discusses the issue of creation and eternity in *Or ha-Shem,* III, i, and he falls considerably short of showing the doctrine of eternity to be "undoubtedly false." The doctrine of eternity would be shown to be false only if the arguments in favor of eternity are answered, and

reason: His assertion that a refutation of the doctrine of eternity constitutes a refutation of the proof from motion is, although correct from one viewpoint, incorrect from another. A successful refutation of the eternity of the world will indeed constitute a refutation of the proof from motion when the proof from motion is advanced by itself as a categorical demonstration of the existence of God; without the premise of eternity, the proof from motion cannot indeed reach an incorporeal first mover. But a refutation of the doctrine of eternity does not affect the approach of Ibn Ṭufayl, Maimonides, and Aquinas. Those philosophers did not intend the proof from motion to stand by itself, but instead proved the existence of God on parallel, alternative tracks, on both the hypothesis that the world is eternal and the hypothesis that it is not. The proof from motion has for Maimonides and the others no function beyond establishing a first incorporeal cause hypothetically—on the first of the two alternatives, that is, the assumption that the world is eternal; on the opposite assumption, a first incorporeal cause would be inferred directly from the world's having come into existence.[86] Crescas' refutation of the doctrine of eternity reveals then, if successful, merely that the proof from motion cannot single-handedly reach an incorporeal first mover. His refutation of the doctrine of eternity does not affect the approach of those for whom the proof from motion is intended to do just part of the job, to establish the existence of God hypothetically, when the eternity of the world is assumed.

Crescas' critique of principle (11), the impossibility of an infinite body. The impossibility of an infinite body is undoubtedly an indispensable premise in the argument showing that the mover of the spheres cannot be a power distributed through a body; for if the body of the universe were infinite it would contain infinite power,[87] and such power should suffice to sustain the motion of the spheres and of the universe as a whole for all eternity.[88] Crescas rejects the premise, but not by offering positive arguments to the contrary; he does not, in other words, undertake to demonstrate that the physical universe is in fact of infinite magnitude. He only refutes the various considerations whereby the "impossibility of an infinite corporeal magnitude" had supposedly been established.[89] Some of those considerations are fully spelled out in Aristotle, while

---

counterarguments are adduced to establish creation. As it turns out, Crescas can discover no counterarguments sufficient, in his view, to establish creation. Of the two sorts of arguments for eternity (cf. above, p. 10), Crescas answers those from the nature of the world, but is hard put to answer those from the nature of the first cause; cf. *Or ha-Shem,* III, i, 4; 5.

[86]Cf. above, n. 57 .

[87]Aristotle, *Physics* VIII, 10, 266b, 6–24, proves that an infinite body *must* contain infinite power.

[88]*Or ha-Shem,* I, ii, 15.

[89]*Or ha-Shem,* I, ii, 1 (2–4). In *Or ha-Shem,* I, ii, 1 (1) (= Wolfson, *Crescas,* pp. 178–191), Crescas refutes, as well, the considerations whereby the "impossibility of an infinite incorporeal magnitude" has supposedly been established. The issue of an infinite *incorporeal* magnitude is not, however, relevant to the proof from motion.

others were developed by the Aristotelian commentators, yet without exception they are anchored in the most basic presuppositions of Aristotelian philosophy.[90] Almost all the counter-considerations in Crescas' critique are adaptations of suggestions that had been made and rejected either by Aristotle himself or by the Aristotelian commentators.

Among the grounds adduced by Aristotle for the impossibility of an infinite body had been the definition of body as that which "is encompassed by one or more surfaces." Since nothing encompassed can be infinite, the definition, so Aristotle's reasoning had gone, entails that every body is finite.[91] Here Crescas makes the simple and obvious response that anyone who accepts an infinite body will reject the Aristotelian definition. To derive the finiteness of body from the definition is to beg the question.[92]

A second ground adduced by Aristotle for the impossibility of an infinite body was based on his theory of physical elements. An infinite body, Aristotle had reasoned, would have to contain either an infinite number of elements or else at least a single element of infinite magnitude. But an infinite body could not, on the one hand, consist of an infinite number of elements; for the natures of the elements are, he postulated, a subject of human knowledge whereas knowledge cannot comprehend an infinite number of items.[93] Nor, on the other hand, could any single element be of infinite magnitude. For the infinite element would overpower all the other, finite elements and destroy their qualities; it would thereby assimilate the other elements to its own nature and give rise to an undifferentiated universe. Moreover, an infinite element would occupy all space in the universe, leaving no room for other elements.[94]

Crescas' response runs thus: Aristotle's reason for not countenancing an infinite number of elements—the infinite is unknowable and hence an infinite number of elements could not be a subject of human knowledge—is not cogent, since there is no justification for assuming that "the principles *qua* principles should be

[90]Cf. Wolfson, *Crescas,* pp. 38–69.

[91]Aristotle, *Physics* III, 5, 204b, 5–6; *Or ha-Shem,* I, i, 1 (2); Wolfson, *Crescas,* pp. 150–151, and n. 57.

[92]*Or ha-Shem,* I, ii, 1 (2); Wolfson, *Crescas,* 190–191, and n. 40. In the note, Wolfson quotes passages from Narboni and Gersonides that may have suggested to Crescas the response he makes; and he quotes two surrejoinders to Crescas which were to be advanced by Isaac ben Shem Ṭob.

[93]On this point, cf. Aristotle, *Physics* I, 6, 189a, 12–13; *Metaphysics* III, 4.

[94]*Or ha-Shem,* I, i, 1 (2); Wolfson, *Crescas,* pp. 150–153. Crescas is following Averroes' *Middle Commentary on the Physics,* which combines Aristotle's *Physics* III, 5, 204b, 10 ff., and *Metaphysics* XI, 10, 1066b, 26 ff., with *Physics* I, 4, 187b, 7–13, and I, 6, 189a, 12–13. The passage in Averroes is cited and translated by Wolfson, *Crescas,* pp. 348–350.

The possibility of a physical universe constituted of several infinite elements does not seem to be covered by the consideration that any infinite element would overpower every finite element, although it is covered by the consideration that an infinite element would occupy all space in the universe. See the passage quoted by Wolfson from Averroes.

known."[95] Aristotle's reason for not countenancing a single infinite element—such an element would overpower the other elements—is also not cogent. For the single infinite element might be wholly devoid of positive qualities just as the celestial spheres, in the Aristotelian system, are wholly devoid of qualities.[96] An infinite element need not, moreover, occupy all space; it might extend infinitely in only one direction leaving room for other elements in the other directions.[97]

A third ground for the impossibility of an infinite body, formulated not by Aristotle but by Averroes,[98] is drawn from the Aristotelian definition of place. The place of any object is, according to Aristotle, the inner surface of the body or bodies surrounding the object.[99] Clearly, if an object has to be circumscribed in order to occupy place, no object occupying place can be infinite.[100]

Crescas responds in two ways. First he notes that even granting the Aristotelian definition of place, the definition fails in the case of at least one object in the universe: The definition cannot be employed to assign the place of the outer celestial sphere or, if we wish, of the universe as a whole. For since nothing, in the Aristotelian scheme, surrounds the outer celestial sphere, the place of the sphere clearly cannot be the inner circumference of a surrounding body. Some sort of *ad hoc* device must be proposed by Aristotelians in order to assign the place of the outermost sphere. And whatever that device should be, it may, Crescas contends, equally be employed to assign the place of an infinite body. For example, if the place of the outermost sphere is understood, exceptionally, to be the outer surface of the body surrounded by the sphere,[101] rather than the inner surface of the body surrounding the sphere, then the same may be maintained regarding the place of an infinite body.[102] Such is Crescas' first response to the contention that the definition of place excludes the possibility of an infinite body.

[95] *Or ha-Shem,* I, ii, 1 (2); Wolfson, *Crescas,* pp. 192–193. In n. 44, Wolfson cites a number of surrejoinders to Crescas.

[96] *Or ha-Shem,* I, ii, 1 (2); Wolfson, *Crescas,* pp. 192–193 and nn. 45, 46. Crescas' response was already anticipated and rejected by Aristotle, *Physics* III, 5, 204b, 23–35.

[97] *Or ha-Shem,* I, ii, 1 (2); Wolfson, *Crescas,* pp. 192–195 and n. 48. In the note, Wolfson shows that Crescas' response was answered, in anticipation, by Averroes, Narboni, and Gersonides; and he shows that it was also later to be answered by Isaac ben Shem Ṭob.

[98] See the passages from Abraham ibn Daud and Averroes which are cited by Wolfson, *Crescas,* pp. 352–355. Wolfson suggests that the argument is a development of Aristotle, *Physics* III, 5, 260a, 2–8, and *Metaphysics* XI, 10, 1067a, 30–33.

[99] *Physics* IV, 4, 212a, 20–21.

[100] *Or ha-Shem,* I, i, 1 (2); Wolfson, *Crescas,* pp. 152–153.

[101] On the different positions taken by Aristotelians regarding the *place* of the outer sphere or the *place* of the entire universe, see Wolfson, *Crescas,* pp. 432–441.

[102] *Or ha-Shem,* I, ii, 1 (2); Wolfson, *Crescas,* pp. 194–195. Crescas means that the place of the entire universe would be the center of the universe, or else the inner part of the universe, around which the universe is positioned. See Wolfson, *Crescas,* p. 195, n. 52.

Secondly, and more fundamentally, Crescas responds that the Aristotelian definition of place leads to so many unacceptable anomalies[103] that a different definition, a definition already weighed and rejected by Aristotle,[104] should be preferred. Place would best be defined, Crescas maintains, as the interval, or the empty void, occupied by an object.[105] And once this preferable definition is adopted, assigning the place of an infinite body is no more difficult than assigning the place of a finite body.[106]

A fourth ground for the impossibility of an infinite body, which had been adduced by Aristotle, is based not on the general definition of place but rather on the Aristotelian theory of natural place, the theory that each of the sublunar elements has a proper place to which it naturally travels and which it naturally occupies. Aristotle had reasoned that every body composed of elements is either heavy, which is equivalent to saying that its natural place is the lower region to the exclusion of the upper region; or else it is light, which is equivalent to saying that its natural place is the upper region to the exclusion of the lower. But an infinite body would be everywhere. It could be neither in the lower region as distinct from the upper, nor in the upper as distinct from the lower, and hence could be neither heavy nor light. That a body should be neither heavy nor light is, however, impossible. And therefore, Aristotle had concluded, an infinite body is impossible.[107] Crescas responds, simply, that an infinite body would in fact be neither heavy nor light. Such a proposition can hardly be objectionable, seeing that the Aristotelian scheme of the universe does recognize bodies that are neither heavy nor light, to wit, the celestial spheres.[108]

A fifth ground for the impossibility of an infinite body, which had been offered by Aristotle and elaborated by Gersonides, is based again on the Aristotelian theory of the natural places of the elements. The reasoning had been: Natural motion consists in the elements' returning to their proper places when removed therefrom. Now an infinite body would have to contain either a minimum of one element of infinite magnitude or else an infinite number of elements. But, on the one hand, a single infinite element cannot be supposed; for an infinite element would occupy all space, would always be present in its proper place, and would therefore never have the opportunity of undergoing natural motion. On the other

[103] *Or ha-Shem,* I, ii, 1 (2); Wolfson, *Crescas,* pp. 194–199.

[104] Cf. Aristotle, *Physics* IV, 4, 211b, 14–29.

[105] *Or ha-Shem,* I, ii, 1 (2); Wolfson, *Crescas,* pp. 194–195, 198–199, and n. 55. Cf. John Philoponus, *Commentary on Physics,* ed. H. Vitelli, *Commentaria in Aristotelem Graeca,* Vol. XVII (Berlin, 1888), pp. 567–569; German translation: W. Boehm, *Johannes Philoponus, Ausgewaehlte Schriften* (Munich, 1967), p. 92.

[106] *Or ha-Shem,* I, ii, 1 (2).

[107] Aristotle, *Physics* III, 5, 205b, 24–31; *Or ha-Shem,* I, i, 1 (2); Wolfson, *Crescas,* pp. 152–153, and nn. 65, 67.

[108] *Or ha-Shem,* I, ii, 1 (2); Wolfson, *Crescas,* pp. 194–195 and n. 49, where Wolfson refers to a similar position taken by Bruno. Aristotle had established that the heavens are neither heavy nor light in *De Caelo* I, 3.

hand, an infinite number of elements also cannot be supposed. For the infinite number of elements would require an infinite number of natural proper places to serve as terminuses for their natural motions, whereas the number of natural places must be finite, since it is determined by two absolute directions, namely, absolute up and absolute down.[109] There is thus, so the conclusion had been drawn, no way in which an infinite body might undergo natural rectilinear motion.

Crescas' response is that the infinite body might after all be composed of an infinite number of elements, and nevertheless the proper natural place of each of the elements could be assigned. An infinite number of natural places might "exist one above the other *ad infinitum*."[110] That is to say, just as both the natural proper place of air and the natural proper place of fire are, on the Aristotelian scheme of the universe, in the upper region, one above the other, so an infinite number of proper places can be conceived in the upper region, one above the other, *ad infinitum*.

The foregoing ground for the impossibility of an infinite body was the contention that such a body could not undergo rectilinear natural motion. That is now complemented by a sixth ground, the contention that such a body could not undergo the other form of natural motion recognized by Aristotelian physics, namely, circular motion. A number of separate considerations came into play here. Circular motion, Aristotle had reasoned, takes place around a fixed center; but an infinite body would, having no extremities, have no precise center, and hence could not undergo circular motion.[111] Further, circular motion implies a spherical body; but an infinite body could not be circumscribed, and hence would not have a spherical shape or any other shape.[112] Again, as an infinite body revolved, a point infinitely distant from the center would travel over an infinite path; but an infinite path is not traversable, and certainly not traversable in a finite time.[113] And, yet again, to assume the circular motion of an infinite body would involve a paradox. Should an infinite radius be drawn from the center and an infinite stationary line drawn parallel thereto, then as the sphere revolved, no point could be designated as the first meeting of the revolving radius with the stationary line. For no matter how distant a point we should propose for the first

[109] Aristotle, *Physics* III, 5, 205a, 10–32; 205b, 31–35; *Or ha-Shem,* I, i, 1 (3); Wolfson, *Crescas,* pp. 156–159. In nn. 96 and 97, Wolfson quotes the formulations of Averroes and Gersonides.

[110] *Or ha-Shem,* I, ii, 1 (3); Wolfson, *Crescas,* pp. 202–203. Crescas' response was suggested, and rejected, by Gersonides. See Wolfson, ibid., nn. 97, 98.

[111] *De Caelo* I, 7, 275b, 12–15; *Or ha-Shem,* I, i, 1 (4); Wolfson, *Crescas,* pp. 174–177. The point that the infinite has no extremities was added by Averroes; cf. the passage referred to by Wolfson in n. 158.

[112] Aristotle, *De Caelo* I, 5, 272b, 17–24; *Or ha-Shem,* I, i, 1 (3); Wolfson, *Crescas,* pp. 172–173 and nn. 145 and 148, where Wolfson cites pertinent passages in Averroes.

[113] Aristotle, *De Caelo* I, 5, 271b, 26—273a, 6; *Or ha-Shem,* I, i, 1 (3); Wolfson, *Crescas,* pp. 169–175 and nn. 132, 140, for diagrams. In Aristotle and in Crescas the argument appears several times, in several different formulations.

meeting of the two lines, we could always suppose a more distant point where the two lines might previously have met. If no point can be designated for the first meeting of the radius with the line parallel to it, the radius would be unable to revolve.[114]

Crescas responds to all these considerations. An infinite body, he maintains, could revolve without a precise center.[115] Further, a revolving body need not be spherical or have any other determinate shape. Therefore, although an infinite body would lack determinate shape, it would not be prevented thereby from undergoing circular motion.[116] Again, no given point in the infinite body would traverse an infinite path. For in an infinite body, the "distance [from the center[117] to points ever more distant on the radius] would increase [infinitely, only] in the manner in which number increases [infinitely],[118] that is to say, without ever ceasing to be limited." In other words, although the radius would extend outwardly without limit, every given point on the radius would remain finitely distant from the center and hence would describe a finite arc as it revolved. Crescas acknowledges that this explanation "is remote from the imagination; nevertheless, reason requires us to accept it."[119] As for the supposed paradox in circular motion, it, according to Crescas, is removed by an Aristotelian observation regarding motion and other types of change. Aristotle had noted that there is strictly "no beginning of change," no "first *when* in which a change occurs,"[120] and "no first *where* to which a change [of place] occurs."[121] Since the beginning of motion can in no instance be demarcated, we are, Crescas maintains, not justified in inquiring about the first point where an infinite revolving radius would meet an infinite line parallel to it.[122]

Finally, a seventh ground for the impossibility of an infinite body, a ground that had been adduced by Aristotle, is based upon the general Aristotelian theory of motion. An infinite body, the reasoning here ran, could not move itself. For only animate beings move themselves; animate beings have objects of sense perception outside them and surrounding them; and what is surrounded is finite.[123]

[114]Aristotle, *De Caelo* I, 5, 272a, 7–20; *Or ha-Shem,* I, i, 1 (3); Wolfson, *Crescas,* 172–173, and p. 387 for a diagram.

[115]*Or ha-Shem,* I, ii, 1 (4); Wolfson, *Crescas,* pp. 214–215 and n. 125, where Wolfson refers to a similar position taken by Bruno.

[116]*Or ha-Shem,* I, ii, 1 (3); Wolfson, *Crescas,* pp. 212–213. In n. 123, Wolfson cites a passage in Averroes which anticipates Crescas' response, and in n. 122, he refers to a passage in Bruno where that philosopher takes a position similar to Crescas'.

[117]As just seen, an infinite body would not in fact have a precise center.

[118]Cf. Aristotle, *Physics* III, 6, 206b, 16–19: The infinite "by way of addition" consists in the possibility of "always taking something *ab extra,* without, however, exceeding every [determinate] magnitude."

[119]*Or ha-Shem,* I, ii, 1 (3); Wolfson, *Crescas,* pp. 206–213.

[120]*Physics* VI, 5, 236a, 14–15.

[121]Ibid., 236b, 15–16.

[122]*Or ha-Shem,* I, ii, 1 (3); Wolfson, *Crescas,* pp. 210–213 and n. 120.

[123]*De Caelo* I, 7, 275b, 25–27; *Or ha-Shem,* I, i, 1 (4); Wolfson, *Crescas,* pp. 176–177. The

Nor could an infinite body be moved by something else. For only a commensurately infinite body would suffice to move the infinite body or act upon it in any way whatsoever. And to assume one infinite body's being acted upon by another is unacceptable for two reasons. In the first place, the other body would, being infinite, possess infinite power; and since the time required to complete an action is always inversely proportional to the power performing the action, an infinite power would produce instantaneous motion, whereas instantaneous motion is an absurdity.[124] In the second place, two actually infinite bodies could not exist together; for, as Averroes completed the train of reasoning, an actual infinite cannot be added to, and hence no actual magnitude at all can coexist with an infinite and add to it.[125]

Crescas responds that an infinite body might in fact move itself without "having objects of sense perception outside, surrounding it."[126] His meaning, apparently is not that the infinite body would be animate, but that it would be moved by a physical power contained within it[127] or else by its own nature.[128] As for the two reasons why an infinite body could not be moved by another infinite body, Crescas responds, in the first place, by proposing a hypothesis going back to John Philoponus. The hypothesis will be examined more fully in later connection,[129] but stated briefly it is this: Although the time required for any given operation is undoubtedly reduced as the power performing the operation is increased, every operation has, intrinsic to itself, a certain minimum "basic time," which is irreducible no matter how great a power might be brought to bear. An infinite power would accordingly perform each operation not instantaneously, but in the irreducible time intrinsic to the operation. Thus, to assume an infinite body containing infinite power would not entail the assumption of instantaneous action.[130] In the second place, Crescas responds that two infinite bodies could after all exist side by side. For infinite bodies need not be infinite in every direction; and the infinite may be added to on its finite side.[131]

The upshot of Crescas' critique of the various grounds whereby Aristotle and his followers had ruled out an infinite body is that an infinite body might in fact

---

consideration that "animate bodies have objects of sense perception outside them and surrounding them, etc." is Averroes' addition. See the passage referred to by Wolfson, ibid., n. 160.

For Maimonides' argument against an animate body's moving itself for all eternity, cf. above, pp. 246–247.

[124] Aristotle, *De Caelo* I, 7, 274b, 33—275b, 4; *Or ha-Shem,* I, i, 1 (3); Wolfson, *Crescas,* pp. 164–169.

[125] Aristotle, *De Caelo* I, 7, 275b, 27–29; *Or ha-Shem,* I, i, 1 (4); Wolfson, *Crescas,* pp. 176–177 and n. 160, where Wolfson refers to the pertinent passage in Averroes.

[126] *Or ha-Shem,* I, ii, 1 (4); Wolfson, *Crescas,* pp. 214–215.

[127] Cf. below, pp. 264–265.

[128] Cf. below, pp. 269–270.

[129] Below, pp. 262–263.

[130] *Or ha-Shem,* I, ii, 1 (3); Wolfson, *Crescas,* pp. 204–205.

[131] *Or ha-Shem,* I, ii, 1 (4); Wolfson, *Crescas,* pp. 214–215; above, n. 97.

exist. Such a body could either be comprised of natural elements of the sublunar kind or else have the nature of the celestial region, the latter possibility, the existence of an infinite celestial region, clearly being intended more seriously.[132] An infinite celestial body would not, Crescas has explained, be strictly spherical nor would it have any other definite shape. It would rotate without having a precise fixed center. It would be free of the qualities of the sublunar elements and would possess neither weight nor lightness. Its place could be assigned in whatever way Aristotelians might choose to assign the place of the outer celestial sphere which, in the Aristotelian system, is finite. And an infinite celestial body could contain infinite power distributed throughout it. It could, consequently, sustain its own circular motion for all eternity.

The object of Crescas' critique in the present section was, as must be kept in mind, to refute just a single premise in the argument intended to show that the motion of the spheres cannot be due to a power distributed through the body of the spheres. The argument had been that since all bodies are finite, every power distributed through a body must be finite; and therefore the eternal motion of the universe cannot be sustained by a power distributed through a body. Crescas, in the stage of his critique we have been examining, has rejected the premise that an infinite body is impossible. If an infinite body is after all possible, an infinite power distributed through a body is also possible. Even granting then that the motion of the universe is eternal and must be sustained by infinite power, the eternal motion of the universe might be sustained through the physical power distributed through the infinite body of the spheres or the infinite body of the universe.

Crescas' critique of the principle that a finite body can contain only finite power. Crescas has, he is confident, refuted two premises in the argument intended to show that the motion of the spheres cannot be due to a power distributed through a body; those premises are the doctrine of eternity and the impossibility of an infinite body. A third premise in the argument is the principle that a finite body can contain only finite power. The premise is again indispensable. For—conceding now the finiteness of the body of the universe, as well as the eternity of the world—if a finite body could contain infinite power, such a body could contain within itself the power needed to sustain motion for an infinite time. The first cause of motion in the universe, even if motion were eternal, could be an infinite physical power distributed through the finite body of the universe.

As is his custom, Crescas undertakes not to disprove the principle at issue, but only to refute the philosophic grounds for it. He does not, in other words, propose

[132] An infinite body comprised of elements of the sublunar kind could be assumed only if the distinction between the sublunar and celestial regions were eliminated; and Crescas does not seem willing to countenance that step seriously. The picture he is suggesting is that of a celestial region, dotted with stars and spheres, which encompasses the sublunar world and extends outwardly without limit.

to show that the principle is false and that finite bodies do in reality contain infinite power. He proposes to show merely that the principle is unfounded and that finite bodies might contain infinite power.

Aristotle's reasoning in support of the principle had been more intricate than the reasoning Maimonides was subsequently to employ, and it is with Aristotle's reasoning that Crescas is concerned. Aristotle had undertaken to reduce the assumption of an infinite power contained within a finite body to absurdity. On the one hand, he had held, an infinite power contained within a finite body could not perform its operation instantaneously, since an instantaneous physical operation is absolutely inconceivable. And, on the other hand, he had contended, an infinite power contained within a finite body could not operate over a finite time span either; for there is no time span that might be assigned for its operation. The inconceivability of instantaneous physical operations was so obvious to Aristotle that he could take it for granted.[133] His efforts were directed towards demonstrating that there is no time span assignable for the operation of an infinite power contained within a finite body.

Aristotle's reasoning ran: "Suppose *z* to be the time required for the infinite power to perform [a certain operation]." Since a finite power would surely require more time, we may suppose "*z* plus *y* to be the time required for a given finite power to perform the same operation." Now the time required to complete a fixed amount of work is inversely proportional to the power applied.[134] Should the given finite power be increased, the time—*z* plus *y*—needed to complete the operation in question would decrease. Eventually, by continuing to "increase the finite power . . . we can reach the level where it performs the operation in time *z*."[135] For example, if *z* plus *y* were twice *z*, by doubling the finite power we would reduce the time required to perform the operation by half. "The finite power would then require the same time that the infinite power was supposed to require." That a finite power and an infinite power should perform the same work in the same time—that they should, in other words, operate at the same velocity—is, however, absurd. And yet, no matter how brief a time span and how rapid a velocity might be supposed for the operation of the infinite power contained within the finite body, no matter how brief *z* might be, the situation would be unchanged. A finite power would in each instance be discoverable which can perform the same operation in the same time, at the same velocity; and a situation of the sort is absurd. To suppose that the infinite power performs the operation in no time, at an infinite velocity, was excluded at the outset as absolutely inconceivable. The conclusion accordingly drawn by Aristotle is that no time span or velocity whatsoever can be assigned for the operation of the infinite power. If there is no time span or velocity in which an infinite power contained within a

[133] Cf., however, *De Sensu* 6, 447a, 1–3.
[134] Cf. *Physics* VII, 5.
[135] The level will be reached when the finite power is multiplied by $\frac{z+y}{z}$.

finite magnitude can act, the supposed power simply could not act. And a power that does nothing is not a power. Thus a finite body can nowise contain an infinite power.[136]

Such were Aristotle's grounds for the principle that a finite body can contain only finite power. The key to Aristotle's reasoning is the rule that the time required for any operation is inversely proportional to the power applied, and Crescas addresses himself to that rule. Crescas does not dispute the empirical fact that as the power performing a given operation is increased, the time required to complete the operation decreases. He maintains, though, that the empirical fact does not necessarily imply that power and time stand exactly in inverse proportion. At least as plausible is a different hypothesis, going back to John Philoponus,[137] which was mentioned earlier in another connection.[138]

The alternate hypothesis is that each given physical operation has, intrinsic to itself, a certain minimum basic time span (*zeman shorshi*) for its occurrence, a time span that is irreducible no matter how great a power may be brought to bear. Stated otherwise, every physical operation has its own intrinsic maximum velocity which cannot be exceeded. When an operation is performed by a finite power, the minimum time span does not suffice, and additional time is required, with the amount of the addition dependent on the power. What is inversely proportional to the power performing a given operation is not the total time required for the operation, but rather the increment over the minimum time. For example, should the power be doubled, what would be reduced by half is not the total time of the operation, but the increment. As long as the power is multiplied finitely, the increment will be divided finitely, with some of it always remaining; and when the power reaches infinity the increment will be reduced to zero. "The infinite

[136] *Physics* VIII, 10, 266a, 24—266b, 6. There is an embarrassing side to Aristotle's argument. The argument appears to prove not merely that no infinite power in a finite magnitude could perform any operation, but more generally that no infinite power whatsoever could, and hence that no infinite power whatsoever can exist. Such, of course, is not the conclusion Aristotle was aiming at. He wished to affirm, on the contrary, that an infinite power does exist, to wit, the infinite power maintaining the heavens in motion, and he traced that infinite power to an incorporeal being. To salvage Aristotle's proof of the principle that a finite magnitude cannot contain infinite power, some sort of distinction would have to be drawn. The inconceivabilty of finite and infinite powers' performing the same amount of work in the same time would have to be restricted in some way to powers in bodies, and more precisely, as Aristotle formulates his proof of the principle, to powers in finite bodies. The power of an incorporeal being could then be subject to different rules, so that it could, with no absurdity ensuing, be assumed infinite and yet able to mete out its activity at the same velocity as a finite power contained in a body. A distinction of the requisite sort, between powers in bodies and powers of incorporeal beings, seems to be adumbrated by Simplicius, *Commentary on Physics*, pp. 1321, 1339; and such a distinction is worked out by Averroes, *Long Commentary on the Physics*, VIII, comm. 79.

[137] For Philoponus, see M. Clagett, *The Science of Mechanics in the Middle Ages* (Madison, 1959), pp. 433–434. The hypothesis was advanced in the Middle Ages by Ibn Bajja and was known to Crescas through Averroes' critique of that philosopher; See Wolfson, *Crescas*, pp. 403–408.

[138] Above, p. 259.

[power] would thus require no time beyond the irreducible minimum in order to produce motion, whereas every finite [power] will require some increment over the minimum."[139]

The hypothesis outlined avoids the twin horns of the dilemma posed by Aristotle. Instantaneous physical operations and infinite velocity, regarded by all as absolutely inconceivable, are avoided, since a finite time span and a finite velocity are assigned for every operation performed by the infinite power contained within a finite body. And the absurdity of a finite power's operating at the same velocity as an infinite power is also avoided. The infinite power and it alone would complete each operation within the irreducible minimum time intrinsic to the operation, whereas every finite power would require an increment over the minimum.[140] Aristotle's proof of the principle that a finite body can contain only finite power consisted in showing that no time span could be assigned for the operation of an infinite power contained within a finite body. By explaining how a time span can be assigned, with no absurdity ensuing, Crescas has refuted Aristotle's proof of the principle.

Crescas' object, we must recall once again, is to invalidate the argument showing that celestial motion and, in general, motion in the universe cannot be due to a power distributed through a body. In the present stage of his critique, Crescas has refuted the Aristotelian grounds for one more premise in the argument, the principle that a finite body can contain only finite power. The effect of his refutation is to permit ascribing infinite power to the finite universe, and hence to permit ascribing to the finite universe sufficient power for sustaining its own motion over an infinite time.[141]

The distinction between finiteness in respect to intensity and finiteness in respect to continuity. Crescas has refuted the grounds for three of the four premises in the argument showing that the ultimate mover of the spheres cannot be a power

[139] *Or ha-Shem,* I, ii, 8; Wolfson, *Crescas,* pp. 270–271.

[140] Crescas, *Or ha-Shem,* I, ii, 8, also proposes a variation of the hypothesis just outlined. A finite power, he writes, may in some instances be conceded to require no increment above the minimum time. In other words, a finite power may be conceded to move some objects in the same minimum time and at the same velocity that an infinite power requires. Yet the finite power will not necessarily be as efficacious. For even acknowledging those instances, we may conceive of a larger object to be moved, an object of such magnitude that the finite power would no longer move it in the minimum time but would now require an increment. The infinite power would, by contrast, never require more than the minimum time to move its objects, no matter how great they might be. According to Crescas, this variation of the hypothesis again invalidates Aristotle's proof of the principle that a finite body cannot contain infinite power. Aristotle's contention, as Crescas understands it, turns on the absurdity of a finite power's being fully as efficacious as an infinite power. In the present variation of the hypothesis, a velocity can again be assigned for the operation of the infinite power without that absurdity's ensuing, for although the finite power would in some instances operate as fast as the infinite power, it would not do so in all instances and hence would be less efficacious.

[141] *Or ha-Shem,* I, ii, 15.

distributed through the body of the spheres. He raises an additional objection: Even conceding the premises—the eternity of the universe, the finiteness of the body of the universe, and the principle that a finite body can contain only finite power—the argument is still invalid. For the argument ignores the distinction between finiteness in respect to *intensity* and finiteness in respect to *continuity*. The distinction between these two senses of finiteness was suggested to Crescas by Averroes, but Crescas utilizes the distinction differently.[142]

"The term *infinite*" he explains, "clearly may be used in two senses, both with respect to intensity and also with respect to time [or continuity]." The principle that a finite body can contain only finite power, granting now that Aristotle had successfully demonstrated the principle, affirms no more than that a finite body cannot contain power "infinite in respect to intensity." A power finite in intensity might, however, be infinite in respect to continuity, and so might suffice to sustain the motion of the heavens eternally.[143] The principle that a finite body contains only finite power therefore does not, Crescas objects, lead to the conclusion that the power distributed through a finite body would be insufficient to sustain eternal motion.[144]

Crescas presumably means the following: Aristotle's reasoning in support of the principle that a finite body can contain only finite power consisted in showing that no time span, in other words, no velocity, can be assigned for the operation of an infinite power contained in a finite body. Whatever time span might be proposed, a finite power could be discovered which would perform the same operation in the same time and at the same velocity;[145] and that would be absurd. But in Aristotelian physics, the intensity, not the continuity, of a power determines velocity.[146] That is to say, increasing the intensity of a power increases the velocity of its operations and decreases the time span required to complete each operation; increasing the time over which the power is applied has, by contrast, no effect on the velocity. To assert that increasing the time over which a power operates decreases the time required to complete the operation would, in the Aristotelian scheme, be nonsensical.[147] Aristotle's reasoning in support of the principle that a finite body can contain only finite power—reasoning which consisted in showing that no time span can be assigned for the operation of an infinite power contained in a finite body—is comprehensible, then, solely in reference to finiteness of intensity. Aristotle can at best be understood to have shown solely that

[142] Cf. below, p. 323. The distinction between infinite in intensity and infinite in continuity had also been drawn by Avicenna, *Najāt,* p. 128.

[143] *Or ha-Shem,* I, ii, 8; Wolfson, *Crescas,* pp. 272–273.

[144] *Or ha-Shem,* I, ii, 15. Cf. the account of Buridan's discussion of the principle in A. Maier, *Metaphysische Hintergruende der Spaetscholastischen Naturphilosophie* (Rome, 1955), pp. 257 ff.

[145] Cf. above, pp. 261–262.

[146] Continuing to apply power merely sustains velocity and does not produce acceleration. Cf. above, nn. 16, 29.

[147] See previous note.

no conceivable time span and no velocity can be assigned for the operation of a power of infinite *intensity* contained within a finite body, since a power of finite intensity could always be discovered which would perform the same operation in the same time and at the same velocity. Aristotle's reasoning nowise explains why a velocity cannot be assigned for the operation of a power of infinite *continuity* contained within a finite body. Aristotle may, of course, have regarded intensity and continuity as convertible, so that a power infinite in one respect would be infinite in the other as well. But he did not affirm, let alone defend, the convertibility of intensity and continuity. Hence Crescas' objection: Even granting the principle that a finite body contains only finite power, the grounds whereby Aristotle supported the principle can establish no more than that a finite body contains power finite in intensity; and a power finite in intensity might very well be infinite in continuity. The body of the spheres and the body of the universe might thus be finite in magnitude; the power contained within the body of the spheres and the body of the universe might be finite in intensity; and yet the power of finite intensity contained within the universe might suffice to sustain the motion of the spheres for all eternity. The ultimate cause of eternal motion may be construed as a physical power contained within and distributed through the finite body of the universe, even if all the premises in the Aristotelian argument intended to exclude the possibility be granted.

Circular motion may be natural to the spheres. Crescas raises yet another objection to the argument whereby Aristotle tried to show that the mover of the spheres cannot be a power distributed through the body of the spheres. The argument, he maintains, can be completely circumvented by construing the motion of the spheres as "natural to the body of the spheres just as rectilinear motion is natural to the elements."[148] Crescas' objection must be read against the background of positions taken by Aristotle and by ancient and medieval philosophers standing in the Aristotelian tradition.

"Nature," a well-known Aristotelian formula affirms, is "a certain principle and cause of [a thing's] being moved and being at rest."[149] The rectilinear motion of the four sublunar elements toward their proper places had, in accordance with the formula, been characterized by Aristotle as natural, inasmuch as the elements have within themselves "a principle of motion."[150] The circular motion of the celestial spheres had similarly been characterized by Aristotle as natural, inasmuch as the spheres too have a "principle of motion in themselves."[151] Yet side by side with the characterization of the motion of the elements as natural, Aristotle

[148] *Or ha-Shem,* I, ii, 8; Wolfson, *Crescas,* pp. 272–273.

[149] *Physics* II, 1, 192b, 21–22.

[150] *De Caelo* III, 2, 301a, 21 ff.; IV, 3, 310b, 24–25. In the former passage, Aristotle speaks of the "natural momentum (ῥοπή)" of the elements for moving up or down.

[151] *De Caelo* I, 2; II, 2, 285a, 30.

had undertaken to establish that the sublunar elements "are not moved by themselves."[152] That is to say, although the elements are moved *owing to* their nature, they are not moved *by* their nature. How the nature of the elements contributes to their motion is not clear in Aristotle.

Convinced as he was that the elements do not move themselves, Aristotle was hard put to discover what, external to the elements, does produce their rectilinear motion. The position he arrived at was that the factor originally bringing a given element into existence should be taken as the cause of the motion of the element to its proper place; for example, the motion of a portion of air to the upper region is to be ascribed to the factor bringing the portion of air into existence. And, Aristotle further explained, in instances where an element has already been brought into existence but is somehow prevented from moving to its proper place, the factor removing the obstacle is to be identified as the cause of motion.[153] These explanations account at best for the initiation of the movement of the elements. In the Aristotelian system, motion has to be sustained as long as it continues, and Aristotle's explanations fail to account for the continuation of the elements' motion as they make their way toward their proper places.[154]

Having traced the natural rectilinear motion of the elements to a cause distinct from them, Aristotle undertook to trace the natural circular motion of the celestial spheres to a cause distinct from the spheres. But since he understood the motion of the spheres to be eternal he did not, in their case, seek a cause that initiates motion. Rather, he sought the cause that sustains the motion of the spheres. In certain passages Aristotle explained that the motion of the celestial spheres is sustained by a soul.[155] In other passages he explained, as we have seen in the present chapter,[156] that celestial motion is sustained by an entity that has no magnitude and is nowise attached to a sphere, in other words, by a purely incorporeal being. Although the two accounts may appear divergent to modern scholars,[157] they were read by the ancient and medieval commentators as complementary.[158]

According to Aristotle, then, the elements and spheres are in motion—the former rectilinearly and the latter circulary—owing to their nature; yet the cause producing the motion of elements and spheres is not their nature but something

[152] *Physics* VIII, 4, 255a, 1 ff.

[153] Ibid., 256a, 1–3; *De Caelo* IV, 3, 311a, 9–11.

[154] Cf. G. Seeck, "Leicht-schwer und der Unbewegte Beweger," in *Naturphilosophie bei Aristoteles und Theophrast,* ed. I. Duering (Heidelberg, 1969), pp. 214–215.

[155] *De Caelo* II, 2, 285a, 29–30; 12, 292a, 20–21.

[156] Above, n. 49; also *Metaphysics* XII, 7.

[157] Cf. H. von Arnim, "Die Entstehung der Gotteslehre des Aristoteles," reprinted in *Metaphysik und Theologie des Aristoteles,* ed. F.-P. Hager (Darmstadt, 1969), pp. 1–74; W. Guthrie, "The Development of Aristotle's Theology," *Classical Quarterly,* XXVII (1933), 162–171; Ross's introduction to his edition of the *Physics,* pp. 94–100. Guthrie finds a less radical development than does von Arnim, and Ross finds a still less radical development than Guthrie.

[158] Alexander already read Aristotle in that way; cf. *Aporiai,* ed. I. Bruns, *Commentaria in Aristotelem Graeca,* Supplementary Vol. II/2 (Berlin, 1892), II, i, p. 3.

external. In the case of the rectilinear motion of the elements, Aristotle identified the cause responsible for the inception of motion, the initiating cause. The motion of a given portion of an element is initiated by the factor bringing the portion of the element into existence, or else by the factor removing the obstacle to its motion. Aristotle's writings reveal no attempt to identify what it is that sustains the motion of the elements as they continue to move. In the case of the motion of the spheres, by contrast, Aristotle had to identify only the cause responsible for the continuation of motion, that is, the sustaining cause. The circular motion of each sphere is sustained—as ancient and medieval philosophers read Aristotle—by an incorporeal being that interacts with the soul of the sphere.

Despite Aristotle's insistence that the sublunar elements do not move themselves but are moved by something external to them, a number of ancient and medieval students of his philosophy did discover a motive factor within the elements. And the motive factor was found to be, appropriately enough, the nature of the elements; nature had, after all, been characterized by Aristotle as a cause and principle of motion, with no exposition of the way in which it is a cause and principle.

The earliest work, as far as I know, that construes the nature of the elements as a motive cause is John Philoponus' commentary on Aristotle's *Physics*. Aristotle's definition of nature—the "principle and cause of [a thing's] being moved and being at rest"—was interpreted by Philoponus as meaning that natural substances such as the elements have the cause of their motion "in themselves." Hence when a heavy or light object is "released," its movement towards its proper place is brought about by its own nature and not by anything outside it.[159] Philoponus probably did not wish, hereby, to reject Aristotle's identification of the factor bringing the element into existence as the initiator of the element's motion.[160] He wished merely to ascribe to the nature of the element a function not treated by Aristotle, namely, the function of sustaining motion once the element exists and is free to move.

A statement to the effect that the elements are moved by their own nature appears, in the Middle Ages, in a corpus of Arabic works attributed to Alexander of Aphrodisias; unfortunately, however, the exact intent there remains obscure.[161] Subsequently, Avicenna, followed by Ghazali in his restatement of Avicenna's philosophy, asserted that the "nature" of the sublunar elements is the "cause" of their motion. And the context wherein Avicenna made the assertion indicates[162]

[159]Philoponus, *Commentary on Physics*, p. 195; cf. S. Pines, "A Refutation of Galen by Alexander," *Isis*, LII (1961), 48. Also cf. Simplicius, *Commentary on Physics*, pp. 1217, 1220, and Pines, ibid., pp. 39–40.

[160]He does eliminate the factor removing the hindrance to motion as in any sense a motive factor.

[161]*Mabādi' al-Kull*, pp. 253–254. Cf. Pines, "A Refutation of Galen by Alexander," pp. 27, 32, 42–48; H. Wolfson, "The Problem of the Souls of the Spheres," reprinted in his *Studies in the History of Philosophy and Religion*, Vol. I (Cambridge, Mass., 1973), p. 29.

[162]Since Avicenna and Ghazali mention no cause apart from *nature*, they apparently recognized no such cause.

that he, followed by Ghazali, was construing the nature of the elements as the exclusive cause of their motion. In other words, each element is not merely sustained in motion, but is set in motion as well, by its own nature.[163] Avicenna acknowledged, of course, that the elements are brought into existence by something different from themselves.[164] Yet he demurred, so it appears, at Aristotle's identification of the factor bringing an element into existence as the initiator of its motion.[165]

Avicenna's position that the nature of each element is responsible for all aspects of the element's motion was not generally adopted. But from the time of Averroes onward, the nature of the elements was repeatedly credited with an active function in the elements' natural motion. The motion of an element was commonly said to be initiated, indeed, by the factor bringing it into existence; but once the element does exist, the factor sustaining its motion until it reaches its proper place was understood to be the element's inner nature. A theory of the sort was offered in divers versions by Averroes,[166] by the Jewish philosopher Isaac Albalag,[167] and by a line of Latin writers extending from Albertus Magnus to the fourteenth-century Scholastics.[168]

Crescas for his part follows the lead of Avicenna in dispensing completely with external factors to explain the motion of the elements to their proper place. What causes each element to move up or to move down to its proper place is, Crescas maintains, "the proper form" of the element, which operates through the "medium of the power . . . called nature."[169] That is to say, the form of the element, operating through the nature of the element, is the sole cause of the element's motion.

Such was Crescas' position on the natural motion of the sublunar elements and his precedents.

There were in addition at least two precedents, both of them conscious departures from Aristotle, for explaining the circular motion of the celestial spheres as due exclusively to the nature of the spheres, thereby dispensing with the soul

[163] Avicenna, *Najāt,* pp. 108–109; Ghazali, *Maqāṣid,* p. 239. Cf. Shem Ṭob's commentary on *Guide,* II, introduction, prop. 17; Wolfson, *Crescas,* p. 673; Pines, "A Refutation of Galen by Alexander," p. 51.

[164] They are brought into existence through a form emanated from the active intellect.

[165] There is a bit of hairsplitting here, since the cause of the existence of the element and of its nature would have to be recognized as an ultimate cause of the motion of the element.

[166] *Epitome of De Caelo,* in *Rasā'il Ibn Rushd* (Hyderabad, 1947), p. 68; *Long Commentary on De Caelo,* in *Aristotelis Opera cum Averrois Commentariis,* Vol. V (Venice, 1562), III, comm. 28.

[167] Isaac Albalag, *Sefer Tiqqun ha-De'ot* (Commentary on Ghazali's *Maqāṣid*), ed. G. Vajda (Jerusalem, 1973), pp. 102–103.

[168] A. Maier, *An der Grenze von Scholastik und Naturwissenschaft* (Rome, 1952), pp. 156–170. Maier finds that until Duns Scotus, the nature of the elements was construed as only an "instrument" of the factor bringing the element into existence. Also cf. J. Weisheipl, "The Principle *Omne quod movetur ab alio movetur,*" *Isis,* LVI (1965), 39–41.

[169] *Or ha-Shem,* I, i, 17; Wolfson, *Crescas,* pp. 298–299 and n. 7. Aristotle, *De Caelo* III, 2, 301b, 18, distinguishes nature from power.

of the spheres and with the incorporeal intelligences. Aristotle, as already seen, had stated that the celestial spheres perform their circular motion thanks to their nature.[170] Undoubtedly with that statement in mind, John Philoponus contended in a nonphilosophic tract that "God" may very well have furnished the heavenly bodies with "a power of movement" analogous to the "momentum"[171] whereby the sublunar elements travel to their proper places.[172] In other words, the power actively moving the spheres may be their nature. Ghazali offered a similar suggestion when he turned from his restatement of Avicenna's philosophy to a critique of Avicenna. The body of the heavens, Ghazali now contended, may "have been created, . . . possessing in itself a factor that necessitates circular motion." The factor in question would be "a principle of motion," that is, a nature,[173] and its character would be "analogous to what the philosophers believe regarding the downward motion of a stone."[174] In these passages, Philoponus and Ghazali exploited the Aristotelian notion of nature to develop an anti-Aristotelian stand. They assumed the creation of the heavens by God. And they contended that, once created, the heavens may well be moved—whether set in motion or only sustained in motion is not made explicit—exclusively by their nature.

In the passages quoted, Philoponus and Ghazali had exchanged philosophic for theological garb and were approaching the issue of celestial motion theistically. Their contention was that once God created the celestial spheres, the nature of the spheres could suffice to move them. Crescas is approaching the issue of celestial motion from a different direction, and as a result he goes further. He, like Aristotle and Maimonides, is addressing the question: What could sustain the motion of the celestial spheres on the assumption that the spheres and their motion are eternal? The theory Crescas advances is that the eternal motion of the spheres can be construed as "natural to the spheres," just as the "rectilinear motion of the elements" is "natural" to the elements.[175] That is to say, just as the rectilinear motion of the sublunar elements to their proper places can, on the position of Crescas' outlined previously,[176] be ascribed exclusively to the nature of the elements, with no role assigned to external factors, so too the circular

[170]Above, pp. 265–266.

[171]Cf. above, n. 150.

[172]*De Opificio Mundi*, ed. G. Reichardt (Leipzig, 1897), p. 29, and also pp. 231–232; German translation: Boehm, *Johannes Philoponus, Ausgewaehlte Schriften*, pp. 333–334, 335–336. Cf. Wolfson, "The Problem of the Souls of the Spheres," p. 26; Pines, "A Refutation of Galen by Alexander," p. 50.

[173]Cf. above, n. 149.

[174] Ghazali, *Tahāfut al-Falāsifa*, ed. M. Bouyges (Beirut, 1927), XIV, §11; English translation in Averroes' *Tahafut al-Tahafut*, trans. S. van den Bergh (London, 1954), p. 290.

[175]*Or ha-Shem*, I, ii, 15; also ibid., I, i, 6; Wolfson, *Crescas*, pp. 236–237. Crescas, ibid., I, ii, 15, acknowledges that the analogy is not exact since "in the case of the elements, to rest in their proper place is natural, whereas in the case of the celestial bodies, to move in their proper place is natural."

[176]Above, p. 268.

motion of the spheres can be ascribed exclusively to the nature of the spheres, and no additional factor need be introduced. Ascribing the motion of the spheres exclusively to their nature would sidestep the argument to the effect that the power contained within the finite body of the universe cannot sustain eternal motion. It may be granted, Crescas is contending, that motion is eternal, that the universe is finite, that the power contained within the body of the universe is also finite, even that the power contained within the universe is finite in respect to continuity. Conceding all this, the argument that a power distributed through a finite body would not suffice to sustain eternal motion can yet be circumvented by construing the motion of the spheres as natural to the spheres. The nature of the spheres is present and operative as long as the form is present,[177] that is, as long as the spheres exist. Hence the nature of the spheres can move the spheres as long as they exist, whether it be for a finite or infinite time. And no incorporeal mover beyond the spheres need be posited.

Resumé. We have been considering the argument intended, in Maimonides' version of the proof from motion, to eliminate one conceivable explanation of the motion of the celestial spheres, the alternative that the cause of celestial motion is a power distributed through a body. The argument had been that no body can contain infinite power, whereas finite power cannot sustain eternal motion.

Crescas undertook to refute three of the premises in the argument. A refutation of the first premise, the doctrine of eternity, would not affect the final outcome. For the function of the argument was to serve as a step in an overall proof of the incorporeality of the first cause, and denying eternity and thereby affirming a beginning of the world leads to the incorporeality of the first cause by another route.[178] The refutation of a second premise, the principle that only finite bodies exist, would invalidate the argument. For if the heavens and the universe as a whole should be infinite, they would contain infinite power, and infinite power would suffice to sustain eternal motion. The refutation of a third premise, the principle that a finite body contains only finite power, would also invalidate the argument. For if a finite body should contain infinite power, that power, again, would suffice to sustain eternal motion.

Having objected that the argument is invalid since two—or, including the doctrine of eternity, three—premises are unfounded, Crescas contends further that the argument remains invalid even when all the premises are granted. Aristotle's reasoning supporting the principle that a finite body contains only finite power would, if correct, establish, no more than that a finite body contains power finite in intensity. Aristotle's reasoning nowise establishes that a finite body contains power finite in continuity. Therefore, even conceding the principle that a

[177] See above, n. 169.
[178] Above, n. 57.

finite body contains only finite power, such power, finite though it is in intensity, may still suffice to sustain eternal motion. Finally, Crescas contends, even if yet another concession is made and the validity of the entire argument is granted, the argument can still be circumvented by identifying the cause of the eternal motion of the spheres as their nature.

(iii) The alternative that the first mover of the spheres is a power attached to, but not distributed through, the body of the spheres

The first conceivable explanation of the movement of the spheres had been the hypothesis that the spheres are moved by another body. The argument whereby that alternative was ruled out in Maimonides' version of the proof from motion and Crescas' reply thereto were examined earlier.[179] The second conceivable explanation was the hypothesis that the mover of the spheres is a power distributed through a body. The argument whereby that alternative was ruled out and Crescas' intricate reply have just been examined. The third conceivable explanation is the hypothesis that the mover of the spheres is a power present within, although not distributed through, the body of the spheres, in effect, a soul of the spheres.

Maimonides' version of the proof from motion had eliminated the third alternative through an argument that ran: The motion of the spheres is by hypothesis eternal. But a power in a body, even if, like a soul, not distributed throughout the body, would undergo motion accidentally; the accidental has the possibility of not existing, and hence is not eternal; the motion of a power attached to a body must, therefore, eventually cease; and the motion of the body inextricably associated with a power whose motion must cease must also eventually cease.[180]

The argument has a flaw even when considered from a purely Aristotelian point of view, inasmuch as it ignores the differing senses of *accident*. Accident, it is true, was defined by Aristotle as "what is present in something . . . but neither necessarily so nor for the most part."[181] The accidental in this sense, it may cogently be reasoned, has the possibility of not existing and, by the principle that every possibility is eventually realized, could not be eternal. Maimonides established that the motion of a power attached to a body is accidental in a certain sense, in the sense that such a power cannot undergo motion in and by itself, or *essentially*—only bodies can—yet is moved by being associated with something that does undergo motion in itself and essentially. To complete Maimonides' argument, the sense in which the motion of the soul is accidental would have to be shown to be identical with, or to imply, the other sense of accidental. Motion that is accidental in the sense of being an accompaniment of essential motion

[179] Above, pp. 243–244, 250–251.

[180] Above, pp. 246–247. See n. 62.

[181] *Metaphysics* V, 30, 1025a, 14–15. Significantly, accident in a second sense, ibid., 31–33, is that which "belongs to each thing by virtue of itself but is not in the essence . . . ; accidents of this sort may be eternal." What is accidental in the second sense obviously does not have the possibility of not existing.

would have to be shown to be accidental in the sense of not being necessary; only then would there be cogency in the contention that the motion of a soul of the spheres must, because it is accidental, have the possibility of not existing and accordingly must eventually cease.

Crescas states an objection along analogous lines. He writes: "What exists by accident has the possibility of not existing" solely in instances "where it does not result necessarily from something existing essentially [and eternally]." By contrast, "accidental things that do result necessarily from essential things" will "be eternal, should the latter be eternal." Now "when we ascribe accidental motion to a soul of a sphere," we do so only insofar as the soul is "connected to the sphere." That is to say, the motion of the soul is accidental only in the sense that a soul is moved not in and by itself, but thanks to its presence in a body; the motion of the soul is not accidental in the critical sense of not being necessary. There is consequently no reason to suppose that the motion of the soul of the sphere has a possibility of ceasing.[182] The soul of a celestial sphere might thus very well move its sphere unceasingly. Since the motion undergone by the soul would be the necessary concomitant of an unceasing essential motion—the motion of the sphere which is produced by the soul itself—it too would be unceasing and necessary. And we would have a situation in which the motion of the soul of the sphere, although accidental in a certain sense, is yet necessary and eternal, and hence a situation in which the soul eternally performs its task of moving the sphere.

(iv) Summary of Crescas' critique of the argument for the incorporeality of the first mover

To establish the incorporeality of the first mover of the universe and complete the proof from motion, three alternative explanations of the movement of the celestial spheres had to be ruled out. The first conceivable explanation of the movement of the spheres was the hypothesis that the mover is a body which is moved by another body, and so on *ad infinitum*. Crescas ostensibly rejects the argument designed to rule out this alternative, by defending the possibility of an infinite number of bodies. In fact, however, Crescas acknowledges that every series of causes can be traced back to a first cause. He thereby acknowledges that philosophic reasoning can establish a first mover which is moved by nothing else, and hence a mover that is not a body, all bodies being moved by something apart from themselves.[183]

The second conceivable explanation of the motion of the spheres was the hypothesis that the first mover of the universe is a power distributed through a body. Crescas contends that two—or three—indispensable premises in the argument ruling out the hypothesis are unfounded; that even granting all the premises,

[182] *Or ha-Shem,* I, ii, 5; Wolfson, *Crescas,* pp. 248–253 and nn. 6, 8. Aristotle, *Physics* VIII, 6, 259b, 28–31, seems to be saying the same.

[183] Above, p. 251.

the argument ruling out the hypothesis remains invalid, since a power finite in intensity might be infinite in continuity; and that even granting the validity of the entire argument, the argument can still be circumvented by identifying the factor moving the spheres as the nature of the spheres.[184]

The third conceivable explanation of the eternal motion of the spheres is the hypothesis that the spheres are sustained in their eternal motion by a soul or souls attached to them. Crescas contends that the argument designed to rule out the hypothesis is again not cogent; for although the soul of the spheres would undergo motion that is accidental in a certain sense, such motion would not be accidental in the specific, critical sense implying noneternity.

The upshot of Crescas' critique is that Aristotelian argumentation cannot establish the correctness of the fourth alternative, the thesis that the motion of the spheres and of the universe as a whole is due to an incorporeal being independent of the spheres. The first mover might be an inanimate power distributed through the body of the spheres, or the nature of the spheres, or the soul of the spheres. The proof from motion thus fails to demonstrate the incorporeality—or transcendence, if one will—of the first mover, and consequently fails as a genuine proof of the existence of God. Crescas' critique incidentally discloses the extent to which the argument or arguments for the incorporeality of the first mover are grounded in the presuppositions of Aristotelian physics. The arguments can be accomplished only by granting, *inter alia,* Aristotle's definition of *body* and of *place,* his theory of the physical elements and natural place, his general theory of motion, and his theory of natural rectilinear and natural circular motion.

(c) The unity of the first mover

Aristotle had established the unity of the first mover in two ways, by inferring the attribute of unity from the attribute of incorporeality, and by inferring the unity of the mover from the unity of the universe and the attendant unity of its motion.[185] Crescas replies explicitly to the former inference and thereby also suggests a reply to the latter.

In the former Aristotle had reasoned: Since "all things that are many in number have matter," the first mover, having no matter cannot be many in number.[186] Crescas starts by rejecting the conclusion on the basis of his earlier refutation of the argument for the incorporeality of the first mover: If the incorporeality of the first mover is not demonstrable, the unity of the first mover obviously cannot be derived from its incorporeality.[187]

But Crescas goes further. He contends that even if the incorporeality of the first mover should be granted, unity still cannot be inferred from incorporeality. Here, as elsewhere, his contention is rooted in positions taken by his predecessors.

[184] Above, p. 270.
[185] Above, p. 248.
[186] Above, n. 24.
[187] *Or ha-Shem,* I, ii, 15 (implied).

Aristotle, at least in certain passages, and medieval Aristotelians universally, recognized a multiplicity of incorporeal beings, the incorporeal movers of the spheres or so-called intelligences.[188] A multiplicity of incorporeal beings could be countenanced only by qualifying the rule according to which objects are differentiated solely through their matter.[189] The qualification best known to Crescas and the one to which he addresses himself is the qualification formulated by Maimonides. As was seen earlier, Maimonides explained that incorporeal beings can be differentiated from each other and enumerated in a single instance, when they stand in the relationship of "causes and effects."[190] Since the incorporeal movers of the spheres form, in Maimonides' picture of the universe, a causal series in which each link emanates the next, his version of the rule regarding the differentiation of beings is compatible with the multiplicity of movers of the spheres: The movers of the spheres, although incorporeal, are differentiated from each other inasmuch as they form a series of causes and effects. And yet, the rule, as qualified by Maimonides, still permits the unity of the first mover to be inferred. For if incorporeal beings are distinguishable only when they stand to each other in the relation of cause and effect, there can exist no more than one incorporeal being that is a cause without being an effect. Maimonides' version of the rule for the differentiation of beings thus accommodates a multiplicity of incorporeal beings, while allowing, so Maimonides thought, the unity of the first mover to be inferred from the first mover's incorporeality.

Crescas demurs. The qualified rule for differentiating objects is compatible, he objects, not merely with a multiplicity of incorporeal beings, but also with a multiplicity of incorporeal first movers. A multiplicity of first movers would be conceivable, should "one be [understood to be] the cause of one effect and another [understood to be] the cause of another effect." That is to say, uncaused causes can be differentiated from each other even though one uncaused cause obviously cannot stand to another in the relation of cause to effect. Uncaused causes can be differentiated on the supposition that each has its own proper effect. To be specific, should the emanationist scheme that had been grafted adventitiously on Aristotle's cosmology be disregarded, the movers of the several celestial spheres would not stand to each other in a causal relation, yet would be differentiated inasmuch as "one is the cause of one sphere and another is the cause of another sphere." Each incorporeal mover would be a first cause in its own right, the several first causes would be differentiated, and the universe would not depend on a single deity.[191] Alternatively, a multiplicity of unconnected physical universes might conceivably exist. Aristotle had, to be sure, undertaken to demonstrate the impossibility of more than one universe;[192] but Crescas, in another

[188]Cf. *Metaphysics* XII, 8; above, n. 25.

[189]Various qualifications of the rule are discussed by Wolfson, "The Plurality of Immovable Movers," pp. 8–13.

[190]Above, n. 70 [principle (19)].

[191]*Or ha-Shem,* I, ii, 15.

[192]*De Caelo* I, 8–9.

connection, answered Aristotle on that issue and defended the possibility of a multiplicity of universes.[193] Should several universes in fact exist, the first cause of each would be differentiated from the first cause of the others by virtue of its having a different effect. Each universe would have its own first cause, and the totality of universes would not depend on a single deity.[194]

The foregoing is Crescas' reply to the argument that the first mover must be one because it is incorporeal. The other Aristotelian argument for the unity of the first mover consisted in inferring the unity of the mover from the unity of the universe and its motion. By defending the possibility of a multiplicity of physical universes, Crescas suggests an appropriate reply to this argument also: If a number of distinct universes exist, a number of first movers would likewise exist, each universe having its own mover.

To summarize: Crescas maintains that the unity of the first mover cannot be inferred from the attribute of incorporeality since the incorporeality of the first mover had not been demonstrated successfully. Furthermore, even granting the incorporeality of the first mover, the attribute of unity still does not follow, since a multiplicity of incorporeal first movers might be differentiated by virtue of having as their effects either different parts of a single universe or independent universes. In addition, Crescas' defense of the possibility of several unconnected universes suggests an answer to Aristotle's other argument for the unity of the first mover, the inference of the unity of the first mover from the unity of the universe.

Crescas' analysis of the proof from motion reveals that the proof is grounded in Aristotle's physics, its presuppositions are the presuppositions of Aristotle's physics, and many of those presuppositions can be overturned from within the Aristotelian system. Moreover, even when all the premises in the proof are granted, the hoped for conclusions can still not legitimately be drawn. As a consequence, the proof completely fails to establish two of the three theses required for a genuine demonstration of the existence of God. It fails to establish that only a single first mover exists and it fails to establish that the first mover or movers are incorporeal or even animate.

4. Another proof from motion

The first thesis in the proof from motion is the existence of a first unmoved mover,[195] and the primary consideration adduced by Aristotle in arriving at a first mover was, as seen earlier, the impossibility of an infinite regress of causes.[196] Aristotle adduced several secondary considerations as well, and among them is

[193] *Or ha-Shem,* I, ii, 1 (4); Wolfson, *Crescas,* pp. 216–217.
[194] *Or ha-Shem,* I, ii, 15.
[195] Above, pp. 241–242.
[196] Above, n. 35.

what has been called the argument from "logical symmetry."[197] That is an argument resting on a principle asserting: When two things are known to exist in combination with each other, and when in addition one of the two is known to exist separately, then the other must likewise be assumed to exist separately. The principle is illustrated by the Greek commentators through the example of honey-water[198] or honey-wine.[199] Honey-water or honey-wine is known to exist as a compound, and honey is in addition known to exist separately; therefore the second component in the compound must, so the thinking went, also be assumed to exist separately.

The necessity of qualifying the principle was pointed out by the Greek commentators. Alexander observed that the principle is not valid in the case of compounds of substance and accident.[200] That is to say, although a given substance and a given accident exist together in a compound, and although, further, the substance may, by its nature, exist independently of accidents,[201] nevertheless the accident may not be concluded capable of separate existence. Simplicius too observed that the principle is not applicable universally and, quoting an expression used by Aristotle in connection with it,[202] he characterized the principle as no more than "reasonable." An exception was discovered by Simplicius in the realm of speech. In speech, vowels and consonants are sometimes pronounced together, and in addition, vowels are sometimes pronounced separately; nevertheless, consonants can never be pronounced separately.[203]

The principle that when one element in a compound exists separately the other must likewise exist separately was, as already mentioned, adduced by Aristotle as a secondary consideration in the course of the proof from motion. Aristotle's precise argumentation remains slightly blurred because of textual uncertainties,[204] but for our purposes the way in which subsequent philosophers read him will suffice. Aristotle was typically understood to have argued: Sense perception testifies in the first place to the existence of something that both undergoes motion and produces motion, and in the second place, to the existence of something that

[197] T. Gomperz, *Greek Thinkers*, Vol. IV (London, 1912), p. 219; Wolfson, "Notes on Proofs of the Existence of God," p. 573.

[198] Alexander, quoted by Averroes, *Long Commentary on Metaphysics*, XII, comm. 35; German translation in J. Freudenthal, *Die durch Averroes erhaltenen Fragmente Alexanders zur Metaphysik des Aristoteles* (Berlin, 1885), pp. 107–108.

[199] Themistius, *Paraphrase of Physics*, ed. H. Schenkel, *Commentaria in Aristotelem Graeca*, Vol. V/2 (Berlin, 1900), p. 223; Simplicius, *Commentary on Physics*, p. 1227.

[200] See n. 198. Aquinas makes a similar statement, *Commentary on Physics*, VIII, §1044.

[201] In fact, material substances are never devoid of accidents, and therefore Alexander's point is weak or at least difficult to bring out. Aquinas, in the passage referred to in the previous note, writes that the principle of logical symmetry is not applicable to accident and substance, since "accident is in substance *per se*," whereas the contrary is not true.

[202] Aristotle, *Physics* VIII, 5, 256b, 23.

[203] Simplicius, *Commentary on Physics*, p. 1227.

[204] See Ross's apparatus and note to *Physics* VIII, 5, 256b, 20–24.

undergoes motion without producing motion. Accordingly it is, in Aristotle's words, "reasonable if not necessary" to conclude that something also exists which produces motion without undergoing motion.[205]

The argument from logical symmetry establishes, at best, an uncaused cause of motion. But no argument for the existence of an uncaused cause would, as we know, reach the level of a complete proof of the existence of God until supplemented by arguments for incorporeality and unity.[206] In Aristotle, the argument from logical symmetry did not require special supplementation, because of the context in which it was employed. It had served as nothing more than a subsidiary consideration for reaching a first mover; and once a first mover had been reached through it or through the primary Aristotelian consideration that causes cannot regress infinitely, Aristotle had completed his overall demonstration through the arguments for the incorporeality and unity of the first mover which were examined in earlier sections of the present chapter.[207]

Alexander and Themistius proposed another procedure by which a first mover could be shown to be incorporeal, explaining that an unmoved mover, being subject to no motion or change, would be free of potentiality, and hence free of matter.[208] Joining this other explanation of the incorporeality of the first mover to the argument from logical symmetry gave rise to a new, not overly profound, proof of the existence of God. An unmoved mover was established through the principle of logical symmetry; and the unmoved mover was found to be incorporeal because free of potentiality.[209] The unity of the unmoved mover still remained to be inferred, but, as was seen in connection with the standard proof from motion, the attribute of unity is derivable from the attribute of incorporeality.[210] The existence of a single, incorporeal, unmoved mover—a deity—could thus be demonstrated.

In the Middle Ages, the argument from logical symmetry is not frequent, but it is used. The corpus of Arabic works attributed to Alexander has it in two guises, neither of which, however, seems intended as a complete proof of the existence of God, their function being rather to exhibit God's attributes. The argument in its first guise follows Aristotle's example and establishes an unmoved mover.[211] The alternate guise looks at purely actual, as distinct from potential,

[205] Aristotle, ibid. Cornford, in the Loeb edition of the *Physics,* Vol. II, p. 336, observes that the passage seems out of place. The argument also appears in Aristotle, *Metaphysics* XII, 7, 1072a, 23–26.

[206] Above, pp. 3, 242.

[207] Above, pp. 244, 248.

[208] Cf. Aristotle, *Metaphysics* XII, 2, 1069b, 24–25.

[209] Alexander, in the passage cited in n. 198; Themistius, *Paraphrase of Metaphysics,* interpreting *Metaphysics* XII, 7, 1072a, 23–26. Arabic text in *Arisṭū 'ind al-'Arab,* p. 14; Hebrew text in *Commentaria in Aristotelem Graeca,* Vol. V/5, ed. S. Landauer (Berlin, 1903), p. 16.

[210] Above, p. 248.

[211] Alexander, *Mabādi' al-Kull,* in *Arisṭū 'ind al-'Arab,* pp. 257–258.

being and runs: Something is discovered to exist which is purely potential, to wit, prime matter;[212] and other things are discovered to exist which are both potential and actual; consequently, something must exist as well which is purely actual.[213]

In Maimonides, the argument from logical symmetry gains the status of a full-fledged proof of the existence of God and is enumerated as the second of four such proofs, the standard proof from motion having been enumerated by him as the first. From "Aristotle," Maimonides adduces the "principle . . . that if something is found composed of two things, and in addition one of those things is found by itself apart from the composite, then the other must also exist apart from the composite." As an illustration, Maimonides cites the example of "honey-vinegar."[214] He thereupon proceeds to apply the principle to the phenomenon of motion. There exist things "composed of mover and moved," that is to say, things producing motion while themselves undergoing motion; and things that are "moved without moving anything"; whence Maimonides concludes that in addition "something must exist which moves without being moved," that is to say, a "first mover." Since what does not undergo motion is "indivisible and incorporeal,"[215] the unmoved mover must be incorporeal.[216] Maimonides' readers are left to infer the unity of the first mover from the attribute of incorporeality; the reasoning whereby that inference can be drawn had already been spelled out by Maimonides in his statement of the standard proof from motion.[217]

In scholastic philosophy a version of the argument from logical symmetry is employed by William of Auvergne. From the presence in the universe of beings that both bestow and receive existence, and of other beings that only receive existence, William concludes that something must exist which only bestows existence.[218] Albertus Magnus, who knew Maimonides' philosophy well, employs the argument from logical symmetry to establish the existence of an unmoved mover. And he characterizes the argument as the "strongest proof of the existence of God."[219]

Crescas' critique

Maimonides had not merely cited the principle underlying the argument from logical symmetry. He had justified it, explaining that if the two components in question were such that "they can exist exclusively in combination, . . . neither

[212]On the Aristotelian physical scheme, prime matter is never in fact found in isolation.

[213]Alexander, *Mabādi' al-Kull*, p. 272.

[214]Cf. above, p. 276.

[215]Cf. above, pp. 247-248, principle (16).

[216]*Guide*, II, 1 (2).

[217]*Guide*, II, 1 (1).

[218]S. Schindele, *Beitraege zur Metaphysik des Wilhelm von Auvergne* (Munich, 1900), pp. 47–49.

[219]Albertus Magnus, *De Causis et Processu* (part of his *Parva Naturalia*) I, 1, 7; 9–10.

of them would under any circumstances be able to exist independently of the other. The fact that one of the two does exist separately proves that there is no necessary interconnection between them. Consequently, the other too must exist separately."[220]

Crescas responds that the logic is faulty. If one of two components in a compound exists separately and accordingly is not bound to the second component, the latter may legitimately be judged to have the "possibility of existing" separately. Nothing, however, justifies concluding that the latter actually does exist separately. Therefore, Crescas maintains, the principle at issue, when applied to motion, will at most permit inferring that an unmoved mover may exist, not that an unmoved mover actually does exist.[221]

A type of composite is moreover pointed out by Crescas—who is now repeating an observation made by Gersonides—where the principle is not at all applicable, where empirical evidence shows that not even the possibility of separate existence can be supposed. In the hierarchy of nature, each lower stage is discovered to exist independently of the one above it, whereas the higher cannot exist except in conjunction with the one below. Plainly, the separate existence of the lower stage, taken together with the combined existence of the lower and higher stages does not permit the conclusion that the higher too exists separately. For example, objects exist which possess the vegetative faculty alone, and others exist which possess both the vegetative and animal faculties; but nothing exists which possesses the animal faculty without the vegetative faculty.[222] Crescas, following Gersonides, is suggesting that the attempted inference of an unmoved mover falls precisely in the area where the principle does not apply. The characteristic of only undergoing motion is at a lower stage than the characteristic of only producing motion. If the principle at issue does not permit inferring the separate existence of a higher stage from the separate existence of a lower stage, the principle does not permit inferring the existence of something that produces motion without undergoing motion from the existence of things both undergoing and producing motion, taken together with the separate existence of things only undergoing motion.[223]

[220]*Guide,* II, 1 (2).

[221]*Or hạ-Shem,* I, ii, 16. For the criticism to hold, Crescas would have to explain why the rule that every possibility is eventually realized does not apply.

[222]Ibid. See Gersonides, *Milḥamot ha-Shem* (Leipzig, 1866), V, iii, 6, p. 267. Gersonides is refuting the use of the argument from logical symmetry to establish the existence of the movers of the spheres and not strictly its use as a proof of the existence of God.

[223]Crescas offers a further criticism, which appears differently in the two editions and in the manuscript at my disposal. I was not able to reach a satisfactory interpretation of any of the versions.

The argument from logical symmetry might also be challenged by maintaining that nothing at all does exist which is moved without moving something else. For objects that seem to be moved without moving anything else do, in fact, move the air around them; and each portion of air moves the portion beyond it.

In fine, the underlying principle of the argument from logical symmetry, which the Greek commentators already insisted must be restricted,[224] is rejected by Crescas on both logical and empirical grounds. And Crescas finds that the argument from logical symmetry, like the standard proof from motion, fails as a demonstration of the existence of God. The standard proof failed to establish the incorporeality and unity of the first mover. The argument from logical symmetry fails to establish even the existence of a first unmoved mover.

[224] Above, p. 276.

IX

Avicenna's Proof of the Existence of a Being Necessarily Existent by Virtue of Itself

1. First cause of motion and first cause of existence

Aristotle's proof from motion, ostensibly at least, sought to establish a first incorporeal cause solely of the motion of the universe, not of the existence of the universe. For its existence, the universe would, in Aristotle's system, be self-sufficient, depending on no further cause outside itself. Plotinus subsequently expounded the conception of a cause of the very existence of the universe,[1] and later Neoplatonist philosophers draw the contrast between a first cause of existence and a first cause of motion. They also introduce the possibility of a cosmological proof establishing a cause of the former sort.

We find the contrast between a first cause of existence and a first cause of motion in Proclus, who draws it in order to distinguish Plato's position from the position of Aristotle and his followers. Plato, according to Proclus, had maintained that the entire universe must depend for its existence on an "efficient" cause, whereas Aristotle and his school failed to recognize an efficient cause of the existence of the universe. Nevertheless, in Proclus' view, Aristotle and the Peripatetics would have approached the correct Platonic position had they only seen the full significance of the proof from motion. Central to the proof from motion is the principle that since every body must be finite, it can contain only finite power. Employing the principle, Aristotle inferred that the power contained in the heavens is insufficient to maintain eternal motion; and inasmuch as the heavens do move eternally, they must depend for their motion on an incorporeal cause outside themselves.[2] That was as far as Aristotle went. But, Proclus contends, the same principle also applies to existence. The principle shows that the

[1]Cf. E. Zeller, *Die Philosophie der Griechen,* Vol. III, Part 2 (5th. ed.; Leipzig, 1923), pp. 530 ff., 550 ff.

[2]Cf. above, p. 244.

power contained in the heavens is insufficient to maintain eternal existence as well as eternal motion. And inasmuch as the heavens do exist eternally, they must depend for their existence, and not merely for their motion, on an incorporeal cause outside themselves. In the reasoning of Aristotle's proof from motion Proclus thus detects the germ of a proof for an incorporeal cause of the very existence of the corporeal, though not of the incorporeal,[3] realm.[4]

Simplicius—following, as he says, a book by his teacher Ammonius,[5] who, it happens, had at one time been a student of Proclus—states the issue in a similar fashion, but with one addition. Simplicius too observes that on the reading of such peripatetics as Alexander of Aphrodisias, Aristotle established a "final and motive cause of the heavens, not an efficient cause [of their existence]." He further contends that the principle that a body can contain only finite power is "clearly" not limited to "motive power." It applies as well to the power "constituting the being" of a corporeal object, and "therefore [the heavens] . . . must receive even their eternal corporeal being from an incorporeal [cause]."[6] So far Simplicius has not diverged from the account given by Proclus. But Simplicius had undertaken a general harmonization of Aristotle with Plato. Accordingly, he goes beyond Proclus and contends that not Aristotle, but his students, failed to see the implications of the proof from motion. Aristotle himself "was in agreement with his master [Plato] in holding God to be . . . an efficient cause of the [existence of the] entire universe and the heavens," and Aristotle intended his proof from motion to establish that thesis.[7] To support his reading of Aristotle, Simplicius has recourse to a subtle, rather unconvincing, exegesis of the Aristotelian text.[8]

In late Greek Neoplatonic philosophy we find, then, the view that Aristotle's proof from motion implied a proof of a first efficient cause of the very existence of the physical universe, although there was a difference of opinion as to whether Aristotle himself was conscious of the implication. The tradition apparently filtered down to Avicenna's time. In a commentary on Aristotle's *Metaphysics* XII, Avicenna criticizes the proof from motion on the grounds that it establishes no more than a first cause of motion; it is "incapable . . . of establishing the one, the True, who is the first cause (*mabda'*) of all existence."[9] Yet, despite his

[3]The principle that a finite body cannot contain infinite power would not apply to incorporeal beings. See, however, below, n. 37.

[4]Proclus, *Commentary on Timaeus*, ed. E. Diehl, Vol. I (Leipzig, 1903), pp. 266, 295. Also cf. the passage from a lost work of Proclus quoted by John Philoponus, *De Aeternitate Mundi contra Proclum*, ed. H. Rabe (Leipzig, 1899), pp. 238–239.

[5]The book was known to Alfarabi. Cf. M. Mahdi, "Alfarabi against Philoponus," *Journal of Near Eastern Studies*, XXVI (1967), 236, n. 9.

[6]Simplicius, *Commentary on the Physics*, ed. H. Diels, *Commentaria in Aristotelem Graeca*, Vol. X (Berlin, 1895), pp. 1362–1363.

[7]Ibid., p. 1360.

[8]Ibid., pp. 1361–1362.

[9]Avicenna, *Commentary on Metaphysics XII*, in *Arisṭū 'ind al-'Arab*, ed. A. Badawi (Cairo, 1947), p. 23.

criticism of Aristotle's proof, Avicenna disputes the position of certain "commentators," including one Abū Bishr, according to whom Aristotle recognized nothing more than a first cause of motion. Avicenna cites a statement of Aristotle's in the *Metaphysics* to the effect that "the heavens depend" on the "first cause (*mabda'*)"; and he explains the statement as an acknowledgment that the first cause of motion is also a first cause of the very "existence (*qiwām*)" of the universe.[10] Avicenna, in other words, understands that although the proof from motion does not establish a first cause of existence, Aristotle did recognize that a complete picture of the universe requires construing the first cause of motion as a cause of existence as well.

Elsewhere, in another collection of philosophic notes, Avicenna adds a refinement. He takes up the following question: If a proper proof of the existence of God must establish a "first cause (*mabda'*) of the existence and substance [of the universe]," why did the earlier philosophers limit themselves to proving a "cause merely of motion?" Avicenna's brief reply is that "openly," indeed, the earlier philosophers demonstrated merely a first cause of motion; yet by "allusion" and by "implication (*bi-l-qūwa*)" they also proved a first cause of the existence of the universe.[11] The present passage still assumes that the proof from motion is unable in itself to establish anything more than a first cause of motion. It adds that Aristotle had not only acknowledged a first cause of existence, but had consciously alluded to and implied—perhaps, but not necessarily, in his proof from motion[12]—another proof which is able to establish a first cause of the very existence of the universe.

Averroes too found a proof of a first cause of existence implied in Aristotle. Aristotle's proof from motion, Averroes explains, is tantamount to a proof of a first cause of existence. He writes: "Should the motion [of the heavens] cease, the heavens [themselves] would cease"; for the essential nature of the heavens is such that they must move, and without motion they could not possess their essential nature as heavens. Furthermore, "should the movement of the heavens cease, the movement of what exists under the heavens would likewise cease."[13] Without the first cause of motion, "the universe as a whole would cease to exist"—at least in the form of a universe.[14] Aristotle's first cause of motion is consequently a first cause of the existence of the universe.[15]

[10]Ibid., p. 26, referring to *Metaphysics* XII, 7, 1072b, 14. An identification of the Abū Bishr whom Avicenna mentions is made by H. Brown, "Avicenna and the Christian Philosophers in Baghdad," *Islamic Philosophy and the Classical Tradition* (Walzer Festschrift), (Oxford and Columbia, South Carolina, 1972), pp. 43–45.

[11]Avicenna, *Mubāḥathāt,* §290, in *Aristū 'ind al-'Arab,* p. 180. Also cf. §264, p. 173.

[12]The medieval philosophers found another proof of the existence of God in *Metaphysics* II; see below, Chapter XI.

[13]Cf. above, p. 242.

[14]It could still presumably exist as inert matter.

[15]*De Substantia Orbis,* IV; cf. Aristotle, *De Caelo* I, 9, 279a, 28–30.

2. The existence of God: a problem for metaphysics

Avicenna's proof of a first cause of the existence of the universe is characterized by him as a *metaphysical* proof, and Avicenna attached considerable significance to that characterization. The notion is elaborated most fully in Avicenna's most comprehensive philosophic work, the *Shifā'*. There Avicenna undertakes to define the "subject matter" (*mawḍū'*) of the science of metaphysics and in particular to show that the subject matter of metaphysics is not the "deity (*innīya allāh*)."[16] He employs three Aristotelian rules regarding the subject matter of any science, and finally arrives at Aristotle's position on what the proper subject matter of metaphysics is.[17] And in the course of his discussion he shows that one task of metaphysics—as distinguished from the subject matter of metaphysics—is the demonstration of the existence of God.

Aristotle had explained that a science (a) does not "demonstrate" the existence of its own subject matter. Astronomy and mathematics, for example, do not "demonstrate" the existence of celestial and mathematical entities, but rather "accept" their existence. Nor (b) does a science demonstrate its "principles," the most fundamental premises from which it reasons; these too it accepts. The procedure of each science is (c) to "examine" various "essential attributes," whose "meaning" is accepted but whose "existence" remains to be proved, and to demonstrate that the attributes do belong to the subject matter. Arithmetic, for example, accepts the existence of its subject matter, number, and proceeds—with the aid of the principles of arithmetic, which it also accepts, together with the aid of subordinate premises, which it derives from the principles—to demonstrate the essential attributes of number.[18]

Using these rules, Avicenna shows that the deity cannot be the proper subject matter of the science of metaphysics. His reasoning is as follows: The existence of God—this Avicenna posits with no explanation—is surely neither "self-evident" nor "unprovable";[19] it must, hence, be "amenable to proof." Since no science apart from metaphysics could possibly prove the existence of God, metaphysics remains the only place were the proof can be undertaken. By Aristotle's rules, however, the existence of the subject matter is not to be demonstrated by a given science, but is rather to be "accepted." Consequently, metaphysics, which does have the task of demonstrating the existence of the deity, must have something other than the deity as its subject matter. Put in another way, since the existence of God is a "subject of inquiry" (*maṭlūb*) of the science of metaphysics, God cannot be the "subject matter" (*mawḍū'*) of metaphysics.[20]

[16] *Shifā': Ilāhīyāt*, ed. G. Anawati and S. Zayed (Cairo, 1960), p. 5; French translation, with pagination of the Arabic indicated: *La Métaphysique du Shifā'*, trans. G. Anawati (Paris, 1978).

[17] The assumption is that the *subject matter* of a given science is not arbitrary, but something to be determined objectively.

[18] *Posterior Analytics* I, 9–10.

[19] Cf. F. Rosenthal, *Knowledge Triumphant* (Leiden, 1970), p. 211.

[20] *Shifā': Ilāhīyāt*, p. 6.

Avicenna was cognizant of an objection that would immediately occur to students of Aristotle. The existence of a first mover had been proved in greater detail in Aristotle's *Physics* than in his *Metaphysics*. How then might it be maintained that demonstrating the existence of God is a task for the science of metaphysics and not for the science of physics? In answering the objection, implied rather than expressly stated by him, Avicenna points out that physics is the science of corporeal entities, whereas the first cause is known even by physics to be incorporeal. The existence of the first cause cannot therefore be a proper subject of inquiry for physics, and when physics does treat of the first cause, it impinges upon an issue not genuinely its own. Physics merely "gives a fleeting idea" of the existence of God, thereby encouraging philosophers to pursue the subject in the discipline proper to it, that is, in the science of metaphysics.[21]

Having established that the existence of the deity is a subject of inquiry of metaphysics, and thereby having ruled out the deity as the proper subject matter of metaphysics, Avicenna follows Aristotle in stating just what the subject matter of metaphysics is. Aristotle had explained that metaphysics "examines the existent *qua* existent and what belongs to it by virtue of itself."[22] Avicenna accordingly concludes, after some additional discussion, that the subject matter, whose existence is simply accepted by the science of metaphysics, is "the existent *qua* existent." The subject of inquiry of metaphysics is, stated broadly, the attributes that "belong to the existent merely by virtue of its being existent."[23] In other words, the task of metaphysics is to demonstrate that certain attributes do indeed belong to the existent solely insofar as it is existent.

Here Avicenna is led to consider a further difficulty. Aristotle had laid down the rule that a science does not demonstrate its own principles, but only accepts them.[24] Avicenna construes the rule as meaning both that a science does not demonstrate the principles from which it reasons—that is to say, principles in the sense of primary premises—and also that it does not demonstrate the existence of the principles of its subject matter—that is to say, principles in the sense of causes of existence.[25] When applied to the science of metaphysics, the rule shows that the "principles" of what exists cannot be part of the subject of inquiry of metaphysics; for were that the case, metaphysics would improperly be demonstrating the existence of the principles of its own subject matter. The subject of

[21]Ibid., pp. 6–7. Alfarabi similarly states that metaphysics has the task of demonstrating the existence of incorporeal beings; cf. Alfarabi, *Iḥṣā' al-'Ulūm,* ed. and trans. A. Gonzalez Palencia (Madrid, 1953), Arabic part, p. 88.

[22]Aristotle, *Metaphysics* IV, 1, 1003a, 20–21, and D. Ross's note in his edition (Oxford, 1924). Cf. Alfarabi, *Aghrāḍ mā ba'd al-Ṭabī'a,* in *Rasā'il al-Fārābī* (Hyderabad, 1931), p. 4; below, p. 313.

[23]*Shifā': Ilāhīyāt,* p. 13.

[24]*Posterior Analytics* I, 10.

[25]In the passage cited in the previous note, Aristotle appears to be making only the first of these two points. But both senses of *principle* are Aristotelian; cf. *Metaphysics* VI, 1, 1021b, 34 ff. Averroes understood Aristotle's rule in the way Avicenna did; cf. below, p. 314.

inquiry of metaphysics must, as just seen, be the "attributes belonging to" the existent insofar as it is existent.[26] These considerations lead to a two-sided difficulty, which is taken up by Avicenna: First, the deity seems to be a principle or cause of everything that exists, and hence a principle of the subject matter of metaphysics; by Aristotle's rule, the existence of God would therefore have to be accepted, and not demonstrated, by metaphysics. Secondly, in any event, the existence of the deity hardly seems to be an attribute belonging to the existent *qua* existent, and therefore hardly an appropriate subject of inquiry for metaphysics.

In resolving the first side of the twofold difficulty, Avicenna notes that God cannot after all be a principle or cause of the entire subject matter of metaphysics. The subject matter of metaphysics is the existent insofar as it is existent, "the existent with no further qualification." And God cannot be the cause of the "existent without qualification," since he clearly is not the cause of absolutely everything that exists, not being a cause of himself.[27] God is the cause of only a portion of existence, the portion that is "caused." Inasmuch as God is the principle of only part of the subject matter of metaphysics, it is, Avicenna contends, legitimate for metaphysics to undertake a demonstration of his existence; for it is legitimate for a science to demonstrate the existence of the principle of part of its own subject matter.

As for the second side of the objection raised by Avicenna—according to which the existence of the deity, if indeed a subject of inquiry of metaphysics, would have to be an attribute belonging to the subject matter—that ceases to be odd on closer inspection. The proof of the existence of God establishes that the property of *not being constituted by anything else* is exemplified in the realm of actual existence. In a sense, therefore, the proof of the existence of God does demonstrate that a certain attribute is applicable to the existent *qua* existent;[28] and the proof of the existence of God turns out to be a proper subject of inquiry for the science of metaphysics.[29]

Avicenna concludes, then, that the subject matter of the science of metaphysics is the "existent *qua* existent." The subject of inquiry of metaphysics can, he writes, be broken down into three subheadings: the attributes of the existent *qua* existent; the ultimate causes of caused existents, the investigation of which, as just shown, can be construed as part of the inquiry into the attributes of existence; and also the principles of the individual sciences.[30]

[26]Cf. above, p. 285.

[27]Avicenna will undertake to prove that God exists *by reason of himself*, but would reject as senseless the description of God as *cause of himself*.

[28]The proof does not, of course, establish that the property in question is applicable to all existence. The case of attributes such as unity and plurality would be analogous. Metaphysics studies the applicability of those attributes and similar pairs to the existent insofar as it is existent, even though neither member of the pair belongs to all existence; cf. Aristotle, *Metaphysics* V.

[29]*Shifā': Ilāhīyāt*, p. 14.

[30]Ibid., pp. 14–15. Cf. Alfarabi, *Iḥṣā' al-'Ulūm*, pp. 87–89. On the third subheading, cf. H.

The foregoing discussion of the subject matter and subject of inquiry of the science of metaphysics, with its justification of a metaphysical proof of the existence of God, is found in Avicenna's *Shifā'*. Avicenna touches upon the question of the metaphysical proof of the existence of God in the *Ishārāt* as well.

There he discusses the method that he pursues in his proof of the existence of God. The proof, he writes, consists in "examining nothing but existence itself"; by "considering . . . the nature (*ḥāl*) of existence," the proof has "existence *qua* existence testify to the first [cause]."[31] The method thus delineated is contrasted by Avicenna with another, whereby the existence of God is established not from a consideration of existence in general, but instead from a consideration of one segment of existence, namely "creation and effect." Although the latter method, which takes its departure from "creation and effect," is recognized by Avicenna as valid, his own method, he insists, is "more certain and more exalted."[32]

The difference between the two methods is stated here in language that is deliberately allusive but easily deciphered. Metaphysics, as was seen, is the science that "examines the existent *qua* existent and what belongs to it by virtue of itself." When Avicenna claims to have constructed a proof exclusively by examining "existence itself," "existence *qua* existence," and "the nature of existence," he plainly means that he has constructed a proof using propositions drawn exclusively from the science of metaphysics. He contrasts his metaphysical proof with the proof that begins with God's "creation and effect" and that reasons back from them to the existence of God as a first cause. Avicenna's intent cannot be that the metaphysical proof uses absolutely no data drawn from God's "creation and effect." For, it will appear, his proof does require at least one datum from the external world;[33] and the parts of the world accessible to man are man himself and physical nature, both of which belong to the realm of "creation and effect." Avicenna's meaning must be that the metaphysical proof considers no properties peculiar to the realm of "creation" but instead considers the attributes of objects belonging to that realm solely insofar as they are existent. As for the proof of the existence of God which is not a metaphysical proof but focuses on God's "creation and effect," it can be nothing other than a physical proof of the existence of God, that is to say, a proof based not on the attributes of existence in general, but on the attributes peculiar to physical existence. The metaphysical proof, Avicenna

Wolfson, "The Classification of Sciences in Medieval Jewish Philosophy," reprinted in his *Studies in the History of Philosophy and Religion,* Vol. I (Cambridge, Mass., 1973), p. 518; and D. Ross's introduction to his edition of Aristotle's *Prior and Posterior Analytics* (Oxford, 1949), p. 64.

[31]The sentence continues: "whereupon he testifies concerning everything in existence posterior to him." This is a Sufi theme; cf. e.g., Ibn 'Aṭā' Īllāh, *K. al-Ḥikam,* trans. V. Danner (Leiden, 1973), §§ 29, 249.

[32]*K. al-Ishārāt wa-l-Tanbīhāt,* ed. J. Forget (Leiden, 1892), p. 146. The French translation by A. Goichon, *Livre des Directives et Remarques* (Beirut and Paris, 1951), gives the pages of Forget's edition.

[33]Cf. below, p. 303.

asserted, is "more certain and more exalted" than the other proof, but the assertion is not supported by an explanation as to where the superiority of the metaphysical proof lies. Scattered statements elsewhere do, however, furnish an explanation.

Avicenna has written that since physics is the science of *corporeal* entities, a physical proof can give no more than a "fleeting idea" of the existence of an *incorporeal* first cause.[34] He has also put aside Aristotle's proof from motion on the grounds that it is incapable of "establishing the . . . first cause of all existence."[35] In a similar vein, he stresses that the metaphysical proof has the advantage of establishing a cause of "every caused existent . . . insofar as it is caused" and "not merely insofar as it is moved or quantitative."[36] These statements indicate at least two virtues of the metaphysical proof: It establishes a cause of the entire universe, incorporeal as well as corporeal, and it establishes a cause of the very existence of the universe. A physical proof, by contrast can at most establish a first cause only of the corporeal realm and, in the instance of Aristotle's well-known proof from motion, it establishes a first cause exclusively of the motion of the corporeal realm.[37]

It is easy to go beyond Avicenna's explicit statements and expand upon the virtues that he could have perceived in a metaphysical proof, rendering it superior to Aristotle's proof from motion. The proof from motion rested on a set of physical principles: Motion in place underlies all other kinds of change;[38] everything moved has the cause of its motion outside itself;[39] nothing can maintain itself in motion unless it is continuously moved by an agent;[40] only circular motion is continuous;[41] only an infinite force can maintain the heavens in motion for an infinite time;[42] the heavens cannot contain an infinite force.[43] Using all the physical principles, Aristotle undertook to establish the existence of an unmoved incorporeal cause solely of the motion of the universe. Avicenna, although not rejecting Aristotle's physical principles, dispenses with them in his metaphysical proof. And yet, without them, he is confident that he can prove the existence of an incorporeal first cause not merely for the motion of the physical universe—that being the only part of the universe in motion—but for the very existence of the entire universe. The metaphysical proof proves more with fewer premises. It travels, or attempts to travel, further with less fuel.

[34] Cf. above, n. 21. [35] Cf. above, n. 9.

[36] *Shifā': Ilāhīyāt,* p. 14.

[37] An argument for the unity of God could, however, show that all incorporeal beings outside the first cause are also dependent on the first cause for their existence. Cf. above, p. 274.

[38] Cf. above. p. 241.

[39] Cf. above, pp. 242–247.

[40] Cf. above, p. 242.

[41] Cf. above, p. 242.

[42] Cf. above, p. 244, n. 49; pp. 245–246.

[43] Cf. above, pp. 244–245.

3. Necessarily existent being and possibly existent being

The metaphysical proof of the existence of God to be examined in the present chapter appears in two of Avicenna's works, at length in the *Najāt* and somewhat obscurely in the *Ishārāt*.[44] In both works, Avicenna analyzes the concepts *necessarily existent being* and *possibly existent being* and uses the analysis to establish the existence of a being *necessarily existent by virtue of itself*. The concepts *necessarily existent being* and *possibly existent being* are also analyzed by Avicenna in two other works, the *Shifā'* and *Dānesh Nāmeh*;[45] but there they serve to define the nature of God, not to establish his existence, and the existence of God is established in a different manner.[46] In discussing Avicenna's analysis of the concepts *necessarily existent* and *possibly existent* I draw upon all of the works just mentioned. But in discussing the particular proof of the existence of God which is the subject of the present chapter I limit myself to the *Najāt* and *Ishārāt*.[47]

The proof of the existence of God as a necessarily existent being probably was suggested to Avicenna by a passage in Aristotle. In the *Metaphysics,* Aristotle presented a version of his proof from motion, whereupon he added a postscript: Since the prime mover "can in no way be otherwise than as it is," it "is an existent . . . of necessity"—an existent of necessity from which "the heavens and nature depend."[48] Avicenna's proof, particularly the fuller version in the *Najāt,* can be understood as starting where Aristotle left off. Avicenna sets aside all the physical arguments leading up to Aristotle's first cause that is an "existent . . . of necessity." He begins afresh by analyzing the concept "existent of necessity" or *necessarily existent,* as he calls it, working out everything contained in the concept as he construes it. Then, adducing a single datum from the external world, he undertakes to establish that something necessarily existent by virtue of itself does in fact exist.

Before, however, entering into any analysis of metaphysical concepts, including *necessarily existent* and *possibly existent,* Avicenna points out that primary concepts cannot strictly be defined at all. Definitions in Aristotelian logic are framed by taking a wider and already known concept, the *genus,* and setting apart a segment of it through a *specific difference*. It follows that primary concepts, which are not "subsumed under anything better known" and hence are part of no genus, cannot be defined. They are rather "imprinted in the soul in a primary fashion,"[49] and must be grasped immediately. The thesis that primary concepts

[44]*Najāt* (Cairo, 1938), pp. 224 ff.; *K. al-Ishārāt,* pp. 140 ff.

[45]*Shifā': Ilāhīyāt,* pp. 37 ff., 343 ff.; *Dānesh Nāmeh* (Teheran, 1937), pp. 105 ff.; French translation: *Le Livre de Science,* transl. M. Achena and H. Massé (Paris, 1955), pp. 136 ff.; English translation: *The Metaphysica of Avicenna,* trans. P. Morewedge (New York, 1973), pp. 18 ff.

[46]Below, Chapter XI.

[47]The most important passages in the *Najāt* and *Shifā'* are translated by G. Hourani, "Ibn Sīnā on Necessary and Possible Existence," *Philosophical Forum,* IV (1972), 74–86.

[48]Aristotle, *Metaphysics* XII, 7, 1072b, 10–14.

[49]*Shifā': Ilāhīyāt,* p. 29.

cannot be defined is developed most comprehensively by Avicenna in connection with the general concepts *existent* and *thing,* and is thereupon applied to the concepts *necessary, possible,* and *impossible;* those concepts too, he writes, cannot be "made known . . . in a true sense."[50]

Ostensible definitions of *necessary, possible,* and *impossible* had been offered, notably by Aristotle, but Avicenna contends that the ostensible definitions lead in a vicious circle. He considers two ostensible definitions of *necessary*: "that which can (*yumkin*) not be assumed [to be] absent (*ma'dūm*)";[51] and that which is such that an impossibility would result if it should be assumed to be other than it is."[52] The first of the two definitions employs the concept *possible* (*mumkin*)—"can" (*yumkin*)—and the second employs *impossible*. When we turn to ostensible definitions of *possible* we find that they in their turn employ either the concept *necessary* or *impossible*. For possible is defined as "that which is not *necessary*; or which is absent (*ma'dūm*), but is such that its existence is not *impossible* if it should be assumed to occur at any time in the future."[53] And ostensible definitions of *impossible,* for their part, employ either *necessary* or *possible*.[54] Attempts to define the triad thus chase one another in a circle.[55] Nevertheless, although primary concepts are not explicable by anything wider and better known, and are consequently inaccessible to true definition, there is, Avicenna understands, a way of presenting them to the man who for some reason does not have them imprinted in his soul. One may "direct attention" to the primary notions and "call them to mind" through a "term or an indication."[56] On that basis, Avicenna ventures an explication of *necessary*: "It signifies certainty of existence."[57]

The distinction between *necessary* and *possible* is employed by Avicenna to distinguish two types of being. A "necessarily existent being" is a being that "perforce exists"; alternatively, it is "such that when it is assumed not to exist, an impossibility results." A "possibly existence being" is a being that "contains no necessity . . . for either its existence or nonexistence (*'adam*)"; alternatively it is "such that whether assumed not to exist or to exist, no impossibility results."[58]

[50]Ibid., p. 35.

[51]Implied in Aristotle, *Prior Analytics* I, 13, 32a, 19–20.

[52]Cf. Aristotle, *Metaphysics* V, 5, 1015a, 34.

[53]Cf. Aristotle, *Prior Analytics* I, 13, 32a, 19–20.

[54]The impossible is "what is necessarily nonexistent," or "what cannot (*lā yumkin*) exist." Cf. Aristotle, *De Interpretatione* 13, 22a, 19–20; 22b, 6.

[55]*Shifā': Ilāhīyāt,* p. 35.

[56]Ibid., p. 29.

[57]Ibid., p. 36.

[58]*Najāt,* pp. 224–225. *Avicennae Metaphysices Compendium,* trans. N. Carame (Rome, 1926), p. 66, is misleading. Carame's translation reads: "Necesse-esse est ens quod si ponatur non esse implicat contradictionem (*maḥāl*). Possible vero esse est illud quod sive ponatur esse sive non esse, non inde oritur repugnantia (*maḥāl*)." The mistranslation of *maḥāl* as "contradictio" suggests that denying the existence of a necessarily existent being involves a logical contradiction. Cf. below, p. 394.

What Avicenna is offering is clearly not strict definitions. Necessarily existent can hardly be defined, in the strict sense, through *impossibility,* since *impossibility* could only be defined through *necessity* or through *possibility,* which in its turn would have to be defined through one of the other two terms. What Avicenna has given are explications of the type that merely "direct attention" to the meaning of concepts and "call them to mind."[59]

The distinction between *necessarily existent* and *possibly existent* is supplemented by Avicenna with a further distinction, suggested by Aristotle. When Aristotle analyzed the term *necessary,* he observed that "for certain things, something else is a cause of their being necessary." But "for some, nothing is [a cause of their being necessary]; rather it is through them that others exist of necessity."[60] That is to say, there is a class of things necessary without having a cause of their being so; and a second class of things necessary through a cause, a cause to be found in the former class. The distinction drawn by Aristotle in regard to *necessary* suggests to Avicenna a distinction between two types of *necessarily existent being*. One can, Avicenna writes, conceive of a being as necessarily existent either by virtue of itself or by virtue of something else. The former would be something "such that because of itself and not because of anything else whatsoever, an impossibility follows from assuming its nonexistence." The latter would be a being "such that it becomes necessarily existent, should something other than itself be assumed [to exist]." The illustrations Avicenna adduces for the latter category are "combustion," which is "necessarily existent . . . when contact is assumed to take place between fire and inflammable material," and "four," which is "necessarily existent . . . when we assume two plus two [to exist]."[61] Now if anything is necessarily existent by virtue of something else, it must—since it will not exist by reason of itself without that other thing—be "possibly existent by virtue of itself."[62] In all, Avicenna thus envisages three categories: (a) the necessarily existent by virtue of itself; (b) the necessarily existent by virtue of another, but possibly existent by virtue of itself; and (c) the possibly existent by virtue of itself which is not rendered necessarily existent by virtue of another.

Here Avicenna makes a point that has escaped some students of his philosophy.[63] Of the three categories, the first two, he states unambiguously, are the only conceivable categories of actual existence. Everything actually existent, including everything "entering existence" in the physical world—as combustion, to take the illustration used by Avicenna—is necessary in one sense or the other.[64] Put in another way, the possibly existent can never actually exist unless rendered

[59] Above, n 56.

[60] Aristotle, *Metaphysics* V, 5.

[61] *Najāt,* p. 225 For Avicenna, it must be stressed, "2 plus 2 equals 4" is not *logically* necessary.

[62] *Najāt,* p. 225.

[63] As, for example, Averroes; cf. below, p. 318. Some modern scholars have also missed Avicenna's point.

[64] Cf. *Shifā': Ilāhīyāt,* p. 37.

necessary by something distinct from itself.[65] Avicenna's division of being hereby differs from, for example, Alfarabi's. Alfarabi applied the designation *possibly existent* to those objects that actually exist, yet have the possibility of not existing and are hence unable to exist forever. In other words, he designated all actual transient objects in the sublunar world as *possibly existent* with no further qualification; and he restricted the designation *necessarily existent* to beings that cannot cease to exist, that is, to eternal beings.[66] Avicenna, by contrast, insists that all objects that actually exist, even transient beings, are to be characterized as *necessarily existent,* and all objects that exist by reason of something else, even eternal beings, are *possibly existent.* Both sets belong to the single category of the *necessarily existent by virtue of another, possibly existent by virtue of itself.* Alfarabi's usage is unquestionably more genuinely Aristotelian than Avicenna's.[67]

In justifying his designation of all actually existent beings as necessarily existent, Avicenna employs a train of thought similar to that underlying arguments from the concept of particularization.[68] The possibly existent can enter the realm of actual existence, he reasons, only if a factor distinct from itself should "differentiate out" existence for it in preference to nonexistence. And once a factor of the sort is present, the possibly existent being perforce exists; for its existence is rendered necessary.[69] The proper way of construing possible existence, according to Avicenna, is therefore to say that during the time the possibly existent actually exists, its existence is necessary, and during the time it does not exist, its existence is impossible. The necessity and the impossibility of its existence are both conditioned, due not to itself, but to the presence or absence of an external condition, which necessitates its existence or nonexistence. Considered in itself, in isolation from the external condition—and only considered in that way—the possibly existent at all times remains possible.[70]

Actual existence is then either: (a) Necessarily existent by virtue of itself; this is something "such that if assumed not to exist an impossibility results," with the proviso that it has its character by reason of itself. Or (b) necessarily existent by virtue of another, but possibly existent by virtue of itself; this is something,

[65]Cf. *Najāt,* p. 226.

[66]Alfarabi, *al-Siyāsāt al-Madanīya* (Hyderabad, 1927), pp. 26–37. There is a suggestion of Avicenna's usage in Alfarabi, *Commentary on Aristotle's De Interpretatione,* ed. W. Kutsch and S. Marrow (Beirut, 1960), p. 192.

[67]For Aristotle, *necessary* and *eternal* are mutually implicative; cf. below, pp. 294, 319.

[68]Cf. above, pp. 161f.; 178f. Averroes saw the Kalam influence on Avicenna in this point. Cf. Averroes, *K. al-Kashf,* ed. M. Mueller (Munich, 1859), p. 39; German translation, with pagination of the original Arabic indicated: *Philosophie und Theologie von Averroes,* trans. M. Mueller (Munich, 1875); *Long Commentary on Physics,* in *Aristotelis Opera cum Averrois Commentariis,* Vol. IV (Venice, 1562), II, comm. 22; *Tahāfut al-Tahāfut,* ed. M. Bouyges (Beirut, 1930), p. 276; English translation with pagination of the original Arabic indicated: *Averroes' Tahafut al-Tahafut,* trans. S. van den Bergh (London, 1954).

[69]Thus all events in the universe occur necessarily.

[70]*Shifā': Ilāhīyāt,* pp. 38–39; *Najāt,* pp. 226, 238.

again, such that if assumed not to exist, an impossibility results, with the proviso that it has its character by reason of another, only inasmuch as "something other than itself is assumed [to exist]."[71] The *necessity* characterizing the two categories of necessarily existent being, is, as already seen, construed by Avicenna as an indefinable primary concept to be grasped by the human mind immediately.[72] As a mere "indication" of the meaning of necessity Avicenna wrote that the term "signifies certainty of existence."[73] The necessarily existent by virtue of itself would accordingly be that which has certainty of existence by virtue of itself; the necessarily existent by virtue of another would be that which has certainty of existence by reason of another. And the impossibility involved in assuming the nonexistence of a necessary being would not be any logical impossibility, but would consist in contradicting the certainty of its existence, the fact that it does exist.[74] If no more can be said about the meaning of *necessarily existent,* it is difficult to see how necessary existence differs from actual existence. Indeed, necessary existence for Avicenna seems simply to be actual existence, with the added understanding that whatever actually exists, exists by necessity, and with the further understanding that *necessity* is a primary concept, the meaning of which must be grasped immediately.

So far, it must be stressed, Avicenna's analysis has been conducted exclusively in the realm of concepts, and he has not committed himself to the existence of anything.[75] He has merely stated that whatever might be assumed to exist would have to be classified as either necessarily existent by virtue of itself or necessarily existent by virtue of another.

4. The attributes of the necessarily existent by virtue of itself

Having established that whatever actually exists is either necessarily existent by virtue of itself or necessarily existent by virtue of another, Avicenna proceeds to analyze the former concept and to set forth its "properties." The analysis of the concept is designed to serve a double function in his proof of the existence of God. It contributes to the argument showing that something necessarily existent by virtue of itself does exist,[76] and in addition it reveals that the entity in question possesses the attributes of a deity. Avicenna's analysis has its sources in Aristotle, Plotinus, Proclus, and Alfarabi, although the pertinent passages in these writers are not connected with a proof of the existence of God.

Aristotle had written that the most fundamental sense of the term *necessary* is "that which cannot be otherwise." And, he continued, anything of the sort "does

[71] Above, n. 61.

[72] Above, pp. 289–290.

[73] Above, n. 57.

[74] This is especially clear in the *Dānesh Nāmeh,* p. 106; French translation, p. 136; English translation, p. 48.

[75] This is clear throughout, and is stated explicitly in *Shifā': Ilāhīyāt,* p. 37.

[76] Cf. below, §5.

not admit of more states than one," and must therefore by "simple." Aristotle further suggested that what cannot be otherwise must be "eternal and immovable."[77] We thus have three attributes belonging to that which is *necessary* in the sense of not being able to be otherwise: It must be simple, eternal, and immovable.

The Neoplatonic passages that lie behind Avicenna's analysis of the concept *necessarily existent by virtue of itself* were mediated though the *Liber de Causis,* a medieval Arabic work falsely attributed to Aristotle, but in fact a paraphrase of Proclus' *Elements of Theology.* In the *Liber de Causis,* the "first cause" is described as the only entity that is "self-sufficient" (*mustaghnīya bi-nafsihā*), and analysis of *self-sufficiency* reveals that the first cause is "simple in the highest degree." The reasoning runs as follows: When an object is not simple but "composite," it "stands in need of something outside itself, or [at least in need] of the elements of which it is composed"; that is to say, a composite entity is dependent upon whatever external factor joins its parts together and, even supposing that no such external factor is required, a composite entity considered as a whole is distinct from its parts and dependent on those parts. If the composite is inescapably dependent upon something distinct from itself and hence not self-sufficient, anything self-sufficient cannot be composite but must be "simple."[78] The first cause, which is known to be self-sufficient, must therefore be simple.

Besides analyzing the implications of the attribute *self-sufficiency,* the *Liber de Causis* also analyzes the implications of *self-subsistence,* which ostensibly has the same meaning.[79] If a "substance" is "self-subsistent" (*qā'im bi-dhātihi*), the analysis now goes, it cannot "be brought into existence (*mukawwan*)." For were it brought into existence, it "would stand in need" of whatever external factor brings it into existence and perfects it; and it would consequently not after all be

[77] *Metaphysics* V, 5, 1015b, 12–15.

[78] *Liber de Causis,* ed. and trans. O. Bardenhewer (Freiburg, 1882), §20, and cf. §27. This reflects Proclus, *Elements of Theology,* ed. and trans. E. Dodds (Oxford, 1963), §127, which in turn goes back to Plotinus, *Enneads,* II, 9, 1. Cf. also Dodds' note to *Elements of Theology,* §§9, 10.

[79] In the original Greek, *self-sufficient* and *self-subsistent* are treated as synonymous terms; cf. Dodds' commentary on Proclus, *Elements,* p. 224. In the Greek, the entities characterized as self-sufficient and self-subsistent stand at a level of existence below the first cause of the universe; Proclus held the strange view that the self-sufficient and the self-subsistent represents a level of existence that both is its own cause and yet also has a cause above it. Cf. ibid., pp. 196, 224. (Avicenna, *Najāt,* p. 225, refutes the thesis that something can be both necessarily existent by virtue of itself and also necessarily existent by virtue of another.) In the Arabic paraphrase, by contrast, the designation *self-sufficient* is explicitly restricted to the first cause. And it is debatable whether *self-subsistent* is synonymous with *self-sufficient* and thus also a designation of the first cause, or whether the *self-subsistent* represents a lower stage in the hierarchy of existence. §20 taken together with §27 suggests the former interpretation, whereas §§24, 25, 28, suggest the latter interpretation. (§20 appears to be a conscious effort by the author of the paraphrase to make the theory more monotheistic than was intended in the original Greek. For that tendency in other Arabic paraphrases and translations of Neoplatonic works, see G. Endress, *Proclus Arabus* [Beirut, 1973], pp. 206–219, 236–237, 240–241.) These questions of interpretation do not affect my present purpose, which is to show how certain attributes were analyzed out of the concepts *self-sufficient* and *self-subsistent.*

self-subsistent.[80] The "self-subsistent" cannot, moreover, be "subject to destruction." For things are destroyed only by being separated from the cause of their existence, whereas the self-subsistent can never become separated from its cause since it is its own cause.[81] Analysis of the concepts *self-sufficient* and *self-subsistent* has thus established that whatever is so described cannot be composite, cannot be brought into existence, and cannot be destroyed. And the *Liber de Causis* has employed a significant thought that was to be applied more systematically by Alfarabi and Avicenna: A composite entity taken as a whole is different from its components and dependent on them for its existence.

Alfarabi, in his turn, undertakes to determine the attributes of "the First," that is, the first being in the hierarchy of existence and the first cause. The First, he contends, can, by definition, "have no cause through which, from which, or for the sake of which" it exists. For if it did have a cause, it would not itself be the first cause.[82] Not having a cause, the First must be "sufficient in itself for its own duration and continued existence." Now anything containing the possibility of being destroyed is plainly not sufficient in itself for its own duration and continued existence. It follows that the First does not have any possibility of being destroyed and hence is eternal.[83]

Furthermore, since the First has no cause, it can in no way be "divisible." To establish this, Alfarabi deploys an argument similar to that wherein Proclus found the self-sufficient to be simple because the composite "stands in need of" its own parts. Alfarabi reasons: The components through which a divisible, composite object "receives its substance" (*tajawhara*) are "causes of the existence" of the object; and since the First can have no cause, it can have no components whatsoever. It must consequently be completely indivisible. It must be free not only of "quantitative" divisibility, but also of the internal composition represented by the parts of a definition and the composition resulting from the joining of matter and form in a corporeal object. The First must be absolutely indivisible, indefinable, incorporeal, and unextended.[84]

There can, Alfarabi continues, be no more than one First existent. If there were two, they would each have to possess a common element by virtue of which they both deserve the common designation *First*; and at least one of the two would have to have an added element peculiar to itself by virtue of which the two entities could be distinguished from one another and enumerated as two. At least one of the two would, in other words, be composite. But, if composite, its parts, by the already familiar argument, would be the cause of its existence. Inasmuch as at

[80] *Liber de Causis,* §24; cf. Proclus, *Elements,* §45.

[81] *Liber de Causis,* §25; cf. Proclus, *Elements,* §46.

[82] Alfarabi, *K. Arā' Ahl al-Madīna al-Fāḍila,* ed. F. Dieterici (Leiden, 1895), p. 5; German translation, with pagination of the original Arabic indicated: *Der Musterstaat,* trans. F. Dieterici (Leiden, 1900).

[83] Ibid., p. 8, and cf. p. 5.

[84] Ibid., pp. 8–9; *al-Siyāsāt,* pp. 14–15.

least one of the two would have a cause and not after all be First, only one First being is conceivable.[85]

Alfarabi derives additional attributes from the concept of the First. Whatever is immaterial consists in pure intellect; therefore the First, being immaterial, must be pure actual intellect.[86] *Truth* designates the degree of existence a thing has; the First, which *ex hypothesi* occupies the highest degree of existence, is accordingly truth par excellence.[87] *Beauty* similarly is understood by Alfarabi as proportional to the perfection of existence a thing has; the First, having the highest degree of existence, will accordingly be of the highest beauty.[88] *Pleasure* consists in perception of beautiful objects; since the First has the most perfect perception of the most perfect beauty, namely himself, he enjoys the highest conceivable pleasure.[89] And since the First is the object of his own love and desire, he is the "prime object of love" and the "prime object of desire."[90] By analyzing the concept Alfarabi has concluded that the First must be uncaused, eternal, indivisible and simple, undefinable, incorporeal, unextended, one, pure intellect, truth, most beautiful, the prime object of love, and possessed of the highest pleasure.[91]

Avicenna's analysis of the concept *necessarily existent by virtue of itself* runs along the same lines. The necessarily existent by virtue of itself, he contends, clearly can "not have a cause." For if it did, "its existence would be by virtue of [that cause]" and not by virtue of itself.[92] Like Alfarabi, Avicenna gives the proposition the broadest application, explaining that the necessarily existent by virtue of itself can have a cause in no sense whatsoever; it cannot even have internal causes, "principles which combine together and in which the necessarily existent consists." The full argument for rejecting internal components of any kind rests on the distinction between a given entity as a whole and the parts of which it is composed. Any composite entity, Avicenna submits, exists by virtue of its parts and not be virtue of itself as distinct from the parts. As a consequence, it exists, considered as a whole, not through itself, but through something else—through the components that constitute it. It is therefore not necessarily existent by virtue of itself; and the necessarily existent by virtue of itself cannot, conversely, be composite.[93]

The implications of the noncomposite nature of the necessarily existent by virtue of itself are drawn by Avicenna almost exactly as Alfarabi drew the implications of the noncomposite nature of, the First. If the necessarily existent by

[85]*K. Arā' Ahl al-Madīna,* p. 6; al-Siyāsāt, p. 14.

[86]*K. Arā' Ahl al-Madīna,* p. 9; *al-Siyāsāt,* p. 15.

[87]*K. Arā' Ahl al-Madīna,* p. 10.

[88]*K. Arā' Ahl al-Madīna,* p. 13; *al-Siyāsāt,* p. 16.

[89]*K. Arā' Ahl al-Madīna,* p. 14; *al-Siyāsāt,* p. 16.

[90]*K. Arā' Ahl al-Madīna,* p. 15; *al-Siyāsāt,* p. 17.

[91]This type of reasoning is also suggested in a text attributed to Alexander of Aphrodisias; cf. *Arisṭū 'ind al-'Arab,* ed. A. Badawi (Cairo, 1947) p. 266.

[92]*Shifā': Ilāhīyāt,* pp. 37–38.

[93]*Najāt,* pp. 227–228.

virtue of itself can contain no parts whatsoever, it is simple in every conceivable manner. It is incorporeal, inasmuch as it is not composed of matter and form. It is unextended and immaterial, inasmuch as it is free of quantitative parts. It is undefinable, inasmuch as it is not composed of genus and specific difference. And it is free of the distinction of essence and existence.[94]

There can, moreover, be only one entity necessarily existent by virtue of itself. To prove the thesis, Avicenna examines, in some detail, the various ways in which things having a common characteristic can nevertheless be distinguished from each other. Basically, though, his contention is that positing two entities both of which are necessarily existent by virtue of themselves would imply that those entities have a cause and, again, that they are composite. The reasoning is: If the intrinsic nature of the "species" necessarily existent by virtue of itself is such as to belong exclusively to a certain given entity, it cannot belong to anything else. If, by contrast, the intrinsic nature of the species necessarily existent by virtue of itself is not such as to belong exclusively to a certain given entity, some added factor, for example, a substratum, must be responsible for the presence of the species in whatever entities happen to have it. On the latter alternative, however, the added factor would be the "cause" of the presence of the species, whereas the necessarily existent by virtue of itself has, *ex hypothesi,* no cause. The intrinsic nature of the species necessarily existent by virtue of itself must therefore be such as to belong exclusively to a certain entity; and only one entity of the sort is conceivable.[95] Furthermore, positing two entities both necessarily existent by virtue of themselves would amount to positing two beings that are similar in one respect, their necessary existence, but different in another, the respect whereby they can be distinguished and enumerated as two. But that situation would be conceivable only if at least one of the two should be composite, containing both the element it has in common with its counterpart and another element whereby it can be distinguished and by reason of which two distinct beings can be enumerated. At least one of the two would, therefore, be composite and, as already seen,[96] not necessarily existent by virtue of itself.[97] It follows, then, both from the uncaused and the noncomposite character of the necessarily existent by virtue of itself that not more than one is conceivable.

Avicenna derives other attributes from the concept *necessarily existent by virtue of itself.* The necessarily existent by virtue of itself must be pure *intellect,* since beings free of matter are pure intellect. It must be *true*; for truth consists in the highest grade of existence, and the necessarily existent by virtue of itself would have the highest grade of existence.[98] It must be *good,* for evil consists in privation, whereas the necessarily existent by virtue of itself has fullness of being

[94]*Shifā': Ilāhīyāt,* pp. 344–348; *Najāt,* pp. 228–229; *K. al-Ishārāt,* p. 144.
[95]*Najāt,* pp. 229–230. Cf. *Shifā': Ilāhīyāt,* p. 349.
[96]Above, p. 296.
[97]*Shifā': Ilāhīyāt,* pp. 43–47, 350–354; *Najāt,* pp. 230–234; *K. al-Ishārāt,* p. 143.
[98]Cf. Aristotle, *Metaphysics* II, 1, 993b, 26–31.

and suffers no privation. It must constitute the highest *beauty,* be the highest *object of desire,* be possessed of the greatest *pleasure,* and so forth.[99]

Avicenna's analysis of the concept *necessarily existent by virtue of itself* thus establishes that anything corresponding to the concept must be uncaused, simple, incorporeal, one, pure intellect, truth, good, most beautiful, an object of desire, and possessed of the greatest pleasure. The analysis mirrors Alfarabi's analysis of the First so consistently that we may be certain Avicenna borrowed Alfarabi's analysis of the First, elaborated it somewhat, and applied it to the concept *necessarily existent by virtue of itself.* The critical thought is, as in Proclus and Alfarabi, that any composite considered as a whole is distinct from its parts and dependent on them for its existence; and therefore, what exists solely by reason of itself can have no parts. Significantly, both Proclus and Alfarabi came to their results without reference to necessity, and in Avicenna too the element of necessity plays no role. Avicenna in effect considers the implications of a thing's *existing by virtue of itself.* He could as well have analyzed and read out the attributes of the *actually* existent by reason of itself instead of the *necessarily* existent by virtue of itself.

5. Proof of the existence of the necessarily existent by virtue of itself

Avicenna does not regard the analysis of the concept *necessarily existent by virtue of itself* as sufficient to establish the actual existence of anything in the external world. He does not, in other words, wish to offer an a priori or ontological proof of the existence of God, but rather a new form of the cosmological proof.

He is careful to define the degree and mode in which the existence of God can be established by philosophy. The existence of God is taken by him to be neither self-evident nor unprovable.[100] Nor can the existence of God be established through a syllogistic "demonstration" (*burhān*).[101] A truly demonstrative syllogism must be framed with propositions that are "prior to," and the "causes" of, the conclusion.[102] It is, more precisely, a syllogism in which the middle term is the *cause* of the presence of the major term in the minor term.[103] Since there is nothing prior to, and the cause of, the presence of actual existence in the necessarily existent by virtue of itself, a *demonstrative* syllogism leading to the

[99]*Shifā': Ilāhīyāt,* pp. 355–356, 367–370; *Najāt,* pp. 229, 245. Cf. Aristotle, *Metaphysics* XII, 7, 1072a, 34–35.

[100]Cf. above, p. 284.

[101]*Shifā': Ilāhīyāt,* p. 348. This was a commonplace. Cf. Alexander, *Commentary on Metaphysics,* ed. M. Hayduck, *Commentaria in Aristotelem Graeca,* Vol. I (Berlin, 1891), p. 686; *Aporiai,* ed. I. Bruns, *Commentaria in Aristotelem Graeca,* Supplementary Vol. II/2 (Berlin, 1892), p. 4; Themistius, *Paraphrase of Metaphysics,* ed. S. Landauer, *Commentaria in Aristotelem Graeca,* Vol. V/5 (Berlin, 1903), Hebrew part, p. 11; Proclus, *Liber De Causis,* §5.

[102]Aristotle, *Prior Analytics* I, 2, 71b, 19–32.

[103]Aristotle, *Posterior Analytics* II, 2; Avicenna, *Najāt,* p. 67; *K. al-Ishārāt,* p. 84.

existence of an entity of that description is impossible. What can, however, be provided, according to Avicenna, is a "proof" (*dalīl*).[104] A "proof" is a syllogism wherein the middle term is the *effect* rather than the *cause* of the presence of the major term in the minor term; it is a chain of reasoning that moves not from the prior to the posterior, but from the posterior to the prior, from the presence of the effect to the existence of the cause. Stated in another way, a strictly demonstrative syllogism establishes both "that" a certain proposition is true and "why" it is true, whereas a "proof" establishes only "that" it is true.[105] A *proof* of the existence of God, as distinct from a strict demonstration, will, we are therefore to understand, reason from the existence of a *possibly existent being* to the existence of *necessarily existent being,* even though the former is the effect, not the cause, of the latter.[106]

As Avicenna constructs his proof, it requires three philosophic principles, each of which he also undertakes to prove. These are (a) the principle of causality; (b) the impossibility of an infinite linear regress of causes—two principles that are fundamental to most cosmological proofs of the existence of God;[107] and (c) the impossibility of a circular regress of causes. Avicenna ingeniously establishes the three principles through an analysis of the concepts *possibly existent by virtue of itself* and *necessarily existent by virtue of itself.* Significantly, the second and third principles are not genuinely needed for his proof; Avicenna has, without quite realizing it, developed a cosmological proof that can dispense with the impossibility of an infinite regress.

(*a*) In formulating his version of the principle of causality, Avicenna employs a distinction between the cause of the "generation" (*ḥudūth*) of an object and the cause of its "maintenance" (*thabāt*) in existence.[108] The cause of generation is more obvious, since no one, Avicenna is certain, can doubt that whenever an object comes into existence, it does so by virtue of something else.[109] But Avicenna could not pursue a first cause of the generation of every possibly existent being, since he believed that some possible beings are eternal and have no cause of generation. Furthermore, by establishing a first maintaining cause he will

[104] *Shifā' : Ilāhīyāt,* p. 6.

[105] Aristotle, *Posterior Analytics* I, 6, 75a, 33–35, and Ross's note; I, 13, 78a, 22 ff.; Avicenna, *Najāt,* p. 67; *Shifā': Burhān,* ed. A. Affifi (Cairo, 1956), pp. 79–80. Also cf. *Encyclopedia of Islam,* (2nd ed.; Leiden, 1960–), s.v. *dalīl*; J. van Ess, *Erkenntnislehre des 'Adudaddīn al-Īcī* (Wiesbaden, 1966), p. 367.

[106] Stated in scholastic terminology, the existence of God is susceptible to a *demonstratio quia* although not to a *demonstratio propter quid.* Cf., e.g., Aquinas, *Summa Theologiae,* I, 2, 2, resp.

[107] Cf. H. Wolfson, "Notes on Proofs of the Existence of God in Jewish Philosophy," reprinted in his *Studies in the History of Philosophy and Religion,* Vol. I, p. 572.

[108] For the conception of a maintaining cause, cf. *De Mundo* VI; Plotinus, *Enneads,* VI, 4, 2; *Theology of Aristotle,* ed. F. Dieterici (Leipzig, 1882), p. 79; English translation: *Plotini Opera,* ed. P. Henry and H. Schwyzer, Vol. II (Paris and Brussels, 1959), p. 245; Proclus, *Elements,* §13, and Dodds' notes.

[109] *Najāt,* p. 236.

establish not merely a first cause that exercised its causality at a moment in the past and withdrew, but, as it were, a stronger deity, a first cause that continually maintains the universe in existence. Avicenna's proof herein parallels Aristotle's proof from motion. The proof from motion similarly sought not a cause that initiates the motion of the spheres, the motion of the spheres being eternal for Aristotle, but a cause that continually moves the spheres,[110] in other words, a maintaining cause of the motion of the spheres.

Avicenna gives his attention, then, to a maintaining cause—not to the cause maintaining a moving object in motion, but, more comprehensively, to the cause maintaining a possibly existent object in existence. He looks at objects of the type he had designated as *possibly existent by virtue of themselves, necessarily existent by virtue of another,* that is to say, objects that actually exist although they are in themselves only possibly existent.[111] Concerning any such object, Avicenna reasons, irrespective of whether it is generated or eternal, we may legitimately ask what maintains it in existence. Since the possibly existent is something that by definition does not exist by virtue of itself, should a possibly existent object actually exist, some factor distinct from it would have to be responsible for its existence. And some factor would have to be present and maintain the possible object in existence as long as the object exists; for even when the possibly existent is already actual it never ceases to be possible by virtue of itself and thereby dependent on something else for its continued existence.[112] Avicenna acknowledges that the maintaining factor may be a component within the total object. For example, the factor maintaining a statute in a given form is the stability of the material from which the statute is made.[113] But a component is still different from the whole[114] so that here too, the factor maintaining the object is distinct from the object considered as a whole. If the component—for instance, the stability of the material—is also possibly existent, inquiry can, of course, legitimately be made regarding the factor maintaining it in existence.

The analysis of the concept *possibly existent by virtue of itself*—to be precise, merely asking what *possibly existent* means—has disclosed that if anything possibly existent should exist, it must at all times depend on a cause distinct from itself to maintain it in existence.

(*b*) The second premise established by Avicenna is formulated by him as the impossibility that "causes go to infinity," the impossibility of an infinite regress of causes. In fact, unlike other philosophers, and unlike his own procedure in another work,[115] Avicenna does not, in the proof we are now examining, directly argue that an infinite regress is, for one reason or another, absurd. Instead he

[110] Cf. above, p. 238.
[111] Cf. above, pp. 291–292.
[112] Cf. above, p. 292.
[113] *Najāt*, p. 237.
[114] Cf. above, p. 294.
[115] Cf. below, p. 339.

first argues for the broader proposition that the totality of actually existent possible beings, "whether finite or infinite,"[116] must depend on a being necessarily existent by virtue of itself; and then, from the broader proposition, he infers the impossibility of an infinite regress as a corollary.

Avicenna is thinking of a situation wherein *z*, for example, is maintained in existence by *y*, which exists simultaneously with it; and wherein *y* is maintained in existence by *x*, which likewise exists simultaneously; *ad infinitum*. To show that a situation of the sort cannot account for the totality of existence, he mentally collects into a single group all possible beings actually existing at a single moment,[117] asks what maintains the group in existence, examines all conceivable answers to the question, and arrives at the alternative he deems correct. His reasoning is: The totality of possibly existent beings, taken as a whole, must be either (α) necessarily existent by virtue of itself or (β) possibly existent by virtue of itself. The former alternative would imply the thesis that the "necessarily existent [by virtue of itself] is composed of possibly existent beings," which would be "a contradiction." Just where the contradiction lies is not made explicit by Avicenna. But granting the conception of *necessarily existent by virtue of itself* as that which does not even have internal factors making it what it is, the contradiction is plain: If something does not have even internal factors making it what it is, it cannot, as Avicenna earlier pointed out, be composite.[118]

Since the totality of possibly existent beings existing together at any moment cannot (α) constitute an entity that is necessarily existent by virtue of itself, there remains (β) the second alternative according to which the totality, taken as a whole, is possible by virtue of itself. But the possibly existent by virtue of itself needs something to maintain it in existence. Hence, on this alternative, "whether the group is finite or infinite," it stands in need of a factor that will continually "provide [it] with existence." The factor, Avicenna assumes, must be either (β1) within the group or (β2) outside it. Assuming that the whole group is (β1) ultimately maintained by one of its own members would, however, be tantamount to assuming that the member in question is a cause of itself. For to be a cause of the existence of a group is "primarily" to be the cause of the individual members; and inasmuch as the supposed cause is itself one of the members, it would perforce be a cause of itself. Yet the supposed cause has already been assumed, as one of the members of the group, to be possibly existent; and the possibly existent is precisely what does not exist by reason of itself. Therefore it could not be the cause of the collection of which it is one member.

If the totality of possibly existent beings cannot (α) form a group that is necessarily existent by virtue of itself, and if, moreover, the ultimate maintaining

[116]Cf. the passage from Aristotle cited below, p. 337.

[117]He has no qualms about treating an infinite number of objects as a single totality.

[118]Ghazali, *Tahāfut al-Falāsifa*, ed. M. Bouyges (Beirut, 1927), IV, §6, understood Avicenna in this way; English translation in *Averroes' Tahafut al-Tahafut*, trans. S. van den Bergh, (London, 1954), p. 161.

cause cannot be (β1) one of its own members, the sole remaining alternative is that what does maintain the totality of possibly existent beings in existence is (β2) something outside the group. And, since, by hypothesis, all possibly existent beings were included inside, anything left outside is not possibly existent, but must be necessarily existent by virtue of itself.

Avicenna should have stopped here. The totality of all actual beings that are possibly existent by virtue of themselves, he has concluded, depends on a being that is necessarily existent by virtue of itself; and that is as much as is required for his proof of the existence of God. But Avicenna wanted an explicit statement of the impossibility of an infinite regress of causes. Once he has established that a linear series, constituted by the possible beings existing together at any moment, must depend on something necessarily existent by virtue of itself, he goes on to infer, as a kind of corollary, that the series must also be finite. The series of possibly existent causes must, he reasons, "meet" its necessarily existent first cause and "terminate" there; and as a consequence, an infinite regress of causes is impossible—a regress, it may be repeated, of the type in which all the links exist together and each link maintains the next in existence.[119]

(*c*) Avicenna understands that his proof requires one more principle, the impossibility of a self-contained regress of causes, a regress that is "circular" and "finite" rather than infinite and linear. A circular regress is a situation wherein *x*, *y*, and *z*, for example, exist simultaneously in the manner that *x* is the cause maintaining *y* in existence, *y* is the case of *z*, but *z* is the cause of *x*. The impossibility of a situation of the sort can, Avicenna explains, be exhibited in two ways. First, the impossibility of a self-contained circular regress can be exhibited by a "similar proof" to that whereby, in establishing his second principle, an infinite linear regress of possibly existent beings was shown to be impossible.[120] Avicenna means that all the links in the supposed circular chain, like the links in a linear chain, would be caused and possibly existent by virtue of themselves; taken as a totality, the chain would remain possibly existent; an additional factor would be required in order to render it necessary and actual; and since the series would have to meet and terminate at the additional factor, it could not after all form a closed circle. Secondly, a self-contained circular regress is shown to be absurd by an argument applying only to it. In the circular regress *x y z*, *x* would be a distant cause of *z*, and *z* would be the immediate cause of *x*. *x* would consequently be a distant cause of itself, which Avicenna regards as absurd. By the same token, *x* would be a distant effect of itself, which is equally absurd. And the point can be made again in a slightly different way, as follows: *x* would be dependent for its existence upon something—*z*—whose existence is posterior to it. But "when the existence of something depends upon the existence of something else that is essentially posterior to the first, the existence of the first

[119] *Najāt*, p. 235; *K. al-Ishārāt*, pp. 141–142.
[120] *Najāt*, p. 236.

is impossible."[121] A self-contained circular regress of causes cannot, therefore, exist.

In addition to these three philosophical principles—the principle of causality, the impossibility of an infinite linear regress of causes, and the impossibility of a finite circular regress of causes—Avicenna leaves the conceptual realm for a single empirical datum in order to accomplish his proof of the existence of God: "There is no doubt that something exists (*anna hunā wujūdan*)."[122] It makes no difference what happens to exist or what the object's peculiar properties might be; for the purpose of his proof Avicenna is concerned with the "existent *qua* existent"[123] and therefore all he needs is the fact that something does in truth exist. Applying the proposition that there are only two conceivable categories of actually existing beings,[124] Avicenna proceeds: "Everything that exists is either necessary [by virtue of itself], or possible [by virtue of itself and necessary by virtue of another]. On the former assumption, a necessarily existent [by virtue of itself] has been established, and that was the object of our proof; on the other assumption, we must show that the existence of the possible [by virtue of itself but necessary by virtue of another] ends at the necessarily existent [by virtue of itself]."[125] That is to say, if the random existent object with which we start is conceded to be necessarily existent by virtue of itself, it may be assigned all the attributes already shown to belong to such a being, and the proof of the existence of God is complete. But the real issue is of course posed by the other alternative, the assumption that the random object with which we start is necessarily existent only by virtue of another, and possibly existent by virtue of itself. A true proof of the existence of God has the task of showing that anything possibly existent by virtue of itself ultimately depends for its actual existence upon something necessary by virtue of itself.

Assuming that the actually existent object we start with is possible by virtue of itself, it must be maintained in existence by something else that exists together with it (principle of causality). The other factor, in turn, must be either necessary by virtue of itself or possible by virtue of itself. If it is assumed to be necessary by virtue of itself, the proof is again at once complete. If, on the contrary, it is assumed to be possible by virtue of itself, it too must depend on a further factor distinct from it and existing together with it. Once again, Avicenna asks whether the new factor is necessary by virtue of itself or possible by virtue of itself. It is inconceivable, he has established, that anything should be maintained in existence

[121] *Najāt*, p. 236; this principle is not brought into the statement of the proof in *K. al-Ishārāt*, pp. 141–142.

Cf. Aristotle, *Physics* VIII, 5, 257b, 13–20; Alexander(?), *Mabādi' al-Kull*, in *Arisṭū 'ind al-'Arab*, p. 259.

[122] *Najāt*, p. 235.

[123] Above, pp. 286–287.

[124] Above, p. 291.

[125] *Najāt*, p. 235.

either by an infinite linear regress of causes or by a circular regress of causes. The series of causes maintaining any given thing in existence must consequently terminate at a being that is necessarily existent by virtue of itself.[126] And the latter may now be assigned all the attributes earlier shown to belong to the necessarily existent by virtue of itself: It is uncaused, simple, and incorporeal; there is only one being answering the description; it consists in pure intellect; it is, in the highest degree, true, good, beautiful, an object of desire, and possessed of pleasure.[127]

The syllogism that encapsules the entire "proof" (*dalīl*)[128] might be summarized thus: Possibly existent beings are traceable to a being necessarily existent by virtue of itself (from the three philosophic principles established by Avicenna).[129] Something exists which is, presumably, possibly existent by virtue of itself (empirical datum).[130] Therefore something exists which is traceable to a being necessarily existent by virtue of itself; and the latter also exists.

6. Questions raised by Avicenna's proof

Underlying all of Avicenna's argumentation is his analysis of the concept *necessarily existent by virtue of itself,* an analysis dependent on an unusual turn of thought. Avicenna contends that whatever truly exists by reason of itself cannot exist by reason even of internal factors making it what it is. As a consequence, the necessarily existent by virtue of itself cannot contain internal factors and must be free of all composition; and carrying the reasoning forward shows that there can exist no more than one such being.[131] The analysis is essential to the entire proof because it helps establish the existence of a being necessarily existent by virtue of itself[132] and also enables Avicenna to assign to the being in question the attributes of a deity.[133] At first blush Avicenna appears to be speculating about the correct meaning of a concept known to him through an immediate intuition of some kind.[134] But if that is in fact all he is doing, one can demur at his intuition and affirm, on the contrary, that a being may be deemed necessarily existent by virtue of itself even when existing by virtue of its parts; and the entire proof constructed upon the analysis of the concept would collapse. Avicenna's procedure can be defended only if it is something other than the taking of one side in a purely verbal or intuitive dispute, a dispute regarding whether the necessarily existent by virtue of itself may or may not properly be said to exist by virtue of

[126] *Najāt*, p. 239; *K. al-Ishārāt*, pp. 141–142.
[127] Cf. above, pp. 297–298.
[128] Cf. above, p. 299.
[129] Above, pp. 299–302.
[130] Above, p. 303.
[131] Above, p. 294.
[132] Above, pp. 301, 302.
[133] Above, pp. 297–298.
[134] Cf. above, p. 290.

its parts. His procedure can be defended only if understood not as seeking to discover the meaning of the term but rather as working from a definition.[135] He must be understood to have arbitrarily defined *necessarily existent by virtue of itself* as that which, taken as a whole, exists solely by virtue of itself, not by virtue even of internal factors making it what it is; and to have analyzed out the implications of the definition.

The heart of Avicenna's proof is the train of reasoning whereby he establishes that possibly existent beings ultimately owe their existence to a being necessarily existent by virtue of itself.[136] As a first step, he rules out the thesis (α) that all possibly existent beings existing at any one moment might, as a totality, comprise a being that is necessary by virtue of itself; in effect, what is being ruled out is that the physical universe as a whole could exist necessarily by virtue of itself. Avicenna does not explain exactly why possibly existent beings cannot add up to a being necessarily existent by virtue of itself; but, as was pointed out, the reason must be that the latter, in Avicenna's usage, can contain no components whatsoever.[137] The properties of the necessarily existent by virtue of itself, it was just seen, are nothing other than the implications of a definition. It is, consequently, because Avicenna has defined necessarily existent by virtue of itself in a certain fashion that a multiplicity of possibly existent beings cannot add up to such a being.

The step ruling out alternative (α) may be read, then, as definitional. The next step runs: If the totality of possibly existent beings does not comprise a being that is necessarily existent by virtue of itself, the totality must be (β) possibly existent by virtue of itself. And since the totality is possibly existent, something must maintain it in existence. Avicenna lays down two alternatives, namely that the totality is maintained either (β1) by one of the possibly existent beings contained within it, or (β2) by something outside; he rejects the former alternative and thereby affirms the latter.[138] Curiously, however, he does not consider a further alternative, which we may call (β3), the thesis that the totality is maintained in existence not be a single component but by all the components together. On this alternative the totality of possibly existent beings—in effect, the entire universe—would indeed be possibly existent in Avicenna's sense; for it would, taken as a whole, exist by reason of something different from itself. Still, it would not exist by reason of anything external to it, but would be possibly existent only inasmuch as it exists by reason of its own components. It would be possibly existent by virtue of itself, necessarily existent by virtue of its components.

If thus challenged, Avicenna might perhaps have replied that each of the possibly existent components would be part of the cause of the existence of the

[135]Not a strict logical definition; cf. above, p. 291.

[136]Above, pp. 301–302.

[137]Above, p. 301.

[138]Above, ibid.

totality; and since the cause of a whole is primarily the cause of the components making up the whole, each of the possibly existent components would be part of the cause of the existence of itself.[139] By a similar argument, as will be recalled, Avicenna rejected the alternative that a single possibly existent being could be the cause of the totality of which it is one component; he reasoned that since the cause of any whole is primarily the cause of the components, the possibly existent being in question would be the cause of itself, whereas the possibly existent is precisely what does not exist by virtue of itself.[140] But to eliminate the new alternative, according to which the components together maintain the totality in existence. Avicenna would have to show that a possibly existent being cannot be even part of the cause of the existence of itself. Until doing so, he cannot rule out the thesis that the totality of possibly existent beings, in other words, the entire universe, exists by virtue of all its components.[141]

The foregoing can be pursued a little further. There is a way to construe the possibly existent totality as maintained in existence through its own components without even conceding that the components would be part of the cause of their own existence. The cause of the existence of each component may be construed as the sum of its subcomponents. The cause of the existence of each subcomponent could, in turn, be construed as the sum of its own subcomponents, and so on, *ad infinitum*. For at the present stage, Avicenna is still entertaining the possibility of an infinite regress; the impossibility of an infinite regress is what he is now attempting to establish.

Avicenna seems, in fine, to have made the following misstep: Having established, with the aid of a definition, that the possible beings existing at any moment must form a possibly existent totality, he fails to consider that the totality might exist not by reason of a single component but by reason of the components together.

The same criticism may be put in a slightly different fashion by returning to a stage in Avicenna's argumentation which was left unquestioned in the preceding paragraphs. Avicenna contended that a series of possibly existent beings must add up either to (α) a totality that is necessarily existent by virtue of itself in the special sense of not containing even internal factors making it what it is, or (β) a totality that is possibly existent by virtue of itself in the sense of being maintained in existence by something—whether internal or external—different from itself taken as a whole. The dichotomy reflects Avicenna's original distinction of only two classes of actual existence.[142] But, one may interject, it would be more

[139] Alternatively, Avicenna might employ an argument from his refutation of a circular regress and reason that the thesis I have suggested would imply that each component would be a partial distant cause—as well as effect—of itself; cf. above, p. 302. But then it would remain to be shown that nothing can in fact be even a partial distant cause of its being maintained in existence.

[140] Cf. above, p. 301.

[141] As an analogy, we may think of a round arch in which each stone is part of the cause maintaining the whole in position, and thereby part of the cause maintaining itself in position.

[142] Above, p. 291.

illuminating here, as well as in Avicenna's original distinction of the classes of actual existence, to draw a trichotomy rather than a dichotomy, and to state instead: All possibly existent beings existing at any moment must add up to a totality that is either (α) necessarily existent by virtue of itself in the special sense of not even having internal causes, (β) possibly existent by virtue of itself in the sense of having an external cause, or (γ) necessarily existent by virtue of itself in the weaker sense of having no external cause, although it might have internal causes. The first alternative, it may be conceded, is impossible: the totality of possibly existent beings cannot be necessarily existent by virtue of itself in the special sense of having no internal causes. But in order to establish the second alternative, it remains to be shown that the totality also cannot be necessarily existent in the weaker sense, in sense (γ). It remains to be shown that a series of possibly existent beings cannot add up to a being necessarily existent in the sense of having no external causes, although it does have internal causes—more specifically, although it has all its components as internal causes. This is another way of putting the objection already suggested in the preceding paragraphs, but it is a formulation giving the gist of what was to be Ghazali's refutation of Avicenna's proof.[143]

7. The version of Avicenna's proof in Shahrastānī and Crescas

A certain awkwardness in Avicenna's argumentation was mentioned earlier. In the course of establishing the second principle required for his proof, the impossibility of an infinite linear regress of causes, Avicenna demonstrates the critical preliminary proposition that the totality of possibly existent beings must depend for its existence on a being that is necessarily existent by virtue of itself; and then, as a corollary, he infers herefrom the impossibility of an infinite linear regress of causes.[144] Avicenna employs the same preliminary proposition when establishing his third principle, the impossibility of a circular regress of causes.[145] After establishing the principles, he goes on to reason that the series of causes maintaining a given possibly existent being in existence cannot regress indefinitely either linearly or circularly but must terminate at, what he had already demonstrated in the preliminary proposition, a being that is necessarily existent by virtue of itself. He uses the preliminary proposition—that all possibly existent beings ultimately depend on a being necessarily existent by virtue of itself—to establish the impossibility of an infinite linear or a circular regress of causes; and he thereupon uses the impossibility of a regress of causes to prove over again what he already had proved in the preliminary proposition.

The circuitous and redundant route followed by Avicenna must have been due to the influence upon him of other proofs of the existence of God; he illogically forced his own proof into the mold of familiar cosmological proofs that do explicitly reject an infinite regress of causes.[146] The proof is obviously simpler and

[143]Cf. below, pp. 371–372.

[144]Above, p. 302.

[145]Above, ibid.

[146]Cf. previous and following chapter; above, n. 107.

more logical when the issue of an infinite regress is set aside. The argument will then run: A totality of possibly existent beings, whether infinite or finite, and whether arranged in a linear or circular series or in any other manner, must depend on a being necessarily existent by virtue of itself; something actually exists; that thing either must be necessarily existent by virtue of itself or must ultimately depend on something necessarily existent by virtue of itself; therefore, a being necessarily existent by virtue of itself must exist. In this form the proof is not merely simpler and more logical; it also reveals its originality, vis-à-vis Aristotelian proofs of the existence of God, in dispensing with the device of tracing a chain of causes back link by link to a first cause. Even in the new form, the objection suggested in the previous section[147] remains to be answered.

The more straightforward formulation of the proof appears in Shahrastānī's restatement of Avicenna's philosophy and in the Jewish philosopher Ḥasdai Crescas. Shahrastānī looks at the totality of "possibly existent" objects making up the universe and argues: the "totality, taken as a totality, and whether finite or infinite, . . . must be either necessary by virtue of itself or possible by virtue of itself." A totality of possibly existent beings cannot be construed as necessarily existent by virtue of itself, since the "necessarily existent" cannot be "composed of possibly existent beings." The totality of possibly existent beings, taken as a totality, must therefore be possibly existent, and must stand in need of something to provide it with existence. If the source of existence "were inside the totality, one of the members would be necessarily existent"; for whatever provides its own existence is necessarily existent by virtue of itself. But by hypothesis all possibly existent beings and only possibly existent beings were included within the totality. Consequently, the ultimate source of existence for the totality of possibly existent beings must lie outside the totality. The source of existence of the possibly existent beings hence cannot be possibly existent, but must be necessarily existent by virtue of itself, possessing all the attributes that are analyzable out of the concept *necessarily existent by virtue of itself*.[148] Shahrastānī's version of the proof, it will be observed, arrives at a being necessarily existent by virtue of itself with no reference to the impossibility of an infinite regress.

Ḥasdai Crescas explicitly argues that an infinite regress of causes is possible.[149] But, he continues: "Whether [the series of] causes and effects should be finite or infinite, there is no escaping a cause for the totality. For if all the members should be caused, they would be possibly existent by virtue of themselves and would require a factor to tip the scales in favor of their existence and against their nonexistence." The factor that "tips the scales" in favor of existence would be the "cause of the totality. . . . This is the deity."[150]

[147] Above, pp. 306–307.
[148] Shahrastānī, *K. al-Milal wal-niḥal*, ed. W. Cureton (London, 1842–1846), p. 376.
[149] Cf. below, p. 365.
[150] *Or ha-Shem*, I, iii, 2. Spinoza observed the character of Crescas' proof; see Spinoza, *Correspondence, Letter XII*.

8. Summary

Avicenna offers a proof of the existence of God as the first cause of the very existence of the universe, in contrast to Aristotle's proof of a first cause merely of the motion of the universe. He proposes to offer precisely a *proof* as distinct from a *demonstration*; a demonstration, in the strict sense, would consist in a syllogism of the type that reveals the cause of the conclusion, whereas the existence of God has no cause. The proposed proof, Avicenna further explains, is undertaken in the discipline of *metaphysics,* since it considers the attributes of the existent solely insofar as existent; it is thereby distinguished from a proof undertaken in the discipline of *physics,* the science that considers the attributes peculiar to one type of existent, namely movable bodies.

The conceptual framework within which Avicenna works is constructed out of elements from diverse sources. His philosophic terminology is Aristotelian. Also genuinely Aristotelian are the definition of the *subject matter* and *subject of inquiry* of the science of metaphysics; the designation of the first cause of the universe as *necessarily existent*; and the distinction between what is *necessary by virtue of itself* and what is *necessary by virtue of another.* From Proclus, and especially from Alfarabi, Avicenna learned the method whereby he derives a set of attributes from the concept *necessarily existent by virtue of itself*; most significantly, he learned that whatever is wholly uncaused cannot even contain internal factors making it what it is.

Avicenna's proof runs as follows: Three categories may be distinguished: (a) the necessarily existent by virtue of itself; (b) the necessarily existent by virtue of another, possibly existent by virtue of itself; and (c) the possibly existent by virtue of itself which is not rendered necessarily existent by something else. But what is actually existent must fall within one or another of the first two categories; in fact, the concept *necessarily existent* as used by Avicenna has little discernible meaning beyond *actually existent*. The object of Avicenna's proof is to show that actual existence is not restricted exclusively to the second category—the possibly existent by virtue of itself, necessarily existent by virtue of another—and that there also exists something necessarily existent by virtue of itself.

From an analysis of the concept *necessarily existent by virtue of itself,* Avicenna derives the attributes of such a being: incorporeality, simplicity, unity, and the like. From an analysis of the concept *possibly existent by virtue of itself, necessarily existent by virtue of another,* he shows that anything of that description must depend on a factor distinct from it to maintain it in existence. From an analysis of both concepts, he shows that the sum of all possibly existent beings cannot constitute a being necessarily existent by virtue of itself, but must depend for its existence on a factor outside it, hence upon a factor that is necessarily existent by virtue of itself. Now some object undoubtedly does actually exist. If the object should be assumed to be necessarily existent by virtue of itself, the proof would immediately be complete. But even if the object in question should be assumed to be possibly existent by virtue of itself, the sum of all objects of the sort, as already shown, must ultimately depend on a being that is necessarily existent by virtue of itself. In either instance, therefore, the existence of a being necessarily existent by virtue of itself is reached. And that being must possess all attributes analyzable out of the concept; it must possess the attributes of the deity. The reasoning can be pursued without reference of the impossibility of an infinite regress of causes and it was so pursued after Avicenna by Shahrastānī and Ḥasdai Crescas. Avicenna himself unnecessarily forced his proof into the form of a proof employing the impossibility of an infinite regress.

The proof became highly influential, eliciting refutations and adaptations. Chapter X will deal with Averroes' critique; Chapter XI, with the utilization of Avicenna's reasoning in arguments for the existence of God from the impossibility of an infinite regress of causes, and also with Ghazali's critique; Chapter XII, with other adaptations.

X

Averroes' Critique of Avicenna's Proof

Averroes appears to have been obsessed by Avicenna's proof of the existence of God from the concepts *possibly existent* and *necessarily existent*. He discusses the proof in at least a dozen works, and the preliminary issues, which he found even more troublesome than the body of the proof, presented difficulties that took him years to solve. His various references to the proof address four issues, of which three are preliminary to the body of the proof and only the fourth concerns the latter.

(1) First Averroes examines Avicenna's characterization of his proof of the existence of God as a metaphysical proof. After bringing to bear the formal Aristotelian rules for determining the subject matter and principles of a science, Averroes concludes that the existence of a first cause cannot after all be a problem for the science of metaphysics. (2) Then Averroes examines Avicenna's division of being, giving particular attention to the designation *possible by virtue of itself, necessary by virtue of another*; and he rejects the designation as unsound. (3) An examination of Avicenna's position on the nature of the celestial spheres, and the effort to clarify his own position on the nature of the spheres pose philosophic problems, which Averroes struggles to solve. (4) Finally, by the side of these preliminary matters, he undertakes a critique of the body of Avicenna's proof. The confrontation with Avicenna leads Averroes to clarify his own philosophic positions; and it reveals something of the character of Averroes the philosopher. One trait was to be expected in the most devoted of Aristotle's commentators: Averroes approaches Avicenna's proof from a rigorously Aristotelian viewpoint and attacks Avicenna whenever he senses a divergence from Aristotle. But other traits might not have been expected. Averroes is not well informed about the positions he is criticizing, and he wavers in his solutions of the philosophic difficulties that arise in the course of his critique. What Averroes is most certain of is that "Avicenna erred exceedingly . . . ,"[1] and even there Averroes wavers.

[1] *Long Commentary on Physics*, in *Aristotelis Opera cum Averrois Commentariis*, Vol. IV (Venice, 1562), I, comm. 83; cited by H. Wolfson, "Averroes' Lost Treatise on the Prime Mover," reprinted in his *Studies in the History of Philosophy and Religion*, Vol. I, (Cambridge, Mass., 1973), pp. 410–411.

1. The proof of the existence of God as a subject for physics

In one work, Averroes gives the following account of Avicenna's method of proving the existence of God: Avicenna proposed to construct a proof of the existence of God through "an examination of the nature (*ṭabīʿa)* of the existent *qua* existent." That method of proving the existence of God was contrasted by Avicenna with another method, which had been employed by the "ancients"; the ancients had taken their departure from "time and motion, . . . things posterior" to the first cause of the universe, and had reasoned back from time and motion to a first cause. Avicenna further contended that his method, which consists in an examination of the "existent *qua* existent," was "more exalted" than the older method.[2]

Averroes' account is a fairly faithful reflection of a passage quoted in the previous chapter wherein Avicenna proclaimed the superiority of his new proof of the existence of God.[3] In the passage in question, Avicenna was alluding to a distinction between a *metaphysical* and a *physical* proof of the existence of God. And Averroes does in fact several times explicitly draw the contrast between Avicenna's proposed metaphysical proof of the existence of God and Aristotle's physical proof.[4] Averroes concedes, moreover, that if the proposed metaphysical proof could indeed "arrive" at the existence of an incorporeal first cause, Avicenna's declaration regarding the superiority of the proof would have been "valid."[5] Averroes finds, however, that a metaphysical proof of the existence of God is intrinsically impossible.

When Avicenna argued that proving the existence of God belongs properly to the science of metaphysics, he applied Aristotle's rules for defining the subject matter, the principles, and the subject of inquiry of any given science. Averroes now cites the Aristotelian rules to establish, on the contrary, that proving the existence of God is not after all a subject for metaphysics. Aristotle, in Averroes' words, explained that "it is impossible for any science to demonstrate the existence of its own subject [matter]." The existence of the subject matter is rather something that each science "accepts,"[6] whereupon the science has the task of demonstrating that its subject matter possesses certain essential attributes.[7] But "incorporeal beings," Averroes continues, "are the subject matter of first philosophy [that is, metaphysics]." And inasmuch as a science does not demonstrate the existence of its own subject matter, "first philosophy" cannot have the task

[2] *Tahāfut al-Tahāfut*, ed. M. Bouyges (Beirut, 1930), X, p. 419; cf. IV, p. 276 and VIII, p. 393. English translation, with pagination of the Arabic indicated: *Averroes' Tahafut al-Tahafut*, trans. S. van den Bergh (London, 1954).

[3] Cf. above, p. 287.

[4] Cf. the passages cited by Wolfson, "Averroes' Lost Treatise," pp. 407–415.

[5] *Tahāfut al-Tahāfut*, X, p. 419.

[6] *Long Commentary on Physics*, I, comm. 83; cited by Wolfson, "Averroes' Lost Treatise," pp. 410–411.

[7] Cf. above, p. 284.

of "proving that incorporeal beings exist."[8] Averroes supports his position by quoting statements of Aristotle's to the effect that the task of metaphysics is to discover "how incorporeal being is disposed and what it is,"[9] to discover "what" the formal principle of the universe is and "whether it is one or many."[10] The implication of the statements, as Averroes reads them, is that metaphysics investigates the "dispositions . . . and essences" of incorporeal beings, thereby investigating the *attributes* of its subject matter. Metaphysics cannot, accordingly, investigate the "existence" of incorporeal beings, since it then would be investigating the *existence* of its subject matter.[11] The existence of incorporeal beings might only be dealt with in a different science.

The statement that "incorporeal beings" constitute the subject matter of metaphysics is problematical, because Averroes also has a different characterization of metaphysics; he describes it as the "universal" science whose "subject matter" is not any one segment of existence, but, more widely, "existence, with no qualification."[12] The differing descriptions of the subject matter of metaphysics do not originate in Averroes; Aristotle too had identified the subject matter of metaphysics as "immovable substance"[13] and again, more widely, as the "existent *qua* existent, universally and not partially."[14] In one passage Aristotle had even combined the two statements on the subject matter of metaphysics, with no clear indication as to how the subject matter of metaphysics can be construed in two disparate ways.[15] Averroes for his part does suggest a harmonization of the two descriptions. Every science, as has been seen, has the task of studying the attributes belonging to its own subject matter. But, Averroes explains, the attributes belonging to the existent solely insofar as it is existent, and not insofar as it is any specific sort of existent, are identical with the attributes of incorporeal beings.[16] Therefore, by studying the attributes of the one, metaphysics at the same time studies the attributes of the other, and both can properly be taken as the subject matter of metaphysics.

[8] *Long Commentary on Physics,* I, comm. 83; cited by Wolfson, "Averroes' Lost Treatise," pp. 410–411. Cf. also Averroes, *Epitome of Metaphysics,* ed. and trans. C. Quirós Rodríguez (Madrid, 1919), I, §9; German translation: *Die Epitome der Metaphysik des Averroes,* trans. S. van den Bergh (Leiden, 1924), p. 4.

[9] *Physics,* II, 2, 194b, 14–15.

[10] *Physics,* I, 9, 192a, 34–35.

[11] *Long Commentary on Physics,* II, comm. 26; cited by Wolfson, "Averroes' Lost Treatise," pp. 412–413.

[12] *Epitome of Metaphysics,* I, §§2, 4.

[13] *Metaphysics* VI, 1, 1026a, 29.

[14] *Metaphysics* XI, 3, 1060b, 31–32.

[15] *Metaphysics* VI, 1. See Ross's note in his edition of the *Metaphysics,* Vol. I, (Oxford, 1924), pp. 252–253. H. Wolfson, "The Classification of Sciences in Mediaeval Jewish Philosophy," reprinted in his *Studies in the History of Philosophy and Religion,* Vol. I, pp. 517–520, cites various medieval statements of the subject matter of metaphysics.

[16] *Epitome of Metaphysics,* I, §4, where Averroes explains that the "general" attributes that "affect sensible objects insofar as they are existent" are attributes "proper to incorporeal objects."

So far, Averroes' criticism of Avicenna is straightforward: Incorporeal beings are the subject matter of metaphysics; no science can demonstrate the existence of its own subject matter; hence Avicenna was mistaken in maintaining that the existence of God can be proved in the science of metaphysics.

Elsewhere, Averroes approaches the issue from a different vantage point. According to another of Aristotle's rules, not only must the subject matter be accepted, rather than demonstrated, by any given science; the "principles" of the science must likewise be "accepted" rather than demonstrated.[17] The term *principles* here is understood by Averroes, as it had been understood by Avicenna, in a double sense. Averroes understands that a science must accept and presuppose the principles, or fundamental premises, from which it reasons, and also the existence of the principles, or causes, of its own subject matter.[18]

The rule comes into play in connection with a commentary on Aristotle's *Metaphysics* XII, regarded by Averroes as part of the genuine commentary of Alexander of Aphrodisias on the *Metaphysics*.[19] In the commentary Alexander, or an unknown pseudo-Alexander, wrote: "It is for metaphysics to demonstrate what the principles of existent things are";[20] physics thereupon "accepts those principles from metaphysics"[21] and "uses" them "as something it does not itself prove but only assumes."[22] "Taken at face value," Averroes concedes, Alexander is asserting that the "student of physics examines movable substance only after having accepted the principles of movable substance from the student of metaphysics; and the latter would have the task of proving the existence of those principles."[23] Several steps in the reasoning of Alexander and Averroes have been left implicit. The subject matter of physics, we are to remember, is movable, corporeal substance. The principles, or causes, of corporeal substances are incorporeal substances, including the deity. By Aristotle's rule, a science cannot demonstrate its own principles; that is to say, it cannot establish the existence of the principles, or causes, of its own subject matter. The science of physics could not, then, demonstrate the existence of incorporeal substances, inasmuch as they are the principles of the subject matter of physics. But if the science of physics cannot demonstrate the existence of incorporeal beings, the existence of incorporeal beings would have to be established in the science of metaphysics and

[17] Cf. above, p. 285.

[18] For the connection between the two senses, see Averroes' comment on *Posterior Analytics* I, 9, 76a, 16–30, in his *Long Commentary* on same, I, comm. 70 and 72.

[19] J. Freudenthal, *Die durch Averroes erhaltenen Fragmente Alexanders zur Metaphysik des Aristoteles* (Berlin, 1885), argues that the commentary known to Averroes was the genuine commentary of Alexander.

[20] *Long Commentary on Metaphysics*, XII, comm. 5; Arabic original: *Tafsīr mā ba'd al-Ṭabī'a*, ed. M. Bouyges (Beirut, 1938–1948), p. 1420.

[21] Ibid., comm. 6; *Tafsīr*, p. 1429.

[22] *Tafsīr*, p. 1420. These passages are cited by Freudenthal, *Die durch Averroes erhaltenen Fragmente*, pp. 72, 74.

[23] *Tafsīr*, p. 1429.

accepted therefrom by the science of physics. And thus metaphysics must have the task of demonstrating the existence of incorporeal beings, including the deity. Such, Averroes concedes, is the ostensible purport of Alexander's statements.

If, however, that is what Alexander meant, he did not, Averroes insists, correctly understand the relationship between physics and metaphysics. First, Averroes has already shown that metaphysics cannot establish the existence of incorporeal beings, inasmuch as incorporeal beings constitute the subject matter of metaphysics, and no science can establish the existence of its own subject matter. Secondly, Aristotle proved the existence of an incorporeal mover in the eighth book of the *Physics,* thereby indicating the science of physics as the discipline for establishing the existence of incorporeal beings.[24] These considerations, Averroes is satisfied, reveal that proving the existence of incorporeal beings is a task for physics.

And yet Aristotle did, as stated in Alexander's commentary, affirm that a science cannot demonstrate its own principles. Averroes feels called upon, therefore, to explain how physics can, despite that rule, establish the existence of the principles, or causes, of the subject matter of physics. He has recourse to the distinction, already met in Avicenna,[25] between *demonstration* and *proof*. The rule that a science does not establish the existence of its own subject matter, Averroes writes, admits no qualification; "a science . . . can in no way establish that its own subject matter exists—neither by a proof nor by a [strict] demonstration."[26] The reason this rule admits no qualification is presumably that nothing at all can be said in a science until at least the subject matter of the science is granted. The other rule, however, by which a science cannot demonstrate, but must accept its own principles, does admit qualification. The rule does not necessarily mean that a science must accept its principles from another source. It means merely that the science cannot establish its principles through "an absolute demonstration [in other words, through a syllogism] that reveals the cause of the existence [of the subject of the conclusion]."[27] A demonstrative syllogism proceeds from cause to effect. As a consequence, if the principles of the subject matter of a given science are to be established through a demonstrative syllogism, the demonstration would have to be framed not in the science itself, but in a prior, more comprehensive science—to be precise, in a twice removed science, one dealing with entities that are the causes and principles of the principles of the subject matter of the given science.[28] That is the reason why no science can give a strict demonstration of its own principles. In the present instance, it happens that not only is physics unable to demonstrate the principles of its own

[24] *Tafsīr,* pp. 1422, 1424–1425.

[25] Above, pp. 298–299.

[26] *Long Commentary on Physics,* II, comm. 26.

[27] *Long Commentary on Metaphysics,* XII, comm. 5; *Tafsīr,* p. 1423.

[28] The subject matter of a given science, *A,* would have as its principles the entities that form the subject matter of a higher science, *B*. The existence of the subject matter of science *B* could then only be demonstrated in science *C,* a yet higher and more comprehensive science.

subject matter; no science whatsoever can provide the demonstration. For no science can treat of the principles and causes of the class of incorporeal beings, inasmuch as incorporeal beings, taken as a class,[29] have no principle or cause; nothing at all exists beyond incorporeal beings.

Although a science cannot establish its own principles through a demonstration, it can, Averroes maintains, establish them through "proofs" (*dalā' il*); for a proof proceeds not from the prior to the posterior, as a strict demonstration does, but rather "from the posterior to the prior."[30] Physical science can therefore legitimately contemplate a *proof* of the principles of its own subject matter. And the posterior phenomenon through which the existence of incorporeal beings may be established turns out to be motion, a physical phenomenon. The proof of the existence of incorporeal beings consequently does belong, so Averroes concludes, to the discpline of physics.[31]

Averroes has reasoned that a science can in no way establish the existence of its own subject matter; metaphysics hence can in no way establish the existence of incorporeal beings including the first cause. A science also cannot establish the existence of its own principles through a demonstration. Hence physics too cannot provide a demonstration of the existence of God. And in any event, the existence of God is nowise amenable to strict demonstration, since a demonstrative syllogism takes its departure from the principles or causes of the conclusion, whereas the first cause of the universe has no cause. Still, although a science cannot provide a demonstration of its own principles, it can provide a proof. Physical science therefore can legitimately reason from a physical phenomenon, namely, motion, back to the existence of a first incorporeal cause of motion.

Returning to Alexander, his commentary on the *Metaphysics,* if "taken at face value," stated that the science of physics accepts its principles from the science of metaphysics. The statement, Averroes is confident, has been shown to be incorrect; for physics does establish the principles of its own subject matter. Alexander must either have been in error or have expressed himself poorly, meaning something different from what he seemed to say. Because of Alexander's "eminence in philosophy,"[32] Averroes prefers to suppose that Alexander was only guilty of imprecise expression, and he suggests a rather far-fetched exegesis in order to rescue Alexander from the stigma of error.[33] As for Avicenna, Averroes has no qualms, and he asserts that Avicenna was unquestionably in error when

[29]Within the class of incorporeal beings, the first cause can be understood as the cause of the existence of the other incorporeal beings (Averroes' precise views on this require elucidation); nevertheless, the class taken in its entirety has no cause.

[30]*Long Commentary on Metaphysics,* XII, comm. 5; *Tafsīr,* p. 1423; cited by Wolfson, "Averroes' Lost Treatise," p. 414.

[31]Ibid. Cf. *Long Commentary on Physics,* II, comm. 22; *Long Commentary on Posterior Analytics,* II, comm. 70, cited by Wolfson, "Averroes' Lost Treatise," p. 411.

[32]*Long Commentary on Metaphysics,* XII, comm. 6; *Tafsīr,* p. 1436.

[33]See *Tafsīr,* pp. 1426, 1429, 1435.

he identified metaphysics as the discipline for proving the existence of God; Avicenna was "misled" by Alexander's statement to the effect that "it is for metaphysics to demonstrate what the principles of existent beings are."[34] Averroes also criticizes Avicenna for failing to recognize the distinction between a demonstrative syllogism and a proof, and thereby misunderstanding Aristotle's rules that a science cannot demonstrate its own principles.[35] He apparently did not know that Avicenna too had been scrupulous about proposing only a proof, and not a demonstration, of the existence of God.[36]

Averroes' criticism of Avicenna may be recapitulated as follows: (a) Avicenna contended that a proof of the existence of God can be framed in the science of metaphysics. Averroes responds that incorporeal beings are the subject matter of metaphysics and a science can nowise establish the existence of its own subject matter. We already know what Avicenna's rejoinder would be. The subject matter of the science of metaphysics, he would contend, is not the deity but the existent *qua* existent. Proving the existence of a being necessary by virtue of itself is tantamount to proving that a certain attribute is exemplified in the existent insofar as it is existent, namely, the attribute of not being constituted by anything else. And it is a legitimate subject of inquiry for a science to establish that its subject matter possesses a certain attribute.[37] (b) Avicenna, in Averroes' reconstruction, also misunderstood the rule stating that a science does not demonstrate its own principles. As a result of misreading Aristotle and Alexander, Avicenna believed that there is no way in which the science of physics can establish its own principles. In response, Averroes maintains that although a science cannot establish its principles through a demonstration, it can do so through a proof, and physics can, accordingly, offer a *proof* of the existence of God. What Avicenna's rejoinder would be is again clear. Physics, he would grant, can provide an argument of sorts for a first cause of the universe, but the physical argument suffices merely to give a "fleeting idea" of the first cause. A complete proof, what Avicenna calls a "more certain and more exalted" proof, though still not a strict *demonstration,* is feasible only in the science of metaphysics.[38] (c) Averroes further understands that the proof of the existence of God has to take its departure from a physical phenomenon and therefore pertains to the science of physics. Avicenna might here reply that the proof of the existence of God may indeed begin with a physical phenomenon. But as he formulates his proof, it immediately disregards all the peculiar properties of the given phenomenon and pays heed solely to the fact that something exists.[39] His proof is thus executed wholly in conformity with the procedure of metaphysics.

[34] *Tafsīr,* p. 1426, cited by Wolfson, "Averroes' Lost Treatise," p. 415; *Tafsīr,* p. 1436.
[35] *Tafsīr,* p. 1423.
[36] Cf. above, p. 299.
[37] Above, p. 286.
[38] Above. pp. 285, 287.
[39] Above, p. 303.

As already seen, Averroes does not deny metaphysics all part in studying incorporeal beings including the first cause; he denies metaphysics a role only in establishing their existence. Averroes' position is that physics and metaphysics "share" the task of "theorizing about the principles of substance," physics establishing the "existence" of incorporeal beings, and metaphysics investigating the "dispositions . . . and essences" of those beings.[40] More precisely, physics adduces the eternity of motion to establish the existence of an eternal cause of motion;[41] it adduces the principle that a finite body cannot contain infinite power to prove that the mover must be incorporeal;[42] and it adduces the principle that what is potential can cease to exist, in order to prove that the mover is exempt from potentiality and affection.[43] Metaphysics should thereupon enter and explain how, within the causes of celestial motion, all "ascend" to a single first cause.[44] Metaphysics must, moreover, establish that the first cause is not merely a cause of motion but is also the efficient, formal, and final cause of the universe;[45] that it is a substance;[46] that it consists in pure intellect;[47] and that it moves the spheres by serving as an object of their desire.[48]

2. Necessarily existent by virtue of another, possibly existent by virtue of itself

Averroes contends that Avicenna erred as well in his division of being into: (a) the necessarily existent by virtue of itself, (b) the necessarily existent by virtue of another, but possibly existent by virtue of itself, and (c) the possibly existent by virtue of itself with no further qualification. In Averroes' account, the three terms were intended by Avicenna to designate three categories of actual existence. *Necessarily existent by virtue of itself,* as Averroes understands Avicenna, designates the eternal "first principle" of the universe—or, according to another passage in Averroes, all the incorporeal movers of the spheres including the first cause. *Necessarily existent by virtue of another, but possibly existent by virtue of itself,* as Averroes understands Avicenna, designates all eternal beings except the first principle, in other words, the celestial spheres and the incorporeal movers of the spheres—or, according to the other passage in Averroes, the celestial spheres alone. And, Averroes understands, "purely *possible beings* [in Avicenna's usage] . . . are those things that are subject to generation and destruction," in

[40]*Long Commentary on Metaphysics,* XII, comm. 5; *Tafsīr,* pp. 1425–1426. Cf. above, p. 284.

[41]*Long Commentary on Metaphysics,* XII, comm. 5; 29; *Tafsīr,* pp. 1423, 1426, 1561–1562; *Epitome of Metaphysics,* IV, §5.

[42]*Tafsīr,* p. 1425; *Epitome of Metaphysics,* IV, §5. Averroes adds certain refinements to this statement, below, pp. 323–325.

[43]*Tafsīr,* p. 1568.

[44]*Tafsīr,* p. 1425. Cf. Aristotle, *Metaphysics* XII, 8.

[45]*Tafsīr,* pp. 1425–1426; *Epitome of Metaphysics,* I, §7.

[46]*Tafsīr,* pp. 1425, 1626.

[47]*Tafsīr,* p. 1626. Cf. Aristotle, *Metaphysics* XII, 9.

[48]*Tafsīr,* p. 1626. Cf. Aristotle, *Metaphysics* XII, 7.

other words, the transient, physical objects constituting the sublunar world.[49] Averroes will object to Avicenna's terminology and its implications, but he accepts the threefold division of actual existence which he discovers in Avicenna. The threefold division goes back to Aristotle[50] and, in Averroes' view, reflects genuine differences in the hierarchy of existence; for it is philosophically illuminating to distinguish the eternal cause that has no cause from eternal beings that do have a cause, and also to distinguish the latter from beings that are not eternal.

In fact, Averroes has misunderstood Avicenna's division of being. Avicenna intended by his formulae to distinguish not three, but only two categories of actual existence: (a) being actually existent solely through itself, termed *necessarily existent by virtue of itself*—in effect, the deity; and (b) actually existent beings, whether eternal or not, existing through something else, termed *necessarily existent by virtue of another but possibly existent by virtue of themselves*—in effect, everything outside the deity. The possibly existent with no further qualification is not, for Avicenna, an additional category of actual existence, but rather a general designation for all things that can exist only through another, whether or not they do happen actually to exist. Avicenna stated unambiguously that the possibly existent cannot enter the realm of actual existence unless and until something renders it necessary.[51] Averroes has thus misinterpreted Avicenna's twofold division of actual existence as a threefold division.

So far, Averroes has no objection to raise, since he approves of Avicenna's division of being as he mistakenly interprets it. The objection Averroes does raise concerns the formula *necessarily existent by virtue of another, possibly existent by virtue of itself,* which Averroes takes to have been Avicenna's designation for the celestial realm. The necessity in the formula is understood by Averroes as tantamount to eternity; Aristotle had explained that "*necessarily* and *always* go together, since what necessarily exists cannot [ever] not exist."[52] Because he takes necessity as tantamount to eternity, Averroes understands the formula *necessarily existent by virtue of another, possibly existent by virtue of itself* as signifying that certain objects have in themselves only a possibility of existing, but are rendered eternal through something else; and he will presently argue that construing anything as possibly existent in itself yet rendered eternal through something else is self-contradictory. Now, we know, Averroes misunderstood Avicenna's formula. *Necessarily* existent in Avicenna's usage was equivalent to *actually* existent, not to *eternally* existent;[53] *necessarily existent by virtue of*

[49]British Museum, Hebrew MS. 27559, p. 304a-b (= *Derushim Tib'iyim,* IX), and *K. al-Kashf,* ed. M. Mueller (Munich, 1859), p. 39; German translation with pagination of the Arabic indicated: *Philosophie und Theologie von Averroes,* trans. M. Mueller (Munich, 1875). The passage that classifies not only the first cause but all the movers of the spheres as necessary by virtue of themselves is *Long Commentary on Physics,* VIII, comm. 79, reflecting Aristotle, *Metaphysics* XII, 1, 1069a, 30.

[50]Reference in previous note.

[51]Above, pp. 291–292.

[52]*De Generatione* II, 11, 337b, 35 f.

[53]In other words, every actually existent being is necessary at the time it exists.

another did not therefore mean eternally existent through another; and the designation *necessarily existent by virtue of another, possibly existent by virtue of itself* was not restricted by Avicenna to the celestial realm. Insofar, then, as Averroes' objection addresses itself to the use of a given formula by Avicenna, it misses the mark, since Averroes misunderstands the meaning assigned to the formula by Avicenna and what the formula was intended to denote.

Averroes' objection can, however, be considered as directed not against the formula but against Avicenna's conception of the nature of the celestial realm. For although the expression *necessarily existent* does not entail *eternally existent* in Avicenna, Avicenna did construe the celestial spheres and the movers of the spheres as eternal and yet, in themselves, possibly existent. And that conception is precisely what Averroes objects to. Averroes' objection is as follows: To say that something is possibly existent is to say that it contains the possibility both of existing and not existing. Consequently, to maintain that something is possibly existent but eternal amounts to maintaining that "something has the possibility of being destroyed without ever undergoing destruction."[54] But should anything contain the possibility of undergoing destruction, the possibility must eventually be realized and the object destroyed; for over an infinite time every possibility is eventually realized.[55] Aristotle had, hence, correctly stated that "whatever exists eternally is absolutely indestructible" and, conversely, what is "destructible is not eternal."[56] When, by contrast, Avicenna maintained that the heavens are eternal, yet have the possibility of both existing and not existing, he was affirming a possibility that is, over an infinite time, never realized. Avicenna's conception of the celestial realm is therefore untenable.[57]

Such is one statement of Averroes' criticism. Elsewhere Averroes puts the objection more sharply. Asserting that something is necessary—that is to say, eternal—and nevertheless possible would, he writes, contravene the law of contradiction. It would amount to assigning contrary attributes to the same thing in the same respect at the same time.[58] In additional passages, Averroes states what is again essentially the same objection in still another form: Avicenna's position amounts to assuming "that the nature of the possible can be transformed into the nature of the necessary [i.e., the eternal,]" an assumption that Averroes regards as a patent absurdity.[59] To summarize Averroes' objection in its different forms: When Avicenna construed the celestial realm as eternal, yet possibly existent, he was assigning to the heavens mutually exclusive attributes.

[54]*Long Commentary on De Caelo,* in *Aristotelis Opera cum Averrois Commentariis,* Vol. V (Venice, 1562), II, comm. 71.

[55]Cf. Aristotle, *Metaphysics* IX, 4, 1047b, 4–5; above, p. 90, n. 30.

[56]Aristotle, *De Caelo* I, 12, 281b, 25; 282a, 22–23. Cf. *Metaphysics* IX, 8, 1050b, 17.

[57]*Long Commentary on De Caelo,* II, comm. 71.

[58]*Tahāfut al-Tahāfut,* VIII, p. 395.

[59]*K. al-Kashf,* p. 39; *Long Commentary on Metaphysics,* XII, comm. 41; Arabic original, p. 1632; *Epitome of De Caelo,* in *Rasā' il Ibn Rushd* (Hyderabad, 1947), p. 36; British Museum, Hebrew MS. 27559, pp. 304b, 307b (= *Derushim Tib'iyim,* IX, and *De Substantia Orbis,* VII); *Epitome of Metaphysics,* III, §27.

3. The nature of the celestial spheres according to Averroes

In the question of the nature of the celestial spheres, as in the question of the science responsible for proving the existence of God,[60] Averroes understands that Avicenna was misled by Alexander of Aphrodisias.[61] At issue now is another Aristotelian principle, the proposition that since the heavens are a finite body, they cannot contain infinite power.[62] That proposition allowed Aristotle to reason that the heavens are unable to maintain themselves in motion over an infinite time, and that the eternal motion of the celestial spheres must therefore be due to an incorporeal mover.[63] The same proposition allowed Proclus and Simplicius to reason that the heavens are unable to maintain themselves even in existence over an infinite time and therefore the very existence of the heavens, not merely their motion, must derive from an incorporeal cause.[64] As put by Simplicius, the principle that a body cannot contain infinite power "clearly" applies not to "motive power" alone, but also to the power "constituting the being" of the heavens; and consequently, the heavens "must receive even their eternal corporeal being from an incorporeal [cause]."[65] Simplicius' contention, as it happens, appeared in the course of a critique of Alexander. Averroes, who mentions neither Proclus nor Simplicius in the present connection, reports the same contention in Alexander's name.

In "several works," Averroes writes,[66] Alexander took up the Aristotelian proposition that the heavens can contain only finite power; and he inferred from it that the heavens cannot possess their infinite, eternal existence through themselves, but must "acquire" their eternal existence "from their immaterial mover."[67] Accordingly, the heavens would, as construed by Alexander, be "destructible in themselves, yet never actually destroyed because of an infinite immaterial power, that is to say, because of their mover."[68] The mistaken interpretation of Aristotle on the part of Alexander, Averroes continues, led Avicenna to his equally mistaken conception. On the model of Alexander, who construed the heavens as destructible in themselves, but never destroyed by reason of their cause, Avicenna supposedly came to his own conception of the heavens as possibly existent by

[60]Cf. above, p. 317.

[61]Averroes also assumes that the Kalam arguments from the concept of particularization were a factor leading to his position. *Tahāfut al-Tahāfut,* IV, p. 276; *K. al-Kashf,* pp. 37, 39; *Long Commentary on Physics,* II, comm. 22.

[62]*Physics,* VIII, 10; *De Caelo* II, 12, 293a, 10–11.

[63]Cf. above, p. 244.

[64]Above, pp. 281–282.

[65]Above, p. 282.

[66]A number of items attributed to Alexander were current in Arabic, some genuine, others excerpted from Proclus' works, yet others of unknown authorship. See J. van Ess, "Fragmente des Alexander von Aphrodisias," *Der Islam,* XLII (1966), 148–168. Item 15 in van Ess's list is entitled *On the Movements of the Spheres* and could be one of the works referred to by Averroes.

[67]*Long Commentary on Physics,* VIII, comm. 79.

[68]*Middle Commentary on De Caelo,* I, Vatican Library, Hebrew MS. Urb. 40, p. 49a. Cf. British Museum, Hebrew MS. 27559, p. 306a.

virtue of themselves, but necessarily—that is, eternally—existent by virtue of another.[69]

Averroes has drawn up the following scheme: Aristotle's principle that bodies cannot contain infinite power led Alexander and Avicenna to conclude that the heavens have in themselves a possibility of not existing. Alexander, as a consequence, construed the heavens as destructible in themselves, indestructible through their cause; and Avicenna construed the heavens as possibly existent in themselves, necessarily and eternally existent through another. But to be destructible in itself, or to have the possibility of existing and hence also of not existing, is incompatible with an object's existing eternally. The position of Alexander and Avicenna therefore contains a contradiction.

John Philoponus too receives a place in the scheme just delineated. The significance of Philoponus' argument for creation from the finite power of the spheres[70] lies, on Averroes' reading, precisely in its recognizing that possible existence and eternity are mutually exclusive. Averroes explains Philoponus' argument as a response to Alexander: Alexander reasoned that since the heavens are finite, they contain only finite power and cannot be responsible for their own eternal existence; they are in themselves liable to destruction; and their eternal existence must be acquired from an incorporeal cause. Philoponus responded that if the heavens do have only finite power and contain in themselves a possibility of not existing, they cannot be eternal, whatever the source of their existence might be; for every potentiality must eventually be realized. The heavens, Philoponus concluded, cannot have existed through an infinite past time, but must instead have come into existence after not having existed.[71] Philoponus' argument from the finite power of the heavens is characterized by Averroes as the strongest argument for creation, and it, together with other difficulties connected with the nature of the heavens, left Averroes "in a quandary for many years."[72]

The acuteness of Averroes' quandary is understandable. It was Aristotle who established the proposition that the celestial spheres contain only finite power. The proposition apparently does imply that the spheres are possibly existent. The position taken by Alexander and Avicenna—that the heavens are possible in themselves, eternal through their cause—is, however, inadmissible, since it assigns the heavens mutually exclusive attributes. And Philoponus' cutting of the knot by affirming that the heavens are not after all eternal is also unacceptable to Averroes; for he endorses Aristotle's arguments for the eternity of the world both

[69] References in the previous two notes. Also, *Long Commentary on De Caelo,* II, comm. 71; *De Substantia Orbis,* III.

[70] See above, p. 91.

[71] *Long Commentary on Physics,* VIII, comm. 79; *Middle Commentary on De Caelo,* I, p. 49a; *Long Commentary on De Caelo,* II, comm.71; *Long Commentary on Metaphysics,* XII, comm. 41, and *Tafsīr,* p. 1628.

[72] *Long Commentary on Physics,* VIII, comm. 79; *Middle Commentary on Metaphysics,* XIII, Casanatense Library, Hebrew MS. 3083, p. 140 (141)b; *De Substantia Orbis,* V.

on their own merit and because of their provenance. The dilemma here outlined is handled by Averroes differently in different works, and disentangling and harmonizing his different treatments is not easy. Still, two general problems dealt with by Averroes may be distinguished, one primarily exegetical, the other philosophical.

(*a*) The exegetical problem regarding the nature of the celestial spheres relates specifically to Aristotle's statement in *Physics* VIII, to the effect that a physical body can contain only finite power,[73] and his statement in *De Caelo* II, to the effect that the heavens can contain only finite power since they are bodies.[74] Aristotle's statements would seem, as Philoponus argued, to imply that the heavens can move and exist for no more than a finite time, and not for all eternity. And yet Aristotle had demonstrated the eternity of the heavens.

Averroes solves the exegetical difficulty by distinguishing two senses of finite power, of which both apply to sublunar bodies, but only one to celestial bodies. Finiteness of power, he writes, may be understood as finiteness in "intensity," which in the instance of motive power means the capability to produce or undergo no more than a finite "velocity." It also may be understood as finiteness in "continuity" or "time."[75] Now every body that performs any "act" whatsoever, including motion, must do so at a finite velocity; for the nature of bodies is such that they cannot perform an act instantaneously. Because of the very nature of bodies, irrespective of whether they are sublunar or celestial, their power of producing as well as of undergoing motion must be finite in respect to *intensity*. Aristotle's statement in *De Caelo* II regarding the finite power of the celestial spheres is interpreted by Averroes as affirming, in conformity with the present sense of finiteness, that the power to produce and undergo motion is, even in the case of the spheres, finite in intensity;[76] and the context of Aristotle's statement there does in fact lend itself to the interpretation.[77]

Finiteness of power in the other sense, in the sense of *continuity* and *time*, is due, writes Averroes, not to the very nature of bodies but to the nature of a certain type of body. And only sublunar bodies are such that their powers must be finite in respect to continuity and time. Averroes explains: Sublunar bodies are compounds of matter and form. The matter and form are fused and are dependent upon one another for their existence. The matter is brought to actuality through the form, and the latter is a material form, capable of existing solely in matter. The power through which a sublunar body continues to exist and move has its

[73] *Physics*, VIII, 10.

[74] *De Caelo* II, 12, 293a, 10–11.

[75] For other instances of the distinction cf. S. Pines, "A Tenth Century Philosophical Correspondence," *Proceedings of American Academy for Jewish Research,* XXIV (1955), 115; H. Wolfson, *Crescas' Critique of Aristotle* (Cambridge, Mass., 1929), pp. 612–613.

[76] *Middle Commentary on De Caelo,* I, pp. 49a–50a, and the other references in n. 80.

[77] See the Oxford translation of the *Works of Aristotle,* Vol. II (Oxford, 1930), *De Caelo* 293a, n. 2.

source in the form, and since the form must be present in the matter, the power too is present "within" the matter. It is divisible, just as matter is divisible; and hence the power contained within a finite sublunar body must, as Aristotle showed,[78] be equally finite—finite not merely in respect to its intensity, but also exhaustible and finite in respect to its continuity. Finiteness of power in this second sense, in respect to continuity, thus follows from the nature of sublunar bodies.

The nature of celestial bodies, Averroes continues, is different from that of sublunar bodies. The substratum and the form of the spheres do not exist together in a compound. Each exists independently of, although in conjunction with, the other. The form of the spheres is not a material form inhering in a material substratum; and the substratum is not in a state of potentiality requiring a form to render it actual. The substratum therefore does not exist through a power that is distinct from it, but contained within it, a power that must accordingly be finite in respect to continuity. The substratum of the spheres does not exist through a power at all; and the argument that a finite body can contain only finite power, and finite power can give rise only to finite existence, does not apply. In Averroes' most frequent formulation,[79] the substratum of the spheres exists "through itself," through its own simple nature, which contains "no contrary" able to bring about the destruction of the substratum. It has, as a consequence, no possibility of not existing, and its continued existence is in no way subject to limitation. As for the motion of the sphere, it is indeed produced by a power, a power issuing from the immaterial form associated with the substratum. But since the form in question is immaterial, the power whereby the substratum moves is not present "within" the finitely extended substratum. Consequently, the power moving the sphere is not finite, and the ability to move continually through time is for the spheres—as it is not for sublunar bodies—unlimited.[80]

The difficulties occasioned by Aristotle's statements concerning the finiteness of power in a body are, then, solved by Averroes in the following fashion: The statement in *De Caelo* II, to the effect that the heavens contain only finite power refers to finiteness in intensity. It means merely that the heavens, like all other bodies, cannot possess a motive power infinite in intensity which would give rise to motion of an infinite *velocity.* By contrast, Aristotle's statement in *Physics*

[78] *Physics* VIII,10, 266a, 24–266b,6. Aristotle's reasoning there is different from that given here by Averroes. Possibly, Averroes was thinking of a train of reasoning offered by Alexander (in Simplicius, *Commentary on Physics,* ed. H. Diels, *Commentaria in Aristotelem Graeca,* Vol. X [Berlin, 1895], p. 1326) and repeated by Avicenna, *Najāt,* (Cairo, 1938), p. 130. See H. Davidson, "The Principle that a Finite Body can Contain only Finite Power," *Studies in Jewish Religious and Intellectual History presented to Alexander Altmann* (University, Alabama, 1979), p. 77.

[79] Another view is given below, pp. 330–331.

[80] *Long Commentary on Physics,* VIII, comm. 79; *Middle Commentary on De Caelo,* I, pp. 49b, 50b–51b; *Long Commentary on De Caelo,* II, comm. 71; *Middle Commentary on Metaphysics,* XIII, pp. 140(141)b–141a; *De Substantia Orbis,* III; British Museum, Hebrew MS. 27559, pp. 305a, 306b. The account in *Long Commentary on Metaphysics,* XII, comm. 41, and *Tafsīr,* pp. 1629–1631, is somewhat different.

VIII, to the effect that physical bodies contain only finite power refers to finiteness in *continuity* and applies exclusively to sublunar bodies. It means that any finite body consisting in a compound of matter and form cannot contain a power infinite in continuity. The spheres, which do exist and move for an infinite time, are to be construed as a different type of a body. On Averroes' interpretation, Philoponus' argument for creation from the finiteness of power in the sphere loses its *point d'appui* in the text of Aristotle, inasmuch as Aristotle is no longer to be understood as saying that the spheres contain a power which is finite in the significant sense of *continuity.*

The principle that a finite body cannot contain infinite power was fundamental to Aristotle's proof of the existence of God from motion. It enabled Aristotle to conclude that the movement of the heavens must depend on an incorporeal first mover beyond the heavens.[81] Now that Averroes has given his own interpretation of the principle, he is led to recast part of Aristotle's proof. Averroes has nothing to add as regards the reasoning leading up to the existence of the celestial spheres, which are eternally in motion.[82] But he does undertake to make the proof more precise from that stage on. He explains: The substratum of the first sphere cannot be assumed to move itself, since nothing at all can move itself.[83] The motion of the substratum of the first of the spheres must be due to a form.[84] The reason that the form moving the sphere must be incorporeal is not to be formulated as the principle that bodies contain only finite power. It rather is to be formulated as the principle that a certain type of body, consisting in a true compound of matter and form, cannot possess power infinite in respect to continuity. The first formulation, which Averroes rejects, would imply that the celestial sphere consists in a compound of matter and material form, and, considered as a whole, contains only a finite power of continued movement and existence. The sphere would hence have the possibility of both moving and not moving, both existing and not existing; and since every possibility must eventually be realized, the spheres, it would follow, cannot exist or move forever. The second formulation, which has the benefit of Averroes' exegesis, leads to another result: Since bodies compounded of matter and form cannot possess power infinite in respect to continuity, and since the heavens do move for an infinite time, the heavens cannot be such a body. The heavens must instead be construed as a body of a completely different type, consisting in the association of a simple matter-like substratum in motion, and an independently existing immaterial form moving the substratum. The matter-like substratum exists *necessarily by virtue of itself,* and the form is a source of infinite power whereby the substratum moves eternally.[85] Inasmuch

[81]Cf. above, p. 244.

[82]Cf. above, pp. 240–242.

[83]*Middle Commentary on De Caelo,* I, p. 52b; British Museum, Hebrew MS. 27559, p. 305b.

[84]Otherwise we would have to posit an infinite series of moving bodies; cf. above, p. 243.

[85]The incorporeal form would thus be the cause of the substratum's existence as a sphere, although not the cause of the existence of the substratum itself.

as the form of the first sphere serves as the eternal mover of the entire universe,[86] Aristotle's proof, with Averroes' nuance, still arrives at an incorporeal first mover. But by denying that the substratum of the sphere contains a form within itself, and by construing the sole source of power for the sphere as an incorporeal form that is independent of, although associated with, the substratum of the sphere, Averroes is confident that he has avoided attributing to the sphere any power that is finite in respect to *continuity.* He thereby can retain Aristotle's proof of an incorporeal first cause while deflecting Philoponus' argument for creation from the finite power of the sphere.[87]

(*b*) Averroes' explanation of Aristotle's statements concerning the finite power of bodies rests, in sum, on a scholastic distinction. The distinction between two senses of finiteness permits an interpretation of Aristotle which avoids attributing to the spheres a power finite in continuity. A philosophical problem remains, however. The thesis that the heavens are possibly existent by virtue of themselves and necessarily, or eternally, existent by virtue of another is untenable, Averroes has argued, inasmuch as it combines contrary attributes in the same thing at the same time. But when Averroes, for his part, affirms that the substratum of the heavens exists by virtue of itself and is kept in motion through an incorporeal form distinct from it, he is adopting a position that also turns out to involve the combined presence of necessity and possibility in the substratum of the heavens. Averroes is quite aware that the problem of the presence of both necessity and possibility in the heavens is going to attach itself to his position too. Unfortunately, the solution to the problem which he most frequently offers—(i), below—is discovered by him, on careful inspection, to be inadequate. A philosophic note of Averroes' offers another solution to the problem—(ii)—but it is questionable as well. An additional philosophic note—(iii)—offers yet a third solution, a solution that brings the discussion to an unexpected denouement.

(*i*) Maintaining that the substratum of the heavens is kept in motion through a factor distinct from it would appear to imply that the substratum in itself has nothing more than a possibility of motion. In some passages Averroes accepts the implication and does affirm that the substratum of the heavens should be construed as having in itself nothing more than a possibility of motion. He draws a distinction between possibility in respect to existence, which the heavens cannot have, and possibility in respect to motion, which they do have. The simple

[86]The outer sphere, it will be recalled, produces the daily movement of the heavens, to which the peculiar movement of each of the spheres is added.

[87]*Middle Commentary on De Caelo,* I, pp. 51b–52b; *Middle Commentary on Metaphysics,* XII, p. 141 (142)a; *Long Commentary on Metaphysics,* XII, comm. 41, and *Tafsīr,* pp. 1633–1634; British Museum, Hebrew MS. 27559, pp. 305b–306a.

For Ḥasdai Crescas' application of the distinction between power which is infinite in intensity and power which is infinite in continuity, see above, pp. 264–265. Some reverberations of the distinction in Scholastic thought are discussed by Anneliese Maier, *Metaphysische Hintergruende der Spaetscholastischen Naturphilosophie* (Rome, 1955), pp. 238–262.

substratum of the spheres, he writes, "has no contrary"; but things are destroyed only by their contrary; consequently, nothing can conceivably destroy the substratum of the spheres, and it "continues [to exist] through itself . . . not through any [added] factor." That is the reason why the substratum of the spheres is to be construed as necessarily existent by virtue of itself. The motion of the spheres, however, "does have a contrary, namely rest." Consequently, the motion of the substratum of the spheres "cannot continue through itself," and the substratum of the spheres has in itself only the possibility, not the necessity of continued movement.[88] This is the reason that an external factor, the incorporeal form, must be posited; it must be posited to explain why the "possibility . . . of rest" in the spheres is "prevented from realization," and why the celestial spheres do enjoy eternal motion.[89]

Averroes submits that the law of contradiction rules out Avicenna's position on the nature of the celestial spheres but not his own position. The law of contradiction rules out an object's being both necessary and possible in a single respect: "Nothing having a single nature" can be "possible by virtue of one factor and necessary by virtue of another. . . ; for possible is the contrary of necessary." Consequently, nothing can be construed, in the manner that Avicenna construed the heavens, as possible by virtue of itself, and necessary, that is, eternal, by virtue of another, all in respect to its existence. But an object may be "necessary in respect to one nature and possible in respect to another." The law of contradiction is not, therefore, contravened when the substratum of the spheres is construed as "necessary in respect to its substance, possible in respect to its motion in place."[90]

Here, then, Averroes has advanced the thesis that the substratum of the spheres is necessary in one respect, in its existence, and possible in a different respect, in its motion. The thesis is highly vulnerable. The movement of the substratum of the spheres, Averroes stresses, is due to a factor distinct from the substratum of the spheres. If the motion of the spheres continues eternally by virtue of such a factor, the motion of the spheres is plainly necessary, or eternal, by virtue of something distinct from it. The motion of the spheres would, as Averroes explicitly writes, be "necessary through another but possibly in itself."[91] A single phenomenon, the motion of the spheres, would, solely in respect to its occurrence or "existence" (*wujūd*),[92] be necessary through another, possible in itself. Or, if one prefers, a single object, the substratum of the spheres, would, in a single

[88]Averroes would have to harmonize his position with Aristotle, *Metaphysics* IX, 8, 1050b, 20.

[89]*Long Commentary on Metaphysics,* XII, comm. 41, and *Tafsīr,* p. 1631. Cf. *De Substantia Orbis,* V.

[90]*Tahāfut al-Tahāfut,* VIII, p. 395; *Long Commentary on Metaphysics,* XII, comm. 41, and *Tafsīr,* pp. 1631–1632. Cf. Aristotle, *Metaphysics* IX, 8, 1050b, 15–18.

[91]*Tafsīr,* p. 1632.

[92]Ibid.

respect, in respect to its motion, be both necessary and possible; it would be necessary through its mover and possible through itself. Averroes has accused Avicenna of fashioning a contradictory thesis when he construed the spheres as both necessary and possible in a single respect.[93] If the criticism is valid, Averroes has now committed the same error.

The point can be made in another way as well. Averroes has described the substratum of the spheres as containing a "possibility . . . of rest," which possibility is "prevented from realization" by the incorporeal mover. The physical universe, Averroes understands, exists in its form as a universe only through the motion of the spheres.[94] If the substratum of the spheres contains a possibility of rest, the physical universe contains the possibility of not existing in the form of a universe. The universe would hence be possibly existent by virtue of itself and, if eternal, necessarily existent by virtue of another—by virtue of the incorporeal beings that move the celestial spheres. Averroes' position is thus seen again to imply that a single object, the physical universe, is both necessary and possible in a single respect, in respect to existence.[95]

Having ruled out Avicenna's position that the heavens are possibly existent by virtue of themselves and necessarily, or eternally, existent by virtue of another, Averroes has been led to a virtually identical position. Two brief, apparently genuine, philosophical notes recognize that the problem of the presence of both possibility and necessity in the spheres has not yet been solved; the two notes offer alternative explanations of the nature of the spheres designed to solve the problem.[96]

(*ii*) The first of the philosophical notes takes up the thesis that since "without . . . their mover" the heavens would not move continually, "their motion in place is possible by virtue of itself, necessary by virtue of their eternal mover."[97] Such was precisely the position arrived at by Averroes in the passages discussed in the previous paragraphs. But now Averroes rejects the thesis because of the difficulty just pointed out: The implication would be that "something [namely, motion] exists, which is possible in itself, necessary through another."[98] A single thing would be both possible and necessary, or eternal, in the same respect.

The new position adopted by Averroes is that the motion of the heavens is not after all possible by virtue of itself, necessary by virtue of its cause. The movement of the heavens is rather to be construed as "necessary by virtue of itself."[99]

[93]Above, pp. 319–320.

[94]Cf. above, p. 283.

[95]The objection is raised by Sh. Falaquera, *Moreh ha-Moreh* (Pressburg, 1837), pp. 63–64.

[96]The two notes are *Derushim Tib'iyim*, IX, and *De Substantia Orbis*, VII. I cite them from British Museum, Hebrew MS. 27559. They also are found in Paris, Bibliothèque Nationale, Hebrew MS. 989; cf. M. Steinschneider, *Die Hebraeischen Uebersetzungen des Mittelalters* (Berlin, 1893), pp. 181–182. Ibn Abī Uṣaybi'a appears to have known them and regarded them as genuine; cf. *'Uyūn al-Anbā'* (Cairo, 1882), II, p. 78.

[97]British Museum, Hebrew MS. 27559, p. 305a.

[98]Ibid., p. 305b.

[99]Ibid.

Averroes does not mean that the substratum of the heavens moves itself; on that supposition, the incorporeal mover could be dispensed with, and Aristotle's proof from motion would no longer establish an incorporeal being beyond the heavens. Averroes categorically rejects any such thought and insists that the substratum of the spheres does stand in need of a mover. But he now contends that a mover of the spheres must be posited "only . . . because every motion requires a mover" and "not . . . because celestial motion is [merely] possible."[100] Accordingly, the motion of the sphere can, although dependent upon something else, be construed as *necessary in itself.*

Averroes praises his solution as "wondrous,"[101] but it seems to be nothing more than verbal legerdemain. The proposition that continual celestial motion depends on a cause external to the substratum of the spheres led Averroes to construe celestial motion as *possible by virtue of itself,* a construction that had unacceptable implications. Averroes here still assumes that continual celestial motion depends on a cause external to the substratum of the spheres, but he alters his terminology and paradoxically calls the motion of the spheres *necessary by virtue of itself*: Celestial motion is necessary by virtue of itself, although dependent on something else. The unacceptable implications of Averroes' theory are disguised by a change of terminology.

(*iii*) The second of the philosophic notes makes yet another attempt to define the nature of the heavens. Averroes, ignoring the distinction he drew elsewhere between the separate senses of *finite,* poses a dilemma, one horn of which repeats the lines of Philoponus' argument: Since the heavens are finite, they can, as Aristotle showed, contain only finite power; if the power of the heavens is finite, the heavens in themselves are only possibly existent; the heavens hence contain the possibility of not existing; every possibility must eventually be realized; whence it would ensue that the heavens cannot be eternal. And yet—here is the other horn of the dilemma—Aristotle had proved conclusively that the heavens are eternal.[102]

According to Averroes' most frequent solution—(i), above—the substratum of the heavens is not possibly existent by virtue of itself, as Alexander and Avicenna held, but rather necessarily existent by virtue of itself. Averroes was nonetheless led to acknowledge that the motion, if not the substratum, of the spheres is both possible and necessary in respect to its existence, a combination of attributes which he had repeatedly rejected as self-contradictory. According to Averroes' second solution—(ii), above—not only the existence of the substratum of the spheres, but also the motion of the spheres is necessary by virtue of itself. Paradoxically, however, Averroes insisted at the same time that the motion of the substratum of the spheres is not produced by itself. In the present

[100] Ibid., pp. 305b–306a.
[101] Ibid., p. 306a.
[102] Ibid., p. 307a-b.

solution, Averroes embraces the theory of Alexander and Avicenna to which he objected so strongly in other works. The substratum of the heavens, he concedes, is in fact possibly existent by virtue of itself, necessarily existent by virtue of its cause.

The human intellect, Averroes explains, sometimes operates with propositions that are illuminating although not true reflections of reality. The geometer, for example, abstracts from the nature of actual physical objects and affirms that a circle will meet a tangential surface at an indivisible point. Yet the physicist informs us that objects in the physical world never do actually meet at an indivisible point. By the same token, the human intellect "judges an effect to be possible in itself," even when the causal nexus is eternal; in the instance under discussion, it judges the celestial spheres to be possible by virtue of themselves, necessary through their cause. But in truth any effect dependent on an eternal cause is eternal and necessary, and hence contains no possibility of not existing; and inasmuch as the substratum of the spheres is dependent on an eternal cause, it too is eternal and necessary, nothing more. The possible existence of the spheres is an unreal distinction drawn by the intellect when the intellect considers the substratum of the sphere in abstraction from its cause. Since *possible existence* is not a real attribute in the substratum of the spheres, the substratum of the spheres can after all be construed as both possibly existent and necessarily existent, without contravening the law of contradiction. And since the possibility of nonexistence is not a real attribute—not a seed, as it were, that must eventually sprout—the possibility of the nonexistence of the spheres need not be realized over an infinite time.[103]

Averroes realizes that the position just stated is that of Alexander and Avicenna. He acknowledges that "what Alexander says . . . is correct"; Alexander recognized that since the heavens contain only finite power they cannot maintain themselves indefinitely and therefore their "continual [existence and movement] . . . must be due to the first cause." "Avicenna's position too was correct." Avicenna rightly recognized that certain necessary and eternal things are "necessary by virtue of themselves," whereas some are "necessary through another, possible by virtue of themselves"; and he did not mean thereby that the latter "contain any possibility at all." On the present explanation, John Philoponus' argument for creation is easily disposed of. When Philoponus argued for creation from the finite power of the spheres, he failed to understand that some "finite powers" are "by their very nature able to receive necessity [and eternity] from something else."[104]

The foregoing solution to the problem of the nature of the celestial spheres is, as far as I can see, explicitly adopted only in the brief philosophic note under consideration. Echoes of it can, however, be discovered elsewhere. Several times Averroes describes the substratum of the sphere as dependent for its very existence

[103] Ibid., pp. 308a–309a.

[104] Ibid., pp. 308b–309b.

on its incorporeal form;[105] by implication, the substratum of the sphere would be possibly existent by virtue of itself, necessarily existent by virtue of its cause. Once when criticizing Avicenna, he observes that Avicenna's position might be defended by taking the possible existence of the spheres as a mere abstraction; there, however, he rejects the suggestion.[106] As for the philosophic note offering the present solution—*De Substantia Orbis,* XII—there seems to be no way of determining whether it is a very early work of Averroes, a very late work, or, for that matter, not even genuine. An apparent volte-face is insufficient to impugn the genuineness of any of Averroes' works, since he at times did openly change his views on philosophic issues. But even if not written by Averroes, the philosophic note under consideration could have been written only by someone steeped in his philosophy. The note suggests that Aristotelian principles lead logically to Avicenna's position on the nature of the celestial spheres. More generally it indicates that Aristotelian principles inescapably lead to puzzles that can be solved only through nominalist interpretations, such as the nominalist interpretation of *possibility.*

4. Averroes' critique of the body of Avicenna's proof

In his discussion of Avicenna's proof of the existence of God from the concepts *possibly existent* and *necessarily existent,* Averroes, as has been seen, gives considerable attention to the question of the metaphysical character of the proof, and the question whether the attribute *possible by virtue of itself* is compatible with *necessary by virtue of another.* Averroes' critique of the body of the proof is almost completely independent of his treatment of those issues.[107] His critique of the body of the proof also occupies him less than the preliminaries did, the reason being that Averroes finds Avicenna's proof, as he understands it, to be acceptable with modifications.

For his critique of the body of the proof, Averroes takes the term *possibly existent* in Avicenna to connote dependence on a cause, and *necessarily existent*—that is to say, *necessarily existent by virtue of itself*—to connote absence of a cause.[108] These constructions were not placed on the terms in Averroes' discussion of the preliminaries of the proof; and Averroes adopts them in the present context from Ghazali, apparently because his critique of the body of Avicenna's proof is based on Ghazali's restatement and critique of the same proof.[109] Averroes, as has already been seen, supposes as well that *possibly existent* with no

[105]*Middle Commentary on Metaphysics,* XIII, p. 140 (141)b; *De Substantia Orbis,* II.

[106]*K. al-Kashf,* p. 39.

[107]He does assume here, as he did in the preliminaries, that *possibly existent* designates, for Avicenna, the generated-destructible objects of the sublunar world; cf. above, p. 318. And his corrected version of Avicenna's proof is, as he required in the preliminaries, a physical proof.

[108]*Tahāfut al-Tahāfut,* IV, p. 276.

[109]Ibid., referring to Ghazali, *Tahāfut al-Falāsifa,* IV, §12, discussed above, p. 371.

further qualification denotes, for Avicenna, the generated-destructible objects of the sublunar world.[110] As for the correct usage of the terms *possibly existent* and *necessarily existent,* Averroes relies, as he always does in matters of terminoloy, on Aristotle. *Possibly existent,* he insists, can only properly mean: that which has the possibility of both existing and not existing—in effect, again, that which is generated and destructible.[111] And *necessarily existent* can only properly mean: that which has no possibility of not existing—in effect, that which is eternal.[112]

In the light of the foregoing, Averroes is able to discover that "one" error in Avicenna's proof was the employment of the term "*possible* equivocally."[113] The purport of the criticism is this: *Possibly existent* means, in Avicenna, that which has a cause; but the term *possibly existent* was in addition employed by Avicenna—so Averroes supposes—as a designation for generated-destructible beings.[114] The class of beings that have a cause is, however, wider than the class of generated-destructible "beings"; for the former extends to the eternal celestial realm, whereas the latter does not. Consequently, Averroes finds Avicenna was guilty of equivocation in his use of the term *possibly existent.*

A second, related error in Avicenna's proof, according to Averroes, is that "dividing the existent primarily into what is possible [in the sense of having a cause] and what is not possible [in the sense of not having a cause] is [an] incorrect [division]"; "it is not a division that comprehends the existent *qua* existent,"[115] and it is not "self-evident."[116] By contrast, the "division of existent beings which is evident by nature" is the division into the "truly possible and the [truly] necessary."[117] Averroes' meaning is unclear,[118] but perhaps his intention is as follows: Sense perception immediately discovers objects to be generated-destructible. Since something corresponding to the concept *generated-destructible* is at once known to exist, the study of existence can legitimately begin with the dichotomy of *that which is generated-destructible* and *that which is not generated-destructible*. Sense perception does not, by contrast, discover that anything has a cause; causes can be inferred only through an act of reasoning subsequent to, and performed upon, the reports of sense perception.[119] Inasmuch as human knowledge cannot at once take hold of either side of the dichotomy *caused* and *uncaused*—neither side being known from the outset actually to exist—beginning the study of existence with that dichotomy is not legitimate.[120]

[110]Above, p. 319.

[111]Cf. *Metaphysics* IX, 8, 1050b, 7–15.

[112]Cf. *De Generatione* II, 11, 337b, 35 ff. Cf. Alfarabi's position, above, p. 332.

[113]*Tahāfut al-Tahāfut,* IV, p. 279.

[114]Cf. above, p. 318–319.

[115]*Tahāfut al-Tahāfut,* IV, p. 279.

[116]Ibid., p. 276.

[117]Ibid., X, p. 418.

[118]Especially unclear to me is Averroes' statement to the effect that Avicenna's division of being does not comprehend "the existent *qua* existent."

[119]Cf. below, p. 342.

[120]This interpretation of Averroes' objection is supported by *Tahāfut al-Tahāfut,* X, p. 418, and by the criticism of Avicenna's proof which he states, immediately below.

Avicenna's proof as a whole fails, Averroes continues, precisely because of its failure accurately to represent the dichotomy of *possible* and *necessary.* Averroes understands Avicenna's proof from the concepts *possibly existent being* and *necessarily existent being,* as nothing other than a typical argument from the impossibility of an infinite regress of causes. Avicenna, Averroes supposes, began by considering beings that are *possibly existent* in the sense of having a cause;[121] he contended that an infinite regress of beings of the sort is impossible; and he concluded that the regress must end at a being necessarily existent by virtue of itself, which is the cause of everything else. Averroes rejects the inference, because, he argues, it is by no means at once evident that the universe does contain beings having a cause. One might, that is to say, grant the impossibility of an infinite regress of causes but reject the first premise in Avicenna's argument. The argument as supposedly framed by Avicenna would not, then, lead to the conclusion that there exists a first cause, but might leave us instead with the proposition that nothing in the universe has a cause.[122]

Averroes thus understands that Avicenna attempted nothing more than a standard proof from the impossibility of an infinite regress and bungled even that by not starting with *possibly existent being* in the correct sense. With a few modifications, Averroes sets the proof aright. He explains: The proof must begin with an appropriate division of being and with the proper usage of *possibly existent.* The existence of possibly existent beings—taking *possibly existent* as tantamount to *generated-destructible*—is attested to by sense perception, inasmuch as objects are continually observed coming into existence and undergoing destruction.[123] Reason thereupon determines that a possibly existent being does require a cause to actualize it, to effect its transition from potentiality to actuality; for nothing can conceivably bring itself to actuality.[124] Further, a series of possibly existent beings that are causes of one another cannot regress infinitely. For, as Aristotle showed, the true cause is always the first cause, and should there be no first cause there would be no true cause and hence no cause at all;[125] "to assume an infinite number of possible causes" would accordingly entail caused beings "without an agent [to effect their actualization]," something that is "absurd."[126] The impossibility of an infinite regress leads to the preliminary conclusion that the series of possibly existent, or generated-destructible, beings must terminate in a being

[121] Averroes apparently understood that this manner of beginning the proof was the basis for Avicenna's claims regarding the proof's metaphysical character; cf. *Tahāfut al-Tahāfut,* X, p. 419. Averroes' corrected version of the proof, given immediately below, begins with a physical phenomenon.

[122] *Tahāfut al-Tahāfut,* IV, p. 280; X, p. 418. Avicenna, it will be recalled, was willing to grant that the random object we begin our investigation with is necessarily existent by virtue of itself; he would, however, insist that all the attributes of the necessarily existent by virtue of itself be then assigned to that object. Cf. above, p. 303.

[123] *Tahāfut al-Tahāfut,* VI, pp. 359–361.

[124] Ibid., VIII, p. 393, and VI, p. 361.

[125] Ibid., IV, p. 274. Cf. below, p. 337.

[126] Ibid., pp. 278–279.

that is not generated, but is necessary, or eternal. A necessary being is not yet a deity, however, and Averroes proceeds, as will be seen in the next chapter, to argue that the series of necessarily existent, or eternal, beings must, for its part, likewise terminate; it must terminate in a being necessarily existent by virtue of itself, and this last will possess the attributes of the deity.[127] The proof Averroes here offers as an emended version of Avicenna's proof from the concepts *possibly existent* and *necessarily existent* makes no use of the argumentation of Avicenna's proof outlined in the previous chapter. Averroes has simply offered his own version of the argument from the impossibility of an infinite regress, employing therein the terms *possibly existent* and *necessarily existent* as he construes them.[128]

Averroes' critique of the body of Avicenna's proof is to an astonishing extent grounded in misinformation. The criticism that Avicenna employed the term *possibly existent* equivocally reflects a misunderstanding of the way Avicenna did use the term and its counterpart *necessarily existent*.[129] The assumption that Avicenna attempted nothing more than a standard argument from the impossibility of an infinite regress of causes reveals that Averroes had no firsthand knowledge of Avicenna's distinctive proof from the concepts *possibly existent being* and *necessarily existent being*.[130] And the contention that Avicenna failed to accomplish his proof because of an incorrect use of the term *possibly existent* does not in any way address Avicenna's proof from the concepts *possibly existent* and *necessarily existent*[131] or, for that matter, the version Avicenna offers elsewhere of the standard argument from the impossibility of an infinite regress.[132]

5. Summary

Averroes must have relied on derivative accounts of Avicenna's philosophy, such as Ghazali's account, in addition to whatever incomplete copies of Avicenna's works might have reached him in Spain.[133] The fragmentary information he had was supplemented through his own reconstruction of Avicenna's reasoning.

Averroes knows that Avicenna proposed to offer a metaphysical proof of the existence of God, and he rejects the possibility of such a proof on the grounds that Aristotelian rules for determining the subject matter and principles of any given science preclude it.[134] He seems not to know that his objections had been

[127]Ibid., and VIII, pp. 393–394. Cf. below, p. 342.

[128]Cf. below, ibid.

[129]Cf. above, p. 319.

[130]Cf. above, pp. 331–333.

[131]Cf. above, n. 122.

[132]That argument of Avicenna's would begin—as, Averroes insists, the argument from an infinite regress must—with objects that are transient and hence clearly do have a cause. Cf. below, p. 340.

[133]In *K. al-Kashf,* p. 42, Averroes writes that no Mu'tazilite writings of substance had reached Spain. In the *Middle Commentary on Metaphysics,* XII, p. 141(142)a, he states that Avicenna's proof from the concepts *possibly existent* and *necessarily existent* appears in the *Shifā'*. In fact it does not appear there; cf. above, p. 289.

[134]Above, pp. 312–313.

anticipated and answered by Avicenna.[135] On a second issue, Averroes generally takes Avicenna's twofold division of actual existence to be a threefold division[136] because he mistakenly supposes that *possibly existent* designated a category of actual existence for Avicenna.[137] In a third issue, Averroes objects especially to the designation of the celestial realm as *possibly existent by virtue of itself, necessarily existent by virtue of another.* Insofar as this objection is directed against the formula *possibly existent by virtue of itself, necessarily existent by virtue of another,* it misses the mark, since Avicenna did not use the formula in the way Averroes supposes he did.[138] Finally, Averroes' critique of the body of Avicenna's proof from the concepts *possibly existent* and *necessarily existent* also misses completely; for Averroes does not correctly understand the concepts and intention of the proof. The most pertinent point made by Averroes—disregarding, now, the formula *possible by virtue of itself, necessary by virtue of another,* which Averroes misunderstood—is that Avicenna construed the celestial realm as eternal, yet in itself only possibly existent. Averroes objects that whatever has only a possibility of existing also has a possibility of not existing; whatever has a possibility of not existing will at some time not exist; and hence possible existence is incompatible with eternal existence. As Averroes pursues the matter, however, he too cannot escape construing the substratum of the spheres as both possible and also necessary, or eternal; and in one philosophic note he finds himself forced to embrace precisely Avicenna's position.[139]

The failure of Averroes' critique is striking. Averroes' objections are based on incorrect information and misunderstandings. And the one serious and pertinent objection that he does raise leads him through a chain of puzzles back to the position of Avicenna which he generally attempts so diligently to refute.

[135]Cf. above, p. 317.

[136]Above, p. 318. Above, p. 332, however, where Averroes follows Ghazali, he recognizes a twofold division of being in Avicenna. Averroes' objection to the effect that Avicenna was guilty of equivocation in using the term *possibly existent* grows out of the confusion between the—genuine—twofold division of being and the—mistaken—threefold division; for *possibly existent* has a different extension in each.

[137]Cf. above, p. 318.

[138]Above, p. 292.

[139]Above, p. 330.

XI

Proofs of the Existence of God from the Impossibility of an Infinite Regress of Efficient Causes

1. The proof from the impossibility of an infinite regress of causes

Aristotle's proof of the existence of God as the first cause of motion and Avicenna's proof of the existence of God from the concepts *necessarily existent* and *possibly existent* each rests on an elaborate apparatus; the former rests on a comprehensive set of physical principles,[1] the latter, on the careful analysis of the critical concepts.[2] An ostensibly more direct proof of the existence of God is suggested by *Metaphysics* II, where Aristotle establishes the general principle that an infinite regress of causes is impossible. The argument suggested by *Metaphysics* II, would run: The universe contains a series of causes; no series of causes can regress indefinitely; therefore, in tracing causes back, a first cause must eventually be reached. The existence of a first cause would thus be inferred from the impossibility of an infinite regress of causes taken together with the principle of causality.[3] This skeletal argument had also appeared in Aristotle's proof from motion and in Avicenna's proof. The proof growing out of Aristotle's *Metaphysics* II—which I shall call the *proof from the impossibility of an infinite regress*—hence has an unmistakable affinity to both those proofs. Certain significant differences should be taken note of, however, if the three proofs are to be properly understood.

As Avicenna formulated his proof from the concepts *necessarily existent* and *possibly existent,* the existence of a being *necessarily existent by virtue of itself* is indeed inferred from the impossibility of an infinite regress of causes. But the inference is artificial and awkward, and the originality of Avicenna's proof lay precisely in its providing a way to establish a first cause without the procedure

[1]Cf. above, pp. 239–248.
[2]Above, pp. 289–292, 296–298.
[3]Aristotle, *Metaphysics* II, 2.

of tracing a series of causes back to a first term in the series;[4] I shall return to the point later.[5] The impossibility of an infinite regress also plays a role in Aristotle's proof from motion, and here the role is essential and indispensable.[6] Yet a difference is still to be discerned. The proof from motion focused its attention on motion and change, particularly motion in place, that being one reason for the large number of physical principles, and for the complicated argumentation, required in the proof from motion. By contrast, the proof that is the subject of the present chapter disregards the several types of effect which can be brought about by an agent; and it borrows nothing from the science of physics. It considers efficient causation in the abstract, and the first cause whose existence it infers is denominated an *efficient cause,* which generally is placed in conscious contradistinction to a *cause of motion*.

The proof we are considering has its origins, as already mentioned, in Aristotle's *Metaphysics* II. The grounds given there for rejecting an infinite regress of causes were that the true cause of whatever occurs in any causal series is always the first cause, and therefore a first cause must in every instance be posited. The thought was developed by Aristotle through the distinction between the final term in a causal series, that is to say, the final effect; the intermediate link or links; and the first term, which activates the intermediates. The trichotomy is valid, Aristotle remarked, irrespective of whether the intermediate members in a given series should be assumed to be "one or more, infinite or finite."[7] Now in any causal series, Aristotle proceeded, the intermediate cause is clearly not the *true* cause of the final effect; for the intermediates are only instruments, dependent upon causes standing behind them. Each causal series consequently has only one *true* cause, namely, the factor standing behind all the intermediate causes and activating them all. If there should be no factor standing behind the totality of intermediate causes, "if there should be no first cause," then there would be no true cause. The causal series would have "no cause whatsoever," which is absurd. But an infinite regress would be precisely a situation wherein all the members preceding the final effect are "equally intermediates." An infinite regress would be a causal series without a first cause and hence without a true cause, which is impossible.[8]

The principle that no causal series can regress without reaching a first term was applied by Aristotle to the four genera of cause which he recognized. That is to say, every series of material causes, final causes, and formal causes, must have a first term. And by the same token, causes in the sense of the "source of movement," or, in other words, *efficient causes*—the two concepts are equivalent

[4] Above, pp. 302, 307–308.

[5] Below, p. 351.

[6] Above, p. 238.

[7] Aristotle ignores the possibility of a causal series' having only two members; no such series is in fact to be found in nature.

[8] *Metaphysics* II, 2, 994a, 1–19.

for Aristotle—also must have a first term.[9] The priority of the first cause hereby established for each series is, as Alexander was to note, not to be understood as *temporal* priority.[10] Aristotle intended to establish a first cause that exists together with its effects and performs its causal function as long as the effects exist.[11]

In the passage in *Metaphysics* II which we are considering Aristotle did not explicitly move from the impossibility of an infinite regress of efficient causes to a proof of the existence of God.[12] But the step is obvious. If every series of causes must have a beginning, and if, as was beyond question for the Aristotelian school, the universe contains a series of efficient causes, that series must begin at a first uncaused efficient cause. The impossibility of a beginningless regress of causes thus suggests a straightforward proof of the existence of God: Every series of efficient causes has a first term; the universe is seen to contain a series of efficient causes; hence, efficient causes can be traced back to a first cause, which is the deity. In Aristotle, *efficient cause* has no meaning apart from "source of motion."[13] Even to the extent, therefore, that the present argument is implicit in Aristotle, it would lead to nothing other than a first source of motion.[14] Like Aristotle's more elaborate proof of the existence of God, it would lead again to a first cause of motion in place, upon which other motions and changes at all times depend; and there would be little justification for differentiating the present proof from the proof of motion.

The argument from the impossibility of an infinite regress was recognized as an independent proof of the existence of God by at least one of Aristotle's Greek commentators, Simplicius.[15] In the Middle Ages, it became a full-fledged proof, in fact the most popular non-Kalam proof, of the existence of God. Typically, the concept *efficient cause* was understood in a wider sense than the Aristotelian *source of motion*; as Avicenna maintained most explicitly, the concept may cover the cause of the very existence of a thing as well as the cause of its change or motion.[16] When *efficient cause* is understood in the wider, un-Aristotelian sense,

[9]Ibid., 3–11.

[10]Alexander of Aphrodisias, *Commentary on Metaphysics*, ed. M. Hayduck, *Commentaria in Aristotelem Graeca*, Vol. I (Berlin, 1891), p. 151, line 2. For the different senses of *priority* distinguished by Aristotle, see *Categories* 12; *Physics* VIII, 7, 260b, 16 ff.

[11]Cf. above, p. 238; below, p. 345.

[12]A suggestion of such a proof may be found in *Metaphysics* II, 2, 994a, 5–7.

[13]Cf. *Physics* II, 3, 194b, 29; E. Zeller, *Die Philosophie der Griechen*, Vol. II, Part 2 (4th ed.; Leipzig, 1921), p. 327, n. 2.

[14]See reference in n. 12.

[15]Simplicius, *Commentary on Physics*, ed. H. Diels, *Commentaria in Aristotelem Graeca*, Vol. X (Berlin, 1895), p. 1359, lines 18–19. Also cf. Plotinus, *Enneads*, IV, 7, 9; Proclus, *Elements of Theology*, ed. and trans. E. Dodds (Oxford, 1962), §11.

[16]Avicenna, *Shifā': Ilāhīyāt*, ed. G. Anawati and S. Zayed (Cairo, 1960), p. 257; E. Gilson, *History of Christian Philosophy in the Middle Ages* (New York, 1955), pp. 210–212; Averroes, *Epitome of Metaphysics*, ed. and trans. C. Quirós Rodriguez (Madrid, 1919), I, §§6–7; German translation: *Epitome der Metaphysik des Averroes*, trans. S. van den Bergh (Leiden, 1924), pp. 3, 149.

the argument from the impossibility of an infinite regress of efficient causes can be employed to establish a first cause of the very existence of the universe—provided, of course, that causes of existence are known to be present in the universe. Here is another feature of the proof, by the side of its simpler argumentation, which rendered it attractive to medieval philosophers: It can, unlike Aristotle's proof from motion,[17] lead to a first cause of existence. A weakness in various formulations of the proof was, however, the failure to show just where causes of existence are discoverable in the universe and how they can be traced back to a first cause of existence.[18]

In working out their proofs, some medieval writers postulated the impossibility of an infinite regress with no explanation. Others explained that an infinite causal regress is impossible because of the nature of the infinite: An infinite causal regress was rejected because, in general, no actual infinite is possible, or else because, in general, no actual infinite can be traversed or added to.[19] But there also were medieval philosophers who, in the spirit of Aristotle, reasoned not from the nature of the infinite, but from the nature of causality. contending that causal series by their nature require a first term.

Reasoning of the last type is offered by Avicenna in his most comprehensive philosophic work, the *Shifā',* a composition that does not have the distinctive proof, examined in an earlier chapter, for the existence of a being necessarily existent by virtue of itself. In the *Shifā',* Avicenna establishes the principle that no causal series, in any of the four genera of cause, can regress indefinitely; and his thinking is basically the same as Aristotle's, although with a certain shift of emphasis: He directs almost all his attention to the thesis that every causal series must have three components—first term, middle link, and final term—and only in passing does he mention why a first term must be posited. Causal series, Avicenna argues, have three components, each of which has its own "peculiar characteristic." The "[final] term" is solely an effect and "not a cause of anything"; the "[first] term" is the cause of everything following it; and the "intermediate" link is the cause of the final term and the effect of the first. These distinctions are valid whether the intermediate link should be assumed to be "one or more than one, . . . finite . . . or infinite." As for the reason why a causal series must indeed have a first term, Avicenna merely observes that the first term is the "absolute" cause of everything subsequent to it, an apparent allusion to Aristotle's contention that the first cause alone is the true cause. In an infinite regress, Avicenna continues, whatever might stand before the final term would be both cause and effect. But to be both cause and effect is the peculiar characteristic of the intermediate link in a causal series, and hence everything standing

[17]Averroes, it should be noted, even in his later works, which no longer distinguish between *efficient cause* and *cause of motion,* did find a way to construe a first cause of motion as a first cause of existence. See above, pp. 283, 318.

[18]See below, p. 345. .

[19]For instances of these positions, see below, p. 363.

before the final term would be included within the intermediate link. Assuming an infinite causal regress thus amounts to assuming a final term in a causal series, an infinite intermediate link, but no first term. It amounts to assuming a causal series lacking one of the requisite constituents, an assumption that is "absurd." Consequently, a beginningless regress is impossible. Avicenna has, without mentioning Aristotle by name, repeated that philosopher's remark to the effect that the intermediate members in a causal series might be "finite . . . or infinite."[20] He expatiates on the import of the remark: Once a first term in a causal series is granted, the intermediate links might hypothetically be construed as infinite. Hypothetically, that is, a causal series might have a first term, a last term, and infinite intermediate members. But further consideration shows the hypothesis to be unreal; for a series that is infinite, yet terminates, is in fact inconceivable, possible only "in speech, not in belief." Avicenna's conclusion, therefore, is that every causal series has a first term, and that the number of links in every causal series is finite.[21]

The foregoing argument, Avicenna understands, provides an appropriate "proof of the finiteness of all genera of causes,"[22] but it is especially applicable to "efficient" causes. When applied to efficient causation, it serves as the backbone of a proof of the existence of God. It permits tracing efficient causes in the universe to a first uncaused efficient cause; and when the first cause is in addition shown to be incorporeal and one, we arrive at the deity.[23]

Thus far, Avicenna's *Shifā'*. The proof of the existence of God in Avicenna's Persian work, the *Dānesh Nāmeh,* is also an argument from the impossibility of an infinite regress of efficient causes, but now Avicenna establishes the critical principle by a different route: He employs a train of reasoning similar to the reasoning developed in his distinctive proof from the concepts *possibly existent* and *necessarily existent*;[24] and yet his argumentation is not the same as in that proof either. His argumentation is more elliptical, and is conducted without mention of the two critical concepts, *possibly existent* and *necessarily existent.*

In the *Dānesh Nāmeh,* Avicenna mentally gathers into a single group all the links in a supposed infinite regress of efficient causes. Since the group, considered as a whole, exists through its components, it can, he submits, "by no means be uncaused." Avicenna entertains no alternative but that the cause of the group lies outside the group.[25] And inasmuch as all the caused links were, by assumption, included inside the group, he concludes that the cause lying outside would be of a different character and would not itself have a cause. The supposedly infinite series would accordingly not be infinite after all, but would terminate at an

[20]Above, at n. 7.
[21]*Shifā': Ilāhīyāt,* pp. 327–328.
[22]Ibid., p. 329.
[23]Ibid., p. 342.
[24]Cf. above, pp. 301–302.
[25]The criticism stated above, p. 305, will apply even more strongly here.

uncaused efficient cause.[26] Having established the impossibility of an infinite regress of efficient causes, Avicenna goes on to infer the existence of a single first efficient cause for the entire universe, which is the deity.[27]

Averroes too employs the argument from the impossibility of an infinite regress. In an early work, the *Epitome of the Metaphysics,* he establishes the key principle in a manner that reflects the language and thinking of Avicenna, including the shift of emphasis evidenced in Avicenna's *Shifā'*: Like Avicenna, Averroes places more emphasis on the thesis that every causal series must have three parts—first term, middle link, and final term—than on the explanation of why a first term must be posited. Causal series, Averroes explains in his *Epitome of the Metaphysics,* have three components, and each of the three has its "peculiar characteristic." The final term "is not a cause of anything"; the middle link is "cause and effect, . . . cause of the last and effect of the first"; the first term is "exclusively a cause." This situation must obtain irrespective of whether the middle link should be "one or many, finite or infinite."[28] To assume an infinite regress therefore amounts to assuming a causal series with a final term, an infinite intermediate link, but no first term. And unacceptable implications would ensue. One unacceptable implication is merely verbal. *Intermediate,* Averroes notes, designates that which stands between two terminal points, and consequently, to assume an intermediate without assuming a first term would be a self-contradiction. Another unacceptable implication is developed on a more philosophical plane. Averroes brings to bear the device, familiar from Avicenna, of mentally grouping all the intermediate links together; and he reasons: Since all the intermediate links are effects, their totality, unless preceded by an uncaused cause, would comprise an effect without a cause. But an effect without a cause is impossible. Consequently, in addition to the intermediate link which is an effect as well as a cause, every causal series must have a first term, which is the cause of everything following it without being itself an effect.[29]

The argument establishing a first term for every causal series is applicable, Averroes continues, to all "four causes," but especially to the "efficient, motive" cause.[30] When applied to the efficient cause, it leads to an "ultimate agent" of the universe,[31] whereby Averroes clearly means a first cause of motion which is by the same token a first cause of the existence of everything in the universe.[32]

[26] Avicenna, *Dānesh Nāmeh* (Teheran, 1937), p. 103; French translation; *Le Livre de Science,* trans. M. Achena and H. Massé (Paris, 1955), pp. 132–133; English translation: *The Metaphysica of Avicenna,* trans. P. Morewedge (New York, 1973), p. 45.

[27] *Dānesh Nāmeh,* p. 115; French translation, p. 150; English translation, p. 59.

[28] Averroes, *Epitome of Metaphysics,* III, §64; German translation, pp. 98–99.

[29] Ibid., III, §65; German translation, p. 99. [30] Ibid., III, §66; German translation, pp. 99–100.

[31] Ibid., III, §71; German translation, p. 102.

[32] See above, p. 283. In his *Middle Commentary on the Metaphysics,* Averroes also employs the trichotomy of first cause, intermediate link, and final effect. There, however, he establishes that a first cause must exist by contending, as Aristotle had done, that the first cause alone is the true cause. See Casanatense Library, Hebrew MS. 3083, p. 3(4) a-b.

In another, later work, the *Tahāfut al-Tahāfut,* Averroes states the reasoning against an infinite regress of efficient causes with a peculiar nuance. He arrives at a first cause not in one, but in two stages.

The *Tahāfut al-Tahāfut* employs the familiar terms *possibly existent* and *necessarily existent,* but Averroes construes the terms differently from Avicenna; for Averroes, *possibly existent* is equivalent to *generated-destructible,* and *necessarily existent* to *eternally existent.*[33] Averroes' reasoning runs: Observation testifies to the existence of objects that are possibly existent, that is to say, subject to generation and destruction. Each object of the sort requires an external cause to bring it into existence. Should the objects in question "extend to infinity" and have no first cause, they would have "no cause"; for Aristotle showed that the only true cause in a series is the first cause. Inasmuch as effects plainly cannot come about without a cause, possibly existent, or generated-destructible, objects cannot regress indefinitely. They must terminate at a being that is the cause of the entire series of possibly existent beings, hence at a being that is not itself possibly existent, or generated-destructible, but rather eternal, or necessary. The necessarily existent being thus reached by tracing back a series of possibly existent beings may conceivably also have a cause, and the impossibility of an infinite regress has to be applied to necessarily existent beings as well. But now, Averroes insists, the principle is to be applied in a fashion different from its application to possibly existent beings. A possibly existent, or generated, being must by its very nature have a cause. Therefore the proposition that only the first cause is a true cause, and without a first cause there would be no causality at all, leads to a single, definitive conclusion, to a first cause of the series of possibly existent beings. Necessary existent, or eternal, being is not, however, such that it must, by its very nature, have a cause. Therefore, in the case of necessarily existent beings, the proposition that without a first cause there can be no cause whatsoever leads, according to Averroes, not to a single definitive conclusion, but instead to a pair of alternatives: Either the supposed series of necessary and eternal beings terminates at a first necessary cause; or else there can have been no causal series to begin with, and the necessary being that was assumed to have a cause did not after all have one. Each alternative, of course, entails the existence of an uncaused cause, which exists by virtue of itself, to wit, either the necessarily existent being at which the series of possibly existent beings terminates, or else the first member of the series of necessarily existent beings. The difference between applying the impossibility of an infinite regress to possibly existent beings and to necessarily existent beings seems trivial, but Averroes stresses its significance and reproves Avicenna for having overlooked it. Averroes' conclusion is that the impossibility of an infinite regress, when correctly applied, demonstrates a first efficient cause, which is necessarily existent by virtue of itself.[34]

[33]Cf. below, p. 379.

[34] Averroes, *Tahāfut al-Tahāfut,* ed. M. Bouyges (Beirut, 1930), IV, pp. 278–279; cf. VIII, pp.

Aquinas' proof of the existence of God from the "nature of the efficient cause" is a further instance of the argument from the impossibility of an infinite regress. In any series of efficient causes, Aquinas explains, the "first [term] is the cause of the middle, and the middle is the cause of the final [term]," irrespective of whether the "middle [link] is many or only one." Now if there "should be no first [term], there would be no final [term] and no middle"; the reason is that the first cause alone is the true cause of everything following it, and "when the cause is removed, the effect is also removed." But assuming an infinite regress of efficient causes amounts to assuming a series without a first cause. Since without a first cause there could be no middle link or final effect, an infinite regress is impossible, and a first term in the series of efficient causes contained in the universe must be posited.[35] Aquinas may well have intended the argument to establish a first cause of the very existence of the universe; but he leaves that unsaid. He does conclude that the "first efficient cause" established by the argument is "what everyone calls God."[36]

Before proceeding, several unstated assumptions should be pointed out which underlie the various versions of the proof from the impossibility of an infinite regress of causes. Without the assumptions the proof will lack cogency.

The proof disregards the different ways in which an "efficient cause" may act upon its effect; what is considered is efficient causation in general. Implied herein is the assumption that whenever operations of one kind stand behind operations of another kind, a single continuous chain is formed, activated by a single causality running from beginning to end. A situation is not countenanced wherein a series of efficient causes may be traced back to an absolutely first cause in the series—let us call it *x*—which is responsible for all activity of a certain character within the series; and yet *x* will be acted upon by something else in a completely different respect that in no way affects its own causal activity. The assumption tacitly made is, on the contrary, that whenever *x* acts in one respect and is acted upon by something else—let us call it *y*—in another respect, its own causal activity is necessarily dependent upon *y*, and *x* is not in fact a first cause. To invent an example, if *x* is the ostensible first term in a series of causes producing changes of color, but itself undergoes a change of temperature, then its own ability to produce a change of color is assumed to depend on the change brought about in its temperature. The first cause of changes in temperature, in our example, is to be construed as the ultimate cause of a single series that embraces changes of both color and temperature. The assumption could be made because

393–394; X, p. 418. English translation, with pagination of the Arabic text indicated: *Averroes' Tahafut al-Tahafut,* trans. S. van den Bergh (London, 1954).

[35]Aquinas, *Summa Theologiae,* I, q. 2, art. 3, secunda via. The point that the first cause is the true cause is made explicitly in *Summa contra Gentiles,* I, 13.

[36]*Summa Theologiae,* I, q, 2, art. 3, secunda via.

the proof from the impossibility of an infinite regress was thought out against the background of Aristotle's physics and Aristotle's proof from motion. In Aristotelian physics, all types of change are taken as forming a single chain, ultimately originating in motion in place; and a primary cause of motion would accordingly be responsible for all the types of change subsequent upon motion.[37] These propositions were indispensable for the proof from motion and were cited explicitly in the formulation of it. The argument from the impossibility of an infinite regress of causes in Avicenna, Aquinas, and others—not including Averroes[38]—disregards the different kinds of causal operation and does not undertake to trace them all back to motion in place. As a result, the argument simply assumes what was justified explicitly in the proof from motion. It assumes that various kinds of causality do form a single causal chain; that a first link in any chain will be responsible for all the several kinds of causality subsequent upon it; and that a first efficient cause must consequently be uncaused in every respect.

A further assumption underlies the proof from the impossibility of an infinite regress of efficient causes. The object of the proof is to go back through a series of causes and effects existing together at any moment and to arrive at a first term whose primacy is not temporal but solely causal. Yet the various versions of the argument do not necessarily take their departure from causes and effects that coexist. The argument typically begins with a random occurrence of change in the universe. The given change, the reasoning runs, must be due to a cause; and the cause may be one that merely initiates the career of the corresponding effect without continuing to sustain the effect, as when an object is fashioned by an artisan or through a natural process.[39] The artisan or natural process might for its part have been brought into existence or to the state wherein it acts by an earlier factor of the same character, in other words, by a factor that does not coexist with, and sustain it. As the series is traced back, however, a stage must be reached—such is the assumption—where causes do not merely initiate the career of their effects. Thenceforth, as we go back, causes coexist with their effects and exercise their causal function during the entire duration of the existence of whatever it is that they happen to produce. By virtue of the present assumption, the first term in the series of efficient causes is understood to exist together with its immediate effect, to exercise its causal function as long as the effect exists, and in an extended—although not wholly exact[40]—sense, to exist together with,

[37]Cf. above, p. 241.

[38]See below, pp. 348–350.

[39]In the Aristotelian system, the cause of a process has to be present as long as the process takes place, just as the cause of motion has to be present as long as motion takes place. But once the process is completed, the cause that sustained the process would not have to be present for the product to be maintained in existence.

[40]For example, an effect may be produced at moment t^1; and its cause may have been brought into existence at moment t by a prior cause, which in its turn is continually sustained by the first cause. Then by virtue of existing together with its immediate effect and continually exercising its

and exercise its causal function upon the series as a whole. The assumption left unstated here is again a reflex from Aristotle. In Aristotelian physics, motion in place stands behind the other kinds of change, and every motion in place does require a cause to sustain it as long as it continues.[41] Aristotle's proof from motion was accordingly able to trace all changes back to a stage where causes coexist with their effects, and perform their causal function continually. In a similar vein, the argument from the impossibility of an infinite regress tacitly assumes that as one goes back through a series of efficient causes, a stage will be reached where causes sustain their effects continually.

Yet a further assumption would have to be made by any philosopher who expected his argument from the impossibility of an infinite regress of causes to establish a cause of the existence of the universe. Such a philosopher obviously would have to assume that the first effect in the chain, even if eternal, depends upon its cause not merely for any change it might undergo but for its very existence. And if the first effect depends upon something prior to it for its existence, later effects may also be presumed to.[42] By virtue, then, of serving as the cause of the existence of the first effect, the first cause will be responsible for the existence of the entire series. The various versions of the argument are unfortunately sometimes vague[43] as to how, in tracing back efficient causes that bring about change, the first cause might be shown responsible for the very existence of what follows it.

2. Unity and incorporeality

The proof of the existence of God from the impossibility of an infinite regress of causes, even granting the unstated assumptions just examined, is not complete. The argument establishes a first term for every series of efficient causes. But the first efficient cause might conceivably be an inanimate object or an animate physical object; and several independent series of efficient causes might conceivably exist, each with its own first cause. Before the argument can be raised to a complete demonstration of the existence of God, the incorporeality and unity of the first cause remain therefore to be established. The need to supplement the argument was recognized by Avicenna, Maimonides, Aquinas, and Averroes.

Avicenna supplements the argument from the impossibility of an infinite regress through the analysis of the concepts *necessarily existent by virtue of itself* and *possibly existent by virtue of itself*. The analysis of those concepts has already

---

causal function, the first cause will have initiated at moment t the subsequent appearance of an effect at moment t^1.

[41] Above, p. 241.

[42] See above, p. 344, n. 39.

[43] Maimonides and Aquinas give no grounds for construing the first cause as a cause of existence, although they understand it to be such. Avicenna, below, p. 346, and Averroes, above, p. 283, do give grounds for construing the first cause as a cause of existence.

been seen to be integral to the argumentation of another proof of Avicenna's, a proof that led precisely to the existence of a being necessarily existent by virtue of itself.[44] In the argumentation of the present proof, the analysis of the concepts has so far played no role; and the present proof has led to nothing other than a first term for any series of efficient causes. But once Avicenna arrives at a first efficient cause, he reasons that such a cause must exist exclusively by virtue of itself, since otherwise it would not be first. As a consequence, a first efficient cause must have all the attributes of a being that exists necessarily by virtue of itself. It must, most importantly, be one, simple, and incorporeal. Furthermore, since there can exist only one being necessarily existent by virtue of itself, everything else will be possibly existent by virtue of itself and will require a cause for its very existence. The series of efficient causes that has been traced back to a first term because of the impossibility of an infinite regress thus turns out to be a series of causes of existence, and the first term in the series is seen to be the cause of the very existence of everything subsequent to it.[45] Avicenna has in effect hereby supplemented the proof from the impossibility of an infinite regress with elements of his other proof.[46] He has invested the "absolutely" first efficient cause with all the attributes analyzable out of the concept *necessarily existent by virtue of itself*; and he has shown that whatever is *possibly existent by virtue of itself* must depend upon the first efficient cause for its existence.

Since Avicenna's proof from the impossibility of an infinite regress now depends for its completion upon the analysis of the concepts *possibly existent* and *necessarily existent,* the proof, it should be noted, will lie open to an objection similar to that raised in an earlier chapter in connection with Avicenna's other proof.[47] The impossibility of an infinite regress would establish a first efficient cause that exists by virtue of itself only in the sense of not having any further efficient cause outside itself; the impossibility of an infinite regress of efficient causes would not establish that the first efficient cause cannot contain even internal factors, for instance, matter and form, making it what it is. And yet Avicenna assigns it the attributes of what is necessarily existent by virtue of itself in the special sense of not containing even internal causes.[48]

Maimonides also recognized the need to supplement the argument from the impossibility of an infinite regress in order to reach a complete proof of the existence of God. His method is to develop an observation of Aristotle's to the

[44]See above, pp. 298–304.

[45]*Shifā': Ilāhīyāt,* pp. 37–47, 342; *Dānesh Nāmeh,* p. 115; French translation, p. 151; English translation, p. 59 f.

[46]The *Shifā'* is generally understood to have been written by Avicenna before the *Najāt.* If that is correct, then the proof from the concepts *possibly existent* and *necessarily existent,* which appears in the *Najāt,* grew out of Avicenna's rethinking his earlier analysis of those concepts, an analysis that had appeared in the *Shifā'.*

[47]Above, pp. 305–307.

[48]See Ghazali's critique, below, p. 373.

effect that the ultimate cause of motion must abide in an eternal state of actuality.[49] Maimonides begins by considering the efficient causation evidenced in the translation of objects from a state of potentiality to a state of actuality. Instances of transitions from potentiality to actuality are, he observes, to be found throughout the physical realm. Every transition to actuality is brought about by a factor external to the object undergoing the change; Maimonides refers his reader to Aristotle for the principle that nothing potential can pass to actuality solely through itself.[50] If the factor bringing about a given transition from potentiality to actuality happens to have undergone a transition of the same sort, it must be dependent on an additional factor. The series of factors cannot regress indefinitely because of the impossibility of an infinite regress of causes; Maimonides directs his reader to Aristotle for the demonstration of this principle too.[51] His conclusion is that a series of factors bringing about transitions from potentiality to actuality must terminate at a first cause that does not pass from potentiality to actuality, hence at a cause that is eternally in a state of pure actuality. Maimonides hereupon reads out the attributes of a purely actual being: It must be immaterial and incorporeal; for matter and potentiality go hand in hand, and a purely actual being is free of potentiality.[52] Inasmuch, moreover, as it is incorporeal, it contains no ground of individuation, and is the only one of its kind.[53] The impossibility of an infinite regress of causes thus leads Maimonides not merely to a first term in a given series of efficient causes, but to a first, eternally actual being with the attributes of the deity—to a single, incorporeal, first cause of all transitions in the universe, although not necessarily of the very existence of the universe.[54]

In the *Summa contra Gentiles,* Aquinas supplements the argument from the impossibility of an infinite regress through an analysis of several concepts associated with *first efficient cause,* including both the concept *necessarily existent by virtue of itself* and the concept *eternally actual being*. He thereby combines the procedures of Avicenna and Maimonides. The first efficient cause established by the impossibility of an infinite regress must, Aquinas reasons, exist necessarily by virtue of itself.[55] Analysis of the concept *necessarily existent by virtue of itself* and assigning the attributes analyzable out of the concept to the first efficient cause discloses that the first cause is simple,[56] incorporeal,[57] and one.[58] Then,

[49]Aristotle, *Metaphysics* IX, 8, 1050b, 4–6; XII, 6, 1071b, 20. Cf. above, p. 59; Avicenna, *Shifā': Ilāhīyāt,* p. 184; and *R. Zaynūn,* pp. 3–4, in *Rasā'il al-Fārābī* (Hyderabad, 1931).

[50]Maimonides, *Guide to the Perplexed,* II, introd., prop. 18.

[51]Ibid., prop. 3.

[52]Ibid., prop. 24.

[53]Ibid., prop. 16.

[54]*Guide,* II, 1(d).

[55]*Summa contra Gentiles,* I, 15. For Aquinas, *necessarily existent* means *eternally existent,* whereas for Avicenna, it means *actually existent*; see below, p. 384. But this difference does not affect the analysis of attributes out of the concept *necessarily existent by virtue of itself.*

[56]*Summa contra Gentiles,* I, 18(c).

[57]Ibid., 20(a).

[58]Ibid., 42(g), (h), (j), (k).

following Maimonides, Aquinas also reasons that the efficient causes behind a given event are all factors bringing about transitions from potentiality to actuality; and since the series of factors undergoing transitions cannot regress indefinitely, the series must terminate in "something that is wholly actual and nowise potential."[59] Having established that the first efficient cause is wholly actual, Aquinas analyzes the key attributes simplicity,[60] incorporeality,[61] and unity[62] out of that concept as well and again assigns them to the first cause. The first cause established by the impossibility of an infinite regress is, through parallel routes, shown to have the attributes of the deity.

Averroes too recognized that the argument from the impossibility of an infinite regress must be supplemented if it is to become a full proof of the existence of God. In the *Epitome of the Metaphysics,*[63] Averroes accomplishes the task by drawing on the proof from motion. The incorporeality of the first efficient cause is, he writes, made "clear in the science of physics," that is to say, through the proof from motion.[64] And once the first efficient cause has been shown to be incorporeal, its unity can likewise be demonstrated. For inasmuch as the beings contained within any single species are distinguishable from each other only through their matter, species of incorporeal beings can contain no more than one member; hence no more than one member of the species of first efficient cause can exist.[65] It deserves noting that since the first efficient cause established in the proof from the impossibility of an infinite regress is assigned the attributes incorporeality and unity with the aid of the more elaborate proof from motion, Averroes' procedure more or less erases the distinction between the two proofs. If the argument from the impossibility of an infinite regress can be completed solely through the proof from motion, and if the latter for its part incorporates the impossibility of an infinite regress as one of its constituent principles,[66] little distinction between the two proofs remains.[67]

The foregoing is Averroes' procedure in his *Epitome of the Metaphysics*. Averroes also employs the argument from the impossibility of an infinite regress in another work, the *Tahāfut al-Tahāfut,*[68] and there, up to a point, he pursues a line analogous to that of Avicenna and of Maimonides. He invests the first cause established by the impossibility of an infinite regress with the attributes of a deity

[59]Ibid., 16(f).
[60]Ibid., 18(a).
[61]Ibid., 20(b).
[62]Ibid., 42(k), (1).
[63]For the first part of the proof, see above, p. 341.
[64]*Epitome of Metaphysics,* III, §72; German translation, p. 102.
[65]Ibid.
[66]Above, p. 238.
[67]Avicenna similarly blurs the distinction between the proof from the impossibility of an infinite regress, and his proof from the concepts *possibly existent* and *necessarily existent*; see below, p. 351.
[68]For the beginning of the proof, see above, p. 342.

through the analysis of a concept, the concept, appropriately enough, of *first efficient cause*. A first efficient cause, Averroes contends, must be completely free of potentiality; if it did contain potentiality, it would require a factor behind it to bring it to actuality and consequently would not be first.[69] Since the first efficient cause is free of potentiality, it cannot have a material substratum and cannot be a body.[70] The first efficient cause cannot be compound even in the sense of having an attribute distinct from its essence. For in every compound the components are welded together through the presence of "unity"; internal unity in the compound must originate in an entity that stands behind the compound and possesses internal unity through its essence, in a being that is "essentially one, that is to say, simple"; and therefore a compound object cannot be a first cause, and, conversely, a first cause cannot be compound.[71] Furthermore, only one uncaused efficient case can exist. For were several to exist, all belonging to the class of uncaused first efficient cause, they would have to contain both a factor in common with each other as well as another factor through which they are distinguished. They would thereby be composite, whereas the first efficient cause must, as just seen, be simple.[72] Through an analysis of the concept *first efficient cause,* Averroes thus shows that anything corresponding to the concept is incorporeal, simple, and one.

But having come this far through analysis of the concept *first efficient cause,* Averroes in the end still maintains that Aristotle's proof from motion cannot be dispensed with. The simple, incorporeal, first cause might conceivably be the heavens—more precisely, the outer celestial sphere. Although the celestial sphere consists in an extended substratum, it is simple in the sense that it does not contain distinguishable components, such as matter and form. Not being composed of matter and form, it is not a body. In fact, it does not even contain matter, at least not the changing matter of the sublunar world; to describe the sphere as *material* is to use the term equivocally.[73] As far as the argument from the impossibility of an infinite regress goes, the eternally actual and simple substratum of the celestial sphere might then be identified as the first efficient cause of the

[69] *Tahāfut al-Tahāfut,* VI, p. 361; X, p. 422. Cf. above, n. 49.

[70] *Tahāfut al-Tahāfut,* ibid.

[71] *Tahāfut al-Tahāfut,* VI, p. 333; cf. X, p. 420. In III, p. 181, Averroes attributes this theory to Aristotle, perhaps on the basis of the image of the army in *Metaphysics* XII, 10; see Themistius' commentary there. For the theory, see Plotinus, *Enneads,* VI, 9, 1–3; Proclus, *Elements of Theology,* §§1–5 and 13; *Liber De Causis,* ed. and trans. O. Bardenhewer (Freiburg, 1882), §23; Kindi, *Rasā'il,* ed. M. Abu Rida (Cairo, 1950), I, p. 132; English translation, with the pagination of the Arabic indicated: *Al-Kindi's Metaphysics,* trans. A. Ivry (Albany, 1974); M. Marmura and J. Rist, "Al-Kindi's Discussion of Divine Existence and Oneness," *Mediaeval Studies,* XXV (1963), 339 and 351; G. Endress, *Proclus Arabus* (Beirut, 1973), p. 242.

[72] *Tahāfut al-Tahāfut,* V, pp. 292–293; VII, pp. 379; 385–386. Cf. above, p. 297. Averroes distinguishes his own argument for the unity of the first cause from Avicenna's argument by pointing to a small nuance which, he insists, Avicenna missed. See especially V, pp. 292–293.

[73] Cf. *Tahāfut al-Tahāfut,* IV, p. 271.

universe. Alternatively, a power present in, and distributed throughout, the spheres might be identified as the first efficient cause.[74] To establish a first cause that is incorporeal and immaterial in the stricter sense of not being extended, a first cause that consists in pure thought, Averroes again has recourse to the proof from motion. He argues that the substratum of the spheres is eternally in motion; since the substratum is finite, it cannot contain within itself the infinite power required for eternal motion;[75] the substratum of the spheres consequently cannot be construed as the first cause. That is the sole reason why the ultimate cause of all activity in the universe must be a nonextended being consisting in pure thought.[76]

To summarize, the preeminent medieval philosophers recognized that the argument from the impossibility of an infinite regress must be supplemented in order to frame a complete proof of the existence of God. Avicenna supplemented the argument by observing that the first efficient cause clearly exists necessarily by virtue of itself; and he assigned to the first efficient cause the attributes of deity analyzable out of the concept *necessarily existent by virtue of itself.* Maimonides applied the impossibility of an infinite regress to transitions from potentiality to actuality and he reached a first cause that is purely actual. Then by analyzing the concept *first actual cause,* he established the requisite divine attributes. Aquinas joined the procedures of Avicenna and Maimonides. Averroes in one work immediately identified the first efficient cause with the first mover established in the proof from motion. In another work he analyzed certain divine attributes out of the concept of first efficient cause; but in the end he concluded there too that the proof can be completed only by identifying the first efficient cause with the prime mover established in the proof from motion. In Averroes, as a consequence, the proof from the impossibility of an infinite regress and the proof from motion are in principle indistinguishable.

The following section will offer additional instances of medieval philosophers supplementing the argument from the impossibility of an infinite regress through the analysis of the concept *first efficient cause* or equivalent concepts.[77]

3. The proof from the impossibility of an infinite regress of efficient causes and the proof from the concepts possibly existent and necessarily existent

I have been assuming that Avicenna's proof from the concepts *possibly existent* and *necessarily existent,* discussed in Chapter IX, is fundamentally different from the proof from the impossibility of an infinite regress of causes, which is the

[74]Ibid., X, pp. 420–421.

[75]See, more fully, above, p. 244.

[76]*Tahāfut al-Tahāfut,* IX, pp. 405–406; X, pp. 422–423.

[77]Cf. also, Albertus Magnus, *De Causis et Processu,* in *Opera Omnia,* ed. A. Borgnet, Vol. X (Paris, 1891), I, 1, 7; 9–10; Duns Scotus, *Opus oxoniese,* I, dist. 2, q. i, in *Duns Scotus, Philosophical Writings,* ed. and trans. A. Wolter (Edinburgh and London, 1962), pp. 39–51.

subject of the present chapter. The justification for distinguishing the two proofs is that the proof from the concepts *possibly existent* and *necessarily existent* allows the existence of God to be inferred without reference to the impossibility of an infinite regress and without the device of going back through a series of causes step by step until a first term in the series is reached. Shahrastānī and Ḥasdai Crescas, it was seen, drew the inference without that device. They left the question of an infinite regress open and did not trace causes back linearly, link by link. Merely from the analysis of the concepts *possibly existent* and *necessarily existent,* taken together with the empirical datum that something actually exists, they arrived at the existence of a being necessarily existent by virtue of itself.[78]

Avicenna, the author of the proof was, however, to such an extent under the influence of arguments from the impossibility of an infinite regress that he forced his new proof into the older mold. The resulting train of reasoning, as was shown in Chapter IX, is circuitous and redundant. Avicenna reasoned that any series of possibly existent beings must depend for its existence on a being necessarily existent by virtue of itself; he inferred therefrom that the series of possibly existent beings cannot regress indefinitely; and from here he concluded again that the series of possibly existent beings must terminate at, and depend for its existence on, a being necessarily existent by virtue of itself.[79] Apart from its circuitousness and redundance, Avicenna's formulation blurs the demarcation between the two proofs.[80] As Avicenna formulates the proof from the concepts *possibly existent* and *necessarily existent*—in his *Najāt* and *Ishārāt*—it establishes the impossibility of an infinite regress of causes maintaining a given object in existence, infers the existence of a being necessarily existent by virtue of itself, and finally assigns to the being in question all the attributes analyzable out of the concept *necessarily existent by virtue of itself.* And as he formulates the proof from the impossibility of an infinite regress—in his *Shifā'* and *Dānesh Nāmeh*—it establishes the impossibility of an infinite regress of efficient causes, identifies the first efficient cause as necessarily existent by virtue of itself, and assigns to the former all the attributes of the latter. The first of the two lines of argumentation is not put forward by him as an independent proof of the existence of God but rather in two texts as a version of, and in two others as a supplement to, the second. It provides an alternative reason for rejecting an infinite regress of causes. And it supplements the argument from an infinite regress by providing a method for assigning divine attributes to the first cause, no matter how the impossibility of an infinite regress and the existence of a first cause might be established.

Not surprisingly, writers subsequent to Avicenna employed the reasoning of the proof from the concepts *possibly existent* and *necessarily existent* as Avicenna himself did, either as a version of, or as a supplement to the argument from the

[78] See above, p. 308.

[79] See above, p. 307.

[80] Similarly, Averroes removed the distinction between the proof from the impossibility of an infinite regress and the proof from motion.

impossibility of an infinite regress. On the one hand, almost all proofs of the existence of God from the impossibility of an infinite regress which appear in Islamic and Jewish philosophy after Avicenna draw, in some measure at least, upon the analysis of the concepts *possibly existent* and *necessarily existent*. And on the other hand, with the aforementioned exception of Shahrastānī and Crescas,[81] philosophers who cite Avicenna's distinctive proof, whether with the intent of accepting or refuting it, do not regard it as an independent method of establishing the existence of God; the proof is always represented as a version of the argument from the impossibility of an infinite regress.

A variety of instances may be cited wherein the two proofs are combined in some fashion or other.

The use of the analysis of *possibly existent* and *necessarily existent* in formulating an argument from the impossibility of an infinite regress may, to begin, be observed in Ghazali's *Maqāṣid al-Falāsifa*, which is a compendium of Avicenna's philosophy. The *Maqāṣid* establishes the impossibility of an infinite regress of efficient causes by mentally collecting together all the links in the supposed infinite regress. The totality, Ghazali reasons, could not "be necessary [by virtue of itself] since it results from individuals (*āḥād*) that are caused, and nothing resulting from what is caused can be necessary [by virtue of itself]." Exactly why "nothing resulting from what is caused can be necessary" is not explained by Ghazali; appropriate reasoning must be supplied, as, for example, the consideration that the necessary by virtue of itself has no components or causes whatsoever making it what it is.[82] Once satisfied that the totality formed by the links in an infinite regress could not be necessarily existent by virtue of itself, the argument proceeds: The totality would have to be "possible and caused"; it would have to have a "cause outside [itself] which is not caused"; it would therefore terminate at its cause, and accordingly not be infinite after all.[83] Having established the impossibility of an infinite regress of causes, Ghazali's argument applies the principle to the physical universe: The physical universe, being composite, is possibly existent by virtue of itself and requires a cause for its existence. An infinite linear regress of causes is, for the reason just given, impossible. A circular regress is likewise impossible since it would involve an object's being both "cause . . . and . . . effect" of the same thing.[84] The physical universe must therefore ultimately depend on a first efficient cause outside itself, a cause that cannot be possibly existent, but must be necessarily existent by virtue of itself.[85] This first efficient cause will possess the attributes analyzable out of the concept *necessarily*

[81]Other possible exceptions are Joseph b. Yaḥya, below, n. 120; Āmīdī, *Ghāya al-Marām* (Cairo, 1971), pp. 14–15.

[82]Cf. above, p. 296.

[83]Ghazali, *Maqāṣid al-Falāsifa* (Cairo, n. d.), p. 127. Page 137 shows that Ghazali is using the term *necessarily existent* as an ellipsis for *necessarily existent by virtue of itself*.

[84]Cf. above, p. 302.

[85]*Maqāṣid*, pp. 146–148.

existent by virtue of itself. It will be one, simple, incorporeal, and the like,[86] and hence possess all the attributes of a deity.

The impossibility of an infinite regress is here established with the aid of the analysis of the concepts *possibly existent* and *necessarily existent,* although the exact reasoning at the critical point remains unclear.[87] The composite nature of the physical universe thereupon shows that the universe cannot be necessarily existent by virtue of itself, but must have a cause. And since the first efficient cause reached by the impossibility of an infinite regress must exist necessarily by virtue of itself, the first cause can be assigned the divine attributes analyzable out of the concept *necessarily existent by virtue of itself*. Ghazali's argument may be viewed as Avicenna's proof from the concepts *possibly existent* and *necessarily existent* forced once again into the mold of an argument from the impossibility of an infinite regress; or, if one prefer, it may be viewed as an argument from the impossibility of an infinite regress supported at each stage by the reasoning of the proof from the concepts *possibly existent* and *necessarily existent*. On either reading, Ghazali plainly did not regard the two proofs as distinct. Formulations similar to Ghazali's are to be found in Abū al-Barakāt[88] and Ibn Nafīs.[89]

There are extant a number of brief unsystematic works, best described as philosophic florilegia, which are mistakenly attributed to Alfarabi but are in fact written in the spirit of Avicenna's philosophy.[90] Several of them contain proofs of the existence of God from the impossibility of an infinite regress of causes, and in each the argument incorporates strands from the proof from the concepts *possibly existent* and *necessarily existent*. One of the compositions carefully distinguishes what is possibly existent by virtue of itself, necessarily existent by virtue of another from what is necessarily existent by virtue of itself. Then a statement is made flatly and with no explanation to the effect that "possible things cannot extend infinitely either as cause and effect, or in a circle; they must arrive at something necessarily existent by virtue of itself." The necessarily existent by virtue of itself hereby arrived at must, of course, possess the divine attributes analyzable out of the concept; and the existence of the deity has with no ado been established.[91] A second of the compositions employs the distinction between the

[86]Ibid., pp. 138–141.

[87]Above, at n. 82.

[88]*K. al-Mu'tabar* (Hyderabad, 1939), III, p. 24.

[89]*Theologus Autodidactus,* ed. and trans. M. Meyerhof and J. Schacht (Oxford, 1968), Arabic part, p. 8; English translation, pp. 43–44. The translation is misleading.

[90]S. Pines, "Ibn Sina et l'auteur de la Risalat al-Fusus fi'l-Hikma," *Revue des études islamiques,* XIX (1951), 121–124, and F. Rahman, *Prophecy in Islam* (London, 1958), p. 21, give reasons for not accepting the attribution of these works to Alfarabi. The three works that I am examining here give an imprecise restatement of Avicenna's positions, and that is the reason I regard them as having been written by lesser thinkers who stood under Avicenna's influence.

[91]*'Uyūn al-Masā'il,* §3, in *Alfarabi's philosophische Abhandlungen,* ed. F. Dieterici (Leiden, 1890). German translation: *Alfarabi's philosophische Abhandlungen aus dem Arabischen uebersetzt,* trans. F. Dieterici (Leiden, 1892), p. 94. The translation I have given takes into consideration another

last term, the middle member, and the first term, in any causal series. "When there is an effect," the argument runs, "it requires a cause." But if the cause should have the "status of an intermediate, irrespective of whether it should be finite or infinite," it can "not exist without the presence of a [first] uncaused term."[92] Consequently, a causal series cannot regress indefinitely but must terminate at a first cause—to be precise, a first cause that "exists together with the effect." The proposition that the "first [cause] must be necessarily existent by virtue of itself," is, hereupon, presented as "primary, not acquired knowledge." The conclusion drawn is that the first cause reached by tracing back a series of causes in the universe is to be assigned all the attributes analyzable out of the concept *necessarily existent by virtue of itself*; and it is the deity.[93] The proofs in both these compositions reason from the impossibility of an infinite regress to the existence of a first cause. And in both, the argument becomes a proof of the existence of God when the first cause is assigned the divine attributes analyzable out of the concept *necessarily existent by virtue of itself*.

A third of the brief compositions mistakenly attributed to Alfarabi offers a slightly more complex argument. It too contends that every causal series must have a "final term that is caused," a "first term that is cause," and a "middle" link. The reasoning proceeds: To assume an infinite regress would amount to assuming a situation wherein the "first [term]" is not in truth a first term but instead has the "status of the middle link, which [by its nature] requires a [prior] term." In other words, to assume an infinite regress would amount to assuming a contradictory situation in which something that cannot have the "status of a middle link" would nevertheless be assigned such status. It follows that "possible beings must terminate at an uncaused being," and an infinite regress of causes is impossible. The argument now goes on to consider "causes and effects existing together." An effect can never "do without . . . its cause," not only in order to come into existence but also during the entire period in which it exists. For if an effect could continue to exist through its own power, without being maintained by anything else, it would "become necessarily existent by virtue of itself after having been possible"; but that is inconceivable because the mere fact of an object's entering the realm of existence does not provide it with an "existence . . . necessary by virtue of itself." Every effect accordingly requires a cause to maintain it in existence as long as it continues to exist. Applying the impossibility of an infinite regress to a series of effects maintained in existence demonstrates that the members of the series must ultimately depend for their existence on an "uncaused being." And "what has no cause" in the strictest sense cannot contain even the

redaction of the text, entitled *al-Da'āwī al-Qalbīya*, p. 2, in *Rasā'il al-Fārābī* (Hyderabad, 1931); English translation in G. Hourani, "Ibn Sīnā on Necessary and Possible Existence," *Philosophical Forum*, IV (1972), 75–76.

[92] *Ta'līqāt*, p. 6, in *Rasā'il al-Fārābī* (Hyderabad, 1931). The text is corrupt.

[93] Ibid., p. 5.

distinction of matter and form; for matter and form are "causes of the existence" of whatever is constituted by them. Thus the existence of a being is established which is absolutely uncaused and incorporeal and which is the ultimate cause maintaining all possible beings in existence; this is the deity.[94] The argument, as will have been observed, began by establishing the impossibility of an infinite regress. The impossibility of an infinite regress is applied to the possibly existent beings existing together in the universe. And a necessarily existent being is inferred which is explicitly assigned one divine attribute—and, by implication, others as well—deducible from the concept *necessarily existent by virtue of itself*. We have an instance of the argument from the impossibility of an infinite regress wherein that principle is applied to causes maintaining a given object in existence, and wherein the first maintaining cause is found to be God through the analysis of the concept *necessarily existent by virtue of itself*.

Not only philosophers in the Aristotelian tradition, but Kalam writers as well employed an argument from the impossibility of an infinite regress of causes. The earlier Kalam proved the existence of God by establishing the creation of the physical universe and inferring the existence of a creator.[95] Conceivably, the cause of creation thus inferred could itself have a cause. In order to arrive at an absolutely first cause, Bāqillānī,[96] Baghdādī,[97] Juwaynī,[98] 'Abd al-Jabbār,[99] and Ghazali in his Kalam writings,[100] all argued that even if the creator of the world should himself have a cause, an infinite regress of causes is impossible; and the series must therefore terminate at a first cause, who is the deity.[101] In ruling out an infinite regress of causes, none of the Kalam writers mentioned brought forward considerations relating specifically to a *causal* regress. Juwaynī, 'Abd al-Jabbār, and Ghazali did adduce Kalam considerations that rule out infinite series in general; Juwaynī and 'Abd al-Jabbār apparently understood, and Ghazali in his Kalam guise made it a matter of principle,[102] that the Kalam grounds ruling out every infinite series likewise exclude an infinite series of causes.

Fakhr al-Dīn al-Rāzī, a Kalam thinker who was strongly influenced by Avicenna takes a new tack, and in one of his works grafts the reasoning of Avicenna's proof from the concepts *possibly existent* and *necessarily existent* on to familiar Kalam proofs of the existence of God. In the work in question, Fakhr al-Dīn al-Rāzī employs the common Kalam procedure of proving the creation of the world and inferring the existence of a creator; and side by side therewith he also employs

[94] *Fī Ithbāt al-Mufāraqāt*, pp. 3–4, in *Rasā'il al-Fārābī*, (Hyderabad, 1931).
[95] Cf. above, pp. 154–162.
[96] *K. al-Tamhīd*, ed. R. McCarthy (Beirut, 1957), p. 32.
[97] *K. Uṣūl al-Dīn* (Istanbul, 1928), p. 72.
[98] *K. al-Irshād* (Cairo, 1950), pp. 25–26, 32.
[99] *Sharḥ al-Uṣūl* (Cairo, 1965), p. 181.
[100] *al-Iqtiṣād fī al-I'tiqād* (Ankara, 1962), p. 35.
[101] Cf. also above, p. 165.
[102] See below, p. 367.

arguments from the concept of particularization which establish a cause of the universe without reference to creation.[103] Once a creator or a cause of the world was established by one means or the other, it remained to be shown that the causes behind the world cannot regress indefinitely. When Rāzī now undertakes to prove that causes cannot regress, he does not posit the impossibility of an infinite regress without explanation, nor does he rest the impossibility of an infinite regress on the Kalam grounds for ruling out every infinite series. Instead he reasons exactly as Avicenna had done once he had set down the empirical datum that "something exists."[104] Rāzī's reasoning goes: We have established that the physical universe has a cause of its existence. If the cause is assumed to be "necessarily existent," that is to say, necessarily existent by virtue of itself, the proof of the existence of God is at once complete. If, on the contrary, the cause of the universe is assumed to be "possibly (*jā'iz*) existent," it too requires a "cause" (*mu'aththir*) of its existence. The series of causes "will either form a circular regress, form a linear regress, or terminate at the necessarily existent." Should the causes form a circular regress, a situation would obtain in which *x*, for example, "precedes" *y* while *y* "precedes" *x*, hence a situation wherein each link ultimately "precedes itself, something that is impossible." Should the causes form a linear regress, the series comprising all the causes would, considered as a "totality, . . . stand in need of each of its components." Since each of the components is "possible" (*mumkin*) and since "whatever stands in need of something possible is itself possible," the "totality would [also] be possible." The "totality would therefore have a cause." The cause of the totality could not be either the totality as a whole or one of the components; for both of those assumptions again entail a "thing's preceding itself,"[105] which is impossible. The cause (*'illa*) of the existence of the totality could only be something lying "outside" the totality. By hypothesis, all the possibly existent causes were included inside the totality; any cause lying outside would therefore be "necessarily existent." The series of causes consequently cannot after all regress indefinitely, but must terminate at a being "necessary [by virtue of itself.]"[106] Through an analysis of the concept *necessary by virtue of itself,* Rāzī further shows that any being answering the description possesses a set of attributes proper to the deity.[107]

The train of thought exhibited here is of some interest. Rāzī presents a faithful restatement of the reasoning of Avicenna's proof from the concepts *possibly existent* and *necessarily existent,* but the reasoning of Avicenna's proof is not offered as an independent proof of the existence of God. Instead, it is appended

[103] *Muḥaṣṣal* (Cairo, 1905), pp. 106–108; and cf. above, pp. 188–189.

[104] Cf. above, p. 303.

[105] (a) If the totality were the cause of all the components, then the totality would precede them. But since a part always precedes its whole, each component precedes the totality. Thus each component would precede itself which is impossible. (b) If a component were the cause of the totality of which it is a part, the component would be the cause of, and would precede, itself.

[106] *Muḥaṣṣal*, p. 108.

[107] Ibid., pp. 43–46, 108.

by Rāzī to Kalam proofs establishing a creator or a cause of the existence of the physical universe; it is appended as grounds for the impossibility of an infinite regress of causes at the point where the Kalam writers themselves had recognized the need to show that the creators or causes of the existence of the universe do not regress.

Once Rāzī accepted Avicenna's argumentation showing that any series of causes must terminate at a being necessarily existent by virtue of itself and possessed of a set of divine attributes, the Kalam reasoning leading up to a cause of the physical universe as a whole could be dispensed with in proving the existence of God. A proof of the existence of God could be formulated taking its departure, like Avicenna's proof, from the empirical datum that something exists, and reasoning back to a being necessarily existent by virtue of itself. Such, it turns out, is the course Rāzī does pursue in another of his many works. There, by the side of Kalam proofs for the existence of God, he offers a proof beginning with the empirical datum that some object exists. The object with which he begins must, Rāzī explains, be either "necessary by virtue of itself or possible by virtue of itself," In the latter case "it requires a cause." A circular regress and an infinite linear regress are ruled out on the grounds that were just examined.[108] "Accordingly," Rāzī concludes, "there is no avoiding the termination [of the series] at a being necessarily existent by virtue of itself" who is the deity.[109] Rāzī has given Avicenna's proof from the concepts *possibly existent* and *necessarily existent,* formulating it, like Avicenna, as an argument from the impossibility of an infinite regress; and the proof is offered by the side of Kalam arguments for the existence of God, and not as an appendage to any of them.

Fakhr al-Dīn al-Rāzī was followed by other Kalam writers, Amīdī and Ījī (fourteenth century), who also present Avicenna's proof as a self-contained method of proving the existence of God.[110] Rāzī and the others hereby bring Kalam thought to a new stage, a stage wherein it abandons its age-old opposition to the ways of the "philosophers" and sanctions a proof for the existence of God in an Aristotelian or, as a Kalam thinker would say, 'philosophic' mode.

I have cited instances in Islamic philosophy where proofs of the existence of God from the impossibility of an infinite regress incorporate, in one manner or another, the reasoning of Avicenna's proof from the concepts *possibly existent* and *necessarily existent.* Instances from medieval Jewish and Christian philosophy can be added.

The Jewish philosopher Abraham ibn Daud (twelfth century), who was well acquainted with Avicenna's philosophy, derives the impossibility of an infinite

[108] Above, p. 356.

[109] *K. al-Arba'īn* (Hyderabad, 1934), pp. 70–84, especially p. 84.

[110] Amīdī, *Ghāya al-Marām,* pp. 13, 14–15; Ījī, *Mawāqif* (Cairo, 1907), VIII, p. 5, taken together with IV, p. 160.

regress of causes from the more general principle that no series whatsoever of ordered objects can extend to infinity. He justifies the more general principle, in its turn, with the aid of the trichotomy of first term, middle link, and final term. The trichotomy was hitherto met in connection with causal series alone;[111] and one would suppose that it can be known to be present in either a causal, or any other kind of series only after the series is somehow demonstrated to possess each of the three components, including, most importantly, a first term. Aristotle, as will be recalled, did furnish grounds for affirming a first term in every causal series. Ibn Daud, for his part, assumes that the trichotomy of first term, middle link, and final term, characterizes a wider type of series than causal series, to wit, all ordered series; and he makes the assumption without proposing any reason for affirming that the three components are in truth characteristic of all ordered series. He simply postulates: "Things that are ordered have a beginning, a single or multiple middle [link], and an end." Given the postulate, the impossibility of any infinite ordered series is easily established. In an infinite series "every member . . . would belong to the middle," and the three requisite constituents of an ordered series would not all be present. The notion of an infinite ordered series is consequently self-contradictory; for whereas an ordered series must have a beginning, middle, and end, an infinite series could have neither beginning nor end.[112]

When Ibn Daud turns from the issue of an infinite regress to the issue of the existence of God, he introduces the terms *possibly existent* and *necessarily existent*; and he is perplexed by the varying ways in which the former term had been employed. Possibly existent, he finds, might denote (α) that which has only a possibility and "potentiality" of existing, "without actually existing now"; or (β) that which exists only for a time, in other words, that which is generated-destructible;[113] or again (γ) that which always exists, but does so "through another" and "not through itself." The third subcategory alone is described by Ibn Daud as both possibly existent by virtue of itself and also "necessarily existent by virtue of another,"—a usage that diverges from Avicenna's.[114] Despite the varying denotations of the term *possibly existent,* Ibn Daud ventures a single comprehensive definition. In the spirit of Avicenna, he defines "the possibly existent" as that "whose existence is dependent on the existence of something else." In parallel fashion, he defines "the necessarily existent [by virtue of itself]" as actual being "which does not receive its existence from anything else."[115]

If, Ibn Daud reasons, everything actually existing were possibly existent, that is to say, dependent for its existence upon something else, the series of causes

[111] Above, pp. 337, 339.

[112] *Emuna Rama* (Frankfurt, 1852), p. 16.

[113] This is the way Averroes, Maimonides, and Aqùinas, use the term. Cf. below, pp. 380, 384.

[114] For Avicenna, everything that actually exists, and not merely that which exists eternally, is possibly existent by virtue of itself, necessarily existent by virtue of another. See above, p. 293; below, p. 378.

[115] *Emuna Rama*, p. 47. Cf. above, pp. 291–292.

existing together and standing behind a given object would be infinite; for "above the existence [of each cause], there would be something else providing it with existence." But an infinite regress of causes is impossible, since causal series are one kind of ordered series, and all infinite ordered series were previously shown to be impossible. Any series of possibly existent beings must, as a consequence, terminate in a first being that is not dependent upon anything else for its existence, a being that is "necessarily existent [by virtue of itself]."[116] Having established a first necessarily existent cause Ibn Daud proceeds, again in the spirit of Avicenna, to derive simplicity, unity, and other divine attributes from an analysis of the concept *necessarily existent by virtue of itself.*[117]

Ibn Daud's proof is an argument from the impossibility of an infinite regress with the impossibility of an infinite regress applied specifically to the series of possibly existent beings; and the first cause in the series gains the character of a deity by being assigned the attributes analyzable out of the concept *necessarily existent by virtue of itself.* We again have an argument from the impossibility of an infinite regress supplemented at key points with elements of Avicenna's proof from the concepts *possibly existent* and *necessarily existent.*

An obscure Jewish writer, Joseph Ibn Aqnin (Joseph b. Yaḥya),[118] follows Avicenna more closely than did Ibn Daud. He distinguishes only two categories of actual existence: the necessarily existent by virtue of itself; and the possibly existent by virtue of itself, necessarily existent by virtue of another. The former is such that "if it is assumed not to exist, an impossibility results, in consideration of itself, not in consideration of another"; the latter is such that "if it is assumed not to exist, an impossibility results [in consideration of something else]."[119] Physical objects, Joseph explains, must be "possibly existent," since they are composite. Every physical object accordingly requires a factor to "tip the scales in favor of existence and against nonexistence," and the factor is needed not merely to bring the possibly existent object into existence, but also to maintain it in existence as long as it actually exists. "Should the factor [in question also] be possibly existent, it too requires a factor to tip the scales [in favor of its existence]." An infinite regress of such factors, all existing together, is impossible because, in the first place, any "infinity of actually existent beings is impossible." But "another reason [why every regress of possibly existent beings must terminate]" is as follows: When all possibly existent beings are treated as a single

[116]Ibid., p. 48.

[117]Ibid., pp. 49–51.

[118]Regarding this author, see D. Baneth, "Joseph ibn Shim'on," *Oṣar Yehude Sefarad,* VII(1964), 11–20; also idem, "He'arot Pilologiyot le-ḥibburo ha-meṭaphisi shel R. Joseph b. Yehuda," *Tarbiẓ* XXVII (1957–1958), 234–239.

[119]Joseph Ibn Aknin (Aqnin) (Joseph b. Yaḥya), *Treatise as to Necessary Existence,* ed. and trans. J. Magnes (Berlin, 1904), p. 21, with the manuscript reading recorded in the apparatus. (I give the pagination of the English translation which is also indicated in the margin of the Hebrew.) For Avicenna's definitions, see above, p. 291.

group, they need "a cause outside them" to maintain them in existence; the external cause "cannot be possibly existent, since all possibly existent beings were [by hypothesis] included within the group"; hence the external cause is perforce "necessarily existent by virtue of itself [and the series must terminate there]."[120] The series of causes standing behind any given possibly existent being must, then, terminate, and the termination must occur at a first cause that is necessarily existent by virtue of itself.[121] To complete his proof Joseph b. Yaḥya reads the attributes of the deity out of the concept *necessarily existent by virtue of itself.*[122]

What Joseph b. Yaḥya has formulated may be described as an argument from the impossibility of an infinite regress, supplemented at every stage by the reasoning of the proof from the concepts *possibly existent* and *necessarily existent.* Or, if one prefer, it may be described as a paraphrase of Avicenna's proof from the concepts *possibly existent* and *necessarily existent,* cast as an argument from the impossibility of an infinite regress of causes, and supplemented by an additional reason for ruling out an infinite regress.

In medieval Christian philosophy an instance of the type of composite argument we are examining is to be found in the twelfth century writer, Dominic Gundissalinus. Gundissalinus is best known for his translations from the Arabic, which include some of Avicenna's works as well as books written in the spirit of Avicenna's philosophy.[123] Like one of the brief works attributed to Alfarabi but in fact written in the spirit of Avicenna,[124] a work that Gundissalinus himself apparently translated into Latin,[125] Gundissalinus posits the impossibility of an infinite regress of causes without explanation.[126] The first cause established by the impossibility of an infinite regress must, he proceeds, be uncaused and eternal, hence "necessarily existent" (*necesse est esse*; *necessarium esse*),[127] that is to say, *necessarily existent by virtue of itself.* And analysis of the concept *necessarily existent* discloses the attributes of unity and simplicity.[128] Taking his departure from the impossibility of an infinite regress and supplementing that principle with the analysis of the concept *necessarily existent,* Gundissalinus thus arrives at the existence of the deity as a single, simple, eternal, uncaused cause of the universe.

[120]I understand these two sentences as "another reason" for the impossibility of an infinite regress of causes. They may be understood, somewhat less naturally, as a self-contained reason for the existence of a being necessarily existent by virtue of itself, in other words, as a self-contained proof of the existence of God, without introducing the issue of an infinite regress.

[121]Ibid., pp. 23, 26–27.

[122]Ibid., p. 28.

[123]See M. Alonso, "Traducciones del Arcediano Domingo Gundisalvo," *Al-Andalus,* XII (1947), 295–338.

[124]Above, n. 91.

[125]Alonso, "Traducciones," p. 319; M. Cruz Hernandez, "El Fontes Quaestionum de Abū Naṣr al-Fārābī," *Archives d'histoire doctrinale et littéraire,* XVIII (1950–1951), 305.

[126]Gundissalinus, *De Processione Mundi,* ed. G. Buelow (Muenster, 1925), pp. 5, 18.

[127]Ibid., pp. 5–7.

[128]Ibid., pp. 10 ff.

Several generations later, William of Auxerre (d. 1231) offers a proof of the existence of God with unmistakable echoes of Avicenna. William does not employ the terms *possibly existent* and *necessarily existent*; but he establishes an uncaused cause in a familiar fashion. He contends: Any "totality of objects that are caused, . . . whether finite or infinite," must itself be "caused"; this is simply postulated. Any totality of caused objects therefore requires a cause of existence "outside itself"; this too is postulated. It follows that should all caused objects in a supposed infinite regress be mentally grouped together, the resulting totality would have to have a cause outside itself; and the cause would itself be uncaused since everything caused was, by hypothesis, included within the totality. Every series of causes therefore terminates at an uncaused cause, and causes are found not to regress indefinitely after all. A circular regress is also shown by William to be impossible.[129] His conclusion is that there must exist a "first principle," which is "God."[130]

William clearly intends his proof as an argument from the impossibility of an infinite regress. The reasoning whereby he rejects an infinite regress parallels the reasoning of Avicenna's proof from the concepts *possibly existent* and *necessarily existent*.[131] William does not, however, explain why a totality of caused beings must itself have a cause, nor why the cause must lie outside the totality. He fails, moreover, to give any grounds for identifying the first cause in a given series of causes as the deity. What we have is Avicenna's proof from the concepts *possibly existent* and *necessarily existent,* reformulated with the terms *caused* and *uncaused*[132] and once again cast as an argument from the impossibility of an infinite regress. And wherever William's argumentation reveals gaps, we may presume that he has failed to repeat the appropriate reasoning that was explicit in Avicenna.

Further instances of an argument from the impossibility of an infinite regress supplemented through the reasoning of the proof from the concepts *possibly existent* and *necessarily existent* are to be found in Albertus Magnus,[133] Duns Scotus,[134] and, as seen in the previous section, Thomas Aquinas.[135]

[129]His reasoning is similar to that of Avicenna's, above p. 302.

[130]William of Auxerre, cited by A. Daniels, *Quellenbeitraege und Untersuchungen zur Geschichte der Gottesbeweise im dreizehnten Jahrhundert* (Muenster, 1909), p. 26, §1. Cf. William of Auvergne, *De Trinitate,* ed. B. Switalski (Toronto, 1976), chap. 2.

[131]The reasoning happens to be particularly similar to that of Avicenna's *Dānesh Nāmeh,* where the terms *possibly existent by virtue of itself, necessarily existent by virtue of another* and *necessarily existent by virtue of itself* are not explicitly used.

[132]That is to say, William uses the term *caused* where Avicenna had used *possibly existent by virtue of itself, necessarily existent by virtue of another*; and he uses the term *uncaused* where Avicenna had used *necessarily existent by virtue of itself.* As a result of the change William loses the implications of the *necessarily existent by virtue of itself* which derive from such an entity's being free even of internal factors making it what it is.

[133]Cf. above, n. 77.

[134]Ibid.

[135]Cf. above, p. 347.

4. Resumé

The arguments that have been examined so far reveal several developments.

(*a*) Among medieval Aristotelians, Aristotle's method of disproving an infinite regress came into play, but with an increasing change of emphasis. The critical consideration for Aristotle was that in every causal series, the first cause is always the true cause; consequently, should there be no first cause, there would be no true cause and no causality at all. Solely in order to bring out the point, Aristotle introduced the distinction between first cause, middle link, and final effect. Avicenna, for his part, made the trichotomy of first cause, middle link, and final effect, central; and only in passing did he explain why a first cause must be posited.[136] An early commentary of Averroes copied Avicenna's example. There Averroes went so far as to offer the purely verbal consideration that an infinite regress is impossible because of the meaning of *intermediate*; simply because of the meaning of the word, he contended, the intermediate member of a causal series must be preceded by a first term.[137] Still, both Avicenna and Averroes did explain in one way or another why all three components must be present in a causal series. By contrast, two of the brief compositions mistakenly attributed to Alfarabi merely posit, with no explanation, the proposition that all causal series must have three constitutents, a first term, a middle link, and a final term.[138] And the Jewish philosopher Abraham ibn Daud extended the proposition, to cover not only causal series, but all ordered series, although the original reason for affirming a first term was related exclusively to causal series. Ibn Daud affirms that the very nature of an ordered series is such that it must have a first term, a middle link, and a final term; and he rules out an infinite causal series as a special case of infinite ordered series.[139]

A new reason for the impossibility of an infinite regress of causes grew out of Avicenna's proof of the existence of God from the concepts *possibly existent* and *necessarily existent*. As one proposition in the proof, Avicenna established that the totality of possibly existent beings must depend on a being necessarily existent by virtue of itself; and he inferred therefrom, as a kind of corollary, the impossibility of an infinite regress of causes.[140] This was not Avicenna's sole reason for rejecting an infinite regress; elsewhere, as has been seen, he restated Aristotle's reason.[141] Ghazali's compendium of Avicenna's philosophy, however, gives the argument from the concepts *possibly existent* and *necessarily existent* as the only grounds for ruling out an infinite causal regress. An infinite regress of causes is impossible, Ghazali's compendium explains, because it would comprise an

[136] Above, p. 339.
[137] Above, p. 341.
[138] Above, p. 354.
[139] Above, p. 358.
[140] Above, p. 302.
[141] And in addition he suggested yet another argument for the impossibility of an infinite regress of causes; see above, p. 127; below, pp. 367–368.

infinite number of possibly existent beings, whereas the totality of possibly existent beings must depend on, and hence terminate at, a being necessarily existent by virtue of itself.[142]

A number of medieval approaches to the impossibility of an infinite regress can now be distinguished. First, some writers affirmed the impossibility of an infinite regress of causes without explanation. Instances are Isaac Israeli,[143] Bāqillānī,[144] Baghdādī,[145] *'Uyūn al-Masā'il*[146] (one of the compositions attributed to Alfarabi), Hillel b. Samuel,[147] Gundissalinus,[148] Alan of Lille,[149] and Alexander of Hales.[150] Secondly, a number of writers ruled out an infinite regress of causes because of the nature of the infinite, contending (α) that no actual infinite is possible, (β) that no actual infinite could be traversed, or (γ) that no actual infinite could be added to. Instances of these positions are (α) Kindi,[151] Juwaynī,[152] Ghazali in one passage,[153] Baḥya ibn Paquda,[154] Averroes in one passage,[155] and Joseph b. Yaḥya;[156] (β) Moses of Narbonne[157] and Witelo;[158] (γ) 'Abd al-Jabbār,[159] Fakhr al-Dīn al-Rāzī,[160] Tūsī,[161] and Ījī.[162] Thirdly, Averroes, in certain passages,[163] and Aquinas[164] followed Aristotle in ruling out an infinite regress because of the nature of causality and causal series; they reasoned that the true cause in a causal series is always the first cause, and therefore a first cause must always be posited. An outgrowth of Aristotle's grounds for ruling out an infinite regress of causes was the thesis that every causal series must by its very nature have a first term, middle link, and final term, a thesis sometimes explained, but sometimes affirmed without explanation. Arguments for the

[142]Above, p. 352.
[143]A. Altmann and S. Stern, *Isaac Israeli* (Oxford, 1958), p. 126.
[144]Above, n. 96.
[145]Above, n. 97.
[146]Above, n. 91.
[147]Hillel b. Samuel, *Tagmule ha-Nefesh* (Lyck, 1874), appendix, commentary on prop. 3, p. 33b.
[148]Above, p. 360.
[149]Alan of Lille, *Opera,* in *Patrologia Latina,* ed. J.-P. Migne, Vol. CCX (Paris, 1855) p. 598b.
[150]Alexander of Hales, *Glossa in Quattuor Libris Sententiarum* (Florence, 1951), I, 40.
[151]Kindi, *Rasa'il,* I, p. 142.
[152]Above, n. 98.
[153]Above, n. 100.
[154]*al-Hidāya* (*Ḥobot ha-Lebabot*), ed. A. Yahuda (Leiden, 1912), I, 5.
[155]Averroes, *Tahāfut al-Tahāfut,* I, p. 27.
[156]Above, p. 359.
[157]Moses of Narbonne, *Commentary on Guide* (Vienna, 1852), Part II, introd., prop. 3. Cf. H. Wolfson, *Crescas' Critique of Aristotle* (Cambridge, Mass., 1929), pp. 226–227.
[158]C. Baeumker, *Witelo* (Muenster, 1908), p. 339.
[159]Above, n. 99.
[160]Above, p. 127.
[161]Glosses to Razi's *Muḥaṣṣal,* p. 108.
[162]*Mawāqif,* IV, p. 167.
[163]Above, p. 341, and n. 32.
[164]Above, p. 343, and n. 35.

impossibility of an infinite regress of causes utilizing the thesis are to be found in Avicenna;[165] in two of the compositions attributed to Alfarabi;[166] in Averroes;[167] Abraham ibn Daud;[168] and Aquinas.[169] Fourthly, Avicenna and followers such as Ghazali in his compendium of Avicenna's philosophy,[170] the commentator Altabrizi,[171] Fakhr al-Dīn al-Rāzī,[172] Amīdī,[173] and Ījī [174] rule out an infinite regress of causes because of the nature of causality, as disclosed by analysis of the concepts *possibly existent* and *necessarily existent*. They reason that any totality of possibly existent beings must itself be possibly existent, and hence must depend for its existence on, and terminate at, something outside the totality.

(*b*) The impossibility of an infinite regress leads to a first term in each series of efficient causes. But it does not show that all series of efficient causes can be traced back to a single common first cause, that the first cause is incorporeal and simple, or that the first cause is the cause of the very existence of the universe. For any of those conclusions to be drawn, the argument from the impossibility of an infinite regress has to be supplemented. The argument from the impossibility of an infinite regress could be supplemented by the proof from motion, as was done by Averroes. In Averroes' procedure, the demarcation between the two proofs is virtually erased, since the proof from motion had in any event already included the impossibility of an infinite regress as one of its premises. The argument from the impossibility of an infinite regress could also be supplemented by assigning to the first efficient cause all the attributes analyzable out of the concept *necessarily existent by virtue of itself*. Various combinations of the argument for a first cause from the impossibility of an infinite regress and the proof from the concepts *possibly existent* and *necessarily existent* have been examined here; they were offered by Avicenna himself, and by followers among Moslems, Jews, and Christians. When the reasoning of the proof from the concepts *possibly existent* and *necessarily existent* was used not only to derive the attributes of the first cause but also to establish the impossibility of an infinite regress, the demarcation between that proof and the proof of the existence of God from the impossibility of an infinite regress would also be virtually erased.

A self-contained proof of the existence of God from the impossibility of an infinite regress of causes was formulated by Maimonides. Maimonides took up a suggestion of Aristotle's and applied the impossibility of an infinite regress to the series of factors leading up to a given transition from potentiality to actuality.

[165] Above, p. 339.
[166] Above, p. 354.
[167] Above, p. 341.
[168] Above, p. 358.
[169] Above, p. 343.
[170] Above, p. 352.
[171] Cited by Wolfson, *Crescas*, pp. 483–484.
[172] Above, p. 356.
[173] Above, n. 110.
[174] Ibid.

He was thereby able to establish a first, eternally actual cause, and by analyzing the concept *eternally actual,* to infer that the first cause must be one, incorporeal, and simple. But Maimonides' version of the proof too is incomplete. For although Maimonides contemplated a first cause of the very existence of the universe, his version fails to establish that the universe does depend for its very existence on its cause.

5. Crescas on the impossibility of an infinite regress

In arguing against an infinite regress of causes, Aristotle remarked that the grounds he adduced for positing a first term are valid whether the intermediate members of the series of causes should be "infinite or finite"; and in the Middle Ages, remarks to the effect that the intermediate links in a causal chain might be finite or infinite recur.[175] Nevertheless, Avicenna's position may be taken as generally accepted. Avicenna insisted that once a first term in a series of causes is posited, the infinity of the intermediate links becomes an empty hypothesis; for if a series has a beginning, it cannot conceivably extend back infinitely.[176]

Crescas takes another tack, maintaining that although the existence of a first cause is indeed demonstrable, nothing excludes construing the intermediate links in the series of causes and effects as infinite. The proof of the impossibility of an infinite regress considered by Crescas is Avicenna's argument that the totality of possibly existent beings must depend for its existence on a being necessarily existent by virtue of itself. And the argument, Crescas understands, does not exclude an infinite series. He explains: The argument is clearly compatible with the supposition that the first cause or a subsequent cause in a series should have innumerable effects. This can be seen, for example, from the circumstance that the argument is indisputably compatible with the supposition that the first cause should give rise to innumerable coordinate effects; what was thought to be ruled out was only an infinite number of effects arranged in a linear series. The argument is also compatible, however, with the supposition that whatever effects the first cause does have, are arranged linearly. Combining the two permissible suppositions[177]—that the first cause may have innumerable effects and that the effects of the first cause may be arranged linearly—shows, according to Crescas, that an infinite series of causes had not after all been excluded. Consequently, he concludes, once granted that every series of causes and effects must depend on an uncaused cause no objection remains to construing the series of causes and effects as infinite.[178]

Crescas' contention, although directed at Avicenna's refutation of an infinite regress of causes, would seem to apply to Aristotle's as well, and Aristotle had

[175] Above, pp. 339, 341, 354.

[176] Above, p. 340.

[177] Crescas should have added the condition that the two suppositions are compatible with one another.

[178] *Or ha-Shem,* I, ii, 3; Wolfson, *Crescas,* pp. 224–227, 490.

in fact admitted as much in the remark quoted a little earlier.[179] Aristotle's grounds for ruling out an infinite regress of causes were that only the first cause is a true cause. Crescas could again reason that once a first cause is granted, the number of links between it and the final effect might conceivably run to infinity.

Such is Crescas' critique of the principle that an infinite regress of causes is impossible. His strictures, whatever their philosophic interest, plainly will not affect any demonstration of the existence of God. Crescas too acknowledges that an uncaused cause must be posited for every series of causes and effects; and that is the only part of the principle which counts in demonstrating the existence of God. Ghazali's critique of the principle affirming the impossibility of an infinite regress is more radical.

6. Ghazali's critique of Avicenna's proof

Ghazali's *Maqāṣid al-Falāsifa,* the *Intentions of the Philosophers,* is a compendium of Avicenna's philosophy and side by side with it Ghazali composed a critique of Avicenna's philosophy entitled *Tahāfut al-Falāsifa,* the *Destruction of the Philosophers.* The proof of the existence of God given in Ghazali's *Maqāṣid* may, as was shown earlier, be read as Avicenna's proof from the concepts *possibly existent* and *necessarily existent* forced, as in Avicenna, into the mold of a proof from the impossibility of an infinite regress; or, alternatively, it may be read as an argument from the impossibility of an infinite regress supported at every stage by the reasoning of the proof from the concepts *possibly existent* and *necessarily existent.*[180] Ghazali's *Tahāfut al-Falāsifa,* the critique of Avicenna's philosophy, naturally enough undertakes a critique of a proof for the existence of God similar to the proof presented by Ghazali in his own compendium of Avicenna's philosophy. Ghazali undertakes to refute a proof from the impossibility of an infinite regress, more specifically, however, a proof wherein the impossibility of an infinite regress is established through analysis of the nature of possibly existent being and necessarily existent being, and wherein the first cause is invested with the attributes analyzable out of the concept *necessarily existent by virtue of itself.*

Ghazali makes three points. He begins (a) by laying down that the sole valid rational grounds for rejecting an infinite regress of causes are arguments that indiscriminately rule out all infinite series, whether they be series of causes or of anything else. It would follow that an infinite regress of causes cannot consistently be rejected by those who advocate the eternity of the world and accept infinite past time. Then (b) Ghazali examines the only argument known to him which attempts to disprove an infinite regress of causes not because of the nature of the infinite in general, but because of the nature of causality. The argument is, of course, Avicenna's proof from the concepts *possibly existent being* and *necessarily existent being*; if successful, Avicenna's reasoning would rule out an infinite

[179] Above, p. 337.
[180] Above, pp. 352–353.

regress of causes alone, without excluding other infinite series. Ghazali determines that the proof from the concepts *possibly existent being* and *necessarily existent being* fails to establish that the totality of possibly existent beings must depend for its existence upon something outside the totality, and Avicenna's reasoning is accordingly found to be incapable of disproving an infinite regress of causes. Finally (c), Ghazali contends that even if Avicenna had successfully established the existence of a first cause necessarily existent by virtue of itself, his proof would still fall short of a demonstration of the existence of God; for the attributes of the deity cannot be analyzed out of the concept *necessarily existent by virtue of itself*. Ghazali's entire discussion is conducted in the guise of an imagined debate between him and a philosopher adversary. The argumentation is intricately dialectical, something typical of the *Tahāfut*. And particularly in the presentation of the first of the three points, Ghazali's thinking is largely implied, not stated.

(*a*) Ghazali's first point commences: The impossibility of an infinite regress of causes is not known "necessarily" and "without a middle [term]"; it is not, in other words, an item of immediate and self-evident knowledge. It can be known, if at all, only through "speculation," that is to say, through logical reasoning. But philosophers such as Aristotle, Alfarabi, and Avicenna cannot consistently employ "speculation" to demonstrate the impossibility of an infinite regress; for since they espouse the eternity of the world, they must affirm that the heavenly bodies have described an "infinite number of revolutions."[181] Ghazali relies on his readers to complete his train of thought. The only valid grounds for rejecting an infinite regress of causes, we are to understand, are arguments that rule out all infinite series, including an infinite series of past revolutions of the celestial spheres; Ghazali has in mind the Kalam arguments from the impossibility of an infinite number which he himself employed elsewhere.[182] The philosophers, however, by advocating the eternity of the world, do affirm an infinite number of past revolutions of the spheres. They must as a consequence relinquish arguments that indiscriminately rule out each and every infinite series, and they are left without any grounds for rejecting an infinite regress of causes.

The issue raised by Ghazali was not new. Alfarabi and Avicenna had tried to explain why arguments against the existence of certain kinds of infinite do not exclude others. Avicenna had even used an argument of the Kalam type to establish the impossibility of an infinite magnitude—line, surface, or body—and had indicated that the same argument may be used to establish the impossibility of an infinite regress of causes.[183] He had, therefore, to show why the argument that he accepts does not imply the finiteness of past time, which he rejects. The

[181]*Tahāfut al-Falāsifa,* IV, §7. English translation in *Averroes' Tahafut al-Tahafut,* trans. S. van den Bergh, (London, 1954), p. 161.

[182]Above, pp. 120, 122.

[183]Avicenna, *Najāt* (Cairo, 1938), pp. 124–125; cf. above, p. 127.

explanation given by both Alfarabi and Avicenna was that the existence of an infinite number of objects is impossible only when two conditions are met, only when the objects exist together at the same time, and when they also have a relative "position" (*waḍ'*) to one another or are "essentially ordered" (*mutarattib al-dhāt*).[184] Arguments against the infinite would now not apply to infinite past time and past events, although they would, for example, rule out an infinite physical magnitude and even an infinite regress of causes.

The explanation given by Alfarabi and Avicenna was known to Ghazali, and he addresses himself to it. For rhetorical effect, Ghazali does not consider the two conditions conjointly as he should have. He does not say that philosophers employ arguments from the impossibility of an infinite number to rule out only infinite series meeting two conditions, namely series whose members both exist together and also have a relative position or essential order. Instead, he first raises the possibility that his philosopher opponent might restrict the arguments to objects existing together. When that route is shown to provide a philosopher no escape from inconsistency, Ghazali raises the further possibility that his philosopher opponent might fall back and restrict arguments from the impossibility of an infinite number to objects having an order. And the second route too is shown to provide no escape from inconsistency. The impression Ghazali undoubtedly wished to create was that his opponent had been driven from pillar to post.

Perhaps, Ghazali writes, alluding to the first of the two conditions stated by Alfarabi and Avicenna, a philosopher opponent will maintain that objects not actually "existing together at the same time" cannot properly be enumerated and so cannot be characterized as infinite, or for that matter even as finite.[185] Arguments against an infinite series would accordingly rule out an infinite series of causes all existing together—as well as an actually existent infinite physical magnitude—yet would not exclude an infinite number of past revolutions of the heavens; for the revolutions of the heavens are not objects existing together and hence are not truly subject to enumeration. And there would be no inconsistency in a philosopher's advocating the eternity of the world while employing arguments to disprove an infinite regress of causes.

Should a philosopher take such a stand, Ghazali responds, he would not rescue himself from inconsistency. Avicenna—and, in Ghazali's opinion, Aristotle, the Aristotelian commentators, and Alfarabi, as well—held that the human soul enjoys individual immortality.[186] The doctrine of immortality taken in conjunction with the doctrine of eternity entails an infinite number of objects existing together, the actual simultaneous existence of an infinite number of immortal human souls. Moreover, Ghazali adds, even should philosophers deny individual human immortality, there would still remain, on their view, the possibility of

[184]Above, p. 128. The conditions are stated as Avicenna formulated them.

[185]*Tahāfut al-Falāsifa,* IV, §8; English translation, p. 162.

[186]For the positions of Alfarabi and Avicenna on immortality, see H. Davidson, "Alfarabi and Avicenna on the Active Intellect," *Viator,* III (1972), 143–144, 172–175.

"supposing" (*qaddara*) objects' coming into existence at every moment of past time and remaining in existence. A philosopher who maintains the eternity of the universe would therefore have either to affirm the actual existence today of an infinite number of immortal souls or at least recognize the legitimacy of *supposing*[187] the existence today of an infinite number of objects. Even conceding that the past revolutions of the heavens are not actually existing objects, capable of being enumerated, the eternity of the world thus still entails the existence—or at least the possibility of supposing the existence—of an infinite number of objects existing together. The first condition, which, according to Alfarabi and Avicenna, must be fulfilled before applying the rule that objects cannot run to infinity, the condition that the objects exist together, is consequently insufficient to allow those who believe in eternity to adduce arguments against an infinite regress of causes.[188]

The second of the two conditions which, according to Alfarabi and Avicenna, must be fulfilled before asserting that objects do not run to infinity is that the objects possess an order. As Avicenna had put it, objects existing together cannot run to infinity when they have "order in respect to position [that is, when they are arranged, for example, in a line], or in respect to nature [that is, when they are related as cause to effect]." Objects, by contrast, which do not have an order, as, for example, "angels and evil spirits," may be of infinite number, Avicenna explained, even should they exist together.[189] With these statements of Avicenna's in mind, Ghazali proceeds a step further in his imagined debate. Perhaps, he writes, his philosopher opponent will retreat to the stand that only when objects also have "an order in respect to position . . . or nature" is it impossible for them to be of infinite number. An infinite number of immortal souls, accumulating from eternity and existing together at the present moment, could be compatible, after all, with arguments against infinite numbers of objects; and the arguments could still legitimately be used by advocates of eternity to disprove an infinite regress of causes.[190] In responding to this new stand, which might be proposed by his philosopher opponent, Ghazali begins by dismissing the distinction between an infinite ordered series and an infinite unordered series as arbitrary; there can be no justification, he insists, for rejecting the former while accepting the latter. But in any event, the immortal human souls accumulating from all eternity would, he contends, form an ordered series, and the same is true of other objects that may be "supposed" to come into existence at every moment of past time. For in each instance, the objects would come into existence successively, thereby arranging themselves in order.[191] Consequently, even granting the second—in addition to

[187]On the import of the contention that an infinite number of objects might be *supposed* to come into existence over an infinite time, cf. above, pp. 123–124.

[188]*Tahāfut al-Falāsifa,* IV, §§8, 19; English translation, pp. 162, 169.

[189]Avicenna, *Najāt,* p. 125.

[190]*Tahāfut al-Falāsifa,* IV, §9; English translation, p. 162.

[191]Ibid., §10. Similarly Shahrastānī, *K. Nihāya al-Iqdām* (Oxford and London, 1934), pp. 26–27.

the first—proposed condition governing arguments from the impossibility of an infinite number, Ghazali's philosopher adversary would still find himself caught in a hopeless predicament. The adversary advocates the eternity of the world while attempting to disprove an infinite regress of causes. In advocating the eternity of the world, he admits an infinite series—or the possibility of an infinite series—of objects existing together and possessing an order, whether it be "human soul, or genie soul, or evil spirit, or angel, or what you will."[192] Since he admits the existence of an infinite ordered series, he cannot accept and employ arguments that exclude every such series, with no exception. And once he relinquishes arguments ruling out every infinite ordered series, he is left with no grounds for rejecting an infinite regress of causes.

To recapitulate Ghazali's argumentation thus far: The only valid grounds for rejecting an infinite regress of causes are arguments that rule out all infinite series without exception; Ghazali has in mind the Kalam arguments from the impossibility of an infinite number which ultimately derive from John Philoponus.[193] Philosophers in the Aristotelian tradition, affirming, as they do, the eternity of the world, cannot employ the Kalam arguments and are unable to disprove an infinite regress of causes. Philosophers will gain nothing by maintaining that arguments from the impossibility of an infinite number apply only when two conditions are met, only when the objects considered exist together and possess order. For by affirming the eternity of the world, Aristotelian philosophers must recognize an infinite series of objects existing together and possessing order, for example, the human souls accumulating from eternity; and inasmuch as they recognize an infinite series of objects existing together and possessing order, they must relinquish arguments ruling out all such series. Even granting, then, the two conditions for applying arguments from the impossibility of an infinite number, the philosophers are still unable to employ the arguments. They are left without grounds for rejecting an infinite regress of causes, and they cannot accomplish the proof of the existence of God from the impossibility of an infinite regress. The upshot of Ghazali's intricate dialectic is a reaffirmation of a fundamental Kalam thesis: The existence of God cannot be demonstrated on the assumption of the eternity of the world. It can be demonstrated only by those who subscribe to the Kalam arguments for creation.

(*b*) Ghazali has been assuming that an infinite regress of causes may be disproved only through arguments from the impossibility of an infinite number. The arguments he has in mind reasoned, for example, from the intrinsic impossibility of an infinite's being added to, and ruled out every infinite series. In taking notice only of these arguments, Ghazali has been disregarding others like Aristotle's, which do not focus on the nature of number and the nature of the infinite in general, but rather on the nature of causal series, in other words, arguments that, because of the nature of causality, specifically rule out an infinite regress of

[192] *Tahāfut al-Falāsifa,* IV, §19; English translation, p. 169.

[193] Cf. above pp. 88–89.

causes. The only argument of the latter type known to Ghazali was the one developed by Avicenna in the course of his proof from the concepts *possibly existent* and *necessarily existent,* and already given in Ghazali's compendium of Avicenna's philosophy, the *Maqāṣid,* as the accepted philosophic grounds for the impossibility of an infinite regress of causes. Ghazali's critique of Avicenna, the *Tahāfut,* now takes up Avicenna's argument. Although Ghazali considers it in connection with an infinite regress of causes, he is in effect refuting Avicenna's entire proof of the existence of God from the concepts *possibly existent* and *necessarily existent,* quite apart from the question of an infinite regress.[194]

Perhaps, Ghazali writes, his philosopher opponent will maintain that the "apodictic demonstration of the impossibility of an infinite regress of causes" is the proof from possibly existent being and necessarily existent being. The "demonstration" to which Ghazali refers is summarized by him as follows: "Every cause is either possible by virtue of itself or necessary [by virtue of itself]. If [a cause is] necessary [by virtue of itself], it does not require a [further] cause [and the series ends there]. If [all causes were] possible [by virtue of themselves], the totality would be possible. But everything possible requires a cause in addition to itself. The totality therefore needs a cause outside itself [and terminates at the external cause]."[195]

To refute the purported demonstration, Ghazali begins by directing his attention to the meaning of "possible," that is to say, the possibly existent by virtue of itself, necessarily existent by virtue of another, and to the meaning of "necessary" that is to say, the necessarily existent by virtue of itself.[196] "Possible [by virtue of itself, necessary by virtue of another]" is, he submits, a "vague" and "incomprehensible" term until properly construed. It can legitimately be construed as nothing more than "what has an [external] cause for its existence." "Necessary [by virtue of itself]" is similarly "vague" and incomprehensible unless construed as nothing more than "what has no [external] cause for its existence." But once possible, that is to say, the possibly existent by virtue of itself, necessarily existent by virtue of another, and necessary, that is to say, the necessarily existent by virtue of itself, are so understood, a collection of possibly existent beings need not be assumed to depend on anything outside the totality. For a series of beings may be taken to be such that each component is "possible in the sense of having a cause distinct from it, whereas the totality is not possible in the sense of having a cause distinct from, and external to it."[197]

[194]That is to say, he is also refuting the argument as Shahrastānī and Crescas had stated it, above, pp. 307–308.

[195]*Tahāfut al-Falāsifa,* IV, §11; English translation, p.163.

[196]Ghazali, I am assuming, correctly understood that *necessarily existent* is tantamount in Avicenna to *actually existent,* and hence actual existence is divided into the *possibly existent by virtue of itself, necessarily existent by virtue of another,* and the *necessarily existent by virtue of itself.*

[197]*Tahāfut al-Falāsifa,* IV, §12; English translation, p. 164. Cf. ibid., X, §5; English translation, p. 252.

The import of Ghazali's statements is this: If the totality of possibly existent beings is itself possibly existent, the totality undoubtedly stands in need of an external cause of its existence. The totality of possibly existent beings need not, however, be possible. It may be necessarily existent by virtue of itself in the sole legitimate sense of the term, namely, the weaker sense of not having an external cause, although allowing of internal causes. Only if the totality of possibly existent beings were demonstrated not to be necessarily existent even in the weaker sense of the term would the totality of possibly existent beings be shown to stand in need of a cause outside the totality; and only then would an infinite regress of possibly existent causes be refuted.[198]

The suggestion that possibly existent beings might add up to a necessarily existent entity cannot, Ghazali continues, be rejected on the grounds that objects of one character never add up to a totality of another character; for many instances can be cited from nature where objects of one character do add up to a totality of another.[199] The suggestion is also not to be rejected by telescoping an illegitimate sense of the term *necessarily existent by virtue of itself* with the legitimate sense, and by reasoning that possibly existent beings cannot add up to something necessarily existent by virtue of itself, inasmuch as the latter is free even of internal factors making it what it is. There is as yet no reason for positing anything necessarily existent by virtue of itself in the peculiar, strong sense of not having even internal factors making it what it is; the existence of an entity of that description is precisely what our proof is attempting to establish. So far there is reason for affirming the existence of something necessarily existent by virtue of itself only in the weaker, and sole legitimate, sense of freedom from an external cause of existence.[200] And the necessarily existent by virtue of itself in the weaker sense may simply be the totality of possibly existent beings. Avicenna's proof hence does not successfully show that the totality of possibly existent beings terminates at a cause outside the totality.

The upshot of Ghazali's critique is once again that his philosopher opponent is unable to accomplish the proof from the impossibility of an infinite regress of causes. An infinite regress of causes can be disproved solely by arguments that rule out every infinite including infinite past time; the existence of God, is consequently demonstrable solely by those who recognize the creation of the world.

(*c*) Ghazali has contended that Avicenna and others who reject the Kalam arguments for creation are left no grounds for ruling out an infinite regress of causes. In particular, he has undertaken to refute Avicenna's proof from the concepts *possibly existent* and *necessarily existent,* and the argument for the impossibility of an infinite regress contained in it. Ghazali makes a further point: Even granted the impossibility of an infinite regress of causes, philosophers who

[198]This is essentially the critique that I suggested, above, p. 307.
[199]*Tahāfut al-Falāsifa,* IV, §14; English translation, pp. 164–165.
[200]Ibid., VII, §§8–9; English translation, p. 225.

accept the eternity of the world still cannot trace efficient causes back beyond the heavens and arrive at a transcendent deity.

Avicenna had discovered the attributes of the deity in the necessarily existent by virtue of itself through his analysis of the concept. The necessarily existent by virtue of itself, he had reasoned, cannot contain internal factors making it what it is; for if it contained internal factors it would, considered as a whole, exist by virtue of them and not exclusively by virtue of itself. It followed that the necessarily existent by virtue of itself must be simple, not composite; free of internal distinctions, notably, the distinction of matter and form; in fine, possessed of a set of attributes appropriate to a transcendent deity.[201] Ghazali responds that the impossibility of an infinite regress of causes establishes at best only the "termination of the regress and nothing else whatsoever." In other words, the impossibility of an infinite regress, if granted, can establish only a first cause with no external cause beyond it, but it can reveal nothing about the inner nature of the first cause. Nor is anything gained by terming the first cause necessarily existent by virtue of itself. For the only legitimate construction to be placed on the term, the only construction that might be justified by the argument from the impossibility of an infinite regress of causes, is: what has no external cause for its existence. And when necessarily existent by virtue of itself is so understood, analysis of the concept reveals nothing about the inner structure of the being that corresponds to the concept.[202] The argument from the impossibility of an infinite regress, even if accepted, consequently fails to prove that the first cause—or, if one will, the necessarily existent by virtue of itself—is simple, not composite,[203] hence free of the distinctions of essence and existence,[204] genus and specific differences,[205] matter and form.[206] The first cause, as far as the argument goes might be a corporeal object.

Furthermore, Ghazali continues, there might exist not one first cause, but many. Avicenna had, in his analysis of the concept, offered two reasons for ruling out a plurality of beings necessarily existent by virtue of themselves. First, if the species necessarily existent by virtue of itself is not intrinsically such as to belong exclusively to a certain given entity, an additional factor would have to be posited to explain why the species is present in each of the several entities that happen to have it; and the additional factor would be a cause, whereas the necessarily existent by virtue of itself can have no cause. Secondly, to assume two beings both of which are necessarily existent by virtue of themselves would require assuming that at least one of the two is composite; for at least one of them would have to contain both the common element rendering it and the other necessarily

[201] Above, pp. 296–297.
[202] *Tahāfut al-Falāsifa,* VI, §9; English translation, p. 190.
[203] Ibid., VII, §8; IX, §3; X, §7; English translation, pp. 225, 242, 253.
[204] Ibid., VIII; English translation, pp. 235 ff.
[205] Ibid., VII, §§8, 13–14; English translation, pp. 225, 229.
[206] Ibid., IX, §§2–3; X, §7; English translation, pp. 241, 253.

existent by virtue of itself, as well as an additional element through which it is distinguished from the other. Since the necessarily existent by virtue of itself was demonstrated not to be composite, no more than a single being answering the description might exist.[207] In reply to Avicenna's first reason, Ghazali insists that the attribute *necessary by virtue of itself* is not a positive element in the entity so characterized. It is merely a negative description, the characterization of something as uncaused, and no additional factor need be posited to explain its presence in the entities that have it.[208] In reply to Avicenna's second reason, Ghazali reiterates that the impossibility of an infinite regress, if granted, fails to show the necessarily existent by virtue of itself to be, in fact, free of composition. Freedom from composition cannot therefore serve as grounds for the unity of the first cause.[209]

Ghazali's contention, then, is that the impossibility of an infinite regress of causes could at most be employed to trace the various series of causes in the universe back, for example, to the heavens. The argument could not be used to infer a simple intelligent being beyond the heavens,[210] nor even to affirm that all causal series terminate in a single celestial sphere. The heavens might, as far as the argument goes, comprise a set of inanimate corporeal first causes each existing in its own right. The philosophers thus have no answer to those—known in Arabic as the "*dahrīya*"—who espouse a naturalistic scheme of the universe.[211]

Earlier Ghazali found that the impossibility of an infinite regress of causes can be demonstrated only through arguments, like those of the Kalam, which rule out every infinite series, including infinite past time. He now concludes that the dependence of the physical universe upon a transcendent cause likewise can be established only through a Kalam argument, to wit, the standard Kalam proof for creation. The Kalam proof, as briefly restated by Ghazali, runs: Whatever "is not free of generated things" is itself "generated, . . . and whatever is generated requires a cause for its generation."[212] The proof demonstrates, Ghazali stresses, not only that the physical universe is generated and consequently dependent for its generation upon a cause.[213] It alone shows the ultimate cause to be incorporeal. For since all bodies are generated—inasmuch as they are not "free of generated things"—the ultimate cause, which is not generated, cannot be a body.[214]

Ghazali's position, in sum, is that neither the existence nor the transcendent nature of the first cause can be established through the philosophic proof from

[207]Above, p. 297.
[208]*Tahāfut al-Falāsifa*, V, §4; English translation, p. 171.
[209]Ibid., V, §9; English translation, p. 174.
[210]Ibid., IV, §5; X, §§1–3; English translation, pp. 161, 250–251.
[211]Ibid., X, §§1, 3; English translation, pp. 250–251.
[212]Ibid., IX, §1; X, §1; English translation, pp. 241, 250.
[213]Ibid. Cf. above, pp. 134, 145.
[214]Ibid., IX, §1; English translation, p. 241.

the impossibility of an infinite regress. Although his objections are directed specifically against the proof from the impossibility of an infinite regress, they apply equally to Avicenna's proof from the concepts *possibly existent* and *necessarily existent* considered apart from the question of an infinite regress.[215] Ghazali has, to his satisfaction, hereby vindicated the Kalam method of proving the existence of God. Creation, he has reaffirmed, must first be established through Kalam reasoning; only from creation can the existence of a single incorporeal creator be inferred.

7. Summary

The present chapter has examined a proof of the existence of God which grew out of Aristotle's *Metaphysics* II, although the proof was not offered by Aristotle as an independent method for proving the existence of God.[216] The proof begins by establishing the impossibility of an infinite regress of causes and at once infers the existence of a first efficient cause. The first cause thus arrived at is not, we must remember, temporally first. It exists together with its effects and is first in the sense that it stands behind all of them.

One virtue of the proof was its directness. A first cause was inferred directly from the impossibility of an infinite regress of causes taken together with the unquestioned presence of causality in the universe. Directness was, however, purchased at a price. Indispensable assumptions were not stated, let alone demonstrated;[217] and the basic argument did not by itself arrive at the existence of God, since the incorporeality and unity of the first cause remained to be proved. That the proof had to be supplemented with grounds for the incorporeality and unity of the first cause was recognized. Averroes supplemented the proof by assimilating it to Aristotle's proof from motion,[218] while other philosophers supplemented it by analyzing the attributes of the deity out of the concept of first efficient cause or out of equivalent concepts.[219]

As for the central principle, the impossibility of an infinite regress of causes, various approaches were pursued. Some medieval philosophers focused on the nature of the infinite in general and reasoned that an infinite regress of causes is impossible since any infinite series of objects is impossible.[220] Others focused, like Aristotle, on the nature of causality and reasoned that causality is such that

[215]Ghazali appeared in many guises, and in another work, where he presents his own position, he does accept the proof. See *Faḍā'iḥ al-Bāṭinīya,* ed. A. Badawi (Cairo, 1964), pp. 82–83; L. Goodman, "Ghazālī's Argument from Creation," *International Journal of Middle East Studies,* II (1971), 75.

[216]Above, p. 338.

[217]Above, pp. 343–345.

[218]Above, pp. 348–349.

[219]Above, p. 346–348.

[220]Above, p. 363.

causes cannot extend back indefinitely;[221] Aristotle had rejected an infinite regress of causes through the consideration that only a first cause can be a true cause and hence every causal series must have a first cause. A new reason for the impossibility of an infinite regress, again turning on the nature of causality, was provided by Avicenna's analysis of the concepts *possibly existent* and *necessarily existent*. Analysis of the concepts revealed that the totality of beings possibly existent by virtue of themselves cannot maintain itself in existence but must depend on, and consequently terminate at, a cause outside the totality.

The impossibility of an infinite regress of causes was not at all required for the proof from the concepts *possibly existent* and *necessarily existent,* although Avicenna did choose to cast that proof in the mold of a proof from the impossibility of an infinite regress.[222] In addition, Avicenna offered another proof of the existence of God from the impossibility of an infinite regress of causes; and here the critical principle was established not through analysis of the concepts *possibly existent* and *necessarily existent* but in a manner virtually the same as that whereby Aristotle had established the principle. In this other proof, the grounds whereby Avicenna supplemented the basic argument and showed that the first cause is incorporeal and one, were, however, borrowed from his analysis of the concept *necessarily existent by virtue of itself.* He observed that the first cause reached through the impossibility of an infinite regress must exist necessarily by virtue of itself; consequently it must possess all the attributes analyzable out of the concept *necessarily existent by virtue of itself.*[223] Avicenna thus blurred the boundary between his proof from the concepts possibly existent and necessarily existent, on the one hand, and the proof from the impossibility of an infinite regress, on the other. He formulated the former as a version of the latter; and he supplemented the latter through analysis of one of the key concepts in the former. A precedent was hereby set for combining the two proofs, and various combinations of the two proofs were offered by subsequent medieval Islamic, Jewish, and Christian philosophers.

The most far-ranging critique of the proof from the impossibility of an infinite regress of causes was drawn up by Ghazali. Ghazali objected, first, that no argument from the impossibility of an infinite regress of causes can be consistently propounded by philosophers who affirm the eternity of the world; since advocates of eternity must admit the existence of certain infinite series they cannot, Ghazali contended, reject others, such as an infinite series of causes. A second objection of Ghazali's was addressed specifically to the argument for the impossibility of an infinite regress which grew out of Avicenna's analysis of the concepts *possibly existent* and *necessarily existent*. Avicenna, Ghazali contended, had failed to show that a series of possibly existent beings cannot constitute a totality which

[221]Above, pp. 337, 363.
[222]Above, pp. 302, 307.
[223]Above, pp. 340, 346.

exists by virtue of itself. A third objection of Ghazali's was that the principle of the impossibility of an infinite regress, even if granted, cannot lead to anything more than the existence of a first term in the series of efficient causes; neither the incorporeality nor the unity of the first cause would be known. The principle consequently is insufficient for a genuine demonstration of the existence of God.

XII

Subsequent History of Proofs from the Concept of *Necessary Existence*

1. Maimonides and Aquinas

Maimonides too offered a proof of the existence of God employing the terms *possibly existent* and *necessarily existent,* a proof described by him as a "demonstration leaving no room for doubt or refutation . . . except for those who are ignorant of the method of demonstration." Its reasoning, according to Maimonides, is "drawn from Aristotle's statements, although put forward by Aristotle for a different purpose."[1] Maimonides' proof can be properly appraised only against the background of Avicenna's distinctive proof from the concepts *possibly existent* and *necessarily existent,* and the proof that Averroes formulated using the same terms.

Avicenna, as will be recalled, operated with a twofold division of actual existence, the dichotomy of *possibly existent by virtue of itself, necessarily existent by virtue of another,* on the one hand, and *necessarily existent by virtue of itself,* on the other; *necessarily* existent in Avicenna was tantamount to *actually* existent. By analysis of the concepts, Avicenna established that the totality of beings possibly existent by virtue of themselves, necessarily existent by virtue of another, must depend for its existence on a being necessarily existent by virtue of itself. He turned to the external world for a single datum, the fact that some object actually exists. And he reasoned that if the object he starts with is not necessarily existent by virtue of itself, it must ultimately depend for its existence on something that is; thus in either case, there exists something necessarily existent by virtue of itself, hence something possessing the divine attributes analyzable out of that

[1] *Guide to the Perplexed,* II, 1(c). Munk suggests that Maimonides is referring to Aristotle's *De Caelo* I, 10.

concept.[2] Avicenna introduced the issue of an infinite causal regress into his proof, but he need not have done so.[3]

Averroes, as will further be recalled, did not correctly understand Avicenna's use of the terms *possibly existent* and *necessarily existent*. He supposed that *possibly existent* with no further qualification designated for Avicenna the generated-destructible realm, and that *possibly existent by virtue of itself, necessarily existent by virtue of another* was restricted to eternal objects having a cause; Averroes could not conceive of *necessary* as tantamount to anything other than *eternal*. In addition, Averroes had no knowledge of the heart of Avicenna's proof, namely, the method of showing through an analysis of the concepts that the totality of possibly existent beings must depend for its existence on a being necessarily existent by virtue of itself.[4] Lacking a correct understanding of what Avicenna had intended in his proof from the concepts possibly existent and necessarily existent Averroes took the proof to be a typical, though poorly executed, argument from the impossibility of an infinite regress of causes. Since Averroes, for his part, operated with a threefold division of actual existence, his own corrected version of Avicenna's proof proceeded to the necessarily existent by virtue of itself, not directly but in stages. In one stage Averroes took his departure from *possibly existent being,* which for him signified generated-destructible being, and arrived at *necessary being,* which for him was tantamount to eternal being; and in a separate stage he passed from *necessarily existent being* to the *necessarily existent by virtue of itself*. His reasoning ran: An infinite regress of causes is impossible. Any series of possibly existent beings can therefore be traced back to a first term which is not possibly existent, that is to say, generated-destructible, but rather necessarily existent, that is to say, eternal. If the necessarily existent being hereby arrived at should not exist through itself, the series of necessarily existent beings can likewise be traced back to a first term, which does exist by virtue of itself. The existence of a first efficient cause, which is necessarily existent by virtue of itself, is thus established. Analysis of the concept *first efficient cause* and, in the end, identifying the first efficient cause with the prime mover established in the proof from motion thereupon shows that the first efficient cause possesses the attributes of the deity.[5] Averroes' tentative derivation of certain divine attributes from the concept of *first efficient cause* was patterned on Avicenna's derivation of divine attributes from the concept *necessarily existent by virtue of itself*. Otherwise, his proof exhibits none of the argumentation of Avicenna's proof from the concepts *possibly existent* and *necessarily existent*. Averroes' proof is simply his own argument from the impossibility of an infinite regress of causes. employing the terms *possibly existent* and *necessarily existent* as he understood them.

[2]Above, pp. 303–304.
[3]Above, pp. 302, 307.
[4]Above, pp. 333–334.
[5]Above, pp. 342, 349.

Maimonides, we find, operates like Averroes with a trichotomy of actual existence. For the purpose of his proof he distinguishes: generated-destructible being; eternal being having a cause;[6] and eternal being that has no cause. Curiously, Maimonides designates and construes each of the three classes in the trichotomy just as Averroes mistakenly supposed Avicenna to have done.[7] The generated-destructible objects of the sublunar world are termed by Maimonides *possibly existent* with no further qualification; eternal being that has a cause is termed *possibly existent by virtue of itself, necessarily existent by virtue of another*; and eternal being that has no cause is termed *necessarily existent by virtue of itself.*[8] Maimonides' use of the formula *possibly existent by virtue of itself, necessarily existent by virtue of another* is particularly noteworthy. Averroes mistakenly took the formula in Avicenna to designate the eternal celestial realm, the realm of beings that although eternal owe their existence to a cause outside themselves; and, as was seen in an earlier chapter, he rejected the formula as self-contradictory.[9] Maimonides understands the formula exactly as Averroes took it to have been intended by Avicenna. Since *necessarily existent* is for Maimonides, like Averroes, tantamount to eternally existent, *possibly existent by virtue of itself, necessarily existent by virtue of another* is understood by him too as a designation for the realm of beings that are eternal, but owe their existence to another. Unlike Averroes, however, Maimonides accepts the formula as legitimate and does not hesitate to use it; he sees no contradiction in the conception of an object as necessary or eternal, and yet, in itself, possibly existent.

Maimonides, then, operates like Averroes with three classes of actual existence, designating them, though, not as Averroes held they should be designated, but as Averroes mistakenly and disapprovingly supposed Avicenna to have designated them. Since Maimonides distinguishes three classes of actual existence, his argumentation, like Averroes', proceeds in stages, passing from generated-destructible, or possible, being to eternal, or necessary, being; and from the latter to a being necessarily existent by virtue of itself. Avicenna, by contrast, had distinguished only two classes of existence for the purpose of his proof, and accordingly had reasoned directly from the possibly existent by virtue of itself—whether it be generated-destructible or eternal—to the necessarily existent by virtue of itself. The reasoning whereby Maimonides passes, in the first stage, from the existence of generated-destructible being to the existence of eternal being is somewhat complicated.

[6]According to *Guide,* I, 71, and II, introduction, prop. 26, Maimonides assumes the eternity of the world only for the sake of argument. He explains that on the contrary assumption, the assumption of creation, the world undoubtedly has a creator, and therefore no further argument is needed to establish the existence of God. Cf. above, pp. 4–5.

[7]Above, pp. 318–319.

[8]Not completely explicit, but clear, from *Guide,* II, introduction, props. 19 and 20; II, 1(c).

[9]Above, pp. 319–320.

It goes: Inasmuch as everything subject to generation is subject to destruction, and vice versa,[10] one of the following alternatives must represent the totality of existence. Actually existent beings must be either (a) such that "none are subject to generation and destruction"; (b) such that "all are subject to generation and destruction"; or (c) such that "some are subject to generation and destruction, and some are not."[11] The first of the alternatives is immediately excluded, since "we perceive many beings coming into existence and undergoing destruction." The second alternative is excluded with the aid of the principle that every possibility is realized over an infinite time.[12] "If," Maimonides writes, "each existent being should be subject to generation and destruction, . . . existent beings taken as a whole must [at some moment] be destroyed . . . ; for what is possible in a species must inevitably occur." In other words, if each existent being had the possibility of undergoing destruction, the possibility of each object's undergoing destruction would have been realized in the infinite past,[13] and the totality of existence would at some past moment in fact have been destroyed. But in that event, "nothing would remain to bring things [back] into existence" and "nothing at all could exist [today]." Obviously things do exist today. At no moment in the past, therefore, can the totality of things have ceased to exist.[14] The second of the alternatives originally posed is hence excluded; it cannot be the case that everything in the universe is generated-destructible. Two of the three alternatives being ruled out, the remaining alternative must be accepted as correct: Some actually existent beings are generated-destructible, while something also must exist which is not.[15]

Maimonides' reasoning contains a questionable link. The hypothesis that every individual thing has the possibility of being destroyed, he has contended, entails that the totality of things would at some moment cease to exist.[16] Clearly, the principle that every possibility is eventually realized justifies the inference that any given thing having the possibility of being destroyed must eventually undergo destruction. Conceivably, however, as Crescas was later to object, things undergoing destruction might not all do so simultaneously; and although each individual thing would eventually be destroyed, something at all times might remain existent.[17] The objection that might, then, be raised against Maimonides' reasoning is: How does it follow from the destructibility of each individual being that at a certain moment all beings would undergo destruction simultaneously?

[10]Cf. Aristotle, *De Caelo* I, 12.

[11]There is an echo here of Aristotle, *Physics* VIII, 3, 253a, 24–25.

[12]See above, pp. 90, 320.

[13]Maimonides, as will be recalled, assumes the eternity of the world only for the sake of argument; cf. above, n. 6.

[14]Cf. Plotinus, *Enneads,* IV, 7, 12, beginning. [15]*Guide,* II, 1(c).

[16]Cf. H. Wolfson, "Patristic Arguments against the Eternity of the World," reprinted in his *Studies in the History of Philosophy and Religion,* Vol. I (Cambridge, Mass., 1973), p. 187.

[17]Crescas, *Or ha-Shem,* I, ii, 17; cf. the passage from Theophrastus quoted by Wolfson, *op. cit.*

There is an answer to the objection which probably reflects Maimonides' intent. At first thought, it might indeed seem that things undergoing destruction might not all do so together, and although each individual object is destroyed, actual existence as a whole would continue. Nevertheless, the possibility remains that objects might all cease to exist at the same moment. And since every possibility must be realized, the last possibility too, the possibility that everything undergoes destruction simultaneously, would eventually have to be realized.[18] The assumption that each individual thing is subject to destruction therefore would after all lead to the conclusion that the totality of actual things must at some moment in the infinite past have ceased to exist.

Maimonides, to the point we have followed him, has completed only the stage of his proof which establishes the existence of something that is not generated-destructible. To arrive at a being that can be identified as the deity, the proof has to be carried through an additional stage. Maimonides continues: By virtue of the principle that every possibility is eventually realized, the being whose existence has so far been established—a being exempt from generation and destruction—can have no possibility of not existing. It is not "possibly existent," but "necessarily existent," which apparently means nothing more than that it is eternally existent. Should the being in question depend for its existence on something else, it would, although not unqualifiedly possibly existent, nonetheless have its "existence and nonexistence possible insofar as it is considered in itself"; and its existence would be "necessary [only] through its cause." It would, in other words, be possibly existent by virtue of itself, necessarily existent by virtue of another. Should the cause, in its turn, likewise be possibly existent by virtue of itself, it also would have a cause for its existence. Ultimately something must be reached which is "necessarily existent by virtue of itself." The reason therefor, the reader is left to understand, is that an infinite regress of causes is impossible; the impossibility of an infinite regress of causes is not explicitly cited by Maimonides here, but is explicitly employed by him in the course of two other proofs of the existence of God.[19] Maimonides has "demonstrated" that there "must exist a being necessarily existent by virtue of itself." "Without that being, nothing at all could exist"; for whatever is not necessarily existent by virtue of itself depends for its existence upon the being that is.[20]

In the stage of his proof just examined Maimonides passed from the existence of an eternal, or necessary, being to the existence of a being necessarily existent by virtue of itself. He goes on to analyze the concept *necessarily existent by virtue of itself* and read out the attributes of such a being. Anything existing exclusively through itself cannot, he argues, have even internal factors making

[18]Cf. the reasoning of Lactantius quoted by Wolfson, "Patristic Arguments," pp. 189–190.

[19]Cf. *Guide*, II, introduction, prop. 3; II, 1, (a) and (d); and Crescas' comment on the proof we are considering, *Or ha-Shem*, I, i, 29.

[20]*Guide*, II, 1(c).

it what it is. It consequently must be simple and incorporeal and can "easily" be demonstrated to be one.[21] "This is the deity."[22] The concluding derivation of divine attributes from the concept *necessarily existent by virtue of itself* is, we should note, open to an objection taken up earlier; for, granting that Maimonides has established a being that is necessarily existent by virtue of itself in the sense of not having any external cause, he seems to have given no justification for assuming that it cannot have even internal factors making it what it is.[23]

Maimonides' proof plainly betrays echoes of Avicenna's proof from the concepts *possibly existent* and *necessarily existent*. Yet no less plainly it is not Avicenna's proof. Maimonides construes *possibly existent* and *necessarily existent* differently. And he does not know the method of establishing, solely through analysis of concepts, that the totality of possibly existent beings must depend for its existence on something necessarily existent by virtue of itself. Maimonides follows Avicenna's argumentation in only a single respect, in deriving from the concept *necessarily existent by virtue of itself* a set of attributes appropriate to the deity. Even here a certain difference is to be observed, inasmuch as *necessarily* existent is tantamount to *eternally* existent in Maimonides, whereas it had been tantamount to *actually* existent in Avicenna; but the difference does not affect the analysis of the concept.

Maimonides' proof is one more version of the argument from the the impossibility of an infinite regress, establishing first, as did Averroes' version, a being that is necessary, or eternal, and then proceeding to the eternal that is uncaused. Maimonides accomplishes the first stage without adducing the principle of the impossibility of an infinite causal regress; but he accomplishes the second and more critical stage with the aid of that principle, which he tacitly takes for granted. The similarity of Maimonides' formulation to Averroes' is, we may conjecture, due to both writers' having derived their incomplete information about Avicenna's proof from similar sources and having reconstructed the proof in similar ways.[24]

One of Thomas Aquinas' celebrated five proofs for the existence of God in the *Summa Theologiae* is characterized by Aquinas as a proof from "the possible and the necessary."[25] It turns out to be an almost point for point restatement of the proof of Maimonides' just discussed.[26] Like both Maimonides and Averroes,

[21]Cf. *Guide,* II, introduction, prop. 21; above, p. 297. [22]*Guide,* II, 1(c).

[23]Cf. above, pp. 305–307, 371–372.

[24]Maimonides, as far as is known, did not become acquainted with Averroes' writings until after he wrote the *Guide*. Cf. H. Wolfson, *Crescas' Critique of Aristotle* (Cambridge, Mass., 1929), p. 323.

[25]*Summa Theologiae,* I, 2, art. 3, third way.

[26]Aquinas' first and second ways also echo two of Maimonides' proofs, namely the first and fourth. Some writers have refused to recognize the almost point for point reflection of Maimonides' third proof in Aquinas' third way. See the literature cited by E. Gilson, *Le Thomisme,* (6th ed.; Paris, 1965), p. 79, n. 40; English translation: *The Christian Philosophy of St. Thomas Aquinas* (New York, 1956), chap. 3, n. 41.

Aquinas proceeds in stages, establishing the existence of something necessary and passing from the necessary to the necessary by virtue of itself; and he accomplishes each stage exactly as Maimonides did. He makes only two changes worth mentioning.[27] He carefully avoids using the formula *possibly existent by virtue of itself, necessarily existent by virtue of another* to designate the class of beings that are not subject to generation and destruction, yet owe their existence to an external cause.[28] Averroes had convinced him that the attributes *possibly existent* and *necessarily existent* are incompatible.[29] And in the course of the second stage of the argument he explicitly cites the impossibility of an infinite regress of causes, a principle left implicit by Maimonides.[30]

Aquinas' *Summa contra Gentiles,* a work dated earlier than the *Summa Theologiae,* offers a parallel argument, not strictly as grounds for the existence of God, but, instead, for the eternity of God[31] and for the dependence of everything else upon Him.[32] Here too Aquinas proceeds in stages, passing first from the existence of "possible beings, . . . that is, generated-destructible beings," to the existence of "necessary" beings, that is, beings which are not generated-destructible; and then from necessary beings having a cause to a being "necessary through itself." The first stage, however, is accomplished in the *Summa contra Gentiles* not as in Maimonides and in Aquinas' own *Summa Theologiae,* but as in Averroes, that is, through adducing the principle of the impossibility of an infinite causal regress. In addition, Aquinas introduces a motif reminiscent of Avicenna. To establish that "possibly existent," or generated-destructible,[33] beings require a cause of existence, he deploys the reasoning of particularization arguments;[34] he submits that inasmuch as a possibly existent being is inclined "equally . . . to . . . existence and nonexistence," some factor must have selected out existence for it in preference to nonexistence. Once he has established that every

[27] Aquinas does not explicitly spell out the three alternatives in the first stage of the argument, but they are implied.

[28] Aquinas, like Maimonides, affirms the creation of the world, yet accepts the hypothesis of eternity for the purpose of proving the existence of God. Even within his creationist system Aquinas finds an application for the term *necessarily existent.* He applies the term to beings, such as angels, which, though not strictly eternal, are not subject to natural generation and destruction. Cf. *Summa Theologiae,* I, 44, 1, obj. 2, and reply; 50, 5, reply to obj. 3. For Aquinas' use of the term *necessarily existent* within the present proof of the existence of God and elsewhere, see M. F., "La preuve de l'existence de Dieu par la contingence dans la Somme Theologique," *Revue de Philosophie,* XXXII (1925), 320–322; P. Brown, "St. Thomas' Doctrine of Necessary Being," *Philosophical Review,* LXXIII (1964), 76–90.

[29] *De Potentia,* q. 5, art. 3, resp.; *Commentary on Physics,* VIII, §§1154–1155.

[30] *Summa Theologiae,* I, 2, art. 3, third way.

[31] *Summa contra Gentiles,* I, 15(d).

[32] Ibid., II, 15(e).

[33] In Aquinas, unlike Avicenna, *possibly existent* with no further qualification designates a class of actual existence. Cf. above, pp. 292, 333, 380.

[34] Above, pp. 161, 292; below, pp. 396–397.

possibly existent being does have a cause that selected out existence for it, Aquinas proceeds through the two now familiar steps: Since causes cannot regress indefinitely, he argues, possible beings must terminate at something necessary, that is to say, something not subject to generation and destruction. The necessary being thus arrived at may itself have a cause. But necessary beings having a cause must also terminate; and they will terminate at a "first necessary" being,[35] a being existing necessarily "through itself."[36]

2. The influence of Avicenna's proof

Through the last few chapters I have been examining the utilization of, and reaction to, Avicenna's proof of the existence of God from the concepts *possibly existent* and *necessarily existent*. An overall summary of the influence of Avicenna's proof should be in order.

Avicenna prefaced his discussion of the existence of God by insisting on the legitimacy and desirability of a metaphysical, as distinct from a physical, proof. The insistence on the feasibility and desirability of a metaphysical proof of the existence of God was not taken up by Islamic and Jewish philosophers. It did, however, enter Scholastic philosophy and is repeated by such philosophers as Henry of Ghent,[37] Duns Scotus,[38] and Suarez;[39] and it has survived into the twentieth century among Neo-Thomists.[40]

Avicenna's proof from the concepts *possibly existent* and *necessarily existent* provided a new method of establishing the existence of God, without reference to the question of an infinite regress.[41] The proof was offered without reference to an infinite regress by Shahrastānī, Crescas,[42] and perhaps others,[43] but not by Avicenna himself. Avicenna instead pressed his proof of the existence of God as a being necessarily existent by virtue of itself into the mold of the old and familiar argument from the impossibility of an infinite regress of causes. In the latter guise, as an argument from the impossibility of an infinite regress, Avicenna's proof was subsequently recorded by Ghazali, and was employed by Abū al-Barakāt, Fakr al-Dīn al-Rāzī, Āmidī, Ibn Nafīs, Ījī,[44] and the Jewish writer Moses

[35]*Summa contra Gentiles,* I, 15(d).

[36]Ibid., II, 15(e). For Aquinas' derivation of divine attributes from the concept *necessarily existent by virtue of itself,* see above, p. 347.

[37]A. Pegis, "A New Way to God: Henry of Ghent (II)," *Mediaeval Studies,* XXXI (1969), 99–101.

[38]Duns Scotus, *Philosophical Writings,* ed. and trans. A. Wolter (Edinburgh and London, 1962), pp. 9–10.

[39]A. Pegis, "A New Way to God: Henry of Ghent (III)," *Mediaeval Studies,* XXXIII (1971), 174–175.

[40]R. Garrigou-Lagrange, *Dieu* (Paris, 1920), §7; R. P. Phillips, *Modern Thomistic Philosophy,* Vol. II (London, 1935), pp. 261–262, 271–272; F. van Steenberghen, *Ontologie,* (3rd ed.; Louvain, 1961), pp. 153–154.

[41]See above, p. 302.

[42]Above, pp. 307–308.

[43]Above, p. 352, n. 81.

[44]Above, pp. 352–353, 357.

al-Lawi. The last mentioned finds Avicenna's proof to be superior to the Aristotelian proof from motion, inasmuch as it does not identify the deity with the mover of the first sphere and consequently permits assigning the deity a rank above the movers of the spheres.[45]

Besides being utilized as a complete demonstration of the existence of God, Avicenna's proof from the concepts *possibly existent* and *necessarily existent* was unraveled, and the several strands were utilized separately. Since the proof usually was treated as an argument from the impossibility of an infinite causal regress, the section bearing specifically on the impossibility of a causal regress could, to begin, be extracted from the total fabric of the proof and used in its own right. New grounds for ruling out an infinite causal regress would thereby be made available alongside of the grounds that had been offered by Aristotle and others; and the impossibility of an infinite causal regress as justified by Avicenna's reasoning could be applied by philosophers to whatever purpose they might wish. Avicenna's argumentation was presented as the grounds par excellence for the impossibility of an infinite regress by Ghazali in his compendium of Avicenna's philosophy, by the commentator Altabrizi, by the Kalam writers Rāzī, Āmidī, and Ījī.[46]

Another strand in the proof which lent itself to separate utilization was Avicenna's analysis of the concept *necessarily existent by virtue of itself*. Avicenna's analysis showed that whatever entity corresponds to the concept will possess a set of attributes worthy of the deity. The procedure of deriving divine attributes from the concept *necessarily existent by virtue of itself* could be, and was, extracted from the total fabric of the proof and employed to supplement other proofs of the existence of God. Most notably, the procedure of analyzing divine attributes out of the concept *necessarily existent by virtue of itself* was used to supplement arguments from the impossibility of an infinite causal regress even when the impossibility of an infinite regress was justified on grounds different from those suggested by Avicenna's proof from the concepts *possibly existent* and *necessarily existent*. The reasoning would be: The first efficient cause reached by tracing back a given series of causes must exist *necessarily* and *by virtue of itself;* it must therefore possess the divine attributes analyzable out of the concept *necessarily existent by virtue of itself,* and may be identified as the deity. Avicenna himself took that line in certain of his works,[47] and the reasoning reappears in several brief works mistakenly attributed to Alfarabi; in Abraham ibn Daud, Joseph b. Yaḥya, and Maimonides; in Gundissilanus, Albert Magnus, Thomas Aquinas, and Duns Scotus.[48]

[45]See excerpts translated by G. Vajda, "Un Champion de l'Avicennisme," *Revue Thomiste*, 48 (1948), 485–486.

[46]Above, pp. 352, 356, 357. For Altabrizi, see Wolfson, *Crescas*, p. 484.

[47]Above, pp. 345–346.

[48]Above, pp. 347, 354, 359, 360, 361, 382–383.

Other strands drawn from Avicenna's proof which were utilized in proofs of the existence of God are the consideration that every possibly existent being requires a factor to tip the scales in favor of its existence;[49] the consideration that every possibly existent being stands in need of something to maintain it in existence as long as it exists;[50] and the very use of the terms *possibly existent* and *necessarily existent* in proofs of the existence of God, even in proofs—like Averroes', Maimonides', and Aquinas'—that had little in common with Avicenna's proof apart from the terms. A measure of the influence of Avicenna's proof is the fact that Necessarily Existent became an accepted synonym for God in Islamic and Jewish theology.[51]

A strand of Avicenna's argumentation was also transformed into a proof for creation. When elucidating the possibly existent by virtue of itself, Avicenna applied to existence the mode of thought animating Kalam arguments from the concept of particularization. That is, while Kalam writers contended that some factor must in every instance be posited to explain why an object possesses a given attribute to the exclusion of another which it might equally possess,[52] Avicenna laid down a similar proposition in regard to existence. He contended that some factor must be posited to explain why a being capable of both existing and not existing, in other words, a possibly existent being, does enjoy actual existence.[53] Avicenna's adaptation of what had originally been a Kalam mode of thought appealed to later Kalam writers; and they readapted it into an argument for creation. Shahrastānī, Fakr al-Dīn al-Rāzī, Āmidī, and Ījī reasoned: Inasmuch as the world is possibly existent, it exists actually only because something "tips the scales" in favor of its existence, thereby bringing it into actual existence. But to suppose that anything can be brought into existence after it already exists would be "absurd." Since the world has had the scales tipped in favor of its existence and has been brought into existence, a period must have preceded when the world was not yet existent. And the world must have been created after not having existed.[54]

[49]Above, pp. 161, 191.

[50]Above, pp. 354, 382.

[51]Cf. Tahānawī, *Kashshāf Iṣṭilāḥāt al-Funūn* (Beirut, 1966), p. 1443, s.v. *wujūb*; E. Ben Yehudah, *Millon ha-Lashon ha-'Ibrit* (Tel Aviv, 1948–59), s.v. *ḥyb*.

[52]Above, p. 178.

[53]Above, pp. 291–292.

[54]Shahrastānī, *K. Nihāya al-Iqdām* (Oxford and London, 1934), pp. 17–21 (somewhat different); Rāzī, *K. al-Arba'īn* (Hyderabad, 1934), pp. 30–31, 40–41; Āmidī, *Ghāya al-Marām* (Cairo, 1971), pp. 250–251, and cf. pp. 258–260; Ījī, *Mawāqif* (Cairo, 1907), VII, p. 227. That nothing can have been brought into existence and yet have existed eternally was a recurrent theme. It appears in Augustine, *City of God,* XI, 4 (tentatively); Philoponus, *De Aeternitate Mundi Contra Proclum,* ed. H. Rabe (Leipzig, 1899), pp. 14, 473; Ibn Ḥazm, *K. al-Faṣl fī al-Milal* Cairo, 1964), I, p. 20 (cf. above, p. 84); Ghazali, *Tahāfut al-Falāsifa,* ed. M. Bouyges (Beirut, 1927), III, §17; Bonaventure, *Commentary on II Sentences,* in *Opera Omnia,* Vol. II (Quaracchi, 1882), d. 1, p. 1, a. 1, q. 2. It is rejected by Aquinas in *De Aeternitate Mundi,* §§6–7.

Avicenna's proof of the existence of God from the concepts *possibly existent* and *necessarily existent* elicited a critique from the Aristotelian standpoint on the part of Averroes, and a critique from the Kalam standpoint on the part of Ghazali. The former attacked the philosophic innovations in the proof insofar as Averroes understood them;[55] the latter attacked the proof as an exemplar of all proofs of the existence of God which countenance the eternity of the world. In the course of his critique Ghazali brought forward the decisive objection that the proof could at most establish only a being necessarily existent by virtue of itself in the sense of not having external causes. Ghazali's objection addresses both the proof considered as a complete demonstration of the existence of God, and also the utilization of key parts of the proof to supplement other proofs of the existence of God. His objection invalidates the proof as a whole by showing that the totality of possibly existent beings might be necessarily existent by virtue of itself and accordingly need have no cause of existence outside the totality. By the same token, his objection disallows the use of Avicenna's argumentation specifically to establish the principle of the impossibility of an infinite regress of causes; for if the totality of possibly existent beings need have no cause outside the totality, there are no grounds for supposing that a series of possibly existent beings must terminate. And, again, Ghazali's objection disallows the use of the proof as a device for identifying the first member in a given series of efficient causes as the deity; for if the necessarily existent by virtue of itself is merely a being that has no external causes, analysis of the concept can reveal nothing about the inner nature of such a being.[56]

It can be stated that Avicenna's argumentation, whether accepted or not, whether even properly understood or not, predominated in efforts by Avicenna's medieval Islamic and Jewish successors to prove the existence of God.

3. Proofs of the existence of God as a necessarily existent being in modern European philosophy

From the time of Descartes, there appears a series of both cosmological and ontological proofs of the existence of God as a necessarily existent being. Although precise filiation cannot be traced, inspiration undoubtedly came from the medieval cosmological proofs, initiated by Avicenna, of the existence of a being necessarily existent by virtue of itself. Descartes and, to a greater extent, Spinoza and Leibniz were after all familiar with the medieval discussions. A review of the proofs appearing in modern European philosophy is appropriate in the context of the present book not merely to show what, in a general way, grew out of the medieval proofs. A comparison of the medieval and modern proofs can also help shed new light on both.

In Avicenna, as has been seen, necessarily existent is tantamount to actually existent, and necessarily existent by virtue of itself, to actually existent by virtue

[55] Above, pp. 331–333.

[56] Above, pp. 370–374.

of itself.[57] Writers like Alfarabi, Averroes, Maimonides, and Aquinas, who were more faithful to Aristotle's usage, employ the term necessarily existent as tantamount to eternally existent, and necessarily existent by virtue of itself as tantamount to eternally existent by virtue of itself.[58] In addition there was Ghazali's contention that necessarily existent—that is to say, necessarily existent by virtue of itself—can coherently mean nothing beyond: free of external cause.[59] For these medieval philosophers, necessarily existent could thus mean either actually existent or eternally existent. And necessarily existent by virtue of itself had three different explications, namely: actually existent by virtue of itself; eternally existent by virtue of itself; free of external cause. As for the denotations of the terms the difference between the two explications of necessarily existent is clearcut: The class of actually existent beings clearly differs from the class of eternally existent beings. The difference in denotation of the three explications of the term necessarily existent by virtue of itself is subtler. A difference can be discerned only by construing the first two explications of the term as covering solely those entities that do not have even internal factors making them what they are, and construing the third explication as covering entities, as well, which do contain such factors.

At times, the aforementioned medieval philosophers use *necessarily existent* as an ellipsis for *necessarily existent by virtue of itself.*[60] The ellipsis was more common among the minor writers who offered adaptations of Avicenna's proof.[61] And in some instances, those minor writers apparently forget that necessarily existent is properly tantamount to actually existent or eternally existent; that the term is not therefore restricted to the deity; and that what does characterize the deity and single him out of the wider class of necessarily existent beings is only the added qualification: [*existing*] *by virtue of itself.* As a result of the imprecision, *necessarily existent* without further qualification sometimes appeared with a meaning for which no philosophic explanation or justification was provided. *Necessarily existent* without further qualification sometimes meant *existing by virtue of itself,* or *uncaused.*

In modern European philosophy *necessarily existent,* not *necessarily existent by virtue of itself,* designates the deity, and *necessarily existent,* without qualification, has the meaning: *existing by virtue of itself* or *uncaused.* As explicitly defined by Leibniz, the necessarily existent is that which exists "through its essence" or "through itself."[62] The modern European use of necessarily existent

[57]Above, p. 293.

[58]Above, pp. 292, 332, 380, 385.

[59]Above, p. 371.

[60]Above, pp. 303, 352.

[61]Above, pp. 356, 359f.; Albertus Magnus, *De Causis et Processu,* in *Opera Omnia,* ed. A. Borgnet, Vol. X (Paris, 1891), I, 1, 9.

[62]*Die Philosophischen Schriften,* ed. C. Gerhardt, Vol. IV (Berlin, 1880), p. 406; English translation: *Philosophical Works of Leibnitz,* trans. G. Duncan (New Haven, 1890), p. 137. Leibniz also recognized *hypothetically* necessary being; see below, p. 396.

to designate the deity presumably grew out of the medieval usage that goes back to Avicenna. But, it is to be remembered, in Avicenna and the careful medieval philosophers who follow him the term necessarily existent does not without further qualification designate the deity. In the careful medieval philosophers, necessarily existent designates a wider class of beings, and the deity is set apart from the other members of the class only insofar as he is the necessary being that exists *by virtue of itself.*

Not every cosmological proof in the history of philosophy employs the term *necessarily existent by virtue of itself* or *necessarily existent*. And yet the very nature of a cosmological proof of the existence of God seems to require that, whether or not a given proof happens to use the term, it must conclude with the existence of a being necessarily existent by virtue of itself. That is to say, every cosmological proof—whether it be Avicenna's proof without the principle of the the impossibility of an infinite regress, the same proof cast as an argument from the impossibility of an infinite regress, any other argument from the impossibility of an infinite regress, the proof from motion, and so on—will undertake to establish the existence of a first cause, which is actually existent by virtue of itself, or eternally existent by virtue of itself,[63] or uncaused. Furthermore, the nature of an a priori or ontological proof of the existence of God would seem to require that whether the proof employs the term necessarily existent or not, it must conclude with the existence of a being necessarily existent by virtue of itself in the foregoing sense and necessarily existent in an additional sense as well. Every ontological proof will undertake to establish the existence of a being that exists by virtue of itself. And in addition, every ontological proof will undertake to establish the existence of a being in regard to whom the proposition 'this being is actually existent' is a *necessary truth,*[64] as the term was defined, again, by Leibniz. According to Leibniz's definition, "when a truth is necessary, its reason can be found by analysis, resolving it into more simple ideas and truths until we come to those that are primary. . . . Truths of reasoning are necessary, and their opposite, impossible."[65] An ontological proof takes its departure from a given concept of the nature of God. Through nothing more than an analysis of the concept, it undertakes either directly to deduce the actual existence of the corresponding object, or else to demonstrate the same indirectly, by showing that a denial of the object's actual existence is logically impossible.[66] The ontological

[63]Leaving aside the implications that *existent by virtue of itself* might have for the internal nature of the entity so designated.

[64]I understand such to be the point made by Anselm in *Proslogion,* chap. 3—where some readers discover a "second" ontological proof—as well as in Anselm's first reply to Guanilo.

[65]*Monadology,* §33. The notion of necesary truth raises issues within the framework of Leibniz' philosophy; cf. e.g., G. Parkinson, *Logic and Reality in Leibniz's Metaphysics* (Oxford, 1965), pp. 69–75, 107–108.

[66]Instances of a direct ontological argument are to be found in Descartes, More, Leibniz, Mendelssohn; instances of an indirect ontological argument are to be found in Anselm and Spinoza.

proof, in other words, undertakes to demonstrate the truth of the proposition 'God, defined as such-and-such, is actually existent' either by analyzing the proposition and resolving it, in Leibniz's words, into more simple ideas, or else by analyzing its opposite—or, to be more precise, its contradictory—and showing the latter to be impossible. The ontological proof thereby undertakes to demonstrate the proposition in question as a *necessary* truth. The necessity embodied in a necessary truth is commonly called *logical necessity.*[67]

A cosmological proof of the existence of God, I have submitted, whether or not it happens to use the term, will always undertake to establish the existence of a being that is necessarily existent in the sense of existing by virtue of itself; and an ontological proof, whether or not it happens to use the term, will always undertake to establish the existence of a being necessarily existent in the same sense, and also necessarily existent in the sense that its existence is logically necessary, capable of being demonstrated as a necessary truth. A further configuration might suggest itself. One might imagine that a cosmological proof too could undertake to establish the existence of a being that not merely exists by virtue of itself, but whose existence is logically necessary as well. On inspection, though, the suggested configuration would appear to be wholly illegitimate. Cosmological proofs take their departure from data drawn from the realm of actual existence, and data of the sort may understandably serve as grounds for inferring the existence of a being having certain functions in the realm of actual existence. But how can data drawn from the realm of actual existence serve as grounds for supposing that there is a concept with certain logical characteristics, a concept such that the actual existence of the corresponding object can be analyzed out as a necessary truth? A cosmological proof attempting to establish the existence of God as necessary in the logical sense of the term would compound the dubiousness of the ontological proof. It would not merely reason from a concept to reality. It would reason from reality to the affirmation that there is a concept with the unusual virtue of allowing one to reason back therefrom to reality. Yet we find that, despite its problematic character, a cosmological proof of the type described was proposed by philosophers in the modern period. Neither Avicenna nor the other medieval philosophers who have been examined in the present book contemplated anything of the kind. The terminology of at least the careful medieval philosophers would not even permit a characterization of the deity as uniquely necessarily existent in the sense that his existence is logically necessary. For, as I have stressed, all the careful medieval philosophers employ the term *necessarily*

[67] Aristotle seems to recognize a similar necessity when he writes, *Prior Analytics* I, 9, 30a, 30, that "it is necessary that man is an animal," animal being part of the definition of man; cf. J. Lukasiewicz, *Aristotle's Syllogistic* (Oxford, 1957), pp. 148–149. More important for Aristotle is what may be called demonstrative necessity, i.e., the necessity whereby the conclusion of a syllogism follows from the premises. Cf. *Prior Analytics* I, 1, 24b, 18–24; *Metaphysics* V, 5, 1015b, 6–9, and D. Ross's note, in his edition (Oxford, 1924). Lukasiewicz questions the value for logic of this *demonstrative* necessity; cf. *Aristotle's Syllogistic,* pp. 11–12, 144–145.

existent to designate a class wider than the deity; and, for them, not necessarily existent, but only the qualification [*existing*] *by virtue of itself* characterizes the deity and sets him apart from other necessary beings.

When the terms *necessarily existent being, necessary being,* and *necessary existence,* appear in modern European proofs of the existence of God, it is often difficult to determine which of two general senses of necessity is intended. It is difficult to determine whether by necessarily existent being, the author of a given proof means a being that exists by virtue of itself, or a being whose existence is logically necessary, analyzable out of a concept as a necessary truth. Some philosophers in the modern period apparently did not distinguish the two senses. Some consciously equated them. The difficulty in ascertaining the precise sense of necessary existence occurs in connection with both ontological and cosmological proofs.

Ontological proofs employing the term *necessarily existent* or a variation of it are found in Descartes, Spinoza, Henry More, Leibniz, perhaps Christian Wolff, Baumgarten, and Moses Mendelssohn. The stage at which the term *necessarily existent being* or *necessary existence* appears in the different arguments varies.

In one instance, at least, necessary existence serves as the starting point. Ontological proofs have taken their departure from a variety of concepts, for example, the concept "God" itself,[68] "that than which nothing greater can be conceived,"[69] the "best,"[70] the "absolutely simple,"[71] "supremely perfect being,"[72] "immeasurably powerful being,"[73] "infinite" being,[74] "substance" par excellence.[75] There is no reason why an ontological proof might not as well address itself to the concept *necessarily existent* and through an analysis thereof undertake to demonstrate the actual existence of a corresponding object;[76] and that procedure was in fact adopted by Leibniz. In a version of the ontological proof which he especially recommends, Leibniz considers "necessary being," in the sense of "being to whose essence existence belongs, or being per se," that is to say, being existing by virtue of itself.[77] The proposition "necessary being exists," Leibniz

[68]Descartes, *Arguments Demonstrating the Existence of God and the Distinction Between Soul and Body, Drawn up in Geometrical Fashion,* prop. 1.

[69]Anselm, *Proslogion,* chaps. 2–3.

[70]Richard Fishacre, cited by A. Daniels, *Geschichte der Gottesbeweise* (Muenster, 1909), p.23; William of Auxerre, cited ibid., p. 26.

[71]Richard Fishacre, cited ibid., p. 23.

[72]Descartes, *Meditations,* V.

[73]Descartes, *Reples to Objections urged against the Meditations,* I; English translation: *Philosophical Works of Descartes,* trans. E. Haldane and G. Ross,Vol. II (New York, 1955), p. 21.

[74]Spinoza, *Ethics,* I, prop. xi.

[75]Ibid.

[76]D. Henrich, *Der Ontologische Gottesbeweis* (Tuebingen, 1960), gives especial attention to ontological arguments taking their departure from the concept of *necessary being*. Heinrich's thesis is that such arguments are less open to crticism than other ontological arguments, yet still contain a fallacy that Kant was able to discern. Cf. also N. Malcom in *The Ontological Argument,* ed. A. Plantinga (New York, 1965), pp. 141–146.

[77]On necessary being in Leibniz, see, further, below, p. 396.

avers, is a proposition whose truth is "evident from the terms"; a being to whose essence existence belongs cannot help but have existence. "But God is such a being. . . . Therefore God exists."[78] The existence of God thus follows at once and by logical necessity from an analysis of the concept *necessarily existent being*.[79] Apparently, analysis of the concept *necessary being* in one sense, in the sense of that which exists by virtue of itself, discloses to Leibniz that the being in question is necessary in a second sense as well, in the sense that its existence is analyzable out of its concept as a necessary truth.

Descartes, the first of the modern philosophers to use the terminology of necessary existence in an ontological proof, does not begin his proof with necessary existence, but uses the term as a sort of bridge to carry his argument to its conclusion. The most succinct formulation of Descartes' ontological proof reads: Since "necessary existence is contained in the concept of God, . . . it is true to affirm . . . that God himself exists."[80] Necessary existence here serves as the connective link between the concept of God and his actual existence; on the grounds that necessary existence is comprised within, and analyzable out of, the concept of God, the actual existence of God is inferred. But exactly what Descartes means by necessary existence remains unclear. The necessary existence he finds to be comprised within the concept of God might be understood either as the deity's existing by virtue of himself, or the deity's being such that actual existence can be deduced from his concept as a necessary truth;[81] and Descartes' own statements are compatible with both readings. When he explicates his ontological argument elsewhere, Descartes contends that in the case of all objects apart from God, "we do not conceive that there is any necessity for the conjunction of actual existence with their other properties." In the case of God, by contrast, "actual existence is *necessarily* and at all times linked to his other attributes," just as the attribute of having "its angles equal to two right angles" is "contained in the idea of the triangle"; and for that reason "it follows certainly that God exists."[82] Here the meaning surely is that the existence of God follows as a necessary truth, by logical necessity, from analysis of the concept of God.[83] But a few paragraphs further on Descartes treats the "necessary existence comprised in the idea of a being of the highest power" as equivalent to its existing

[78]*Die Philosophischen Schriften*, IV, p. 359; English translation, p. 50. For other features of Leibniz's proof and for issues raised by it, see Parkinson, *Logic and Reality in Leibniz's Metaphysics*, pp. 77–84; N. Rescher, *Philosophy of Leibniz* (Englewood Cliffs, 1967), pp. 66–68, 148–149, 152.

[79]M. Mendelssohn, *Morgenstunden*, chap. 17, gives a similar proof. Cf. A. Altmann, "Moses Mendelssohn's Proofs for the Existence of God," *Mendelssohn Studien*, II (1975), 26.

[80]*Arguments Demonstrating the Existence of God*, prop. 1. Cf. *Principles of Philosophy*, I, xiv.

[81]The context does not help, nor does *Arguments*, axiom 10.

[82]*Replies to Objections urged against the Meditations*, I; English translation, II, p. 20. Cf. *Principles of Philosophy*, I, xiv; xv.

[83]Cf. *Meditations*, V; English translation: *Philosophical Works of Descartes*, trans. E. Haldane and G. Ross, Vol. I (New York, 1955), pp. 181–182.

"by its own power."[84] Here he is using the term *necessary existence* approximately in the sense of *existing by virtue of itself.* There are accordingly textual grounds for reading Descartes' ontological proof as inferring the actual existence of God from the presence of necessary existence in the concept of God in either of the two senses of necessary existence. And, of course, there is the strong possibility that Descartes did not differentiate the two senses.[85]

The seventeenth-century English philosopher and theologian Henry More offers an ontological proof that is more diffuse than Descartes' but is worked out entirely in the same spirit. More finds necessary existence to be one of the perfections comprised within the notion of "absolutely perfect" being;[86] and inasmuch as the "notion and nature of God implies in it *necessary existence,*" More concludes "that God does exist."[87] Once again necessary existence is the link allowing the passage from a concept of the nature of God to the conclusion that God actually exists. As for the meaning of *necessary existence,* the term is bracketed by More with "stable" and "immutable" existence,[88] and is glossed as a "habitude to existence."[89] That is fairly close to defining the necessarily existent as what exists by virtue of itself. But, More adds, "necessity" is also a "logical term." It "signifies . . . a connection between the subject and predicate" which cannot be "dissevered, . . . so that [subject and predicate] make *axioma necessarium.*" As a consequence, "necessity does not only signify the manner of existence in that which is necessary, but also [signifies] that it does actually exist and could never possibly do otherwise."[90] More finds, in a word, that the concept of *absolutely perfect* being contains necessary existence in two senses, both in the sense of a certain "immutable" manner of existence, and also in the sense that God's existence is a logically necessary truth. And More explicitly infers the actual existence of God from the presence of necessary existence in both senses within the concept of *absolutely perfect* being.[91]

[84]*Replies to Objections,* I; English translation, II, p. 21.

[85]Note that whatever the meaning of the necessary existence which serves as a bridge to the conclusion of Descartes' ontological proof, the proof—like all ontological proofs—must conclude with a being that is necessarily existent in both senses of the term. Cf. above, p. 392.

[86]*Antidote versus Atheism* (4th ed.; London, 1712), Book I, chap. iv, §1.

[87]Ibid., chap. viii, §1.

[88]Ibid., chap. iv, §6.

[89]Ibid., Scholia on Book I, chap. viii, §3. The Oxford English Dictionary explains *habitude* as a "manner of being or existing" and again as an "inherent or essential character."

[90]Ibid., Book I, chap. viii, §4.

[91]The conclusion that God actually exists because the concept of God contains necessary existence in the sense of *logically* necessary existence, is not a genuine inference. The contention could amount to no more than this: The concept of God is such that actual existence follows therefrom as a necessary truth; therefore the actual existence of God necessarily follows therefrom. Descartes can of course be read as not strictly proving the existence of God in his ontological proof, but rather as merely inspecting the concept of God and discovering that existence is comprised therein. Cf. Wolfson, *Spinoza,* Vol. I (Cambridge, Mass., 1948), pp. 168–169; N. Smith, *New Studies in the Philosophy of Descartes* (London, 1952), pp. 305–307; M. Gueroult, *La preuve ontologique de Descartes* (Paris, 1955), pp. 24–25, 42–44, 48.

Not only did the terminology of *necessary existence* serve as the point of departure for an ontological proof and as a bridge from a given concept to the actual existence of the deity; there also are instances where necessary existence explicitly appears in the conclusion of an ontological proof. A prime instance is Spinoza. It would, Spinoza reasons, be "absurd" to suppose that "existence" is not involved in the essence of God, or Substance; consequently, "God [or Substance] necessarily exists."[92] Again there is a question as to what is meant by necessary existence. At the least Spinoza is asserting that since a denial of the proposition 'God is actually existent' would be logically "absurd," the proposition is a necessary truth. The necessity in the conclusion "God necessarily exists" is accordingly logical necessity.[93] The conclusion probably intends something more, however. Necessary existence is, for Spinoza, equivalent to a thing's being "cause of itself"[94] and having its "essence involve existence."[95] In addition, to be "cause of itself" and to have its "essence involve existence" is understood by him to be equivalent to a thing's being such that its "nature cannot be conceived unless existing";[96] the thinking apparently is that because the essence involves existence, the essence cannot be conceived of except in actual existence. The two general senses of necessary existence are, it thus seems, consciously equated by Spinoza. By being the "cause of itself" or existing through itself, and hence having its essence involve existence, a thing is also such that actual existence follows from its concept as a necessary truth.[97] The conclusion of Spinoza's ontological proof—"God necessarily exists"—may therefore be read as affirming that God exists through Himself as well as that the existence of God is a logically necessary truth. And, what is more, the conclusion "God necessarily exists" may be understood as affirming both those things indistinguishably.[98] Spinoza also offered an argument which, although differing from all the familiar cosmological proofs, is characterized by him as an "a posteriori" proof; and it, like the a priori, or ontological, proof,[99] concludes with the necessary existence of God.[100] If the suggestion that has been made regarding the conclusion of Spinoza's ontological argument is correct, the conclusion of his a posteriori argument presumably affirms the existence of a being that is necessarily existent in the two now indistinguishable senses, namely, a being that not only exists through itself, but whose existence is logically necessary.

[92] *Ethics,* I, prop. xi; cf. prop. vii.

[93] Cf. also Spinoza's *Cogitata Metaphysica,* I, 3.

[94] *Ethics,* I, prop. xxiv.

[95] *Ethics,* I, props. vii and xxiv; *Cogitata Metaphysica,* I, 1.

[96] *Ethics,* I, def. 1; cf. axiom vii.

[97] Cf. Wolfson, *Spinoza,* I, p. 160.

[98] The necessary existence in *Ethics,* I, prop. xxxv, can be understood in the same fashion.

[99] Spinoza offers several ontological arguments.

[100] *Ethics,* I, prop. xi, third proof. The peculiar argument is essentially this: If finite being actually exists, infinite being must, *a fortiori,* actually exist.

I have cited instances wherein an ontological argument takes its departure from the concept *necessarily existent,* wherein an ontological argument employs necessary existence as a connecting link to pass from a concept of God to the conclusion that God actually exists, and wherein an ontological argument explicitly concludes with the necessary existence of God. Ontological arguments explicitly concluding with the necessary existence of God merely express something that every ontological argument must undertake. The ontological proof must, by its very nature, undertake to establish the existence of a being both necessarily existent insofar as it exists by virtue of itself and also necessarily existent insofar as its existence is deducible from its concept as a necessary truth. But no ontological argument need go as far as Spinoza went, and treat those two senses of necessary existence as equivalent and indistinguishable.

Modern philosophy also knows of a series of cosmological arguments designed to establish the existence of a *necessary being.* As already mentioned, Spinoza offered an "a posteriori" proof concluding with the necessary existence of God. Despite the label a posteriori, it is not, however, a truly cosmological proof, and Leibniz is responsible for introducing into modern philosophy a truly cosmological proof of the existence of a necessary being. Leibniz works with a dichotomy of actual existence which is curiously similar to the dichotomy met in Avicenna. As will be recalled, Avicenna had maintained that all actual existence must be either possibly existent by virtue of itself, necessarily existent by virtue of another, or else necessarily existent by virtue of itself. Leibniz, for his part, maintains that all actual existence must be either "contingent" or "necessary." Yet contingent beings, so Leibniz explains, are also necessary in a certain qualified sense. They are subject to "physical or hypothetical necessity"; that is to say, they are subject to the necessity consisting in things' happening in the world "such as they do," because the nature of the world is "such as it is."[101] Unqualifiedly "necessary" being, by contrast, is that which enjoys "absolute or metaphysical necessity" and carries the "reason of its existence within itself."[102]

The principle of sufficient reason, one of "two great principles" of human reasoning,[103] teaches Leibniz that nothing exists or occurs without a sufficient reason "inclining" it towards existence[104] and determining why it should exist at

[101] *Principles of Nature and of Grace,* §8; in *Die Philosophischen Schriften,* ed. C. Gerhardt, Vol. VI (Berlin, 1885), p. 602; English translation: Leibniz, *Selections,* trans. P. Wiener (New York, 1951), pp. 527–528. *On the Ultimate Origin of Things,* in *Die Philosophischen Schriften,* ed. C. Gerhardt, Vol. VII (Berlin, 1890), p. 303; English translation: *Selections,* ed. Wiener, p. 346. See also *A Collection of Papers which Passed between the Late Learned Mr. Leibnitz and Dr. Clarke in the Years 1715 and 1716* (London, 1717), V, §5.

[102] *On the Ultimate Origin of Things,* p. 303; English translation, p. 346; *Principles of Nature and Grace,* §8.

[103] *Monadology,* §§31–32.

[104] *On the Ultimate Origin of Things,* p. 303; English translation, p. 346.

all, as well as why it should exist "so and not otherwise."[105] Now, when contingent things are explained by other contingent things, and contingent states of the world are explained by other contingent states of the world, "no progress is made, go as far as one may," towards discovering the sufficient reason of their existence. The effort to explain why things exist will not be furthered a whit, "however infinite may be" the series of contingent beings and states of the world brought into the explanation. "For the same question always remains," namely, why contingent beings exist at all and why in one form rather than another.[106] The "complete" and sufficient reason can be found only "outside the series of contingent things . . . in a substance that is their cause, a substance that is a necessary being carrying the reason of its existence within itself." Consequently, Leibniz concludes, a necessary being, a being that is "absolutely" or "metaphysically necessary," must be posited, and "this is what we call God."[107]

The similarity of Leibniz's proof to Avicenna's is striking. Leibniz distinguishes two classes of necessary being, which parallel the two classes of necessary being distinguished by Avicenna. Everything actually existent must, for both writers, be necessary in one sense or the other. The reason for the existence of contingent, physically necessary being—the possibly existent by virtue of itself, necessarily existent by virtue of another in Avicenna's terminology—must be sought outside of it, in a factor that "inclines" towards (Leibniz) or "selects out" (Avicenna) existence.[108] Leibniz leaves the issue of an infinite regress open, as Avicenna's proof allowed him to do.[109] And like Avicenna, although without Avicenna's argumentation, or, it seems, any argumentation at all, Leibniz finds that the series of contingent beings, taken as a whole, cannot conceivably contain the sufficient reason for its own existence.

Leibniz's cosmological proof was repeated and elaborated by his followers, and the version offered by Christian Wolff evinces an even more striking resemblance to Avicenna. *Necessary being* is defined by Wolff as that which has the "sufficient reason of its existence in its own essence,"[110] and hence exists "through itself."[111] Like Avicenna, Wolff requires one and only one empirical datum to establish that something corresponding to the concept actually exists. The empirical datum adduced by Wolff is the actual existence of an individual human soul. Wolff's argument runs: Everyone has immediate knowledge of the existence of his soul.[112] Since nothing at all exists "without a sufficient reason why it should

[105] *Principles of Nature and Grace*, §7.

[106] Ibid., §8; *Monadology* §37.

[107] *On the Ultimate Origin of Things*, p. 303; English translation, p. 346; *Principles of Nature and Grace*, §8; *Monadology*, §§36–38.

[108] See above, p. 292.

[109] Above, p. 302.

[110] *Ontologia* (Frankfurt, 1736), §309.

[111] *Theologia Naturalis* (Frankfurt, 1739), I, §§31, 33.

[112] Cf. Wolff, *Psychologia Empirica* (Verona, 1736), §§14, 21. The Cartesian echo is obvious.

be rather than not be, . . . our soul" must have a "sufficient reason" for its existence. The reason for the existence of an individual soul is to be sought either in the soul itself or in something else; and in the latter event, the reason for the existence of the other thing is also to be sought either in itself or in something else again. The sufficient reason for the existence of everything in the series is reached only when we arrive at a being "that has the sufficient reason of its own existence in itself," only when we arrive at a "necessary being."[113] The attributes of necessary being are established by Wolff, much as they had been by Avicenna, through an analysis of the concept.[114] Since necessary being carries the sufficient reason of its existence in itself, since it exists through itself, it has no possibility of not existing; it is consequently eternal.[115] It cannot be composite but must be simple; for whatever is composite is subject to generation and destruction, whereas necessary being, as just seen, has no possibility of ever not existing.[116] It is not extended; for whatever is extended is composite.[117] There can, furthermore, be no imaginable grounds for distinguishing two beings, both of which exist by virtue of themselves, and possess all the attributes associated with a thing's existing by virtue of itself;[118] hence only one might exist.[119] Analysis of the concept thus demonstrates the eternity, simplicity, unextendedness, and unity, of the necessary being whose existence is established through the cosmological argument. The physical universe cannot be this necessary being, since the physical universe is composite, whereas necessary being has been shown to be simple.[120] The human soul, though simple,[121] also cannot be the necessary being, since the soul is subject to the influence of the external world and therefore does not exist by virtue of itself.[122] The necessary being reached by seeking the sufficient reason of a single actually existent object must be transcendent, a transcendent being possessed of a full set of divine attributes analyzable out of the concept of *necessary being*; and it is the deity.[123]

In presenting Leibniz's cosmological proof of the existence of a necessary being, I omitted one particular. Leibniz, as was seen, concludes with the existence of a "necessary being" that carries "the reason of its existence within itself," and thereby enjoys what Leibniz calls absolute or metaphysical necessity. But absolute or metaphysical necessity is defined by Leibniz as the condition such that

[113] *Theologica Naturalis,* I, §24. Cf. *Ontologia,* §322.
[114] Cf. *Theologia Naturalis,* I, §6.
[115] Ibid., §§35–39.
[116] Ibid., §§47, 49.
[117] Ibid., §48.
[118] These attributes include *intellect, power,* and *goodness*; see below.
[119] *Theologia Naturalis,* I, §1107.
[120] Ibid., §50.
[121] Ibid., §60.
[122] Ibid., §59.
[123] Ibid., §67.

the "opposite implies contradiction and logical absurdity";[124] and whatever exists by absolute or metaphysical necessity would accordingly be such that to deny its existence is logically impossible. Leibniz understands, in other words, that the necessary being reached by his cosmological proof is necessarily existent not only in the sense that it exists through itself and is uncaused in contrast to everything else, which depends on it for existence; it also is necessary in the sense that the proposition stating its existence is a logically necessary truth. The expression "carrying the reason for its existence within itself" turns out to combine both senses; it means that necessary being exists by *reason* of itself, and that, by the same token, analysis of its concept will reveal the *reason* for its existence. In all likelihood, Leibniz's thinking was—as Spinoza's also probably was—that if something exists by virtue of itself, existence belongs to its essence; and as a consequence, actual, objective existence may be analyzed out of the essence.

The conclusion of Leibniz's cosmological proof is intended to be read, then, as, I proposed, the conclusion of Spinoza's ontological proof and "a posteriori" proof may be read. It is to be read as affirming the existence of a being that is necessary in the supposedly indistinguishable senses of existing by virtue of itself and having logically necessary existence. The conclusion of Wolff's cosmological proof is to be read in the same way. For Wolff construes necessary existence as equivalent to a thing's existing "through itself" and having the "sufficient reason of its existence in its own essence,"[125] but also as equivalent to "absolutely necessary . . . existence."[126] And something is "absolutely necessary" when, "considered in itself," it is such that its opposite is "impossible" and "involves a contradiction."[127] The cosmological arguments of Leibniz and Wolff, in other respects so strikingly similar to Avicenna's proof, are seen now to differ therefrom in a critical respect. When they conclude with a necessary being, they understand that being's existence to be *logically* necessary.

Hume and Kant, in their critiques of the cosmological proof, each had in view an argument in the tradition of Leibniz and Wolff. Hume's critique is directed specifically against a cosmological argument formulated by the English philosopher Samuel Clarke. Clarke had, in the spirit of Leibniz, inferred that the changeable and dependent objects in the universe must have their "ground or reason of existence" in an eternal being which exists "independently"[128] and is "self existent, that is, necessarily existent."[129] But, Clarke adds, "the only idea we can

[124] *On the Ultimate Origin of Things,* p. 304; English translation, p. 349; *A Collection of Papers between Leibnitz and Clarke,* V, 4. A similar definition of *absolutely necessary* appears in Aquinas, *Summa Theologiae,* I, 19, 3. But Aquinas does not maintain that God is an absolutely necessary being.

[125] Above, nn. 110, 111.

[126] *Ontologia,* §§309, 320.

[127] Ibid., §302.

[128] *Demonstration of the Being and Attributes of God* (London, 1705), §2.

[129] Ibid., §3.

frame of self-existence" is existence "by an absolute necessity." Absolute necessity is explicated by him as necessity which "must be antecedent in the natural order of our ideas to our supposition" that a given thing exists; and again as necessity which "must antecedently force itself upon us whether we will or no, even when we are endeavoring to suppose that no such being exists." It follows for Clarke that "the only true idea of a self-existent or necessarily existing being is the idea of a being, the supposition of whose not existing is an express contradiction."[130] Clarke's cosmological proof thus arrives at a being that is necessarily existent in the sense of existing "independently" and by virtue of itself; and thereupon, as a sort of appendix to the proof, the same necessarily existent being, which exists by virtue of itself, is also construed by Clarke as absolutely necessary, as such that its existence is logically necessary

A proof exactly like Clarke's is presented for critical scrutiny in Hume's *Dialogues concerning Natural Religion* when the floor is yielded to the conservative participant in the dialogue. The conservative participant characterizes his method of establishing the existence of God as the "simple and sublime argument a priori." The argument he offers has three steps, the first two of which are not a priori reasoning at all, but correspond instead to Clarke's cosmological proof, or, more precisely, to the body of Clarke's proof. Then the third and final step in the argument restated by Hume concludes that in order to explain the existence of the world "we must . . . have recourse to a necessarily existent being who carries the reason of his existence in himself; and who cannot be supposed not to exist without an express contradiction." The third step combines the conclusion of the body of Clarke's proof, according to which there must exist a being that is *necessary* in the sense of existing by virtue of itself, with the appendix to Clarke's proof, according to which the necessary being thereby established must also be such that its existence is logically necessary. When Hume passes to his critique of the proof he had presented for consideration, he states what is by now a celebrated objection: "Whatever we conceive as existent we can also conceive as nonexistent. There is no being, therefore, whose nonexistence implies a contradiction."[131] Hume's objection would be cogent if directed against a genuine a priori or ontological proof. It also is cogent in its context, directed against the peculiar, hybrid cosmological proof that proposes to establish a being whose existence is logically necessary. The objection is not, however, applicable to the body of Clarke's proof, taken without the appendix. Nor does it address the medieval cosmological proofs that conclude with the existence of a being necessarily existent by virtue of itself; for the medieval cosmological proofs know nothing of logically necessary existence.

No less celebrated than Hume's objection is Kant's objection to the effect that the cosmological proof ultimately reduces itself to an ontological proof. Kant too

[130]Ibid.

[131]*Dialogues Concerning Natural Religion,* IX.

had a specific cosmological argument in mind. Alexander Baumgarten, a follower of Leibniz and Wolff, had advanced parallel cosmological and ontological arguments for the existence of God. The first of Baumgarten's pair of arguments recasts Leibniz's cosmological proof into an argument from the impossibility of an infinite regress.[132] And it arrives at a being whose existence is "absolutely" necessary; absolute necessity is defined by Baumgarten as a condition such that the "opposite is in itself impossible."[133] Side by side with his cosmological argument, Baumgarten offered an ontological argument.[134] The latter takes its departure from the concept of "most perfect being," which is shown to be equivalent to "most real being."[135] From an analysis of the concept of most perfect being, Baumgarten deduces the actual existence of a corresponding object,[136] as well as the presence therein of all compatible "perfections" or "realities"[137] including the attribute of "necessary . . . existence."[138] Two parallel proofs are, then, offered by Baumgarten, a cosmological proof establishing the existence of an "absolutely" necessary being, and an ontological proof starting with the concept of "most perfect" or "most real being," and establishing an actually existent being possessed of all compatible perfections including necessary existence.

Kant's critique presents the following syllogism for consideration: "If anything exists, an absolutely necessary being must also exist. Now I at least exist. Therefore an absolutely necessary being exists." Cosmological argumentation, Kant insists, can go no further than the bare syllogism; for "what properties [necessary] being must have, the empirical premise cannot tell us." As a consequence, human reason is led to "abandon experience altogether and endeavors to discover from mere concepts what properties an absolutely necessary being must have." The only means human reason can find for pouring content into *absolutely necessary being* is to identify it with *most real being*—being possessing the fullness of perfection. In order, however, to show that most real being is in truth identical with the necessary being established by the cosmological argument, human reason must first analyze the concept of most real being and derive necessary existence from it; only by showing most real being to be a necessarily existent being and in fact the only one, will the human reason be able to identify it with the necessarily existent being established by the cosmological argument. The absolutely necessary being whose existence is established through the cosmological argument hence acquires meaning only on the assumption that necessary existence can also be analyzed out of the concept of most real being. But to assume that necessary existence is analyzable out of the concept of most real being is,

[132] *Metaphysica* (Halle, 1779), §§375, 381, 854. Cf. above, Chapter XI.
[133] Ibid., §§102, 109.
[134] Ibid., §856.
[135] Ibid., §806.
[136] Ibid., §810.
[137] Ibid., §812.
[138] Ibid., §823.

Kant contends, to assume that the concept of most real being can legitimately serve as the point of departure for an ontological argument. The complete cosmological proof is accordingly discovered to embody a latent ontological proof.[139]

For our purposes, two separate moments are to be noted in Kant's exposition of the course that human reason must inevitably pursue. In the first place, so Kant understands, the necessary being established in the cosmological proof must be shown to be identical with most real being, and the attributes of most real being must be transferred to necessary being; otherwise the necessary being established in the cosmological proof would remain without content. In the second place, the identification of necessary being with most real being can be accomplished only if necessary existence, and hence actual existence, is analyzable out of the concept of most real being. In each of these two moments, Kant's exposition of what human reason inevitably must do is apparently his reading of what Baumgarten in effect did do. But in each instance, the procedure that Kant discovers in Baumgarten and characterizes as the inevitable course taken by human reason is not the only possible course for reason to take.

In the first place, philosophers did not, historically, see themselves compelled to identify necessary being with most real being in order to provide the former with content. Christian Wolff, for example—even though he construed part of the meaning of necessary being as the possession of logically necessary existence—derived a set of divine attributes from the analysis of the concept necessary being taken solely in the sense of what exists by virtue of itself. He thereby assigned a set of divine attributes to the necessary being arrived at in the cosmological proof without identifying necessary being with most real being.[140] As was seen in earlier chapters, Avicenna and other medieval philosophers also undertook to derive a set of divine attributes from the concept *necessarily existent by virtue of itself*; they certainly had no thought of identifying necessary being with most real being in order to provide the former with content.[141] Kant then does not take into account the procedures in fact employed by philosophers from Avicenna to Christian Wolff. Those philosophers were satisfied that they could analyze a set of divine attributes directly out of the concept *necessarily existent by virtue of itself* or the concept *necessary being;* and they thereby raised the necessarily existent being arrived at in the cosmological proof to the level of the deity without the device of identifying necessary being with most real being.[142] That is not, of course, to say that the analysis of attributes out of one concept or

[139] *Critique of Pure Reason,* A 604–608/B632–636; English translation: *Immanuel Kant's Critique of Pure Reason,* trans. N. Smith (London, 1933), pp. 508–511.

[140] Above, p. 398. Spinoza had done the same. See his *Correspondence,* Letter 40(35).

[141] Above, pp. 296–297, 347, 352–353, 382–383.

[142] Kant even recognizes the propriety of analyzing divine attributes out of a concept. He concedes that the proposition "God is omnipotent" is a "necessary judgment" inasmuch as "omnipotence cannot be rejected if we posit a deity, that is, an infinite being; for the two concepts are identical." *Critique of Pure Reason,* A595/B623; English translation, p. 502.

the other had been successful; a weakness in the reasoning whereby Avicenna and his followers deduced divine attributes from the concept of the necessarily existent by virtue of itself was pointed out earlier.[143]

The second moment in Kant's exposition of the inevitable course pursued by human reason is more interesting. Kant is presupposing the indistinguishability of two separate senses of necessary existence, namely: existence by virtue of itself, and logically necessary existence; and he takes the necessary being arrived at in the cosmological proof as having necessary existence in the two indistinguishable senses. Necessary being, Kant's exposition runs, can be identified with most real being only if the latter is shown to possess, and to be the only thing to possess, necessary existence. Since no distinction is envisaged between the two senses of necessary existence, most real being can be identified with necessary being only if it contains necessary existence in both senses. Whereupon, Kant's criticism ensues: The cosmological proof can be completed only if necessary existence, including necessary existence in the logical sense, is analyzable out of the concept of most real being; to assume that necessary existence in the logical sense is analyzable out of most real being—that is, to assume that the actual existence of most real being is logically necessary—is to credit an ontological proof with most real being as the starting point; and the completed cosmological proof embodies a latent ontological proof.

Kant has in view a peculiar form of cosmological proof, a cosmological proof purporting to conclude, like the cosmological argument of Leibniz and his disciples, with a being that is necessarily existent in the two supposedly indistinguishable senses of the term. If the necessary being in the conclusion of the cosmological proof is necessarily existent in both senses, the identification of necessary being with most real being plainly can be effected only by analyzing necessary existence in the double sense out of most real being. Hence, given Kant's exposition of the cosmological proof, his criticism is cogent. Given the indistinguishability of the two senses of necessary existence, and granting as well the need to identify necessary being with most real being,[144] the cosmological proof can indeed be completed only by showing that the actual existence of most real being is logically necessary.[145]

Kant's exposition has, however, disregarded, another way of construing the necessary being in the conclusion of the cosmological proof. Necessary being

[143]Above, pp. 306–307, 373.

[144]The need to identify necessary being with most real being may be implicit in Baumgarten, but I could not find it stated explicitly there.

[145]I would suggest, though, that Kant's criticism is overly subtle. For if the absolutely necessary being in the conclusion of the cosmological proof possesses *logically* necessary existence, then the legitimacy of logically necessary existence, and, by that token, the possibility of an ontological proof is already affirmed. In other words, the cosmological proof of Leibniz and Baumgarten embodies a latent ontological proof in its conclusion quite apart from the procedure of identifying necessary being with most real being.

may be construed in the single sense of *what exists by virtue of itself*. Cosmological proofs concluding with a being *necessarily existent by virtue of itself* in the single sense were, we know, current in the Middle Ages. A cosmological proof could even be competed, if required or desired, by identifying necessarily existent being with most real being. To effect the identification, most real being would not have to be shown to contain necessary existence in the two supposedly indistinguishable senses. Most real being would have to be shown to contain necessary existence only in the single sense of existing by virtue of itself, with no intimation of, and no place for, logically necessary existence. We would have a cosmological proof concluding with a necessarily existent being; content could even, if needed, be poured into the concept of necessarily existent being through identifying it with most real being; yet the identification would be effected with no suggestion that the actual existence of most real being is logically necessary and with no hint of an ontological proof.

Kant's critique, in sum, was directed against a cosmological proof with two peculiar characteristics: The necessary being in the conclusion of the cosmological argument scrutinized by Kant acquires content only through its identification with most real being; and the necessary being in the conclusion of the cosmological argument is understood by Kant to possess necessary existence in two indistinguishable senses, to wit, existence by virtue of itself, and logically necessary existence. Having thus expounded the cosmological proof, Kant objected that analyzing necessary existence out of most real being amounts to proving the existence of God ontologically; and therefore a compete cosmological proof cannot help but embody a latent ontological proof. As for the first peculiarity of the cosmological proof in Kant's exposition, our historical survey revealed that philosophers were, at least to their own satisfaction, able to provide content to the necessary being in the conclusion of the cosmological proof without identifying necessary being with most real being. Insofar as Kant's critique turns on the need for identifying necessary being with most real being, his critique does not take into account cosmological arguments contemplating no such identification. As for the second peculiarity of the cosmological proof in Kant's exposition, the identification of necessary being with most real being, even if required, might be effected without assuming that most real being possesses logically necessary existence. For our historical survey revealed that medieval cosmological proofs arrive at a necessarily existent being in the single sense of *that which exists by virtue of itself*. A being necessarily existent by virtue of itself in the single sense could, if required, be identified with most real being, without assuming that most real being possesses logically necessary existence, and hence without in any way crediting the validity of an ontological proof.

The objections made by Hume and Kant which have been examined here do not, therefore, address themselves to Avicenna's proof of a being necessarily existent by virtue of itself nor to similar proofs. That is far from saying, of course, that those proofs are satisfactory. If one were to seek objections pertinent to

Avicenna's proof in Hume and Kant, one might best cite the rhetorical question "why may not the material universe be the necessarily existent Being,"[146] a question already put and developed by Ghazali;[147] and, perhaps the comprehensive critique that "all attempts to employ reason in theology in any merely speculative manner are altogether fruitless and by their very nature null and void,"[148] a much more radical rejection of metaphysical methods than the rejection met earlier in Averroes.[149]

4. Summary

From the time of Descartes on, the term *necessarily existent* and related terms appear repeatedly in ontological and cosmological proofs of the existence of God. Given instances of the ontological proof have expressly referred to necessary existence at different stages of their argumentation.[150] But all ontological arguments, whether expressly using the term or not, must undertake to demonstrate the existence of a being that is necessary both in the sense of existing through itself and in the sense that the proposition stating its actual existence is a logically necessary truth. Spinoza's ontological proof apparently undertakes to go further than that. Spinoza, it seems, identified the two senses of necessary existence. And accordingly his ontological proof not merely arrives at the existence of a being that is necessary in both senses, but also treats the two senses as equivalent and indistinguishable. Spinoza, in addition, offered an "a posteriori" proof, arriving at the necessary existence of God and, presumably, likewise affirming that the deity possesses necessary existence in the two supposedly indistinguishable senses of the term.

Leibniz offered parallel cosmological and ontological proofs for the existence of God and concluded, even in his cosmological proof, with the existence of a being that is necessarily existent in the two supposedly indistinguishable senses of necessary existence. Leibniz's cosmological proof differs herein from the medieval cosmological proofs leading to a being necessarily existent by virtue of itself; for the medieval cosmological proofs had no thought of logically necessary existence and aim only at the existence of a being that exists through itself, uncaused. Hume and Kant, in their critiques of the cosmological proof, consider a form of the proof which does purport to establish a being whose existence is logically necessary. The critiques of Hume and Kant examined in the previous section are cogent solely vis à vis that pèculiar form of the proof. Their criticisms do not address themselves to, and do not refute the medieval cosmological proofs.

[146]Hume, *Dialogues,* IX.

[147]Above, p. 373.

[148]Kant, *Critique of Pure Reason,* A636/B664; English translation, p. 528.

[149]Above, p. 317.

[150]As a starting point, above, p. 392; as a bridge, above, p. 393; in the conclusion, above, p. 395.

5. Concluding remark

Avicenna's proof from the concepts *possibly existent* and *necessarily existent* and Leibniz's cosmological proof have found champions into the twentieth century. The Muslim modernist Mohammad Abduh expounds an adaptation of Avicenna's proof.[151] Neo-Thomists employ a "proof from contingency" which they mistakenly suppose to be genuinely Thomistic.[152] Other philosophers, such as Copleston[153] and R. Taylor,[154] espouse adaptations of Leibniz's proof. A critique in the spirit of Hume is also still advanced by contemporary philosophers. Russell, for example, contends that any cosmological proof arriving at a necessarily existent being must *ipso facto* be invalid; for such a proof pretends to establish the existence of a being whose existence is logically necessary, whereas existence simply cannot be logically necessary.[155] The criticism is, we have seen, cogent only when directed against versions of the proof which do propose to arrive at being that enjoys necessary existence in the peculiar sense of logically necessary existence.

[151]Mohammad Abduh, *R. al-Tawḥīd* (Cairo, 1966); chap. 2, p. 26; English translation: *The Theology of Unity,* trans. I. Musa'ad and K. Cragg (London, 1966). Mohammad Abduh reasons: "The totality of existing possible [beings] is possible by virtue of itself, and everything possible requires a cause to give it existence." The cause of the totality cannot be the totality itself; for that would "imply the thing's preceding itself." The cause cannot be one part; for that would again "imply the thing's preceding itself." "It follows that the cause [of the totality of possible being] is outside the totality . . . and [hence] necessary." Cf. above, p. 301. In the sequel, Mohammad Abduh establishes the attributes of the necessarily existent (by virtue of itself) through analysis of the concept.

[152]Garrigou-Lagrange, *Dieu,* §38; G. Joyce, *Principles of Natural Theology,* (2nd ed.; London, 1924), pp. 78 ff.; Phillips, *Modern Thomistic Philosophy,* Vol. II, pp. 285–287; E. Mascall, *He Who Is* (London, 1943), pp. 46–52, 65–68, 75. Cf. van Steenberghen, *Ontologie,* pp. 148–154, who gives the "proof from contingence" but realizes that it is not Aquinas' "third way."

[153]"The Existence of God—, a Debate between Bertrand Russell and Father F. C. Copleston, S.J.," in B. Russell, *Why I am not a Christian* (London, 1957), pp. 145–146.

[154]R. Taylor, *Metaphysics* (Englewood Cliffs, N.J., 1963), pp. 87–88, 90–94.

[155]"The Existence of God, a Debate," p. 146. Restated by J. Smart, "The Existence of God," in *New Essays in Philosophical Theology,* ed. A. Flew and A. MacIntrye (London, 1955), pp. 38–39.

Appendix A
Two Philosophic Principles

Here I wish to sketch the history of two Aristotelian principles that have appeared in several of the preceding chapters but have not been followed from beginning to end at any one spot.

1. The principle that an infinite number is impossible

Aristotle and Alexander of Aphrodisias, a faithful follower of Aristotle, had advanced grounds for the impossibility of an infinite number.[1] Their grounds were subsequently adapted by John Philoponus to a highly un-Aristotelian end: He developed them into a proof for the creation of the world. The eternity of the world, Philoponus reasoned, would entail an infinite series of past events, and since an infinite series is ruled out by the impossibility of any infinite number, the number of past events cannot be infinite and the world must have a beginning.[2] Philoponus' proof of creation, which took several forms, had wide currency. It was taken up by the Kalam school of philosophy,[3] passed into the thought of medieval Jewish and Christian philosophers of varying orientations,[4] and even reached eighteenth-century German rationalism.[5] Argumentation similar to that whereby Philoponus proved the creation of the world was also utilized in the Middle Ages for a separate purpose, to demonstrate the finiteness of all bodies.[6] Medieval Aristotelians endorsed the use of arguments from the impossibility of an infinite number to rule out an infinite body.[7] And yet, since Aristotle had proclaimed the eternity of the world, they perforce rejected the use of the arguments to rule out infinite past time. Applying the impossibility of an infinite number in one area but not in another might seem inconsistent, and the medieval

[1] Above, pp. 87–88.
[2] Above, pp. 87–89.
[3] Above, pp. 117–124.
[4] Above, pp. 120, 121, 122.
[5] Above, p. 120.
[6] Above, pp. 125–127.
[7] Above, pp. 126, 128.

Aristotelians had recourse to a typical medieval device for removing the apparent inconsistency: They drew a distinction. They discovered a difference between divers types of infinite, a difference permitting them to conclude that not all types are covered by the impossibility of an infinite number. What is impossible, they contended, is an infinite series of objects existing together at the same time and arranged in an order. Since the parts of a body do exist together and have an order, an infinite body is ruled out, but since past events do not exist together, an infinite series of past events remains unaffected. The medieval Aristotelians were accordingly confident that they had committed no inconsistency in rejecting the possibility of an infinite body while affirming the possibility of an infinite series of past events and infinite past time.[8]

The distinction drawn by the medieval Aristotelians between divers types of infinite is reviewed by Ghazali as part of his comprehensive critique of "the philosophers." Ghazali notes that still a further infinite—by the side of infinite past events and an infinite body—plays a central role in Aristotelian philosophy: The Aristotelian proofs of the existence of God accomplish their task only through adducing the impossibility of an infinite regress of causes.[9] The Aristotelians are therefore in the situation of linking one fundamental tenet of their system—the existence of a first cause—to the impossibility of one infinite, while, on the contrary, linking another fundamental tenet—the eternity of the world—to the possibility of another. That, Ghazali objects, is the grossest inconsistency; and, he insists, the Aristotelians cannot be allowed to pick and choose. They cannot be allowed to adduce the impossibility of an infinite number when their purpose is served, when they are ruling out an infinite regress of causes and proving the existence of God, if they do not let the impossibility of an infinite run its logical course when their purpose is not served and creation follows. The Aristotelians will gain nothing if they respond that an infinite series of objects is impossible only when the objects exist together in time and are arranged in an order; and hence that an infinite series of causes existing together is, like an infinite body, an impossibility, whereas infinite past events are possible. Even if the distinction between different types of infinite be granted, an eternal universe does entail the existence of an infinite series of objects existing together and having an order, those objects being the immortal human souls that would have accumulated through infinite past time.[10] The eternity of the universe consequently entails an infinite number precisely of the type that the Aristotelians held is impossible when they were ruling out an infinite regress of causes, as well as an infinite body. If, then, the Aristotelians refuse to accept the argument that the world cannot be eternal

[8] Above, pp. 128–129.

[9] Above, pp. 238, 337.

[10] See above, p. 369, where Ghazali explains that his objection stands even if an infinite series of human souls has not in actuality accumulated from all eternity; for such an infinite series can still be "supposed."

because eternity entails an infinite number of objects existing together and arranged in an order, they cannot retain their own argument that a first cause must exist because causes cannot regress infinitely.[11]

The Aristotelian principle that an infinite number is impossible is thus transformed in Ghazali's hands, as it had been in the hands of Philoponus, into weaponry for subverting a key doctrine of the Aristotelian system. Philoponus had developed the principle into a proof of creation, thereby subverting the Aristotelian doctrine of the eternity of the world. Ghazali now contends that as long as the Aristotelians refuse to accept the proof of creation from the impossibility of an infinite number and stubbornly maintain the eternity of the world, they will be unable to rule out an infinite regress of causes and will have to abandon their proofs of the existence of God.

2. The principle that a finite body contains only finite power.[12]

A key premise in Aristotle's proof of the existence of an incorporeal prime mover is the principle that a finite body can contain only finite power. The principle permitted Aristotle to reason that since the physical universe is finite, its power is finite and would not suffice to sustain motion over an infinite time; consequently, the eternal motion of the heavens and of the universe as a whole must be due to an incorporeal mover beyond the universe.[13] The scope of the principle was extended by Proclus, who, in his turn, reasoned that the finite power of the physical universe would not suffice to sustain even existence over an infinite time; and consequently, the very existence of the physical universe, not merely its motion, must be due to an incorporeal cause beyond the universe.[14] Philiponus thereupon developed Proclus' reasoning into another proof of creation, by the side of his proof from the impossibility of an infinite number. Whatever is of finite power, Philoponus argued, cannot maintain itself in existence over an infinite time no matter what the source of its existence might be; for whatever is of finite power has the possibility of not existing, and every possibility must, by virtue of an additional Aristotelian principle, be realized over an infinite time. Consequently, the finite physical universe, being possessed of finite power, cannot have been in existence for an infinite time under any circumstance, but must have had a beginning in the finite past.[15] Philoponus' proof of creation from the finite power of finite bodies was known in the Middle Ages, although it enjoyed less currency than his proof from the impossibility of an infinite number; and it may

[11] Above, pp. 367–370.

[12] Discussed more fully in H. Davidson, "The Principle that a Finite Body Can Contain only Finite Power," *Studies in Jewish Religious and Intellectual History Presented to Alexander Altmann* (University, Alabama, 1969), pp. 75–92.

[13] Above, p. 244.

[14] Above, pp. 89–90, 281–282.

[15] Above, pp. 91–93.

also have engendered new proofs.[16] A response to it was essayed by the arch-Aristotelian, Averroes.

The task facing Averroes was to explain why the principle that a finite body can contain only finite power does not lead to the finiteness of the universe's existence. The explanation once again turns, in the typical medieval fashion, on a distinction, the distinction now between two senses of finite power. Philoponus, Averroes contends, failed to recognize the separate senses of finite power and did not realize that the sense in which the power of the heavens is finite is different from the sense that excludes efficacy for an infinite time. The two senses of finite power distinguished by Averroes are finiteness in respect to *continuity* and finiteness in respect to *intensity.* Bodies that are genuine compounds of matter and form, that is, bodies whose form is present within their matter, have, Averroes writes, power finite in both senses, in continuity as well as intensity. Aristotle's proof of the incorporeality of the prime mover argues, on Averroes' interpretation, that since a body compounded of matter and form would have power finite in respect to continuity, the heavens must be a different sort of body. The heavens must be a body that is not a genuine compound, a body that consists rather in the association of a simple, self-subsistent, material substratum with an independent incorporeal form. If that is the character of the heavens, their power, Averroes understands, will indeed be finite in respect to *intensity,* but not so in respect to *continuity.* And the finite power of the heavens, not being finite in the critical sense of continuity, will not preclude—contrary to Philoponus' supposition—the heavens' existing eternally. Aristotle's proof of the existence of a prime mover, as read by Averroes, still arrives at an incorporeal being that sustains the motion of the heavens, to wit, the incorporeal form associated with the substratum of the heavens.[17] But the finiteness of the power of the heavens does not imply the finiteness of the existence of the heavens and, through them, of the existence of the universe as a whole.[18]

When Crescas reflected on the distinction drawn by Averroes, he discovered that by employing the distinction in a more straightforward manner, a charge similar to the one Averroes laid to Philoponus can also be laid to Aristotle. Aristotle's reasoning, without the benefit of subtle exegesis, had been that since finite bodies can contain only finite power, the eternal motion of the heavens must be due to an incorporeal being beyond the body of the universe. Crescas' response is that Aristotle too failed to recognize the separate senses of finite power; and he did not for his part notice that the sole sense in which he could show the power of bodies to be finite, namely, in respect to intensity, is different from the sense, namely, in respect to continuity, which would entail the inability

[16] Above, pp. 323–325.

[17] Averroes explains that the substratum of the heavens cannot move itself because nothing at all can move itself.

[18] Above, pp. 325–326.

to sustain motion for an infinite time. Granting, then, that the power of any finite body can be shown to be finite, it cannot be shown finite in the critical sense, the sense of *continuity,* which is required by the proof for the existence of an incorporeal prime mover. Until the power of the physical universe is known to be finite in that sense, the finiteness of the power contained by the body of the universe will not permit Aristotle's conclusion that eternal motion must be due to an incorporeal being distinct from the physical universe. The Aristotelian proof of the existence of God as the prime mover accordingly collapses.[19]

The outcome is a fine irony. One of the premises upon which Aristotle's proof of the existence of the prime mover rests suggests to Philoponus a refutation of Aristotle's proof of eternity. And Averroes' attempt to defend the Aristotelian doctrine of eternity suggests to Crescas a refutation of the proof for the existence of God which the premise was originally intended to serve. Aristotle's proof of the existence of God turns out to contain the seed of its own refutation, the conceptual tools for the refutation being provided by the distinction Averroes devised in defending Aristotle.[20]

[19]Above, pp. 264–265.

[20]I leave open the question whether Crescas' critique addresses Averroes' reinterpretation of the proof from motion.

Appendix B

Inventory of Proofs

I. Proofs of Eternity of World	
1. From nature of physical world	
General response	pp. 30–33
(a) From the nature of matter	pp. 13–16
Response	pp. 33–35
(b) From concept of possibility	pp. 16–17
Response	pp. 36–39
(c) From nature of motion	pp. 17–24
Response	p. 39
(d) From nature of time	pp. 24–27
Response	pp. 39–45
(e) From the vacuum	p. 27
Response	pp. 45–46
(f) From nature of celestial spheres	pp. 28–29
Response	p. 46
2. From nature of God	
(a) No moment could have been chosen for creation	pp. 51–56
Response	pp. 68–76
(b) From unchangeability of cause	pp. 56–61
Response	pp. 76–79
(c) From God's eternal attributes	pp. 61–67
Response	pp. 79–85
II. Proofs of Creation of World	
(1) From impossibility of infinite number	pp. 87–89, 95–101, 107–109, 117–125
Responses	pp. 127–133
(2) From principle that a finite body can contain only finite power	pp. 89–93, 101

Response	pp. 322–326
(3) From accidents	pp. 103–104, 134–146
(4) From composition	pp. 102, 148–151
(5) From concept of particularization	pp. 189, 191, 197–201
(6) From matter and form of celestial region	pp. 203–206, 211–212
(7) From Arabic Aristotelian theory of emanation	pp. 208–209
(8) From rule that what has cause of existence must have come into existence	pp. 84, 190–191, 209–211, 387
III. Proofs of Existence of God	
(1) From creation	pp. 154–172, 374–375
(2) From composition	pp. 147–152
(3) From design	pp. 151–152; Chap. VII.
(4) From concept of particularization	pp. 187–189
(5) From motion	
(a) First proof	pp. 237–249
Response	pp. 249–275
(b) Second proof	pp. 275–278
Response	pp. 278–280
(6) From concepts *possibly existent* and *necessarily existent*	Chap. IX; pp. 385–386, 396–399
Response	Chap. X; pp. 370–375
(7) From impossibility of an infinite regress of causes	Chap. XI; pp. 378–385
Response	pp. 367–375
(8) Ontological proofs	pp. 392–395
Response	pp. 399–405

Primary Sources

Aaron ben Elijah. *'Eṣ Ḥayyim*. F. Delitzsch, ed. Leipzig, 1841.

'Abd al-Jabbār. *K. al-Majmū' fī al-Muḥīṭ bi-l Taklīf*. J. Houben, ed. Beirut, 1965.

———. *Sharh al-Uṣūl*. Cairo, 1965.

Abduh, Muhammad. *K. al-Tawḥīd*. Cairo, 1966. English translation: *The Theology of Unity,* I. Musa'ad and K. Cragg, trans. London, 1966.

Abravanel, Isaac. *Mif'alot*. Venice, 1592.

Abravanel, Judah. *See* Leone Ebreo.

Abū al-Barakāt. *K. al-Mu'tabar.* Hyderabad, 1939.

Abū Qurra, Theodore. *Mīmar fī Wujūd al-Khāliq*. L. Cheikho, ed. *al-Mashriq,* XV (1912), 757–774. German translation: *Des Theodor Abu Kurra Traktat Ueber den Schoepfer,* G. Graf, trans. Muenster, 1913.

Alan of Lille. *Opera*. In J.-P. Migne, ed., *Patrologia Latina,* Vol. CCX. Paris, 1855.

Albalag, Isaac. *Sefer Tiqqun ha-De'ot* (Commentary on Ghazali's *Maqāṣid*). G. Vajda, ed. Jerusalem, 1973.

Albertus Magnus. *Commentary on II Sentences*. In *Opera Omnia,* A. Borgnet, ed., Vol. XXVII. Paris, 1894.

———. *De Causis et Processu*. In *Opera Omnia,* A. Borgnet, ed., Vol. X. Paris, 1891.

———. *Physics*. In *Opera Omnia,* A. Borgnet, ed., Vol. III. Paris, 1890.

Albo, Joseph. *Sefer ha-'Iqqarim*. I. Husik, ed. Philadelphia, 1946.

Alexander of Aphrodisias. *Aporiai*. I. Bruns, ed. *Commentaria in Aristotelem Graeca,* Supplementary Vol. II/2. Berlin, 1892.

———. *Commentary on Metaphysics*. M. Hayduck, ed. *Commentaria in Aristotelem Graeca,* Vol. I. Berlin, 1891.

———. *Fragments*. In J. Freudenthal, *Die durch Averroes erhaltenen Fragmente Alexanders zur Metaphysik des Aristoteles*. Berlin, 1885.

[Alexander of Aphrodisias?] *Mabādi' al-Kull*. In A. Badawi, ed., *Arisṭū 'ind al-'Arab*. Cairo, 1947.

Alexander of Hales. *Glossa in Quattuor Libris Sententiorum*. Florence, 1951.

Alfarabi, *Aghrād mā ba'd al-Ṭabī'a*. In *Rasā'il al-Fārābī*. Hyderabad, 1931.

———. *Commentary on Aristotle's Peri Hermeneias* (*De Interpretatione*). W. Kutsch and S. Marrow, eds. Beirut, 1960.

———. *Epitome of Prior Analytics*. Published as *K. al-Qiyās al-Ṣaghīr.* M. Türker, ed. *Revue de la Faculté de Langues, d'Histoire et de Géographie de l'Université d'Ankara,* XVI/3–4 (1958), 244–286.

———. *Iḥṣā' al-'Ulūm*. A. Gonzalez Palencia, ed. and trans. Madrid, 1953.

———. *K. Arā' Ahl al-Madīna al-Fāḍila*. F. Dieterici, ed. Leiden, 1895. German translation: *Der Musterstaat,* F. Dieterici, trans. Leiden, 1900.

———. *al-Siyāsāt al-Madanīya*. Hyderabad, 1927.

[Alfarabi?] *al-Da'āwā al-Qalbīya*. In *Rasā'il al-Fārābī*. Hyderabad, 1931.

———. *Fī Ithbāt al-Mufāraqāt*. In *Rasā'il al-Fārābī*. Hyderabad, 1931.

———. *R. Zaynūn*. In *Rasā'il al-Fārābī*. Hyderabad, 1931.

———. *Ta'alīqāt*. In *Rasā'il al-Fārābī*. Hyderabad, 1931.

———. *'Uyūn al-Masā'il*. In *Alfarabi's philosophische Abhandlungen,* F. Dieterici, ed. Leiden, 1890. German translation: *Alfarabi's philosophische Abhandlungen aus dem Arabischen uebersetzt,* F. Dieterici, trans. Leiden, 1892.

Amīdī. *Ghāya al-Marām*. Cairo, 1971.

al-Anbārī, Jibrīl b. Nūḥ b. Abī Nūḥ. *K. al-Fikr wa-l-I'tibār.* Aya Sofia Library, MS.4836/2.

Anselm. *Proslogion*.

Aquinas, Thomas. *Commentary on Metaphysics*.

———. *Commentary on Physics*.

———. *Commentary on Sentences*.

———. *Compendium of Theology.*

———. *De Potentia*.

———. *Summa contra Gentiles*.

———. *Summa Theologiae*.

Aristotle. *De Generatione*.

———. *De Partibus Animalium*.

———. *Metaphysics*.

———. *Physics*.

———. *Posterior Analytics*.

See also Pseudo-Aristotle.

Ash'ari. *K. al-Luma'*. In *The Theology of al-Ash'arī,* R. McCarthy, ed. and trans. Beirut, 1953.

———. *Maqālāt al-Islamīyīn*. H. Ritter, ed. Istanbul, 1929–1930.

———. *Risāla ilā ahl al-Thaghr. Publications of the Theological Faculty, Istanbul,* II/8 (1928), 80–108.

Athanasius. *Contra Gentes*. R. Thomson, ed. Oxford, 1971.

Augustine. *De Civitate Dei*.

Averroes (Ibn Rushd). *De Substantia Orbis*. In *Aristotelis Opera cum Averrois Commentariis,* Vol. X. Venice, 1562.

———. *Derushim Ṭib'iyim*. H. Tunik, ed. Ph.D. dissertation, Radcliffe College, 1956.

———. *Epitome of De Caelo*. In *Rasā'il Ibn Rushd*. Hyderabad, 1947.

———. *Epitome of De Generatione*. In *Rasā'il Ibn Rushd*. Hyderabad, 1947. Hebrew translation: *Commentarium Medium et Epitome in Aristotelis de Generatione et Corruptione Libros, Textum Hebraicum,* S. Kurland, ed. Cambridge, Mass., 1948. English translation: *Averroes' Middle Commentary and Epitome on Aristotle's De Generatione*. S. Kurland, trans. Cambridge, 1958.
———. *Epitome of Metaphysics*. Published as *Compendio de Metafísica*. C. Quirós Rodríguez, ed. and trans. Madrid, 1919. German translation: *Die Epitome der Metaphysik des Averroes,* S. van den Bergh, trans. Leiden, 1924.
———. *Epitome of Physics*. In *Rasā'il Ibn Rushd*. Hyderabad, 1947.
———. *K. al-Kashf*. M. Mueller, ed. Munich, 1859. German translation: *Philosophie und Theologie von Averroes,* M. Mueller, trans. Munich, 1875.
———. *Long Commentary on De Caelo*. In *Aristotelis Opera cum Averrois Commentariis,* Vol. V. Venice, 1562.
———. *Long Commentary on Metaphysics*. Arabic original: *Tafsīr mā ba'd al-Ṭabī'a,* M. Bouyges, ed. Beirut, 1938–1948.
———. *Long Commentary on Physics*. In *Aristotelis Opera cum Averrois Commentariis,* Vol. IV. Venice, 1562.
———. *Middle Commentary on De Caelo*. Vatican Library, Hebrew MS. Urb. 40.
———. *Middle Commentary on Metaphysics*. Rome, Casanatense Library, Hebrew MS. 3083.
———. *Middle Commentary on Physics*. Oxford, Bodleian Library, Hebrew MS. Neubauer 1380 = Hunt. 79.
———. *Middle Commentary on Porphyry's Isagoge and on Aristotle's Categoriae*. H. Davidson, ed. and trans. Cambridge, Mass., and Berkeley, 1969.
———. *Tahāfut al-Tahāfut*. M. Bouyges, ed. Beirut, 1930. English translation: *Averroes' Tahafut al-Tahafut,* S. van den Bergh, trans. London, 1954.
Avicenna (Ibn Sīnā). *Commentary on Metaphysics XII*. In A. Badawi, ed., *Arisṭū 'ind al-'Arab*. Cairo, 1947.
———. *Dānesh Nāmeh*. Teheran, 1937. French translation: *Le Livre de Science,* M. Achena and H. Massé, trans. Paris, 1955. English translation: *The Metaphysica of Avicenna,* P. Morewedge, trans. New York, 1973.
———. *De Anima*. F. Rahman, ed. London, 1959.
———. *K. al-Ishārāt wa-l-Tanbīhāt*. J. Forget, ed. Leiden, 1892. French translation: *Livre des Directives et Remarques,* A. Goichon, trans. Beirut and Paris, 1951.
———. *Mubāḥathāt*. In A. Badawi, ed., *Arisṭū 'ind al-'Arab*. Cairo, 1947.
———. *Najāt*. Cairo, 1938.
———. *Shifā': Burhān*. A. Affifi, ed. Cairo, 1956.
———. *Shifā': Ilāhīyāt*. G. Anawati and S. Zayed, eds. Cairo, 1960. French translation: *La Métaphysique du Shifā',* G. Anawati, trans. Paris, 1978.
[Avicenna?] *De Caelo*. In *Opera*. Venice, 1508.
Baghdādī. *K. Uṣūl al-Dīn*. Istanbul, 1928.

Baḥya ibn Paquda. *al-Hidāya (Ḥobot ha-Lebabot)*. A. Yahuda, ed. Leiden, 1912.

Basil. *Hexameron*.

———. *Sermon on Deuteronomy*.

al-Baṣīr, Joseph. *al-Muḥtawī*. Cited in P. Frankl, *Ein Mu'tazilitischer Kalām aus dem 10. Jahrhundert*. Vienna, 1872.

Bāqillānī. *K. al-Tamhīd*. R. McCarthy, ed. Beirut, 1957.

Baumgarten, A. *Metaphysica*. Halle, 1779.

Bazdawī, (Pazdawī). *K. Uṣūl al-Dīn*. H. Linss, ed. Cairo, 1963.

Boethius. *Consolation of Philosophy*.

Bonaventure. *Commentary on II Sentences*. In *Opera Omnia*, Vol. II. Quaracchi, 1882.

Bruno, Giordano. *De la Causa*.

Cicero, Marcus Tullius. *De Natura Deorum*.

———. *Tusculan Disputations*.

Clarke, Samuel. *Demonstration of the Being and Attributes of God*. London, 1705.

Clement. *See* Pseudo-Clement.

Corpus Hermeticum. A. Nock and A. Festugière, eds. Paris, 1945–1954.

Crescas, Ḥasdai. *Or ha-Shem*.

Descartes, R. *Arguments Demonstrating the Existence of God and the Distinction between Soul and Body, Drawn Up in Geometrical Fashion*.

———. *Replies to Objections Urged Against the Meditations*.

Duns Scotus. *Opus Oxoniense*. In *Duns Scotus, Philosophical Writings*, A. Wolter, ed. and trans. Edinburgh and London, 1962.

Duran, Simon. *Magen Abot*. Livorno, 1785.

Euclid. *Elements*.

Galen. *Compendium Timaei Platonis*. P. Kraus and R. Walzer, eds. London, 1951.

———. *De Usu Partium*.

Galileo, G. *Discorsi e dimostrazioni matematiche intorno a due nuove scienze*. In *Opere*. Florence, 1898. English translation: *Two New Sciences*, S. Drake, trans. Madison, 1974.

Gersonides, Levi. *Milḥamot ha-Shem*. Leipzig, 1866.

Ghazali. *Faḍā'iḥ al-Bāṭinīya*. A. Badawi, ed. Cairo, 1964.

———. *Iḥyā' 'Ulūm al-Dīn*. Cairo, 1937.

———. *al-Iqtiṣād fī al-I'tiqād*. Ankara, 1962.

———. *Maqāṣid al-Falāsifa*. Cairo, n.d.

———. *al-Risāla al-Qudsīya*. Published as *Al-Ghazali's Tract on Dogmatic Theology*. A. Tibawi, ed. and trans. London, 1965.

———. *Tahāfut al-Falāsifa*. M. Bouyges, ed. Beirut, 1927. English translation in *Averroes' Tahafut al-Tahafut*, S. van den Bergh, trans. London, 1954.

[Ghazali?] *al-Ḥikma fī Makhlūqāt Allāh*. Cairo, 1908.

Gregory of Nyssa. *De Anima et Resurrectione*.

Gundissalinus. *De Processione Mundi*. G. Buelow, ed. Muenster, 1925.
Hadassi, Judah. *Eshkol ha-Kofer.* Eupatoria, 1836.
Hallevi, Judah. *See* Judah Hallevi.
Hillel ben Samuel. *Tagmule ha-Nefesh*. Lyck, 1874.
Hume, D. *Dialogues Concerning Natural Religion*.
———. *Treatise of Human Nature*.
Ibn Abī Uṣaybi'a. *'Uyūn al-Anbā'.* Cairo, 1882.
Ibn Aknin (Aqnin), Joseph (Joseph b. Yaḥya). *Treatise as to Necessary Existence*. J. Magnes, ed. and trans. Berlin, 1904.
Ibn 'Atā' Illāh. *K. al-Ḥikam*. V. Danner, trans. Leiden, 1973.
Ibn Daud, Abraham. *Emuna Rama*. Frankfurt, 1852.
Ibn Ezra, Abraham. *'Arugat ha-Ḥokma. Kerem Ḥemed,* IV (1839), 1–5.
Ibn Ḥazm. *K. al-Faṣl fī al-Milal*. Cairo, 1964. Spanish translation: *Abenházam de Córdoba y su Historia Crítica de las Ideas Religiosas,* M. Asín Palacios, trans. Madrid, 1928.
Ibn al-Nadīm. *K. al-Fihrist*. G. Fluegel, ed. Leipzig, 1871–1872.
Ibn Nafīs. *Theologus Autodidactus*. M. Meyerhof and J. Schacht, eds. and trans. Oxford, 1968.
Ibn Rushd. *See* Averroes.
Ibn Ṣaddiq, Joseph. *ha-'Olam ha-Qaṭan*. S. Horovitz, ed. Breslau, 1903.
Ibn Sīnā. *See* Avicenna.
Ibn Suwār. *Fī an Dalīl Yaḥyā al-Naḥwī 'alā Ḥadath al-'Ālam awlā bī-l-Qubūl min Dalīl al-Muktakallimīn aṣlan*. In A. Badawi, ed., *Neoplatonici apud Arabes*. Cairo, 1955. French translation in B. Lewin, "La Notion de muḥdaṯ dans le kalām et dans la philosophie," *Orientalia Suenica,* III (1954), 88–93.
Ibn Ṭufayl. *Ḥayy ben Yaqdhān*, L. Gauthier, ed. and trans. Beirut, 1936. English translation: *Hayy Ibn Yaqzān*, L. Goodman, trans. New York, 1972.
Ījī. *Mawāqif*. Cairo, 1907.
Ikhwān al-Ṣafā'. *Rasā'il*. Beirut, 1957.
Jābir ibn Ḥayyān. *Textes Choisis*. P. Kraus, ed. Cairo, 1935.
Jaḥiẓ. *K. al-Ḥayawān*. Cairo, 1938.
[Jaḥiẓ?] *K. al-Dalā'il wa-l-I'tibār 'alā al-Khalq wa-l-Tadbīr.* Aleppo. 1928.
———. *K. al-'Ibar wa-l-I'tibār*. London, British Museum, MS. Or 3886.
Jeshua b. Judah. *Bereshit Rabba* (Commentary on Genesis). In M. Schreiner, *Studien ueber Jeschu'a ben Jehuda*. Berlin, 1900.
Job of Edessa. *Book of Treasures*. A. Mingana, trans. Cambridge, Engl., 1935.
John of Damascus. *De Fide Orthodoxa*.
Judah Hallevi. *Kuzari*.
Juwaynī. *al-'Aqīda al-Niẓāmīya*. Cairo, 1948. German translation: *Das Dogma des Imām al-Ḥaramain al-Djuwaynī,* H. Klopfer, trans. Cairo, 1958.
———. *K. al-Irshād*. Cairo, 1950.
———. *K. al-Shāmil*. Alexandria, 1969.
———. *Textes apologétiques de Ǧuwainī (Luma')*. M. Allard, ed. Beirut, 1968.

Kant, Immanuel. *Critique of Pure Reason.*
Khayyāṭ. *K. al-Intiṣār.* A. Nader, ed. and trans. Beirut, 1957.
Kindi. *Rasā'il.* M. Abu Rida, ed. Cairo, 1950. English translation: *Al-Kindi's Metaphysics,* A. Ivry, trans. Albany, 1974.
Lactantius. *De Opificio Dei.*
———. *Institutiones.*
Leibniz, G. W. *A Collection of Papers which Passed between the Late Learned Mr. Leibnitz and Dr. Clarke in the Years 1715 and 1716.* London, 1717.
———. *Monadology.*
———. *On the Ultimate Origin of Things.*
———. *Principles of Nature and Grace.*
———. *Theodicée.*
Leone Ebreo. *Dialoghi d'Amore.* Bari, 1929. Hebrew translation: *Wikkuaḥ 'al ha-Ahaba.* Lyck, 1871. English translation: *The Philosophy of Love,* F. Friedeberg-Seeley and Jean H. Barnes, trans. London, 1937.
Maimonides, Moses (Moses ben Maimon). *Guide to the Perplexed.*
Māturīdī. *K. al-Tawḥīd.* Beirut, 1970.
Māwardī. *A'lām al-Nubūwa.* Cairo, 1971.
Mendelssohn, M. *Morgenstunden.*
Miskawayh. *al-Fawz al-Aṣghar.* Cited in Kh. Abdul Hamid, *Ibn Miskawaih, A Study of His Al-Fauz Al-Asghar.* Lahore, 1946.
More, Thomas. *Antidote versus Atheism.* 4th ed. London, 1712.
Mufaḍḍal b. 'Umar. *K. al-Tawḥīd.* In al-Majlisi, *Biḥār al-Anwār.* Teheran, 1956–1972.
Narboni, Moses. *Commentary on Guide.* Vienna, 1852.
Nasafī. *Creed.* In E. Elder, *A Commentary on the Creed of Islam.* New York, 1950.
Nemesius. *De Natura Hominis.*
Philo Judaeus. *Legum Allegoria.*
———. *De Specialibus Legibus.*
Philoponus, John. *Commentary on De Generatione.* H. Vitelli, ed. *Commentaria in Aristotelem Graeca,* Vol. XIV/2. Berlin, 1897.
———. *Commentary on Physics.* H. Vitelli, ed. *Commentaria in Aristotelem Graeca,* Vol. XVI, Berlin, 1887; Vol. XVII. Berlin, 1888.
———. *De Aeternitate Mundi contra Proclum.* H. Rabe, ed. Leipzig, 1899.
———. *De Opificio Mundi.* C. Reichardt, ed. Leipzig, 1897. German translation: *Joannes Philoponus, Ausgewaehlte Schriften,* W. Boehm, trans. Munich, 1967.
Plato. *Laws.*
———. *Parmenides.*
———. *Republic.*
———. *Timaeus.*
Plotinus. *Enneads.*

Proclus. *Commentary on Timaeus.* E. Diehl, ed. Leipzig, 1903–1906. French translation: *Commentaire sur la Timée,* A. Festugière, trans. Paris, 1966–1967.

———. *Elements of Theology.* E. Dodds, ed. and trans. Oxford, 1963.

———. *Liber de Causis.* O. Bardenhewer, ed. and trans. Freiburg, 1882. Also in A. Badawi, ed., *Neoplatonici apud Arabes.* Cairo, 1955.

Pseudo-Aristotle. *De Mundo.*

Pseudo-Clement. *Recognitiones.*

Rāzī, Abū Bakr ibn Zakarīyā. *Opera Philosophica.* P. Kraus, ed. Cairo, 1939.

al-Rāzī, Fakhr al-Dīn. *K. al-Arba'īn.* Hyderabad, 1934.

———. *Muḥaṣṣal.* Cairo, 1905.

Saadia. *Commentary on Sefer Yeṣira.* Published as *Commentaire sur le Séfer Yesira,* M. Lambert, ed. and trans. Paris, 1891.

———. *K. al-Amānāt wa-l-I'tiqādāt.* S. Landauer, ed. Leiden, 1880. English translation: *Book of Beliefs and Opinions,* S. Rosenblatt, trans. New Haven, 1948.

Sextus. *Adversus Physicos.*

Shahrastānī. *K. al-Milal wa-l-Niḥal.* W. Cureton, ed. London, 1842–1846. German translation: *Religionspartheien und Philosophenschulen,* T. Haarbruecker, trans. Halle, 1850–1851.

———. *K. Nihāya al-Iqdām.* A. Guillaume, ed. Oxford and London, 1934.

Simplicius. *Commentary on De Caelo.* I. Heiberg, ed. *Commentaria in Aristotelem Graeca,* Vol. VII. Berlin, 1894.

———. *Commentary on Physics.* H. Diels, ed. *Commentaria in Aristotelem Graeca,* Vol. X. Berlin, 1895.

Spinoza. *Cogitata Metaphysica.*

———. *Correspondence. Letter XII.*

———. *Ethics.*

Themistius. *Paraphrase of Metaphysics.* Arabic translation in A. Badawi, ed., *Arisṭū 'ind al-'Arab,* Cairo, 1947. Hebrew translation: S. Landauer, ed. *Commentaria in Aristotelem Graeca,* Vol. V/5. Berlin, 1903.

———. *Paraphrase of Physics.* H. Schenkel, ed. *Commentaria in Aristotelem Graeca,* Vol. V/2. Berlin, 1900.

Theodoret. *De Providentia.* In J.-P. Migne, ed., *Patrologia Graeca-Latina,* Vol. LXXXIII. Paris, 1864.

Theology of Aristotle. F. Dieterici, ed. Leipzig, 1882. English translation in Plotinus, *Opera,* P. Henry and H. Schwyzer, eds., Vol. II. Paris and Brussels, 1959.

Ṭūsī. *Glosses* to Rāzī's *Muḥaṣṣal.* Cairo, 1905.

William of Auvergne. *De Trinitate.* B. Switalski, ed. Toronto, 1976.

Wolff, Christian. *Ontologia.* Frankfurt, 1736.

———. *Psychologica Empirica.* Verona, 1726.

———. *Theologia Naturalis.* Frankfurt, 1739.

Xenophon. *Memorabilia.*

Index of Philosophers

Aaron ben Elijah: cites proofs for eternity from nature of world, 13–15, 17, 24, 26, 29; from nature of God, 54, 59, 60; answers proofs for eternity from nature of world, 31, 35, 36, 38, 39, 46; from nature of God, 73, 77; proofs for creation, 121, 142–143; cites Aristotle's proof from motion, 237

'Abd al-Jabbār: cites proofs for eternity from nature of world, 15, 27; from nature of God, 55, 60, 62; answers proofs for eternity from nature of world, 30, 34, 45; from nature of God, 78, 79, 81; proofs for creation, 120, 126, 139–140, 143, 144, 156–157, 189; inferring creator from creation, 75, 163, 355; proofs for divine attributes, 165, 166, 167–169, 171–172, 234; proof of existence of accidents, 182–183; impossibility of infinite regress, 363

Abravanel, Isaac: cites proofs for eternity from nature of world, 14, 16, 17, 22, 23, 25, 26, 27, 29; from nature of God, 54, 59, 60, 63, 67; answers proofs for eternity from nature of world, 36, 38, 39, 43, 44, 45, 46; from nature of God, 82

Abravanel, Judah. *See* Leone Ebreo

Abū al-Barakāt: proofs for eternity, 54; answers refutations of proofs for eternity, 71–72, 75; cites proofs for creation, 122, 141; proof of existence of God, 353

Abū al-Hudhayl: Kalam proof for creation, 134, 139

Abū Bishr, 283

Abū Hāshim, 140

Abū Qurra, 148, 150

Aḥmad ibn Sulaymān, 224

Alan of Lille, 363

Albalag, 268

Albertus Magnus: cites proofs for eternity from nature of world, 13, 17, 22, 24, 26, 29; from nature of God, 53, 54, 59, 64; answers proofs for eternity from nature of world, 31, 35, 38, 39, 45, 46; proof for creation, 120; natural motion of elements, 268; proofs of existence of God, 278, 361

Albo, 53

Alexander of Aphrodisias: impossibility of infinite number, 88, 407; proof of existence of God from logical symmetry, 276, 277; subject matter of physics and metaphysics, 314, 316–317. *See also* 321, 330

[Alexander of Aphrodisias?], 26, 237, 267

Alexander of Hales, 363

Alfarabi: proof for eternity from nature of world (time), 43–44; cites Kalam proofs for creation, 105, 134–137, 139, 143; impossibility of certain infinites, 128, 130, 367–370; emanation, 206; division of being, 292; analysis of concept of *first being*, 295–296; and Philoponus, 93

[Alfarabi?], 148, 215, 237, 353–355, 360

Altabrizi, 127, 386

Āmidī: cites proofs for eternity from nature of world, 17, 25; from nature of God, 56, 62; answers proofs for eternity from nature of world, 37, 42; from nature of God, 72, 79–80, 84; proofs for creation, 120, 141; proofs for divine attributes, 169, 170; inferring deity from creation,

Āmidī (*continued*)
75; particularization arguments, 180, 191; Avicenna's proof, 357, 385–387
Ammonius, 282
Aquinas, Thomas: approaches to proving existence of God, 4–5, 253; classifying proofs for eternity, 11; cites proofs for eternity from nature of world, 13, 15, 17, 24, 25, 26, 27, 29; from nature of God, 53, 58–59, 63, 64, 66; answers proofs for eternity from nature of world, 32, 35, 36, 37–38, 39, 42, 44, 45, 46; from nature of God, 73, 74, 77, 82, 83; cites proofs for creation, 120, 121, 133; proof from motion, 238, 240; proof from impossibility of infinite regress, 343, 344, 347–348, 361, 363, 383–384, 386
Aristotle: proofs for eternity from nature of world, 13, 16, 17–20, 24, 27, 28, 34, 43; from nature of God, 52; impossibility of infinite number, 87, 407; finite body contains only finite power, 89, 244, 409; impossibility of infinite regress, 131, 165–166, 337–338; proof from design, 217; proof from motion, 237–240, 344–345; proof from logical symmetry, 275–277; senses of *accident*, 271; meaning of *first cause*, 281; meaning of *necessary*, 293–294. *See also* 114–115, 284, 368. *See* Crescas, Ḥasdai, critique of proof from motion
Ash'ari: Kalam proof for creation, 105, 135, 136, 137; proof from composition, 148–149; proof for unity from mutual interference, 166–167
Athanasius, 150, 151
Augustine: cites proofs for eternity from nature of God, 60, 61; answers proofs for eternity from nature of God, 68–70, 73. *See also* 64, 71, 176, 194
Averroes: proofs for eternity from nature of world, 13, 17, 21–22, 25, 26, 27, 29, 36–37, 43-44; from nature of God, 54, 55, 58, 63, 76; cites proofs for creation, 141, 143, 145–146; refutes Kalam reasoning, 143, 169, 192; proof from design, 229–230; proof from motion, 237, 240, 283, 325, 349–350; proof from impossibility of infinite regress, 341–342, 348–349, 379; critique of Avicenna, 311–335, 379, 388; finite power of finite bodies, 321–331, 409–410; accidental and essential series, 131–133; subject matter of physics and metaphysics, 312–318; natural motion of elements, 268. *See also* 94, 128, 147, 255, 363
Avicenna: proofs for eternity from nature of world, 13, 16, 25, 26, 29; from nature of God, 53, 58, 62, 66; cites proofs for creation, 119–120, 122; impossibility of certain infinites (method of application), 126–127, 128–129, 367–370; emanation, 207; proof from motion, 237, 282–283; his peculiar proof of existence of God, 149, 214–215, 281–307, 309–310, 336, 351, 378, 385–386; proof from impossibility of an infinite regress, 339–341, 345–346, 351. *See also* 93, 161, 162, 172, 267–268, 361. *See* Ghazali, critique of Avicenna's proof

Baghdādī: proofs for creation, 126; Kalam proof for creation, 140, 143, 144; inferring creator from creation, 156, 161, 163, 178; proofs for divine attributes, 165, 168; proof for existence of accidents, 182; impossibility of infinite regress, 165, 355, 363
Baḥya ibn Paquda: proof for creation, 120; proofs from composition, 152–153; inferring creator from creation, 155, 163; proofs for divine attributes, 165, 168, 170; proofs from design, 224, 227–228, 235
Bāqillānī: answers proof for eternity from nature of God, 55; Kalam proof for creation, 135–140, 143; inferring creator from creation, 75, 156, 160, 163, 177–179, 182; divine attributes, 165, 166, 167, 171–172; proof of existence of atoms, 180–182, 183–184; proof from design, 233; impossibility of an infinite regress, 355, 363
al-Baṣīr, Joseph: Kalam proof for creation, 141; inferring creator from creation, 157; proofs for divine attributes, 165, 168, 171–172, 235
Baumgarten, Alexander, 392, 401
Bazdawī: cites proof for eternity from nature of world, 15–16; Kalam proof for creation, 141; inferring creator from creation, 156, 161

Bonaventure: cites proofs for eternity from nature of world, 23, 26, 27; from nature of God, 59, 66; answers proofs for eternity from nature of world, 39, 45; from nature of God, 76–77, 83; proof for creation, 120
Bruno, Giordano, 121

Chrysippus, 175, 220
Cicero, 217–220
Clarke, Samuel, 399–400
Cleanthes, 220
Corpus Hermeticum, 166
Crescas, Ḥasdai: cites proofs for eternity from nature of world, 11, 14, 22, 23, 25, 26, 27, 29; from nature of God, 53, 54, 59, 60, 63, 67; answers proofs for eternity from nature of world, 31–32, 35, 36, 39, 44, 45; from nature of God, 78; infinity, 121, 125, 127, 253–260, 365–366; critique of proof from motion, 90, 240, 249–275; finite power of finite bodies, 90, 260–265, 410–411; critique of proof from logical symmetry, 279–280; version of Avicenna's proof, 308, 351, 385

Descartes, 388, 392–394
Duran, Simon, 232–233
Duns Scotus, 361, 385, 386

Galen, 154, 217
Galileo, 121
Geber. *See* Jābir corpus
Gersonides, Levi: cites proofs for eternity from nature of world, 13, 17, 22, 23, 25, 26, 27, 29; from nature of God, 54, 55, 59, 60, 63, 67; answers proofs for eternity from nature of world, 31–33, 39, 42–43, 44, 45, 46; from nature of God, 74, 78, 80, 83; accepts proof for eternity of matter, 35; proofs for creation, 120, 121, 122, 123, 209–212; proof from design, 231–232; infinite body, 256–257
Ghazali: approaches to proving existence of God, 3–5, 174; cites proofs for eternity from nature of world, 17, 26; from nature of God, 53, 55–56; answers proofs for eternity from nature of world, 37, 41–42; from nature of God, 71, 194–197; inferring creator from creation, 3–5, 75, 154, 161, 162; proofs for creation, 120, 122, 123–124; infinity, 127, 129–130, 166, 367–370, 408–409; Kalam proof for creation, 141, 145; proofs for divine attributes, 165–166, 171–172, 174; emanation, 207; proofs from design, 226–227, 234; natural motion of elements, 267–268; of spheres, 269; critique of Avicenna's proof, 307, 366–375, 388; impossibility of infinite regress, 352–353, 355, 363, 385, 386. *See also* 331
Gundissalinus, 360, 363

Hadassi, Judah, 168
Hallevi, Judah. *See* Judah Hallevi
Henry of Ghent, 385
Hillel ben Samuel, 363
Hume, David, 176, 216, 399–400, 404–405

Ibn ʿAdī, Yaḥyā, 94. *See* Yaḥyā(?)
Ibn Aqnin, Joseph, 141, 359–360, 386
Ibn Daud, Abraham, 127, 237, 357–359, 386
Ibn Ezra, Abraham, 141, 157, 165
Ibn Ḥazm: cites proofs for eternity from nature of world, 15; from nature of God, 58, 62, 64, 65–66; answers proofs for eternity from nature of world, 34; from nature of God, 76, 80, 81, 84; proofs for creation, 120, 122, 123; inferring creator from creation, 163; proof from design, 225–226
Ibn Nafīs, 353, 385
Ibn al-Rawandī, 118
Ibn Ṣaddiq, Joseph, 141, 163, 169
Ibn Suwār, 94, 135–137, 144
Ibn Ṭufayl: approaches to proving existence of God, 4–5, 253; proofs for eternity from nature of world, 25; from nature of God, 54; proof for creation, 120–121; proof for finiteness of spatial magnitudes, 127; Kalam proof for creation, 141; proof from motion, 237
Ījī: cites proofs for eternity from nature of world, 13, 17, 25; from nature of God, 56, 62, 66, 75; answers proofs for eternity from nature of God, 72–73; proof

Ījī (*continued*)
for creation, 121; Kalam proof for creation, 141; proof from composition, 149–150; inferring creator from creation, 155, 162; proofs for divine attributes, 165, 168, 171, 172, 235; particularization, 179, 189, 191; Avicenna's proof, 357, 385–387
Ikhwān al-Ṣafā', 152, 187, 195, 224–225
Isaiah, Book of, 217
Iskāfī, 117
Israeli, Isaac, 363

Jābir corpus, 15, 30, 34, 237
Jāḥiẓ, 233
[Jāḥiẓ?], 219–220, 221–223
Jeshua ben Judah, 141, 157, 166, 171–172
Job of Edessa, 150–151
John of Damascus, 150–152, 166
Judah Hallevi: proofs for creation, 120, 122; Kalam proof for creation, 141; inferring creator from creation, 161; proofs for divine attributes, 165, 171–172; emanation, 207; proofs from design, 228–229, 235
Juwaynī: cites proofs for eternity from nature of world, 15; answers proofs for eternity from nature of world, 30, 34; inferring creator from creation, 75; proofs for creation, 120, 190; Kalam proof for creation, 140–141, 144–146; particularization, 161–162, 177, 179, 180, 182–183, 184–185, 187–188, 190; proofs for divine attributes, 165, 166, 168, 170, 171–172, 234; proof for existence of accidents, 182–183; infinite regress, 355, 363

Kant, Immanuel, 51, 120, 176, 400–405
Khayyāṭ, 118, 119
Kindi: proofs for creation, 106–115, 123, 125–126, 148; inferring creator from creation, 155, 163. *See also* 153, 363

Leibniz, 176, 388–393, 396–399
Leone Ebreo: cites proofs for eternity from nature of world, 14, 24, 26, 29; from nature of God, 62; answers proofs for eternity from nature of world, 35, 38, 39, 45, 46

Maimonides, Moses: approaches to proving existence of God, 4–5, 253; classification of proofs for eternity, 10–11; cites proofs for eternity from nature of world, 13, 17, 23, 24, 29; from nature of God, 53, 54, 59, 60, 63–64; answers proof for eternity from nature of world, 31, 33, 34–35, 36–37, 38, 39, 42, 46; from nature of God, 73, 77, 82; cites proofs for creation, 120, 121, 128, 131, 141, 145, 149, 192–194; cites proofs for divine attributes, 169, 170; his own proofs of creation, 197–201, 203–209; emanation, 207–209; proof from motion, 237–238, 240–249; proof of existence of God from logical symmetry, 278–279; proof from impossibility of an infinite regress, 346–347, 364–365, 380–383, 386; version of Avicenna's proof, 378, 380–383, 386
Māturīdī: proof from composition, 151; inferring creator from creation, 156, 160, 163; proofs for divine attributes, 167, 171–172; proof from design, 224
Māwardī, 141
Mendelssohn, Moses, 120, 392
Miskawayh, 237
More, Henry, 392, 394
Moses al-Lawī, 385–386
Mu'ammar, 184
Muḥāsibī, 224
al-Muqammiṣ, David, 171–172

Narboni, Moses, 23, 363
Nasafī, 141
Nāṣir-i-Khosraw, 120, 141
Naẓẓām, 117–119, 125, 151

Parmenides, 51–52
Pazdawī. *See* Bazdawī
Philo, 218, 220
Philoponus, John, 7, 9; answers proofs for eternity from nature of world, 30, 39, 40, 43; from nature of God, 69–70, 84, 176; proofs for creation, 86–116, 119-123, 126, 130, 134, 139, 142, 146–147, 322, 329,

407, 409; matter and motion of heavens, 204, 269; nature of elements, 267
Plato, 2, 40, 92–93, 146–147, 162, 217, 281–282
Plotinus, 281, 293
Posidonius, 219
Proclus, 7, 11; proofs for eternity from nature of world, 26, 29; from nature of God, 51, 54–55, 58, 60, 61–62, 64–65, 69–70; cause of existence of universe, 89–90, 281–282, 321, 409; proofs from composition, 147. *See also* 162–163, 293

Qāsim b. Ibrāhīm, 224
Quran, 166, 227, 229

Rāzī, Abū Bakr ibn Zakarīyā, 12, 14
al-Rāzī, Fakhr al-Dīn, 180; cites proofs for eternity from nature of world, 17, 25; from nature of God, 54, 56, 62, 66–67; answers proofs for eternity from nature of world, 37, 42; from nature of God, 72, 80–82; inferring creator from creation, 75; proofs for creation, 120–122, 387; proof for finiteness of spatial magnitude, 127; proof from composition, 149; inferring creator from creation, 75, 154–155, 157–159, 162; proofs for divine attributes, 165, 171, 172, 235; particularization arguments, 180, 188–189, 190–191; proof from impossibility of infinite regress, 355–357, 385

Saadia: cites proofs for eternity from nature of world, 15, 27; answers proof for eternity from nature of world, 34; from nature of God, 70; proofs for creation, 95–107, 118, 123, 148; Kalam proof for creation, 103–104, 134, 137, 143; inferring creator from creation, 155–156, 163–164; proof for divine attributes, 167, 170
Shahrastānī: cites proofs for eternity from nature of world, 17, 26, 29; from nature of God, 51, 58–59, 62; answers proofs for eternity from nature of world, 37, 42; from nature of God, 78, 79, 85; proofs for creation, 120, 122, 126, 127; Kalam proof for creation, 141, 142, 145, 387; proofs from composition, 148–149; proofs for divine attributes, 168, 170; inferring creator from creation, 188; proof from motion, 237; version of Avicenna's proof, 308, 351, 385
Sijistānī, 93
Simplicius: as source for Philoponus, 86, 88, 89, 90, 92; cause of existence of universe, 282, 321; proof from impossibility of infinite regress, 338
Spinoza, 120, 388, 392, 395
Suarez, 385

Taftāzānī, 168
Thābit ibn Qurra, 121–122
Themistius, 18, 277
Theodoret, 219–222
Ṭūsī, 38, 73–74, 141, 145, 155

William of Auvergne, 278
William of Auxerre, 361
Witelo, 363
Wolff, Christian, 392, 397–398

Xenophon, 216–217, 221

Yaḥyā(?), 40

Zeno, 97, 118

Index of Terms

Arabic

^{c}adam, 25, 290
āḥād, 352
aḥkam, 233
^{c}azīma, 75
^{c}azm, 76
bā^{c}ith, 54
burhān, 197, 298
da^{c}āwā, 140
dahr, 15, 34
dahrīya, 15, 374
dā^{c}ī, 54, 73
dā^{c}ī al-ḥāja, 62
dā^{c}ī al-ḥikma, 62
dalīl, 299, 304, 316
falak, 225
fi^{c}l, 115
(fkk) yanfakk, 136, 140
ghā'ib, 30, 156
ghayr wujūdī, 38
ḥadath, 95
ḥadd, 104
(ḥādith) ḥawādith, 102
ḥāl, 54, 287
ḥasharāt, 226
hayūlā, 17
ḥaẓẓ, 143
ḥudūth, 299
ihmāl, 223
^{c}illa, 55, 157, 166, 181, 356
imkān, 16, 17
innīya, 110, 111, 284
i^{c}tibār, 235
i'tilāf, 114
jā'iz, 356
jā'iz al-wujūd, 162
jam^{c}, 114
jawād, 61
jūd, 61
(juz') ajzā' mu'allafa, 114
(jwhr) tajawhara, 295
khāṭir, 54
(khlw) yakhlū, 136
lazima, 208
li-annahu, 66
mabda', 282, 283
mādda, 17
ma^{c}dūm, 290
ma^{c}nā, 55, 137, 140, 142, 180, 181, 182, 184
māni^{c}, 54
maqdūr, 169
maṭlūb, 284
mawḍū^{c}, 284
mizmār, 222
(mkn) yumkin, 290
mu'aththir, 66, 356
mudabbir, 224
muḥdith, 224
mukawwan, 294
mukhaṣṣiṣ, 161, 182
mumkin, 162, 290, 356
muqtaḍī, 54, 182, 187
murajjiḥ, 56, 162
murakkab, 111
mustaghnī, 294
mutarattib al-dhāt, 368

mutaṣāᶜidan, 107
qadarīya, 135
qaddara, 369
qādir, 169
qā'im bi-dhātihi, 294
qiwām, 283
(qdm) yataqaddam, 136, 140
(qṭᶜ) yaqṭaᶜ, tuqṭaᶜ, 107
qūwa, 17, 115, 283
(sbq) sābiq, yasbuq, 136
shāhid, 30, 156
ṣiḥḥa, 168, 178
(slsl) yatasalsal, 165
ṣuᶜūd, 107
ṭabīᶜa, 312
tadbīr, 221
tafakkur, 234
tamānuᶜ, 166
taqdīr, 37, 41, 168, 221
ṭārī, 54
tarkīb, 111, 114
tawḥīd, 164
thabāt, 299
uṣūl, 140
waḍᶜ, 368
(wjd) awjada, 34
wujūd, 162, 327
wujūdī. See *ghayr wujūdī*

Hebrew

mezeg, 212
na'ot, 232
nimus, 63, 232
ᶜinyan, 142, 143
(ᶜry) yitᶜareh, 143
to'ar, 33
zeman shorshi, 262

Latin

mutatio, 32
necesse est esse, 360
necessarium esse, 360
novo consilio, 52

Greek

ἀνω, 88
γνώμη, 217
δύναμις, 36, 37
δυνατόν, 37
ὑπάρχειν, 18